D0705392

1 1 01
FANSHAWE COLLEGE
BOOKSTORE
110 95

RELIABILITY For Technology, Engineering, and Management

Paul Kales
University of Massachusetts

Prentice Hall
Upper Saddle River, New Jersey   Columbus, Ohio

Library of Congress Cataloging-in-Publication Data

Kales, Paul.
Reliability : for technology, engineering, and management / Paul Kales.
p. cm.
Includes bibliographical references and index.
ISBN 0-13-485822-0
1. Reliability (Engineering) 2. Reliability. I. Title.
TA169.K35 1998
620'.00452—dc21 97-1545
CIP

Cover art © Paul Kales
Editor: Stephen Helba
Production Editor: Rex Davidson
Design Coordinator: Karrie M. Converse
Cover Designer: Rod Harris
Production Manager: Laura Messerly
Production Supervision: Bookworks
Marketing Manager: Debbie Yarnell

This book was set in Times Roman by The Clarinda Company and was printed and bound by R. R. Donnelley & Sons Company. The cover was printed by Phoenix Color Corp.

Simon & Schuster/A Viacom Company
Upper Saddle River, New Jersey 07458

Printed in the United States of America

10 9 8 7 6 5 4 3 2 1

ISBN: 0-13-485822-0

Prentice-Hall International (UK) Limited, *London*
Prentice-Hall of Australia Pty. Limited, *Sydney*
Prentice-Hall of Canada, Inc., *Toronto*
Prentice-Hall Hispanoamericana, S. A., *Mexico*
Prentice-Hall of India Private Limited, *New Delhi*
Prentice-Hall of Japan, Inc., *Tokyo*
Simon & Schuster Asia Pte. Ltd., *Singapore*
Editora Prentice-Hall do Brasil, Ltda., *Rio de Janeiro*

PREFACE

Reliability is the branch of quality assurance that deals specifically with functionability upon demand, or, as presented in first chapter of the text, assuring that the users of a product or service will get from that product or service what they want when and whenever they want it. As such, reliability addresses the essence of the customer-focus aspect of quality assurance: having the customer experience not only product satisfaction but continued value from the product every time it is used. Yet, since its introduction more than 40 years ago, the reliability discipline has been restricted to manufactured-product applications, particularly in industries under some form of government regulation. But these techniques offer competitive benefits to nonregulated and service industries as well. Because of this more universal opportunity to incorporate reliability techniques into quality assurance efforts, some university technology and management programs have recently seen the need to include reliability engineering practices into their curricula, either as a stand-alone course or as part of a quality control or quality management course.

For the past three years I have taught a one-semester reliability course to juniors and seniors in an engineering technology program. I developed the course by adapting material I had previously used to teach a two-semester reliability engineering course at the graduate level, attempting to maintain the substance while refraining from dependence on advanced probability theory and higher mathematics. In selecting a textbook for my course, I found that among the many fine books on the subject, most developed the material through mathematical presentations beyond the scope of my curriculum.

Consequently, I progressively relied on supplementary notes to whatever textbook I did adopt during a given year. This book evolved from those notes.

Although the intent of this book has been to provide a suitable text for an undergraduate course in technology, engineering, or business curricula, it has also been structured to serve as a text for a one- or two-semester graduate course, as a text to accompany a short industrial seminar or workshop, as an introduction to reliability engineering practices, and as a reference manual for practicing reliability engineers. The text is presented in three major parts, the first of which, "Book One: The Reliability Objective," describes the ways in which reliability is defined and specified for various types of system operation; the second part, "Book Two: Measuring and Evaluating Reliability," addresses methods of designing reliability into a product or service function; and the third, "Book Three: Reliability Assurance and Improvement," covers techniques for proving and improving the reliability of a system after it has been designed and put into operation. Appendix A, which is optional for the more mathematically inclined readers, supplies probability theory used for deriving relationships in the body of the text. Acknowledging the benefit of reliability practices to nonregulated and service applications, reference to government standards has been kept to a minimum, and examples involving health care, banking, and other service industries are included. To accommodate service systems—that is, systems not necessarily composed of hardware—the more general term *restorability* is used in place of what other documents refer to as *maintainability*. So rather than repairing a system that is down or faulty, we restore it.

The entire text is organized to separate the hows from the whys, so that it can be understood by readers with limited technical training while serving the purposes of readers requiring the analytical framework. The main body of the text presents the techniques of reliability and restorability aimed at readers who are fluent in algebra and understand the notion of probability. It presents the topic within the context of total quality and can be understood by all engineering and technology students as well as business students with a background in quality assurance or the total quality concept. Appendix A and sections within the text that are designated with an asterisk comprise an optional supplement to the main text, with explanations and derivations aimed at readers with backgrounds in probability theory, matrix algebra, and at least a year of calculus. The intention of this optional supplementary material is to enhance a reader's understanding of the applicability and restrictions inherent to the techniques without extraordinary mathematical rigor. It is, nevertheless, possible for readers to skip over all the asterisked sections and Appendix A and gain full comprehension of all techniques, follow the examples, and do the exercises at the end of each chapter. The few exercises requiring the supplementary material are also highlighted by asterisks. I recommend omitting the asterisked sections and exercises whenever this book is used for an undergraduate elective course, a course in an engineering technology or management curriculum, or for brief industrial training seminars.

The reliability and availability solution tables in Appendix C make it possible to solve the MTBF (mean time between failures) and availability of complex systems through fairly simple mathematics. These tables enable individuals to tackle solutions that may otherwise be beyond their analytical capacity. I developed Tables C1 through C10 more than 20 years ago to provide a method for quickly computing the MTBF of alternative design approaches. The tables have been useful aids during design review meetings and trade-off evaluations. However, they also allowed me to include math modeling for the MTBF of fairly intricate systems when I planned courses for ASQC or for industrial organizations, where I could not anticipate the backgrounds of the participants. I developed Tables C11 through C16, which supply availability solutions, specifically for this text. Appendix C is my mechanism for including complex reliability modeling in an undergraduate course or seminar where mathematical sophistication is to be avoided. Where Appendix C tables are used, the exercises of Chapters 9 and 10 can be accomplished without reading the asterisked sections.

The 13 chapters of the main text, omitting all asterisked sections and exercises, can be used for a course in reliability for college juniors or seniors enrolled in any engineering or technology program or enrolled in any business school program where students are familiar with the concept of total quality. Prerequisites are mathematics through algebra and an introduction to quality assurance or to total quality. A course that introduces probability or statistics is a possible but not necessary prerequisite.

The 13 chapters, in their entirety, of the main text and Appendix A can be used for a two-semester senior or graduate course in any business, engineering, or science discipline. Prerequisites are mathematics through integral calculus, matrix algebra (possibly), probability and statistics, and an understanding of the concept of total quality. When teaching this course, I have used Appendix A and Chapters 1 through 7, 11, and 12 for the first semester and Chapters 8, 9, 10, and 13 for the second semester.

I have used the first 10 chapters, omitting the asterisked sections and exercises, as supplementary reading for a four-day industrial seminar on reliability and have included the material from Chapters 11 through 13 for a five-day seminar. These short sessions serve as an introduction and survey of the reliability discipline and can be tailored by the instructor to be specific to an industry or organization. The book has also been written with individual study in mind, whereby readers can introduce themselves to the subject and can, by electing to include or exclude the asterisked sections, choose to make use of or to avoid analytical discussion. I also hope that individuals practicing reliability will find this book to serve as a useful reference through ready access to the various techniques, through the examples given, and through the collection of tables provided in Appendices B and C. In fact, in the reliability tradition, I hope that the users of this book will find value in it every time they choose to use it.

I wish to acknowledge the following reviewers of the manuscript: Harry F. Cullen, Palomar College; Donna Summers, University of Dayton; and William Winchell, Ferris State University.

While this book was in production, many of the military standards and handbooks referenced were in the process of revision or being merged into other government documents. I am thankful to Mike DiFranza of Mitre Corporation for helping me track these changes. For helping me field test the manuscript of this book as a course text, I want to express appreciation to my 1997 Reliability Engineering class, in particular, George Zantos, who eliminated many failure modes by uncovering flaws and inconsistencies within the text and exercises. I appreciate the efforts of Stephen Helba and the editorial staff at Prentice Hall who guided me through the authorship process and produced the book well ahead of the anticipated publication date. I also want to thank Lisa Garboski of bookworks for supervising all the intricacies of editing and proofing and for turning an ambitious production schedule into reality. Finally, I thank Judy Kales for encouraging me to periodically retreat into the tranquil and orderly world of textbook writing during those hectic years when we were immersed in the chaos of major house renovation.

Paul Kales

CONTENTS

BOOK TWO: Measuring and Evaluating Reliability 29

BOOK THREE: Reliability Assurance and Improvement 241

BOOK ONE
The Reliability Objective

1

I WANT WHAT I WANT WHEN I WANT IT!

1.1 INTENT

What does it take to be successful in a business enterprise, a professional career, or, for that matter, any endeavor? An obvious answer is to have that endeavor become desired, sought after, or necessary. This status of desirability is achieved through customer satisfaction. What does it take to satisfy a customer? As an occasional customer, I can answer that it takes nothing short of giving me what I want when I want it. Whenever we purchase a product or service, it is with the intent of using it for some purpose, and we expect it to fulfill that purpose whenever we need it. If we are in the business of providing a product or service, our primary concern has to be that it serves each customer's needs whenever that customer uses it.

Reliability is a discipline that has been developed to address this concern. The overall strategy of this discipline is to identify weaknesses or events that can prevent the customer from getting what he or she wants, to reduce the likelihood of such events, and to provide a fast and easy recovery if such an unfortunate event occurs. The intent of this text is to provide the reader with techniques to apply this strategy for assuring customer satisfaction. It recognizes that reliability is best achieved through the direct efforts of the product or service provider and has been written to integrate these techniques into technology and management activities.

1.2 ORIGINS

Reliability practices can be traced back to the late 1940s or early 1950s and to the apprehensions of air-traffic controllers. Perhaps it all started with one who earlier in his career performed his job by looking for landing aircraft through field glasses from the vantage point of his control tower. Now, with the benefit of newly developed electronics, he was able simultaneously to control the takeoffs and landings of many aircraft, and he was able to perform faster, with ever-improving accuracy and safety. But what if his equipment should fail him? And what if such a failure should happen at a particularly inopportune time?

Such anxieties would have been shared by many others, not only in this profession but in other fields, people whose work had become increasingly dependent upon the satisfactory performance of complex equipment. These people recognized that it was not enough to expect equipment to meet their satisfaction upon delivery. This equipment had to perform when needed. And there were some instances in time that were more critical than others. There were instances when the consequences of a failure could be far more devastating than at other times. Surely any piece of equipment is to be expected to be down occasionally, to be in need of maintenance. But how do we assure ourselves that these downtimes do not coincide with periods of critical demand? Or even periods that are simply inconvenient?

1.3 THE AGREE REPORT

With the introduction of solid-state electronics, equipment became more complex, and individuals within the electronics industry recognized that these concerns of their users were not adequately addressed by the quality control practices of their day. They began using the term *reliability* to refer to the capability of a product not only to meet the customer's needs but to function according to the customer's needs *when the customer need occurred.* U.S. Government regulatory agencies, specifically the Department of Defense and the Federal Aviation Authority, saw the need to establish reliability requirements. So, in 1952 the Department of Defense funded a study group, composed of suppliers and users of electronics equipment, to develop an approach to dealing with reliability issues. This group, identified as AGREE (Advisory Group on Reliability of Electronic Equipment), 5 years later issued a report that became the basis of a new discipline, reliability engineering.

The AGREE report ultimately evolved into a military standard on Reliability, MIL-STD-785, first issued by the Department of Defense in 1965. This document and an accompanying set of operating documents on reliability engineering were soon specified on most defense contracts. Because of its origins within the electronics industry, the emphasis in these early days was on reliability of electronic components and equipment. During the late 1950s the Air Force established a comprehensive study on the reliability of electronic components at their Rome Air Development Center (RADC) in Rome, New York. The RADC database, used to produce MIL-HDBK-217, *Reliability Prediction for Electronic Systems,* is the most respected and commonly used handbook for reliability prediction.

1.4 RELIABILITY ENGINEERING

Of course, it was recognized that items other than electronics were responsible for delivering what the customer wants when he or she wants it. So RADC—and other organizations as well—soon developed databases and produced handbooks for reliability prediction of nonelectronic parts. These handbooks included reliability of mechanical parts, hydraulic components, and others that could impact the overall reliability of a complex system of that day. A generation later, computer software became an issue, for software problems were as likely as, or perhaps more likely than, hardware failures to be the source of our failure to get what we want when we want it. As a result, a variety of techniques for dealing with **software reliability** began to appear during the 1970s, and a new area of specialization within the practice of reliability engineering was born.

Despite our best efforts, there will be those occasions when a system will not provide what we want when we want it. We deal with that unfortunate reality (1) by trying to identify the ways in which our system can disappoint us, (2) by designing features within our system to reduce the expected frequencies of such occasions, and (3) by designing features to provide a timely recovery. The first two of these activities are regarded as **reliability engineering,** whereas the third was known as **maintainability engineering.** In the early 1960s maintainability engineering was established as a discipline of the assurance sciences in recognition of the need to provide recoverability features within a system. Because of the close affinity between reliability and maintainability engineering, they are often indistinguishable, and their combined activities were soon referred to as R/M (reliability/maintainability) engineering. In 1972 what had been the Annual Symposium on Reliability (which had been meeting since 1966) became the Annual Reliability and Maintainability Symposium.

1.5 RESTORABILITY

Through the years many people have confused maintainability with maintenance engineering. The former, as just described, deals with the designing of recoverability capabilities into a system, whereas the latter deals with methods of repairing faulty equipment or of keeping equipment in good working condition. Because maintenance is a subset of maintainability, this confusion can be expected to continue. For this reason and because this book deals with reliability of service as well as that of physical systems, this text avoids the use of the term maintainability, which, through the years, has been associated with the recoverability of equipment. Instead this book uses the more general term *restorability* to describe the third activity, that of providing timely recovery to any type of system. In designing system restorability, our aim may be to minimize the frequency of the interruptions, to minimize the expected length of each interruption, or to minimize the human effort required to respond to an interruption. Another restorability objective is to avoid having interruptions during critical demand periods.

1.6 GOVERNMENT REGULATIONS

During the late 1950s and early 1960s reliability engineering was practiced primarily where mandated by the DOD or the FAA. Reliability specifications by NASA soon followed. During the 1970s the newly established NRC (Nuclear Regulatory Commission) published its own set of reliability standards relating to nuclear power plant systems. In the 1980s the Food and Drug Administration was assigned the responsibility of regulating medical devices, and they joined the group of government regulatory agencies that impose reliability requirements on manufacturers.

1.7 RELIABILITY AND TOTAL QUALITY

Up to this point, reliability had addressed primarily safety issues. However in the late 1980s and early 1990s competitive issues began to emerge as reliability engineering techniques began to be applied to commercial products, such as copy machines, personal computers, and automobiles. Reliability engineering practices became incorporated into the *total quality* initiatives of some organizations, and some reliability engineering methods began to be required as part of ISO-9000 certification. As a consequence there is now an emanating interest in applying reliability engineering methods to service-rendering systems as well as to manufactured equipment.

As with any of the *assurance sciences,* rather than having reliability features delegated to a consulting specialist, they are best incorporated by product or service designers and providers, the individuals with the ultimate responsibility for assuring that the users get what they want when they want it.

BIBLIOGRAPHY

1. AGREE Report. 1957. *Reliability of Military Electronic Equipment.* Arlington, Va.: Department of Defense.
2. ARINC Research Corp. 1964. *Reliability Engineering.* Upper Saddle River, N.J.: Prentice Hall.
3. MIL-STD-785B, 1980. *Reliability Programs for Systems and Equipment.* Arlington, VA: Department of Defense.

EXERCISES

1.1. Describe what is meant by the terms (a) quality assurance, (b) quality control, and (c) reliability engineering. Discuss the differences among these terms.

1.2. Obtain an organization chart from a local company and explain how the reliability engineering function fits into their organizational structure. Describe where, if at all, in the organization the following functions are performed:

a. Reliability estimation and verification
b. Failure analysis
c. Process control
d. Procurement control
e. Reliability/design/cost trade-off studies

1.3. In answering Question 1.2, you may interview friends in industry, fellow classmates, or work associates from your own company. Once you obtain the information, review it on your own and describe and discuss any changes you would recommend to make that company's reliability engineering function more effective.

1.4. Repeat these exercises at various stages as you progress through the book, and discuss any changes in your recommendations.

2

DEFINING RELIABILITY

2.1 INTENT

In Chapter 1, we learned that the discipline known as reliability was developed to provide methods for assuring that a product or service will function when it is required to do so by its user. These methods consist of techniques for determining what can go wrong, how we can prevent it from going wrong, and, if something does go wrong, how we can quickly recover and minimize the consequences. This chapter presents the definitions of common terminology applied to the techniques and provides an understanding of how reliability is regarded under various circumstances.

2.2 SYSTEM

In this text, we shall regard any product or service that is used by a consumer as a **system.** Accordingly, let us consider a system to be an arrangement of related material or nonmaterial items that behave as a whole body, a body whose purpose is to perform some functions or services. Under this definition a system can be a physical assemblage of items (parts) to form a functional piece of equipment or it can be an established sequence of items (that is, a procedure) for performing a service. Hence, as a product a system can be anything from a tape dispenser to a personal computer or a power plant, any product from which a consumer has usage demands. As a service, a system can be a medical procedure, a process for having a restaurant meal served, or the Internal Revenue Service.

2.3 RELIABILITY

We can now define *reliability* in terms of a generic item.

> **The *RELIABILITY* of an item is the *probability* that the item will**
> **perform a *specified function***
> **under *specified operational* and *environmental conditions,***
> ***at* and *throughout a* specified *time.***

The first thing to notice in this definition is that reliability is a *probability,* so we are dealing with the laws of random chance as they appear in nature. Indeed, occurrences of inopportune interruptions in functionability or service in a system are random events, the expected frequencies of which we aim to reduce.

The next thing to notice is that the definition depends on a *specified* function, operating conditions, environment, and time. So before we can deal with reliability, the producer (or provider) and the user must reach formal agreements on what the product or service is to do, how the user is to use the product or service (that is, how he or she will operate the product or receive or apply the service), the range of environments under which the product or service is expected to perform satisfactorily, and the instant or duration in time that the performance of the product or service is demanded.

2.4 DEMAND TIME

The definition of reliability allows for the specification of demand time to be either an instant in time or a time interval. In actuality, the demand time may be a sequence of instances (or periods or cycles) or it may be a series of intervals. That depends, of course, on the type of system or service. How we apply the definition of reliability to an actual product or service depends heavily upon the nature of the demand time. Figure 2-1 illustrates the four types of demand times that can be applied to an item's performance. Each of these suggest a different way of regarding the reliability of an item.

When the demand time of an item's performance is either a discrete instant in time or a series of instances, we shall use the term **cycle-dependent performance.** A cycle-dependent performance may be **one-shot** or **repeated cycles.** Examples of one-shot items are a mouse trap, the cancellation of a postage stamp, or a match. The one-shot item (product or service) is expected to operate just once and at one instance in time, the demand time. A postage stamp needs to be canceled just once, at the point in time that the piece of mail is being processed at the post office. A mouse trap also needs to operate once, at the time the mouse arrives. A one-shot item is absolutely useless if it functions at any time other than the demand time. A toggle switch is an example of a repeated cycle item. It has a series of demand instances, and it must function every time it is demanded for, say, turning on a light.

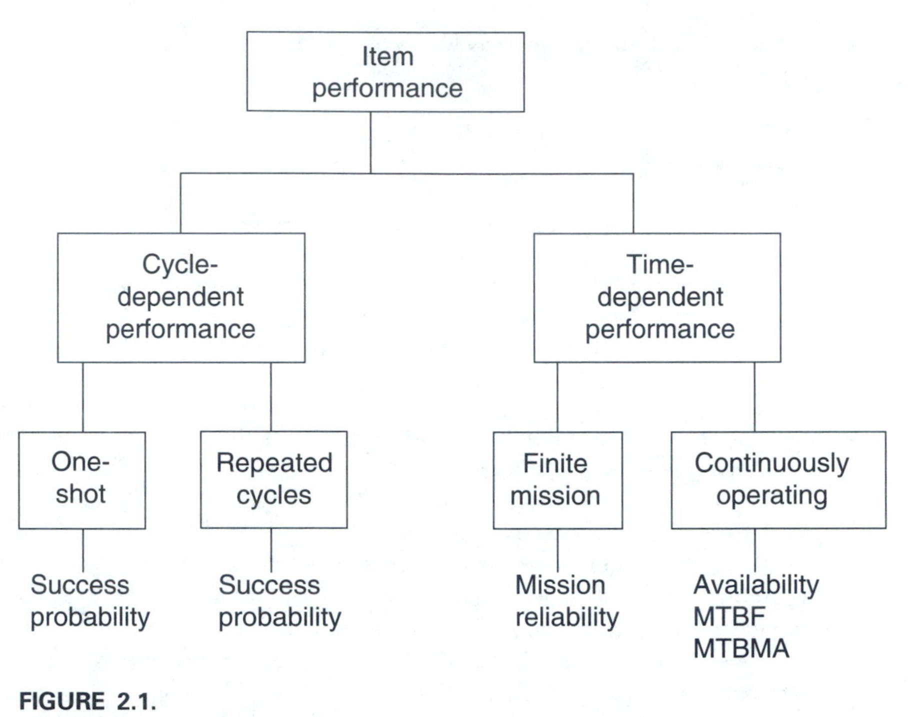

FIGURE 2.1.
Reliability parameters.

If the demand time of an item's performance is a time interval or is continuous, we describe the performance as **time dependent.** A time-dependent performance may be for a *specified mission* or may be *continuously operating*. Examples of a specified-mission operation are the launching of a satellite or a haircut. Examples of a continuously operating item are a refrigerator, a power-generating station, or the telephone company's directory assistance service. Time-dependent items are expected to operate throughout their demand intervals without interruption or, in the case of continuously operating items, all the time.

2.5 ONE-SHOT ITEMS

The reliability of a one-shot item is best described as its success probability—that is, the probability that the item will perform as expected (as specified, under specified operating and environmental conditions) at the specific instance of the item's demand. Consider, as an example, a wooden match. If you, the user, strike the match and it lights, *success* has been achieved. On the other hand, if it fails to light (that is, it is incapable of lighting no matter how many times you strike it), a *failure* occurs.

Regarding the definition of reliability in this example, the *item* is the wooden match, the *specified function* is the lighting of the match, the *specified environmental conditions* consist of typical ambient conditions (dry, no wind, etc.), the *specified operating conditions* consist of the accepted methods for striking a match, and the *specified time* is your demand time (that is, the point in time that you choose to strike the match).

The reliability of the wooden match is its success probability under the specified environmental and operating conditions, and it can be expressed as

$$R = P(S) = \frac{S}{N} = \frac{S}{S + F} \tag{2.1}$$

where $P(S)$ = success probability,
N = number of sequential trials of match strikes
S = number of successes
F = number of failures

In other words, if we struck millions of similar types of matches, matches that were produced through the same process, the ratio of successful lightings to total trials would estimate the success probability.

2.6 REPEATED CYCLES

As with one-shot items, the reliability of an item that functions through repeated cycles is also its success probability. As an example, if you, the user, drop a coin into a postage stamp machine, the reliability of that stamp machine is the probability that the machine provides the expected stamp for you.

To examine the elements of the definition of reliability through our stamp machine example, the specified function is that the machine produces the stamp that it indicates it will—that is, the proper denomination or a specific commemorative stamp as pictured on the face of the machine. A further specified function may be that the machine advertises that it will make change, in which case the correct amount of change is also expected.

The specified operational conditions would be presented as instructions for operating the machine; for example, pennies may be restricted or the machine may or may not take paper money. If you, the user, violate the specified operating conditions by, say, dropping in a penny or by inserting a $10 bill into the paper money slot when the machine is specified to make change for only $1 and $5 bills, then the machine is not expected to perform. Further operating conditions may be that the machine is not to be used if it indicates that it is out of stamps or that only correct change is to be used if the machine indicates that it is out of change.

Specified environmental conditions may include maintenance or reloading (of stamps or change) instructions and restrictions for the location of the machine (e.g., indoors only). It would be up to the maintainer of the stamp machine (the post office or store) to provide the specified environmental conditions. Finally, the specified time is that instant in time when you choose to insert your coin(s) or bill into the machine.

The response to your inserting of the coin(s) or bill will be either a success or failure. If the machine provides the stamp you ordered and the correct change, if applicable, the functioning is considered a success. In other words, you got what you wanted when you wanted it. If, on the other hand, you fail to get a stamp, or you get the wrong stamp, or you get the wrong change, it is a failure. If you fail to get the right stamp or change but specified operating or environmental conditions were violated, the function cannot be considered a success or failure.

For millions of transactions with our postage stamp machine leading to success or failure, the machine's success probability can be estimated by the number of successes divided by the number of trials. So the reliability is

$$R = P(S) = \frac{S}{N} = \frac{S}{S + F} \tag{2.2}$$

where $P(S)$ = success probability
N = number of trials (either success or failure)
S = number of successes
F = number of failures

2.7 TIME-DEPENDENT ITEMS OF SPECIFIED MISSION

When an item's function is time dependent and it is needed throughout a time interval—but only throughout that specified time interval—its reliability is known as **mission reliability,** expresed as

$$R = R(t) \tag{2.3}$$

where $R(t)$ = reliability defined as a function of time t, (i.e., the reliability for the specified duration t)
t = specified mission time duration

An example is a satellite launching rocket, which must operate long enough to launch the satellite but need not operate beyond that. So if the time to launch is, say, 30 min, the rocket must operate without any failures during that 30-min mission, after which time we do not care about it. Hence, its mission reliability is $R = R(t) = R(30 \text{ min})$. We deal with determining mission reliability in the next chapter.

As another example, consider having dinner at a restaurant. The mission in this case is the interval from the time you enter the restaurant until you leave for your next activity (or, possibly, until your body has satisfactorily digested all the restaurant's food). The reliability is then the probability that your restaurant experience is failure free throughout the defined mission. Reliability in this example is certainly a function of the systems within the restaurant. But it is concerned only with results during your demand interval, not before it and not after it.

For some types of services, such as a paint job, the mission time may be defined by the agreed-upon warrantee period. The painter may guarantee that the exterior paint job on your house will be effective for 3 years, and that becomes the mission time.

2.8 CONTINUOUSLY OPERATING ITEMS

Now let us consider the type of item that is required to operate continuously, forever or for the duration of a long, extended service life. Here we are talking about products such as a personal computer or your home's heating system and services such as electric power or the postal system. Because there are no finite missions associated with the operation of such items, mission reliability is not the best way to define or specify reliability requirements. Nevertheless, duration reliability, $R(t)$, values are sometimes used, regardless of the fact that $R(t)$ is relatively meaningless in such situations. For example, we may consider the probability that electric power is provided without incident during the hours of 5:00 P.M. to 8:00 P.M.—in other words, the interval reliability for a fictitious mission within our infinite demand time.

More useful reliability parameters for continuously operating items are the **mean time between failures (MTBF)** and the **availability.** The MTBF tells us how frequently, on the average, we can expect our item to experience an outage. The availability tells us the proportion of the time that we can expect our item to be operating. Associated with these characteristics are the **mean downtime (MDT),** the average time that it takes to return to an operating state after an outage has occurred, and the **outage rate,** the complement of availability, or the portion of the time that we can expect our item to be down.

Availability and outage rate are related as follows.

$$\text{availability} = 1 - \text{outage rate} \tag{2.4}$$

In the simplest possible situation an item's availability is

$$A = \frac{\text{MTBF}}{\text{MTBF} + \text{MDT}} \tag{2.5}$$

and an item's outage rate (OR) is

$$\text{OR} = \frac{\text{MDT}}{\text{MTBF} + \text{MDT}} \tag{2.6}$$

In later chapters you will see that for complex items, Equations 2.5 and 2.6 are not necessarily true.

Associated with continuously operating items are the **mean time between maintenance actions (MTBMA)** and the **mean time to restore (MTTR).** The MTBMA is the average frequency of (preventive or corrective) maintenance actions upon the item, whether or not the maintenance actions impose or are in response to an outage. The MTTR is the average time taken up by maintenance activities, whether or not the item is inoperable during the maintenance activities. If the item is always out during maintenance, then MTBMA = MTBF and MTTR = MDT. You will see in later chapters, however, that the MDT may be composed of a logistics delay time and an administrative downtime in addition to maintenance downtime.

2.9 ITEMS IN STANDBY

Instead of a continuously operating item, we may be talking about an item that is continuously operable, that is, inactive in a standby mode. Examples are a sprinkler or fire alarm system, a fire department, and, perhaps, your automobile insurance. Usually such an item, as in our examples, is associated with safety protection. Standby items may also serve as backup to active items within an operating system, ready to come on-line if the primary item deactivates. For a standby item, the MTBF is the mean time between events that render the item inoperable, the MTTR and MDT are as just defined, and the availability is the portion of the time that the item is operable, or in a state of operational readiness. In fact, the availability of a standby item is often referred to as **operational readiness.**

For a safety system, such as a building sprinkler system, we use the term **dependability** to designate the system's readiness to activate upon demand and then to remain operating throughout a required demand period. So, if the system is required to detect a fire, turn on an alarm and the sprinklers, and keep them going without interruption for a period of 30 min, for example (considered to be enough time for the fire Department to arrive and take control), the operational readiness is the system's available in the standby state. Its ability to operate throughout the 30-min demand period is its 30-min mission reliability in the active state. Hence the dependability is

$$D(t) = A \times R(t) \tag{2.7}$$

2.10 RELIABILITY PARAMETERS

To summarize, the best way to specify the reliability of an item depends upon how the item is expected to function. The nature of the item's functioning interprets the form of the demand time. As we have seen, demand time may be an instant in time, a time interval (a mission), continuously (all the time), or continuously in a pair of modes (standby and active). When demand time is an instant or a series of instances, success probability is used. When demand time is an interval, mission reliability is used. When demand time is continuous, we use MTBF and availability, possibly along with MTBMA, MTTR, and MDT. For continuously operating items normally in a standby mode we also use dependability. Until we refine their definitions in the chapters to come, the following will be used as our preliminary working definitions:

Success probability is the chance that an item that operates discretely will function as specified in a given trial.

Mission reliability is the probability that an item will operate without failure throughout a specified interval.

MTBF is the expected average time between failure events that cause the item to go down.

MTBMA is the expected average time between events requiring maintenance, whether or not the item is down during the maintenance.

MTTR is the expected average time to restore an item after a failure event or as a routine, scheduled maintenance action, whether or not the item is down during the restoration.

MDT is the expected average downtime for restoration activities.

Availability is the fraction of time that an item is actually operating (if it is an active item) or operable (if it is a standby item).

Operational readiness is the availability of a standby item while in standby.

Dependability is the probability that a standby item will be operable upon demand and then will remain operating, without failure, throughout its required demand time.

BIBLIOGRAPHY

1. ARINC Research Corp. 1964. *Reliability Engineering.* Upper Saddle River, N.J.: Prentice Hall.
2. I. Bazovsky. 1963. *Reliability Theory and Practice.* Upper Saddle River, N.J.: Prentice Hall.
3. P. D. T. O'Connor. 1991. *Practical Reliability Engineering,* 3d ed. New York: John Wiley.
4. C. O. Smith. 1976. *Introduction to Reliability in Design.* New York: McGraw-Hill.

EXERCISES

2.1. Which parameter or parameters are most appropriate for specifying the reliability of each of the following?

a. An emergency generator in a hospital
b. An engraver
c. A mousetrap
d. An office copy machine
e. A flashbulb

2.2 Give examples of products or services whose reliability would properly be specified by each of the following parameters:

a. MTBF and MTTR
b. Success probability
c. Dependability

3

COMPUTING RELIABILITY PARAMETERS

3.1 INTENT

Reliability, as presented in the first two chapters, deals with the likelihood that a product or service will perform for its user when the user demands it. The product or service is considered to be a *system,* which in turn is made up of constituent elements. These elements, or *items,* may be physical components of a piece of equipment or procedural steps of a service process. We have just seen in Chapter 2 that, depending on the *type* of performance of a particular item, a specific *parameter* is used for defining that item's reliability. This chapter introduces the **mathematics of reliability.** The reliability parameters are defined mathematically, and examples are given for computing values of reliability parameters for a variety of systems. Readers following the mathematical derivations (asterisked paragraphs) may wish to refer to Appendix A for a review of probability and statistics.

*3.2 RELIABILITY AS A PROBABILITY FUNCTION

Referring to the definition of reliability given in Section 2.3, we observe that *reliability* is a probability function; as such it possesses all the properties of probability functions (See Section A2.1 of Appendix A).

For cycle-dependent operations, where reliability is defined as success probability $P(S)$, the following are true:

1. $0 \leq P(S) \leq 1$.
2. $P(S) = 1$ implies certainty of success.
3. $P(S) = 0$ implies certainty of failure.
4. Let $Q = P(F) =$ unreliability. $R = P(S)$ and $Q = P(F)$ are *mutually exclusive* and *exhaustive*. Hence $R + Q = 1$.
5. If R_A is the reliability of item A, R_B is the reliability of item B, and items A and B operate *independently*, then $R_{AB} = R_A \times R_B$. Otherwise, $R_{AB} = R_{A|B} \times R_B = R_{B|A} \times R_A$.

For time-dependent operations, where reliability is a function of time, or $R = R(t)$, the following hold:

1. $0 \leq R(t) \leq 1$.
2. $R(t) = 1$ implies certainty of success.
3. $R(t) = 0$ implies certainty of failure.
4. If $Q(t)$ is the unreliability at the time t (i.e., the probability of a failure prior to the time t), because success and failure are *mutually exclusive* and *exhaustive* events at any time t, $R(t) + Q(t) = 1$ for all values of t.
5. If A and B operate *independently*, $R_{AB}(t) = R_A(t) \times R_B(t)$. Otherwise, $R_{AB}(t) = R_{A|B}(t) \times R_B(t) = R_{B|A}(t) \times R_A(t)$.

3.3 FAILURE RATES

Reliability for any of the types of systems identified in Figure 2.1 depends either upon successful events or successful operation within a given time. So reliability may be regarded as the absence of failure: either the absence of failure events in the case of items with cycle-dependent performance or the absence of failures during operation of items with time-dependent performance. Accordingly, the reliability of any type of item is a function of the item's **failure rate.**

One-shot items and other types of cycle-dependent items have failure rates designated in **failures per cycle.** Since the reliability of a cycle-dependent item is the success probability, or the rate of successes per cycle, reliability becomes

$$R = 1 - f \tag{3.1}$$

where f is the failure rate in failures per cycle.

EXAMPLE 3.1 A billing process that generates monthly statements has a failure rate of 3.5 per 10 million cycles. What is the reliability of that process?

$$f = 3.5 \text{ per } 10^7 \text{ cycles} = 3.5 \times 10^{-7} \text{ per cycle}$$
$$R = 1 - f = 0.99999965$$

Items with time-dependent performance have failure rates $\lambda(t)$ in terms of **failures per hour.** Of particular interest is the situation where the failure rate of an item is considered constant with time, that is, operation of the item does not increase its failure rate. A failure rate that is not dependent on time is designated by λ. (Some texts refer to the time-dependent failure rate, $\lambda(t)$, as the *hazard function.*)

*3.4 RELIABILITY AS A FUNCTION OF TIME

For time-dependent operations we can generalize by stating that at some point in time we have a reliability $R(t) = 1.0$. Let us define that point in time as $t = 0$. Eventually, if there is no recoverability, a fail state is reached. So we can conclude that $R(\infty) = 0$. Furthermore, we can intuitively state that $R(t)$ approaches zero asymptotically as t approaches infinity. Hence, for a simple nonmaintained system, the $R(t)$-versus-t relationship is expected to be as illustrated in Figure 3.1. Now, let us mathematically derive the actual relationship for the nonrecoverable condition.

Suppose we put a quantity N_0 identical items into operation at the time $t = 0$. Let $N_S(t_A)$ be the number of survivors at the time $t = t_A$, and let $N_f(t_A)$ be the number that fail prior to the time $t = t_A$. If the items are nonrecoverable—that is, failed items are not replaced—$N_S(t_A) + N_f(t_A) = N_0$ for all values of t_A. So we can generalize that $N_s(t) + N_f(t) = N_0$ for all values of t. Also, for all values of t,

$$R(t) = \frac{N_s(t)}{N_0} = \frac{N_s(t)}{N_s(t) + N_f(t)} = 1 - \frac{N_f(t)}{N_0} = 1 - Q(t) \tag{3.2}$$

Since there is no replacement of failed items, we can conclude that successive points in time $t_1, t_2, t_3, \ldots$, where $t_1 < t_2 < t_3 < \cdots$, the quantity of survivors $N_s(t_i)$ is nonincreasing:

$$N_0 \geq N_s(t_1) \geq N_s(t_2) \geq N_s(t_3) \geq \cdots \tag{3.3}$$

From Equation 3.2,

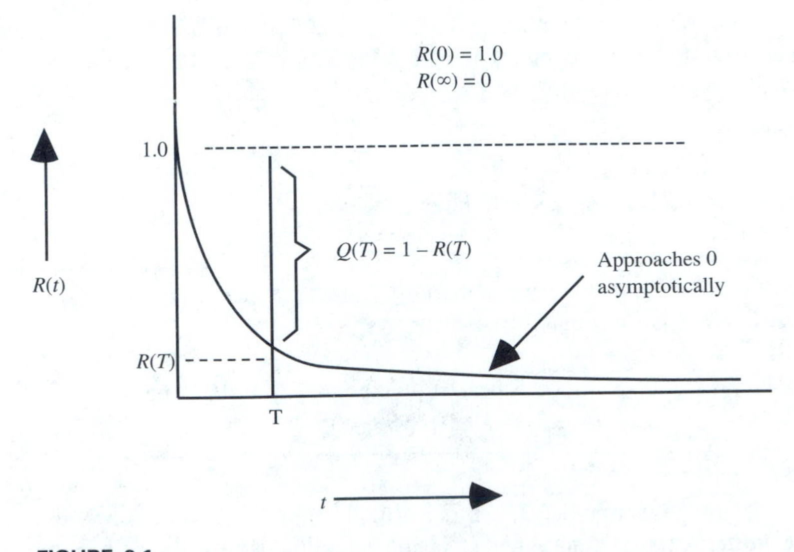

FIGURE 3.1
Reliability versus time for a time-dependent nonrecoverable system.

$$R(0) = N_s(0)/N_0 = N_0/N_0 = 1$$
$$R(t_1) = N_s(t_1)/N_0$$
$$R(t_2) = N_s(t_2)/N_0$$
$$\vdots$$
$$R(t_i) = N_s(t_i)/N_0 \tag{3.4}$$

Hence, as illustrated in Figure 3.2,

$$R(0) \leq R(1) \leq R(2) \leq R(3) \leq \cdots$$

and

$$R(t_i) \leq 1.0 \qquad \text{for all } t_i \tag{3.5}$$

The events governing reliability are *failures*. So let us define the **failure density function, $f(t)$,** such that the probability of a failure occurring at or before the time $t = t_A$ is

$$F(t_A) = \int_{t=-\infty}^{t_A} f(t)\,dt = \int_{t=0}^{t_A} f(t)\,dt \tag{3.6}$$

Because we established $t = 0$ as that point in time when our components are set into operation and we are certain that they are all operating, $R(0) = 1.0$, and there can be no negative values of t. Thus, the lower limit of the integral in Equation 3.6 becomes $t = 0$.

By Equation 3.6, $F(t_A)$ is the unreliability at time t_A

$$F(t_A) = Q(t_A) = 1 - R(t_A) \tag{3.7}$$

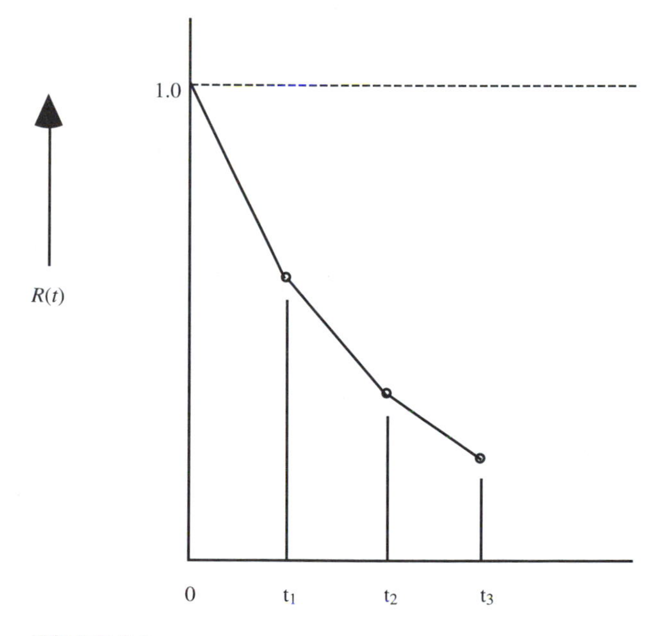

FIGURE 3.2
Reliability versus operating time.

Since Equation 3.7 holds for any t_A, we can generalize that $F(t) = 1 - R(t)$ for all values of t. Also note that $F(t)$ is the cumulative distribution of the time-to-failure (TTF) distribution. (It can easily be shown that $f(t)$, the failure density function, is actually a density function, by proving $\int_{t=0}^{\infty} f(t)\, dt = 1.0$.)

Figures 3.3 and 3.4 illustrate the relationships we have been discussing. Figure 3.3(a) shows the relationship between the quantity of survivors, $N_s(t)$, and time. By dividing the quantity of survivors by the initial quantity, N_0, we have the relationship between time-dependent reliability $R(t) = N_s(t)/N_0$ and time in Figure 3.3(b).

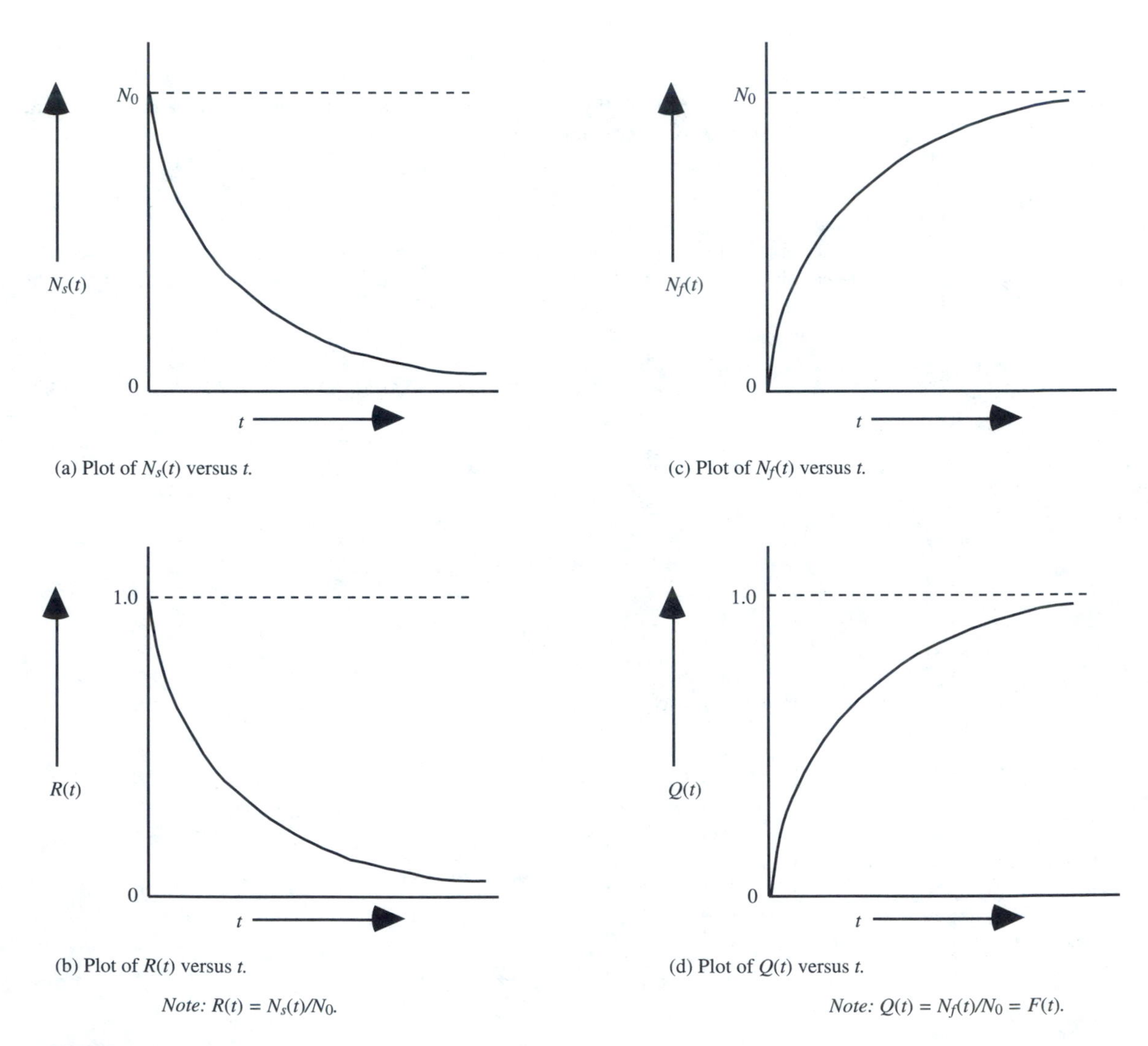

FIGURE 3.3
Development of the relationships among reliability, time-to-failure distribution parameters, and time.

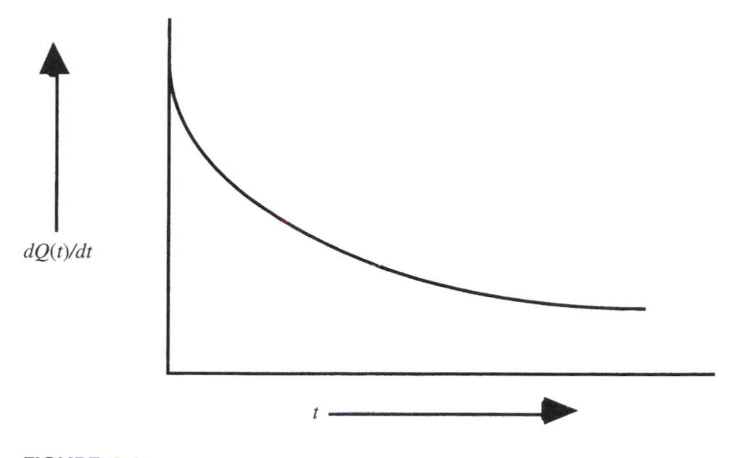

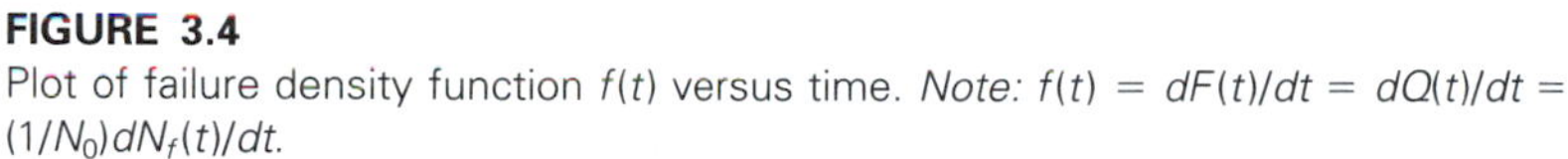

FIGURE 3.4
Plot of failure density function $f(t)$ versus time. *Note:* $f(t) = dF(t)/dt = dQ(t)/dt = (1/N_0)dN_f(t)/dt$.

Figure 3.3(c) relates the failed quantity $N_f(t) = 1 - N_s(t)$ to time, and by dividing the failed quantity by the initial quantity, N_0, we have the relationship between unreliability $Q(t) = 1 - R(t) = N_f(t)/N_0$ and time in Figure 3.3(d). As derived earlier, $Q(t) = F(t)$, the cumulative distribution function of the TTF distribution.

The failure density function—that is, the density function of the TTF distribution—can be developed by differentiating $F(t)$ with respect to time:

$$f(t) = \frac{dF(t)}{dt} = \frac{dQ(t)}{dt} = \frac{1}{N_0}\frac{dN_f(t)}{dt} \tag{3.8}$$

This result is illustrated in Figure 3.4, which is the derivative of Figure 3.3(d).

In Equations 3.6 and 3.7 we see that the unreliability relates to the failure density function:

$$Q(t_A) = \int_{t=0}^{t_A} f(t)\,dt \tag{3.9}$$

Consequently, reliability relates to the failure density function:

$$R(t_A) = 1 - Q(t_A) = 1 - \int_{t=0}^{t_A} f(t)\,dt = \int_{t=t_A}^{\infty} f(t)\,dt \tag{3.10}$$

By differentiating reliability, $R(t)$, with respect to time, we obtain the relationship

$$\frac{dR(t)}{dt} = \frac{d}{dt}\int_{t=t_A}^{\infty} f(t)\,dt = f(\infty) - f(t_A) = 0 - f(t_A) \tag{3.11}$$

from which we conclude

$$f(t_A) = -\frac{dR(t_A)}{dt}, \quad \text{or, more generally,} \quad f(t) = -\frac{dR(t)}{dt} \tag{3.12}$$

Also recall that (Equation 3.2)

$$R(t) = \frac{N_s(t)}{N_0} = \frac{N_s(t)}{N_s(t) + N_f(t)} = 1 - \frac{N_f(t)}{N_0}$$

from which

$$\frac{dR(t)}{dt} = -\frac{1}{N_0}\frac{dN_f(t)}{d(t)} \equiv -f(t) \qquad \textbf{(3.13)}$$

and once again we conclude

$$f(t) = -\frac{dR(t)}{dt} \qquad \textbf{(3.14)}$$

Figure 3.5 illustrates the relationship among reliability, unreliability, and the failure density function. As seen in Equation 3.10, the reliability at time T is the area under the failure density function curve between $t = T$ and $t = \infty$ (i.e., the probability that the failure takes place beyond $t = T$). As seen in Equation 3.9, the unreliability at time T is the area under the failure density function curve between $t = 0$ and $t = T$ (i.e., the probability that the failure takes place prior to $t = T$).

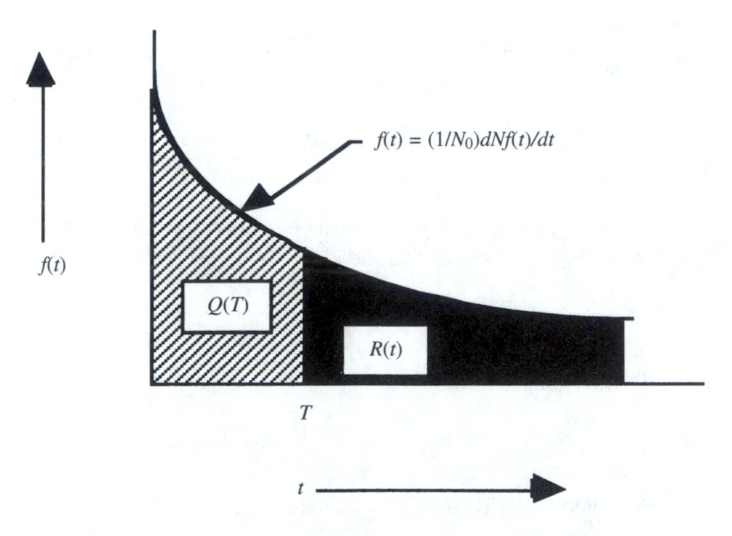

FIGURE 3.5
Plot of the failure density function $f(t)$, illustrating both reliability and unreliability as integrals of $f(t)$:

$$Q(T) = \int_0^T f(t)\,dt$$

$$R(T) = \int_T^\infty f(t)\,dt$$

$$Q(T) + R(T) = \int_0^\infty f(t)\,dt$$

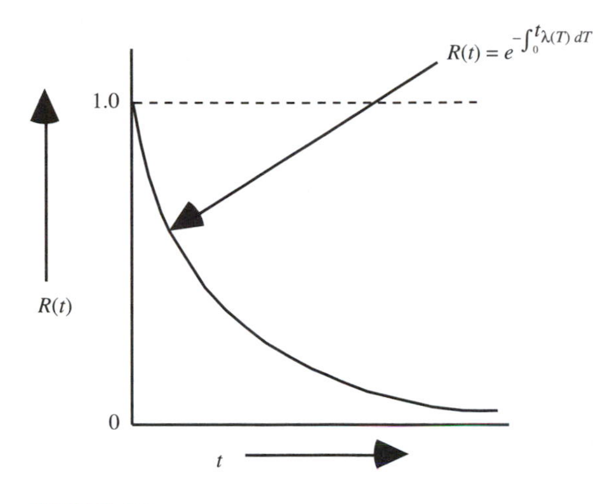

FIGURE 3.6
Reliability as a function of instantaneous failure rate.

*3.5 FAILURE RATE AS A FUNCTION OF TIME

We define **instantaneous failure rate λ(*t*)** as the instantaneous rate at which items are failing—that is, the ratio of the increase in failure quantity at a given time t with respect to time T to the quantity of survivors at time t:

$$\lambda(t) \equiv \frac{dN_f(t)/dT}{N_s(t)} = \frac{1}{N_s(t)}\left[-N_0\frac{dR(T)}{dT}\right] = -\frac{N_0}{N_s(t)}\frac{dR(T)}{dT} = -\frac{1}{R(t)}\frac{dR(T)}{dT} \tag{3.15}$$

where T is the variable time. Therefore,

$$\lambda(t)dT = -\frac{dR(T)}{dT} = -d[1\text{n}\, R(T)] \tag{3.16}$$

Integrating Equation 3.16 between the limits 0 and t,

$$\int_{T=0}^{t} \lambda(T)\, dT = -1\text{n}\, R(T)\Big|_{T=0}^{T=t} = 1\text{n}\, R(0) - 1\text{n}\, R(t) = 1\text{n}(1.0) - 1\text{n}\, R(t) = -1\text{n}\, R(t) \tag{3.17}$$

Hence,

$$R(t) = 3^{-\int_0^t \lambda(T)\, dT} \tag{3.18}$$

Now we can further clarify the relationship between time-dependent reliability and time that we intuitively developed in Figure 3.1. (See Figure 3.6.)

*3.6 CONSTANT FAILURE RATE

If the failure rate is assumed to be constant, $\lambda(t) = \lambda$, then the relationship between reliability and failure rate becomes

$$R(t) = e^{-\int_0^t \lambda(T)\, dT} = e^{-\lambda \int_0^t dT} = e^{-\lambda T} \tag{3.19}$$

or, more generally,

$$R(t) = e^{-\lambda t} \tag{3.20}$$

where λ is the constant failure rate. The failure density function becomes

$$f(t) = -dR(t)/dt = -de^{-\lambda t}/dt = -e^{-\lambda t}d(-\lambda t)/dt = \lambda e^{-\lambda t} \tag{3.21}$$

Notice that Equation 3.21 is the *probability density function* of the *exponential probability law* (see Table A.2), which, from Table A.2, Quality Assurance Applications, we expect to govern the *TTF* distribution when we have constant failure rate.

Also consider that, for constant failure rate λ, $R(t)$ is the probability of having zero failures during time t and the rate of failure per time t units is a constant, λt. For this situation the *Poisson Probability Law* with parameter $\mu = \lambda t$ applies (see Table A.1):

$$P(x) = e^{-\mu}(\mu^x/x!) = e^{-\lambda t}(\lambda t)^x/x! \tag{3.22}$$
$$R(t) = P(0) = e^{-\lambda t}(\lambda t)^0/0! = e^{-\lambda t}$$

which is identical to Equation 3.20. What we have essentially done in Equations 3.19 and 3.20 is to derive the probability density function and probability mass function for the Exponential and Poisson Probability laws, respectively.

Figures 3.7 and 3.8 reproduce Figures 3.5 and 3.6 for the special case where we have constant failure rate λ, where the failure density function is $f(t) = \lambda e^{-\lambda t}$. Computing reliability and unreliability for a mission time $t = T$ by integrating the failure density function, we get the following expected results:

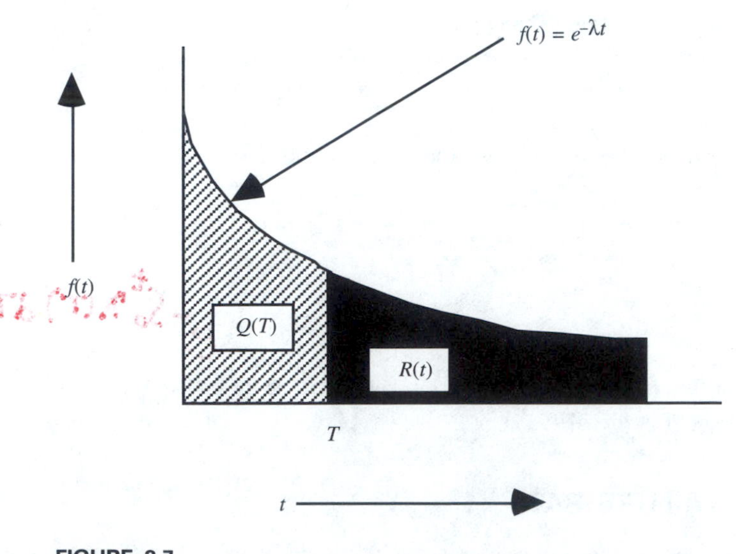

FIGURE 3.7
Failure density function versus time for constant failure rate λ.

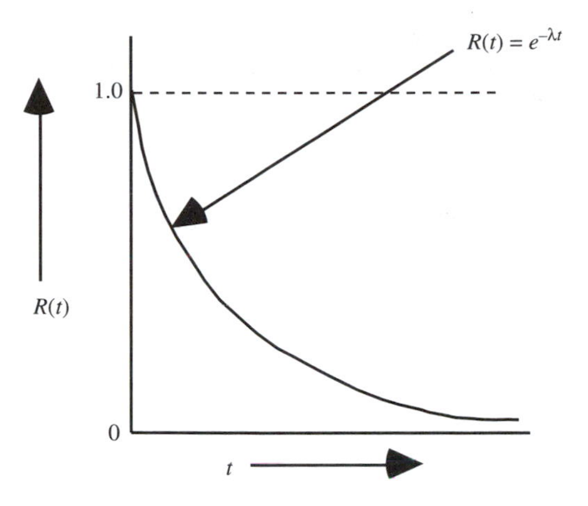

FIGURE 3.8
Reliability versus time for constant failure rate λ.

$$\text{reliability } R(T) = \int_{t=T}^{\infty} f(t)\,dt = \lambda\int_{t=T}^{\infty} e^{-\lambda t}\,dt = -1[e^{-\lambda t}\Big|_{t=T}^{t=\infty} = e^{-\lambda t} - e^{-\infty} = e^{-\lambda t} \tag{3.23}$$

$$\text{unreliability } Q(T) = \int_{t=0}^{T} f(t)\,dt = \lambda\int_{t=0}^{T} e^{-\lambda t}\,dt = -1[e^{-\lambda t}\Big|_{t=0}^{t=T} = e^{-0} - e^{-\lambda t} = 1 - e^{-\lambda t} \tag{3.24}$$

3.7 MISSION RELIABILITY

Section 3.5 shows the mission reliability, $R(t)$, for a specified mission time t to be

$$R(t) = e^{-\int_0^t \lambda(T)\,dT} \tag{3.25}$$

where $\lambda(T)$ is the time-dependent failure rate. However, if—*and only if*—an item has a *constant failure rate,* Section 3.6 shows its mission reliability for a specified mission time t to be

$$R(t) = e^{-\lambda t} \tag{3.26}$$

where λ is the failure rate.

Figures 3.6 and 3.8 show these relationships between *reliability* and *failure rate.* As can be seen in these illustrations, reliability decreases with time. At some point in time an item is known to be performing as specified (refer to the definition of reliability in Chapter 2). Convention refers to that point in time as the time $t = 0$. We shall refer to the $t = 0$ condition as that point in time when an item is set into operation and is known to be performing its specified function under specified operational and environmental conditions. Hence, the reliability at $t = 0$ is 1.0. As time increases, the reliability decreases, and the item ultimately will fail. So at time infinity the reliability is zero:

$$R(0) = 1.0 \tag{3.27}$$

$$R(\infty) = 0 \tag{3.28}$$

EXAMPLE 3.2 A type of power supply used in a computer is assumed to have a constant failure rate of 8.5 failures per million hours. What is its reliability within the system for a 5000-h mission?

$$\lambda = 8.5 \times 10^{-6}/\text{h} \qquad t = 5000 \text{ h}$$
$$\lambda\, t = 5000 \times 8.5 \times 10^{-6} = 0.0425$$

$R(t) = e^{-0.0425}$, which can be solved on your calculator or using Table B8 in Appendix B as

$$R(t) = e^{-0.0425} = 0.958$$

---

*3.8 MEAN TIME TO FAILURE

Allow MTTF to represent the mean time to failure of a set of identical items put into operation at some point in time $t = 0$, where t is a variable representing time to failure (TTF). Then MTTF is the expected value of t, or the mean of the TTF distribution. In Section A1.7 we see that the mean of a continuous distribution is $\mu = \int_{x=-\infty}^{\infty} xf(x)\,dx$. So for the variable t, which has no negative values, the mean is

$$\text{MTTF} = \int_{t=0}^{\infty} tf(t)\,dt = -\int_{t=0}^{\infty} t\frac{dR(t)}{dt}\,dt = -\int_{t=0}^{\infty} t\,dR(t) \tag{3.29}$$

Solving this equation through integration by parts and recognizing that $R(t)$ approaches zero as t approaches infinity,

$$\text{MTTF} = +\int_{t=0}^{t=\infty} R(t)\,dt - [tR(t)]\Big|_{t=0}^{t=\infty} = \int_{t=0}^{\infty} R(t)\,dt \tag{3.30}$$

*3.9 MTTF AS A FUNCTION OF FAILURE RATE

MTTF is a constant, the mean of the TTF distribution. The general solution of MTTF as a function of the variable failure rate $\lambda(t)$, from Equations 3.19 and 3.30, is

$$\text{MTTF} = \int_0^{\infty} R(t)\,dt = \int_0^{\infty} e^{-\int_0^T \lambda(t)\,dT}\,dt \tag{3.31}$$

When—*and only when*—failure rate is a constant, or $\lambda = \lambda(t)$,

$$\text{MTTF} = \int_0^{\infty} R(t)\,dt = \int_0^{\infty} e^{-\lambda t}\,dt = \frac{1}{\lambda} \tag{3.32}$$

As used throughout this text, MTTF refers to the mean time to the failure from the time $t = 0$, that point in time when our item is put into operation. If somebody wished to predict the mean time to failure from some other time, say τ, then $R(t)\,dt$ would have to be integrated from a lower limit of τ rather than from 0.

We have limited our discussion thus far to nonrecoverable items. If we have an item that is restorable or replaceable, the expression *mean time between failures* (MTBF) is applicable. The

MTBF is a constant; it is the expected time between successive failure events over the service life of the item. If we impose the condition that every time the item fails, we restore it completely—that is, we restore it to the $t = 0$ state—then we can use MTBF and MTTF interchangeably. In other words the mean time between failures is the same as the mean time to the first failure (the MTTF) under the stated conditions.

3.10 MEAN TIME BETWEEN FAILURES

From the discussions in Sections 3.8 and 3.9, we can conclude that the **mean time between failures (MTBF)** of an item is

$$\text{MTBF} = \int_{t=0}^{\infty} R(t)\, dt \tag{3.33}$$

If and only if the item has constant failure rate, its MTBF is

$$\text{MTBF} = 1/\lambda \tag{3.34}$$

The MTBF of an item is always a constant, whether its failure rate is constant or not, and Equation 3.34 holds *only* for constant failure rate.

EXAMPLE 3.3 Compute the MTBF for the power supply of Example 3.2.

Since the failure rate is constant,

$$\text{MTBF} = 1/\lambda = 1/8.5 \times 10^{-6}/\text{h} = 118{,}000 \text{ h}$$

3.11 MEAN DOWNTIME (MDT)

As defined in Chapter 2, the MDT of an item is the expected downtime needed to restore the item to full operation ($t = 0$ conditions) when it fails.

3.12 AVAILABILITY

The availability of an item was defined in Chapter 2 to be the fraction of time that the item is actually operating or operable. So, availability (A) is

$$A = \frac{\text{MTBF}}{\text{MTBF} + \text{MDT}} \tag{3.35}$$

EXAMPLE 3.4 If the MDT of the power supply of Examples 3.2 and 3.3 is 2 h, find its availability.

$$A = \frac{\text{MTBF}}{\text{MTBF} + \text{MDT}} = \frac{118{,}000 \text{ h}}{118{,}000 + 2 \text{ h}}$$

$$= \frac{118{,}000}{118{,}022} = 0.99998$$

*3.13 COMPLEX SYSTEMS

Larger restorable systems with a high degree of complexity tend to behave as though they have constant failure rates at the system level. This statement is particularly true if parts that exhibit wearout during operation (e.g., mechanical parts) are on a planned replacement schedule. The more complex the system, the more it appears to behave as though its failure rate is constant. So we usully can consider that for a complex, restorable system, where restorations upon failure are always to the $t = 0$ conditions, MTBF $= 1/\lambda$, where λ represents the system failure rate.

If a large system is composed of N serial items, the failure of any of which comprises a system failure, the system reliability is

$$\begin{aligned} R_{\text{sys}}(t) &= R_1(t) \times R_2(t) \times \cdots \times R_N(t) \\ &= e^{-\lambda_1 t} \times e^{-\lambda_2 t} \times \cdots \times e^{-\lambda_N t} = e^{-t\Sigma\lambda} \end{aligned} \tag{3.36}$$

and

$$\text{MTBF} = 1/\Sigma\lambda_i \tag{3.37}$$

This discussion continues in Chapter 9 on reliability math modeling. Meanwhile, let us turn our attention to situations where we have failure rates that are not necessarily constant.

*3.14 INCREASING FAILURE RATE

Recall from Equation 3.14 that

$$f(t) = -\frac{dR(t)}{dt} = -\frac{d}{dt}\left[e^{-\int_0^t \lambda(T)\,dT}\right] = \lambda(t)e^{-\int_0^t \lambda(T)\,dT} \tag{3.38}$$

Let us assume that $\int_0^t \lambda(T)\,dT$ has a solution of the form $(At)^B$. Then

$$\lambda(t) = \frac{d}{dt}\int_0^t \lambda(T)\,dT = \frac{d}{dt}(At)^B = A^B\frac{d}{dt}t^B = BA^Bt^{B-1} \tag{3.39}$$

Hence,

$$f(t) = BA^Bt^{B-1}e^{-(At)^B} \tag{3.40}$$

which, if we substitute β for B and μ for A, we recognize as the probability density function for the Weibull Probability Law (see Table A.2). So we can conclude that the Weibull Probability Law governs the TTF distribution regardless of whether the failure rate $\lambda(t)$ is constant or not. In fact, if $\lambda(t)$ is increasing with time, the parameter β is greater than 1.0. If $\lambda(t)$ is decreasing with time, $\beta < 1.0$, and if $\lambda(t)$ is constant, $\beta = 1.0$. As we observe in Appendix A, when $\beta = 1$ in the Weibull Probability Density Function, it becomes the density function for the Exponential Probability Law, which governs the TTF distribution for constant failure rate.

Therefore, we can use the Weibull Probability Law for making predictions when we are dealing with a situation where we suspect we have an increasing failure rate. When we are making predictions for electrical or mechanical hardware, failure rates are either constant or increasing. Some reliability analysts use the Weibull Probability Law for making predictions for a supposed "burn-in" period, where the rate of failures is assumed to decrease. However, there is really no reason to assume that behavior during this very questionable phase (if it is, indeed, a phase) follows the Weibull or any other probability law.

When the question of burn-in arises in a reliability engineering class, students are usually asked to reflect on the difference between screening and burn-in testing. The difference is that with the former, we know what we are looking for, whereas with the latter we do not. We just operate equipment for a period of time, hoping that any inherent problems will manifest themselves during that period. Since with burn-in we are not dealing with some known phenomenon, how can we expect a varying outcome governed by some probability law? As a consequence, this text deliberately avoids making burn-in predictions.

The decreasing failure rate is, nevertheless, a circumstance we may encounter with service functions as a consequence of a continual improvement process.

BIBLIOGRAPHY

1. I. Bazovsky. 1963. *Reliability Theory and Practice.* Upper Saddle River, N.J.: Prentice Hall.
2. P. D. T. O'Connor. 1991. *Practical Reliability Engineering,* 3d ed. New York: John Wiley.
3. M. L. Shooman. 1968. *Probabilistic Reliability: An Engineering Approach.* New York: McGraw-Hill.
4. C. O. Smith. 1976. *Introduction to Reliability in Design.* New York: McGraw-Hill.

EXERCISES

3.1. A given type of pressure switch has an estimated failure rate of 24×10^{-6} failures per cycle. What is its reliability?

3.2. A postage vending machine has a failure rate of one failure per 25,000 cycles. What is its reliability?

3.3. An electronic calculator has an estimated failure rate of 34×10^{-6} failures per hour. Assuming constant failure rate, what is its MTBF? Again assuming constant failure rate, what is its reliability for a 100-h mission? A 1000-h mission? A 10,000-h mission?

3.4. A receiver has an estimated failure rate of 100×10^{-6} failures per hour. Assuming constant failure rate, what are its MTBF and its reliability for a 100-h mission? Again assuming constant failure rate, what is its reliability for a 1000-h mission? A 2000-h mission? A 5000-h mission? A 10,000-h mission?

3.5. Plot $R(t)$ versus t for the receiver of Exercise 3.4.

3.6. A computer has an MTBF of 10,000 h. Assuming constant failure rate, what is its reliability for a 10,000-h mission?

3.7. What percentage of the computers of the type described in Exercise 3.6 would you expect to run for 10,000 h without failure?

3.8. An air-conditioner condenser has an MTBF of 100,000 h. Assuming constant failure rate, what is the probability that it will survive an operating period equal to its MTBF without failure?

3.9. Using the results of Exercises 3.6 through 3.8, what conclusions can you draw regarding the reliability of an item for a mission equal to its MTBF when the item has constant failure rate? In other words, for constant failure rate, what is $R(t = \text{MTBF})$? Is this what you would intuitively expect? Explain.

3.10. Suppose an item with a constant failure rate has an MTBF of 1000 h. For what mission time is its reliability 50%? In other words, find t such that $R(t) = 0.50$.

BOOK TWO Measuring and Evaluating Reliability

4

RELIABILITY PREDICTIONS

4.1 ESTIMATING PARAMETERS FROM OBSERVED DATA

In Chapter 3 we saw how to compute reliability parameters from assumed failure rates. In practice, failure rate assumptions do not just drop out of the sky that way. We have to predict failure rates from observed field or laboratory data according to the statistical analysis method described in Appendix A. This chapter shows how to use such data to make failure rate estimates in which we can have confidence when using them to make our reliability predictions.

4.2 CONDITIONS FOR THE PREDICTION

Referring once again to the reliability definition of Chapter 2,

> **The *RELIABILITY* of an item is the *probability* that the item will perform a *specified function***

> **under** ***specified operational*** **and** ***environmental conditions,***
> ***at*** **and** ***throughout a specified*** time.

The data used for estimating failure rates, in order to be valid, *must* be acquired under the intended operational and environmental conditions. When it is impossible to simulate the intended usage and environments in a laboratory or to get sufficient field data under suitable conditions, it is common to make analytic adjustments. This chapter also deals with the necessity to sometimes make such adjustments.

4.3 CYCLE-DEPENDENT PERFORMANCE

We saw in the last chapter that for cycle-dependent performance, the reliability is the success rate, which is 1 minus the failure rate: $R = 1 - f$. If a cycle-dependent item is activated in its intended operational and environmental state for N repeated cycles and it experiences S successes and F failures (note that $S + F = N$), then the observed success rate is $s = S/N$ and the observed failure rate is $f = F/N$. The best estimate of reliability is our *observed* success rate. But how confident are we in the adequacy of that estimate? Obviously, the more cycles we observe, the higher our confidence. That is, the larger N is, the higher is our confidence.

*4.4 CONFIDENCE ESTIMATES FOR SUCCESS PROBABILITY

The success probability of an item whose operation is cycle dependent is governed by the Binomial Probability Law. The α confidence estimate of the success probability is that value of R_α such that

$$P(x > F) = \alpha = 1 - P(x \leq F) = 1 - \sum_{x=0}^{F} \binom{N}{x} R_\alpha{}^{N-x}(1 - R_\alpha)^x \tag{4.1}$$

where N is the number of cycles observed and F is the number of failures among the N cycles. Hence, based on the experience of F failures in N cycles, there is a probability α that the reliability (success probability) is at least R_α.

Table B.5 (Appendix B) is a table of the 50% confidence estimate of the reliability, $R_{0.50}$, versus N and F:

$$0.50 = 1.0 - \sum_{x=0}^{F} \binom{N}{x} R_{0.50}{}^{N-x}(1 - R_{0.50})^x \tag{4.2}$$

Table B.6 (Appendix B) is a table of the 75% confidence estimate of the reliability, $R_{0.75}$, versus N and F:

$$0.75 = 1.0 - \sum_{x=0}^{F} \binom{N}{x} R_{0.75}{}^{N-x}(1 - R_{0.75})^x \tag{4.3}$$

Table B.7 (Appendix B) is a table of the 95% confidence estimate of the reliability, $R_{0.95}$, versus N and F:

$$0.95 = 1.0 - \sum_{x=0}^{F} \binom{N}{x} R_{0.95}{}^{N-x}(1 - R_{0.95})^x \tag{4.4}$$

Confidence estimates other than 50%, 75%, or 95% can be determined from other sources or by using Equation 4.1. For values of N larger than those contained in the tables, the normal approximation to the binomial may be used, provided that $F > 4$. To use the normal approximation set the mean μ equal to $N(S/N)$ and the standard deviation σ equal to $\sqrt{N(S/N)\ (F/N)}$, which are seen in Table A.1 to be the mean and standard deviation, respectively, of the binomial. (S is the number of successes and F is the number of failures in N observed cycles.) Then the α confidence estimate, R_α, is determined by

$$NR_\alpha = \mu - z_\alpha \sigma = N\left(\frac{S}{N}\right) - z_\alpha \sqrt{N\left(\frac{S}{N}\right)\left(\frac{F}{N}\right)} \tag{4.5}$$

from which

$$R_a = \frac{S - z_\alpha \sqrt{SF/N}}{N} \tag{4.6}$$

From Table B.1,

$$z_{0.50} = 0.000$$
$$z_{0.75} = 0.675$$
$$z_{0.95} = 1.645$$

Hence, the 50% confidence estimate is

$$R_{0.50} = \frac{S}{N} \tag{4.7}$$

The 75% confidence estimate is

$$R_{0.75} = \frac{S - 0.675\sqrt{SF/N}}{N} \tag{4.8}$$

The 95% confidence estimate is

$$R_{0.95} = \frac{S - 1.645\sqrt{SF/N}}{N} \tag{4.9}$$

4.5 SUCCESS PROBABILITY CONFIDENCE TABLES

Tables B.5, B.6, and B.7 provide confidence estimates of success rates derived from various quantities of cycles. They provide 50%, 75%, and 95% confidence estimates, respectively. A 90% confidence estimate of success rate, for example, is one for which we are 90% certain that the actual success rate is at least the estimated value.

EXAMPLE 4.1 Suppose we observe 30 cycles of a pressure switch, of which there are two failures: $N = 30$, $F = 2$.

The best estimate of the reliability is

$$R = 1 - F/N = 1 - 2/30 = 0.933$$

The 50% confidence estimate (Table B.5) is

$$R_{0.50} = 0.912$$

The 75% confidence estimate (Table B.6) is

$$R_{0.75} = 0.873$$

The 95% confidence estimate (Table B.7) is

$$R_{0.95} = 0.805$$

Hence, although our observations show a 0.933 success rate, there is a 50% chance that it is at least 0.912, a 75% chance that it is at least 0.873, and a 95% chance that it is no worse than 0.805. The confidence estimate we decide to use depends on how conservative we want to be when making our predictions. When the consequences of overestimating success rate are severe, we may want to use a high confidence estimate (90 or 95%). Otherwise, a 50% estimate may be sufficient.

When the number of cycles N is larger than the N values in Tables B.5, B.6, and B.7, Equations 4.7, 4.8, and 4.9 may be used to get 50%, 75%, and 95% confidence estimates of success probability if the number of failures is five or more.

EXAMPLE 4.2 Suppose we wish to assess the success probability of patient food service in a hospital. A success is a meal that is delivered on time, as ordered and according to dietary instructions. Out of 1500 monitored meal deliveries, 6 were unsatisfactory in some way. Estimate the reliability of food service at that hospital at the 95% confidence level.

$$N = 1500, \qquad F = 6, \qquad S = 1494$$

The best estimate of the reliability is $R = S/N = 1494/1500 = 0.996$. Table B.7 does not provide estimates for $N = 1500$ and $F > 4$. Therefore, Equation 4.9 applies:

$$R_{0.95} = \frac{S - 1.645\sqrt{\dfrac{SF}{N}}}{N} = \frac{1494 - 1.645\sqrt{\dfrac{(1494)(6)}{1500}}}{1500} = 0.993$$

Our observed data showed a 0.996 success rate, and on the basis of that data we are 95% confident that the actual reliability of the food service is no less than 0.993. (There is a 5% chance that the actual reliability is less than 0.993.)

---

*4.6 CONFIDENCE ESTIMATES FOR MTBF, CONSTANT FAILURE RATE

More sophisticated texts on statistical inference show that the upper and lower confidence estimates of the constant failure rate for the Exponential Probability Law can be determined through the Chi-Square Probability Law as follows: For the upper confidence,

$$\lambda_{\alpha} = \frac{\chi^2_{\alpha,2c+2}}{2T} \tag{4.10}$$

and for the lower confidence,

$$\lambda_{\underline{\alpha}} = \frac{\chi^2_{1-\alpha,2c}}{2T} \tag{4.11}$$

where α is the confidence estimate, T is the operating time, c is the number of failures during the time T, and $\chi^2_{a,b}$ is the chi-square statistic at level a and b degrees of freedom. We are interested in confidence estimates of the MTBF, which is the reciprocal of the constant failure rate λ when the Exponential Probability Law applies, i.e., whenever we can assume constant failure rate. Also, we are interested in the worse case, that is, the lower α confidence estimate of MTBF, which is the reciprocal of the upper α confidence estimate of λ:

$$\text{MTBF}_{\alpha} = \frac{2T}{\chi^2_{\alpha,2r}} \tag{4.12}$$

where T = total operating time
c = number of failures during the time T
$r = c$ if the operating time is failure terminated
or $c + 1$ otherwise

4.7 TIME-DEPENDENT PERFORMANCE: MTBF ESTIMATES

The best estimate of MTBF from observed laboratory or field data consisting of total operating time T, during which there were c failures, is T/c. Equation 4.12 can be used for obtaining the α confidence estimate,

where $(\chi^2)_{\alpha,2r}$ is the chi-square statistic (Table B.2)

r is the number of failures, c, if the observed time ended in a failure the last failure occurred at the time T)

$r = (c + 1)$ if the last failure occurred prior to the time T (which is usually the case)

EXAMPLE 4.3 Suppose we wish to estimate the MTBF of the power supplies described in the examples in Chapter 3 on the basis of accumulated field data, assuming that we have 204,683 equipment hours of data and that two failures occurred during that time,

$$T = 204{,}683 \text{ h}, \qquad c = 2$$

The best estimate is MTBF $= T/c = 204{,}683$ h/2 $= 102{,}342$ h. If we want to make a 95% confidence estimate of the MTBF, because we must make the assumption of time (rather than failure) termination of the operating time, $r = c + 1 = 3$, so $2r = 6$. Therefore,

$$\text{MTBF}_{0.95} = \frac{2T}{(\chi^2)_{0.95,2r}} = \frac{2(204{,}683)}{(\chi^2)_{0.95,6}}$$

From Table B.2, where $a = 0.95$ and $2r = 6$, the chi-square statistic is 12.6. So

$$\text{MTBF}_{0.95} = 2(204{,}683)/12.6 = 32{,}489 \text{ h}$$

Thus, although the observed field data had a failure on the average of once every 102,000 h, on the basis of that data we are 95% certain that the MTBF is no less than 32,500 h. (There is a 5% chance that the actual MTBF is less than 32,500 h.) We assume, furthermore, that the field operation was under the operational and environmental conditions applicable to the predictions.

EXAMPLE 4.4 Determine the 60% confidence estimate of the MTBF for the power supply of Example 4.3.

$$\text{MTBF}_{0.60} = \frac{2T}{(\chi^2)_{0.60,2r}} = \frac{2(204{,}683)}{(\chi^2)_{0.60,6}}$$

From Table B.2, where $a = 0.60$ and $2r = 6$, the chi-square statistic is 6.21. So

$$\text{MTBF}_{0.60} = 2(204{,}683)/6.21 = 65{,}920 \text{ h}$$

Thus, based on the field data, there is a 40% chance that the actual MTBF is less than 65,900 h. Reviewing the results of Examples 4.3 and 4.4, we can state that there is a 5% chance that the actual MTBF is less than 32,500 h, a 35% chance that it is between 32,500 and 65,900 h, and a 60% chance that it is more than 65,900 h. As discussed previously, the confidence level used when making our prediction depends upon the criticality of the consequence of a statistical error. When we wish our prediction to be extremely conservative, we use a high confidence level, such as 90% or 95%; when we wish to be more realistic we use a confidence level of about 50% or 60%.

4.8 FAILURE RATE ESTIMATES

For constant failure rates, the best estimate from test of field data is

$$\lambda = c/T \tag{4.13}$$

where c is the number of failures during total operating time T.

From Equations 4.10 and 4.11, the upper α confidence estimate λ_α is

$$\lambda_\alpha = \frac{(\chi^2)_{\alpha,2r}}{2T} \tag{4.14}$$

and the lower a confidence estimate is

$$\lambda_{\underline{\alpha}} = \frac{(\chi^2)_{1-\alpha,2c}}{2T} \quad (4.15)$$

where $(\chi^2)_{\alpha,2r}$ is the chi-square statistic (Table B.2)

r is the number of failures, c, if the observed time ended in a failure (the last failure occurred at the time T)

or $r = (c + 1)$ if the last failure occurred prior to the time T

EXAMPLE 4.5 Determine upper and lower 90% confidence estimates for a device that experienced 11 failures in 495,000 device operating hours in the field. Assume *constant failure rate.*

$$T = 495{,}000 \qquad c = 11 \qquad r = 12$$

$$\lambda_{0.90} = \frac{(\chi^2)_{\alpha,2r}}{2T} = \frac{(\chi^2)_{0.90,24}}{2(495{,}000)} = \frac{33.2}{990{,}000} = 33.5 \times 10^{-6} \text{ per hour}$$

$$\lambda_{\underline{0.90}} = \frac{(\chi^2)_{1-\alpha,2c}}{2T} = \frac{(\chi^2)_{0.10,22}}{2(495{,}000)} = \frac{14.0}{990{,}000} = 14.1 \times 10^{-6} \text{ per hour}$$

Notice that this device experienced 11 failures in 495,000 h, or a failure per 495,000/11 = 45,000 h, which is a failure rate of 22.2×10^{-6} per hour. We want to use this field experience to estimate a failure rate in which we have sufficient confidence to be able to make a valid prediction. As discussed before, when we wish to be extremely conservative we use an upper 90% or 95% confidence estimate. If we want to make as accurate a statement as possible about the actual failure rate, we use a confidence interval, as in this example. When our aim is to make a somewhat realistic prediction, we usually use an upper confidence estimate of 60%, which as we shall see, is favored by most failure rate data source handbooks.

The results of this example indicate that, on the basis of the reported field experience, a best estimate of the device failure rate is 22.2×10^{-6} per hour, there is a 90% probability that the actual failure rate is more than 14.1×10^{-6} per hour, and there is a 90% probability that the actual failure rate is less than 33.5×10^{-6} per hour. Therefore, there is an 80% probability that the actual failure rate is between 14.1×10^{-6} and 33.5×10^{-6} per hour. This is called the 80% *confidence band.* A confidence band is typically used when we are deriving a failure rate estimate from laboratory test data and wish to determine whether or not we have enough data. In this case, since we have too wide a band to make a very accurate estimate at the 80% level, we may wish to gather more data.

EXAMPLE 4.6 Suppose we test the devices of Example 4.5 for another 505,000 device operating hours and experience another 11 failures. This means we have 22 failures in 1,000,000 h, or 22×10^{-6} per hour, the same best estimate we had with the original 495,000 h. But the 80% confidence band (upper and lower 90% confidence estimates), where

$$T = 1{,}000{,}000 \qquad c = 22 \qquad r = 23$$

is

$$\lambda_{0.90} = \frac{(\chi^2)_{\alpha,2r}}{2T} = \frac{(\chi^2)_{0.90,46}}{2(1{,}000{,}000)} = \frac{58.7}{2{,}000{,}000} = 29.3 \times 10^{-6} \text{ per hour}$$

$$\underline{\lambda_{0.90}} = \frac{(\chi^2)_{1-\alpha,2c}}{2T} = \frac{(\chi^2)_{0.10,44}}{2(1{,}000{,}000)} = \frac{32.5}{2{,}000{,}000} = 16.2 \times 10^{-6} \text{ per hour}$$

Notice that we have to interpolate to get the chi-square statistics in Table B.2. (An interpolation is also necessary in Example 4.7.)

So we are now 80% confident that the actual failure rate is between 16.2 and 29.3 $\times 10^{-6}$ per hour, a much tighter band than we had before. In other words, the additional operating hours with the same results reinforced our confidence in our estimate.

EXAMPLE 4.7 Suppose we test the same devices until we accumulate another 1,000,000 device operating hours and that we experience another 22 failures. Then

$$T = 2{,}000{,}000 \qquad c = 44 \qquad r = 45$$

$$\lambda_{0.90} = \frac{(\chi^2)_{\alpha,2r}}{2T} = \frac{(\chi^2)_{0.90,90}}{2(2{,}000{,}000)} = \frac{101.1}{4{,}000{,}000} = 25.3 \times 10^{-6} \text{ per hour}$$

$$\underline{\lambda_{0.90}} = \frac{(\chi^2)_{1-\alpha,2c}}{2T} = \frac{(\chi^2)_{0.10,88}}{2(2{,}000{,}000)} = \frac{73.4}{4{,}000{,}000} = 18.3 \times 10^{-6} \text{ per hour}$$

So, the 2,000,000 device-hours of data provide a sufficiently tight confidence band to predict an accurate failure rate of 0.00002 per hour (we are 80% confident that the actual failure rate is between 18 and 25 $\times 10^{-6}$ per hour). It is clear that as the operating time increases, the confidence band becomes smaller.

A failure rate handbook, which typically uses upper 60% confidence estimates, would list this device's failure rate as 23.2 $\times 10^{-6}$ per hour.

$$\lambda_{0.60} = \frac{(\chi^2)_{\alpha,2r}}{2T} = \frac{(\chi^2)_{0.60,90}}{2(2{,}000{,}000)} = \frac{92.7}{4{,}000{,}000} = 23.2 \times 10^{-6} \text{ per hour}$$

EXAMPLE 4.8 A 500-bed hospital, with an average occupancy rate of 80%, accumulates data on patient infections acquired as a by-product of hospitalization (that is, infections that were not present or incubating at the time of admission). Over a 12-week period (84 days), 40 such infections were reported. Assuming a constant rate of occurrence at that hospital, what is the risk of such an infection to a patient hospitalized for a 3-week period? We use 60% confidence estimates for our predictions.

The observation period is 84 $\times$ 24 = 2016 h, during which time there are an expected 0.80 $\times$ 500 = 400 patients. So

$$T = 2016 \times 400 = 806{,}400 \text{ patient-hours}$$

Also, the number of incidents is $c = 40$, and, because we have time-terminated data.

$$r = c + 1 = 41$$

The best estimate of infection per patient-hour is

$$\lambda = 40/806{,}400 = 49.6 \times 10^{-6} \text{ per hour}$$

The 60% confidence estimate of the patient-infection rate is

$$\lambda_{0.60} = \frac{(\chi^2)_{\alpha,2r}}{2T} = \frac{(\chi^2)_{0.60,82}}{2(2{,}806{,}400)} = \frac{84.6}{1{,}612{,}800} = 52.4 \times 10^{-6} \text{ per hour}$$

The $(c^2)_{0.60,82}$ value was found by interpolation from Table B.2.

Assuming a constant rate of patient infection, the patient reliability over a 3-week (504-h) period is

$$R(t) = e^{-\lambda t} = e^{-0.0000524 \times 504} = 0.974$$

which is the probability that a patient is hospitalized for a 3-week period without acquiring a hospital-induced infection. Hence the risk of infection to a patient for 3 weeks of hospitalization is

$$Q(t) = 1 - R(t) = 1 - 0.974 = 0.026$$

4.9 FAILURE RATE PREDICTION TABLES

Failure rate handbooks traditionally report device failure rates in *failures per million hours* (or 10^{-6} *failures per hour*). During the 1950s, when few devices achieved the low magnitudes of failure rate that were later to be experienced, handbooks often reported failure rates in *failures per hundred thousand hours* (or 10^{-5} *failures per hour*). So, in these older sources you may see a reference to the archaic term percent per thousand hours (which, of course, is the same as 10^{-5} per hour). With some of today's electronic parts achieving increasingly lower failure rates, more recent failure rate sources have started to report their estimates in failures per 10^8 h.

4.10 FAILURE RATES FOR ELECTRONIC PARTS

The most comprehensive failure rate data source available is MIL-HDBK-217E, the *Military Handbook on Reliability Prediction of Electronic Equipment,* published by the Rome Air Development Center (RADC) in Rome, New York. The laboratories at RADC have been testing and compiling electronics component-failure data since the late 1950s. The results are the basis of the periodically updated handbook, which provides failure rate estimates for all types of components under a variety of operational and environmental conditions.

Originally, MIL-HDBK-217E listed 60% confidence estimates of all failure rates. However, as operating time accumulated through the years, the confidence estimates converged (as described in Example 4.7, so that *confidence* is no longer an issue in MIL-HDBK-217E. MIL-HDBK-217E also provides mathematical models for making operational (stress) and environmental adjustments to all listed failure rates. Methods for making such adjustments are discussed at the end of this chapter.

4.11 IMPACT OF NONOPERATING PERIODS

The plethora of failure rate data accumulated by RADC applies to electronic components during operation. However, reliability analysts, in making their predictions, are often just as interested in failure rates of components that are not operating. Spare components frequently experience an extended shelf life before being assigned to a system. Also, fault-tolerant systems, or systems with built-in redundancy, have within them components that are not active all the time. Duty cycling of systems or subsystems (discussed in a later chapter) also causes components to be inactive during part of a system's active life. Consequently, during the 1980s MIL-HDBK-217E added a section for failure rates for electronic components in dormant storage and in a standby state.

Nonoperating electronic components have an inherent failure rate far less than their active failure rates (often less than 1%). On the other hand, mechanical components could have dormant failure rates that are higher than their active failure rates. Consider, for example, a valve, which can freeze in the shut position if not exercised occasionally. At this time there are no recognized sources correlating active and dormant mechanical part failure rates. It is up to the reliability analyst to deal with duty cycling and standby functions of mechanical parts when evaluating the operating conditions under which the failure rate data are derived.

4.12 NONELECTRONIC PARTS

RADC also published a handbook called *Failure Rates for Nonelectronic Parts* (NPRD-3). Because the failure rate estimates in NPRD-3 are based upon a far less comprehensive data base than those of MIL-HDBK-217E, the format is different and more demanding on the analyst. The tables list parts and part types alphabetically in their first columns. The second columns list environments under which the failure data were derived (e.g., ground, ground-mobile, airborne, helicopter, etc.). The next two columns list the data bank information, i.e., part-hours of operation or number of part-cycles and the numbers of failures experienced. Finally, failure rate estimates are provided in failures per 10^6 hours or cycles. Best estimates as well as the 90% confidence bands are provided in the tables.

4.13 OTHER FAILURE RATE HANDBOOKS

The Government-Industry Data Exchange Program (GIDEP) provides for its members, among other services, handbooks on failure and restoration rates. The source of the data used in the GIDEP handbooks is failure information voluntarily submitted by member manufacturers and users of products. There are some industry-specific failure rate handbooks, such as Bell-Core for communications and the Edison Electric Industry (EEI) Handbook for the power industry. As service industries start to apply reliability analysis techniques, there will be need for data exchange and failure rate–estimation handbooks for food service, banking, hospitals, and others.

4.14 EFFECTS OF ENVIRONMENT AND STRESS

Occasionally data used for estimation of failure rates are derived under operating environments and stress levels that are different from those intended for the equipment for which the predicton is being made. Sometimes this difference is by design. For example, accelerated testing may deliberately be used to get an early estimate without extensive hours of expensive testing. In such instances, the failure rate estimates must be adjusted for the environmental difference. A well-known analytical technique for making the adjustment is the *Arrhenius model,* described here with an example of an operating temperature adjustment.

The Arrhenius model is a relationship between degradation and temperature, empirically derived by the Swedish physicist Arrhenius around 1880. He found a linear relation to exist between the logarithm of the degradation rate of a substance and the reciprocal of the absolute temperature of its environment. The relation is inverse; for example, as temperature increases, degradation rate (DR) increases:

$$\log \text{DR} = a - b(1/T) \tag{4.16}$$

where

a is the log DR intercept
b is the slope of the log DR–versus–$(1/T)$ line
T is the absolute temperature

This relation can also be expressed as

$$\text{DR} = Ae^{-(b/T)} \tag{4.17}$$

where $A = e^a$, which is another constant.

The Arrhenius model is backed up by more than a century of additional empirical data and is a simplified version of the Eyring model, which has been used successfully for years. It has been shown to be a valid technique for establishing a failure rate–versus-temperature relationship, which can be used as a model for accelerated reliability testing. The Arrhenius model can also be used to describe the effect other environmental stresses have upon failure rate.

To use the Arrhenius model for relating environmental temperature to failure rate, substitute the failure rate λ for DR in Equation 4.17:

$$\lambda = Ae^{-(b/T)} \tag{4.18}$$

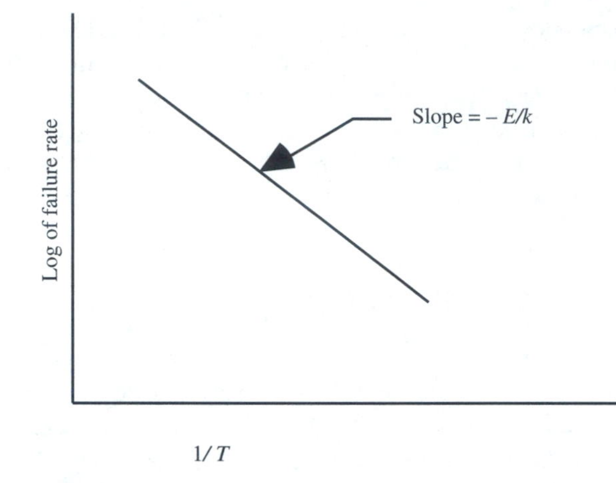

FIGURE 4.1
Relationship between a component failure rate and operating temperature, in K, according to the Arrhenius model: $\lambda = Ae^{-(E/kT)}$, where λ = failure rate, A = scaling constant, E = activation energy, k = Boltzmann's constant, and T = absolute temperature.

The slope of the log λ–versus–$(1/T)$ line has been found to be proportional to the activation energy E of the material of component on test:

$$E = kb \tag{4.19}$$

and k has been found to be *Boltzmann's constant,* 8.614×10^{-5} eV/K. Hence,

$$\lambda = Ae^{-(E/kT)} \tag{4.20}$$

as illustrated in Figure 4.1.

4.15 ACCELERATED TESTING

Figure 4.1 shows the expected relationship between the operating temperature of a component and its failure rate. As the temperature goes up, the expected failure rate goes up exponentially. So, on a logarithmic plot you would see a linear relationship between the logarithm of failure rate and the reciprocal of T (as $1/T$ goes up, log λ and λ go down). To apply this idea to accelerated testing, suppose we want to develop a failure rate estimate at a reasonable confidence level for a given component, but we want to do it in less time than it would ordinarily take.

If we test at higher than operating temperature, we would induce a higher-than-expected number of failures in a given operating time. Consequently, in that time we would achieve more confidence in the failure rate estimate. But that estimate would be too high, and it would have to be adjusted to the lower temperature applicable to the component's operation. The adjustment can be made according to the Arrhenius model, but to use this model we need estimates of the test constant A and of the *activation energy* of the component. Both these values can be estimated by testing at multiple temperatures (at least three) and then plotting log λ versus $1/T$.

Suppose, for example, that we wish to estimate the failure rate for a component that will be operated at 25°C and that we expect a constant failure rate in the magnitude of 10 to 100 failures per million hours. Obviously, we will need many component-hours of testing to get, say, a 60% confidence estimate. So, if we use the Arrhenius model to accelerate the test, we will need to test at three temperatures. Since it is not a good idea to extrapolate data on a linear plot (Figure 4.1), we shall use the operating temperature to do some of the testing. Then we shall select two other test temperatures that are reasonable and higher than the operating temperature. Let us say that we decide to test at 25°, 75°, and 150°C.

Now let us suppose that we have 100 components to test. We expect the highest failure rate at the 150°C test and the lowest at the 25°C test. To equalize the test time and the number of failures, we thus allocate more of our available components to the lower-temperature tests. We may decide to test 50 components at 25°C, 30 at 75°C, and 20 at 150°C. Based on the test results we make our 60% confidence estimates of failure rates (using Equation 4.14) at all three temperatures, and we plot them on a log $\lambda_{0.60}$–versus-temperature graph (Figure 4.2). We probably would use a log-scaled graph paper to be able to plot the failure rates directly instead of plotting the logs of failure rates. (Also, notice that the temperature scale on the horizontal axis of Figure 4.2 runs backward because the relationship in Figure 4.1 is actually log λ versus $1/T$).

Knowing that the Arrhenius relationship is linear, we then best fit a straight line through the three points plotted on the graph. The best-fit line can be determined using the linear regression technique described in Chapter 6. Now Figure 4.2 can be used to

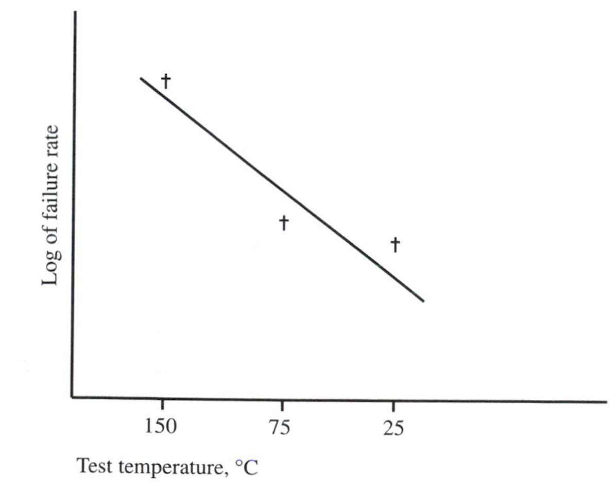

FIGURE 4.2

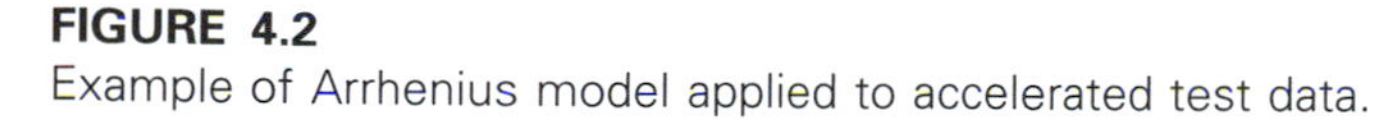
Example of Arrhenius model applied to accelerated test data.

estimate failure rates for any operating temperature. Estimates are valid for operating temperatures between 25°C and 150°C, because, as stated before, extrapolation of data is not good practice.

4.16 ENVIRONMENTAL AND STRESS FACTORS APPLIED TO MIL-HDBK-217E

A refinement of the Arrhenius model, known as the **Eyring model,** is used by RADC for dealing with stress and environmental adjustments to failure rate estimates. This is accomplished in much the same manner as the example given in the preceding paragraph. A typical failure rate–estimation model in MIL-HDBK-217E is in the form

$$\lambda_p = \lambda_b(\pi_E \times \pi_A \times \pi_Q \times \pi_R \times \pi_S \times \pi_C \times \cdots) \tag{4.21}$$

where
λ_p = the part failure rate estimate in failures per 10^6 hours
λ_b = the base failure rate estimate in failures per 10^6 hours
π_E = the environmental factor
π_A = the application factor
π_Q = the quality factor
π_R = the power-rating factor
π_S = the stress factor
π_C = the complexity factor

The π factors are all environmental and operational adjustments to a base failure rate λ_b: The environmental factor adjusts for the environment (e.g., laboratory, factory, highway vehicle, airborne, etc.); the application factor adjusts for the various uses of the part, the quality factor adjusts for fabrication materials, the amount of part screening, and other variables affecting quality; the power-rating factor adjusts for the power rating (watts) of the part; the stress factor adjusts for the amount of derating in the parts application; and the complexity factor (sometimes there is more than one) deals with adjustments for the electronic complexity of the device.

A typical failure estimation model for a microelectronic device in MIL-HDBK-217E is

$$\lambda_p = \pi_Q \pi_L (C_1 \pi_T \pi_V + C_2 \pi_E) \tag{4.22}$$

where
λ_p = the part failure rate estimate in failures per 10^6 hours
π_Q = the quality factor
π_L = the learning factor (the maturity of the design and part history)
C_1 = the failure rate (in failures per 10^6 hours) based on the circuit complexity
C_2 = the failure rate (in failures per 10^6 hours) based on the package complexity
π_E = the environmental factor

MIL-HDBK-217E used the Eyring model to develop temperature and stress factors. Other factors were developed subjectively, using methods such as the *Delphi technique,* which factors in the expert judgments of many users of the components.

Whereas reliability analysts in many fields of application do not at this time have the luxury of failure rate–estimation handbooks as comprehensive as MIL-HDBK-217E, it is necessary to be able to estimate failure rates from field and test data and, for that matter, from data in some failure rate handbooks. The next chapter deals with methods for evaluating failure rate data.

BIBLIOGRAPHY

1. I. Bazovsky. 1963. *Reliability Theory and Practice.* Upper Saddle River, N.J.: Prentice Hall.
2. P. D. T. O'Connor. 1991. *Practical Reliability Engineering,* 3d ed. New York: John Wiley.
3. M. L. Shooman. 1968. *Probabilistic Reliability: An Engineering Approach.* New York: McGraw-Hill.
4. C. O. Smith. 1976. *Introduction to Reliability in Design.* New York: McGraw-Hill.
5. MIL-HDBK-217E. 1986. *Reliability Prediction for Electronic Systems.* Arlington, Va.: Department of Defense.
6. MIL-HDBK-251. 1978. *Electronic System Reliability—Design Thermal Applications.* Arlington, Va.: Department of Defense.
7. L. J. Finison, K. S. Finison, and C. M. Bliersbach. 1993. "The Use of Control Charts to Improve Healthcare Quality," *Journal for Health Care Quality* (January/February), pp 9–23.

EXERCISES

4.1. A remotely controlled camera successfully photographed its subject 22 times in 24 attempts. Determine the 75% confidence estimate of the reliability for this system.

4.2. In words, express what is meant by the 75% confidence estimate of the reliability in Exercise 4.1.

4.3. Fifty sonabuoys are tested by dropping them into the ocean to see if they can detect a planted decoy. Forty-eight of them successfully detect the decoy, and the other two fail to do so. Determine the best estimate of the reliability of this model of sonabuoy based on these test results. Estimate the reliability at 95% confidence.

4.4 An embosser creates 36 successful images in 40 attempts. Based on this information, determine the best estimate, the 50% confidence estimate, and the 95% confidence estimate of its reliability.

4.5. Fifty computers of a given model are tested until 100,000 operating hours are accumulated among them. During that time six failures occur. Whenever a failure occurs, the computer is restored, and its operation continues. Based on these test results, compute the best estimate of MTBF for this computer model. Compute 60%, 80%, 90%, and 95% confidence estimates of MTBF based on the same test results.

4.6. A system is tested until the fifth failure occurs, at which time the test is stopped. Suppose the fifth failure occurs at 1257.3 equipment operating hours. Determine the best estimate, the 60% confidence estimate, and the 90% confidence estimate of the system MTBF.

4.7. A system is tested for 1000 h with no failures occurring. Determine the 60%, 80%, 90%, and 95% confidence estimates of the MTBF based on this test result.

4.8. Suppose 1000 integrated circuits of a given type are each operated for 1000 h. During the 1,000,000 part-hours of operation, four failures occur among them. Upon failure a part is replaced, and the test is resumed. Based on the four failures in 1,000,000 part-hours, determine a 60% confidence estimate of the failure rate of this type of integrated circuit for this operation. Compare the 60% confidence to the best estimate.

4.9. Use the best estimate, the 60% confidence estimate, and the 95% confidence estimate of the MTBF of the computer in Exercise 4.5 to determine its reliability for a 10,000-h mission. Assume constant failure rate.

4.10. In Exercise 4.7, use the 60% confidence estimate of MTBF to compute the probability that this system can operate another 1000 h without failure. Assume constant failure rate. Do the same using the 95% confidence estimate.

5

EVALUATING DATA FOR FAILURE RATE ESTIMATION

5.1 INTENT

The purpose of the seven chapters in Book 2 is to present methods for evaluating the capability of our system to meet the reliability objectives established in Book 1. We want to determine how likely it is that our system will satisfy our customers' demands at the demand times. The technique introduced in Chapter 3 is to determine system reliability through the estimates of failure rates of the system's constituent elements (or components). The previous chapter introduced methods for estimating component failure rates, including the use of failure rate handbooks. In this chapter we examine methods for failure rate estimation from existing component-failure data. We shall concentrate on the evaluation of reliability data for time-dependent operation, which, as we have already seen, is the more challenging.

5.2 RELIABILITY VERSUS OPERATING TIME

Suppose we have some time-dependent failure data to evaluate, where

at $t = 0$,	N_0 items are operating
at $t = t_1$,	$N_f(t_1)$ have failed and $N_s(t_1)$ survive
at $t = t_2$,	$N_f(t_2)$ have failed and $N_s(t_2)$ survive
⋮	

at $t = t_i$, $N_f(t_i)$ have failed and $N_s(t_i)$ survive

.
.
.

at $t = t_F$ N_0 have failed and *none* survive

We can plot $R(t)$ versus t *nonparametrically* (ignoring the relationships of Chapter 3) by plotting $N_s(t)$ versus t.

EXAMPLE 5.1 Suppose five items are put into operation and all are properly operating at time $t = 0$. We then allow them to continue and survey them every 100 h to see how many are surviving. Failed items are not replaced or restored. At the end of 100 h there are four survivors, at the end of 200 h there are three, at the end of 300 h there are still three, and at the end of 400 h there are two. The two survivors continue to survive for the next three surveys (at 500, 600, and 700 h), and at the end of 800 h there is only one survivor. The lone survivor continues to survive for the next three surveys and is observed to have failed by 1200 h. Plot $R(t)$ versus t directly from these data.

$$
\begin{aligned}
N_0 &= 5 & R(0) &= N_0/N_0 = 5/5 = 1.00 \\
N_s(100\text{ h}) &= 4 & R(100\text{ h}) &= N_s(100\text{ h})/N_0 = 4/5 = 0.80 \\
N_s(200\text{ h}) &= 3 & R(200\text{ h}) &= N_s(200\text{ h})/N_0 = 3/5 = 0.60 \\
N_s(400\text{ h}) &= 2 & R(400\text{ h}) &= N_s(400\text{ h})/N_0 = 2/5 = 0.40 \\
N_s(800\text{ h}) &= 1 & R(800\text{ h}) &= N_s(800\text{ h})/N_0 = 1/5 = 0.20 \\
N_s(1200\text{ h}) &= 0 & R(1200\text{ h}) &= N_s(1200\text{ h})/N_0 = 0/5 = 0.00
\end{aligned}
$$

See Figure 5.1.

EXAMPLE 5.2 Use the plot in Example 5.1 to estimate the reliability for a 600-h mission.

From the plot, $R(600\text{ h}) \sim 0.3$.

---

*5.3 FAILURE DENSITY FUNCTION VERSUS OPERATING TIME

If we wish to plot the failure density function, we refer to Equation 3.8:

$$f(t) = \frac{1}{N_0} \cdot \frac{dN_f(t)}{dt} \tag{5.1}$$

Thus, we can estimate $f(t)$ from our data by using the difference equation:

$$
\begin{aligned}
f(t) &= \frac{1}{N_0} \cdot \frac{\Delta N_f(t)}{\Delta t} \\
&= \Delta N_f(t)/(N_0 \times \Delta t)
\end{aligned} \tag{5.2}
$$

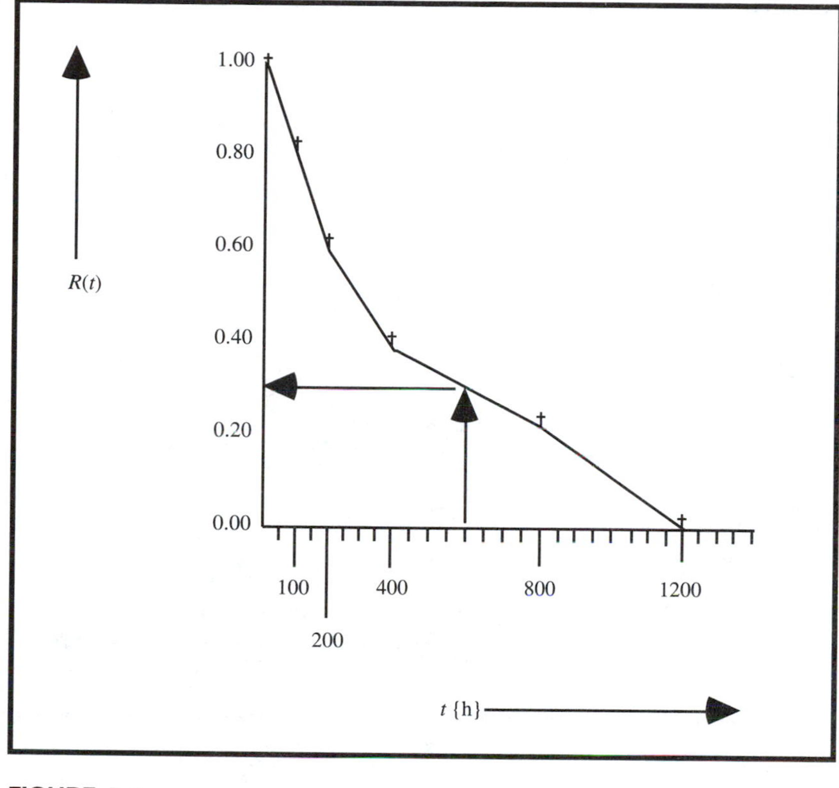

FIGURE 5.1
$R(t)$-versus-t plot for Example 5.1.

EXAMPLE 5.3

Plot $f(t)$ versus t for the data of Example 5.1. Use this failure density function plot to estimate the reliability for a 600-h mission.

Interval	Δt	$N_s(t)$ in	$N_f(t)$ in	$N_f(t)$ out	$\Delta N_f(t)$	$f(t)$
0–100 h	100 h	5	0	1	$1 - 0 = 1$	$1/(5 \times 100\text{ h}) = 0.002/\text{h}$
100–200 h	100 h	4	1	2	$2 - 1 = 1$	$1/(5 \times 100\text{ h}) = 0.002/\text{h}$
200–400 h	200 h	3	2	3	$3 - 2 = 1$	$1/(5 \times 200\text{ h}) = 0.001/\text{h}$
400–800 h	400 h	2	3	4	$4 - 3 = 1$	$1/(5 \times 400\text{ h}) = 0.0005/\text{h}$
800–1200 h	400 h	1	4	5	$5 - 4 = 1$	$1/(5 \times 400\text{ h}) = 0.0005/\text{h}$

See Figure 5.2.

From Equation 3.9, the unreliability for a 600-h mission is the area under the $f(t)$ curve from $t = 0$ to $t = 600$ h (see Figure 3.5). So the unreliability is

$$Q(600\text{ h}) = (100)(0.002) + (1000)(0.002) + (200)(0.001) + (200)(0.0005) = 0.70$$

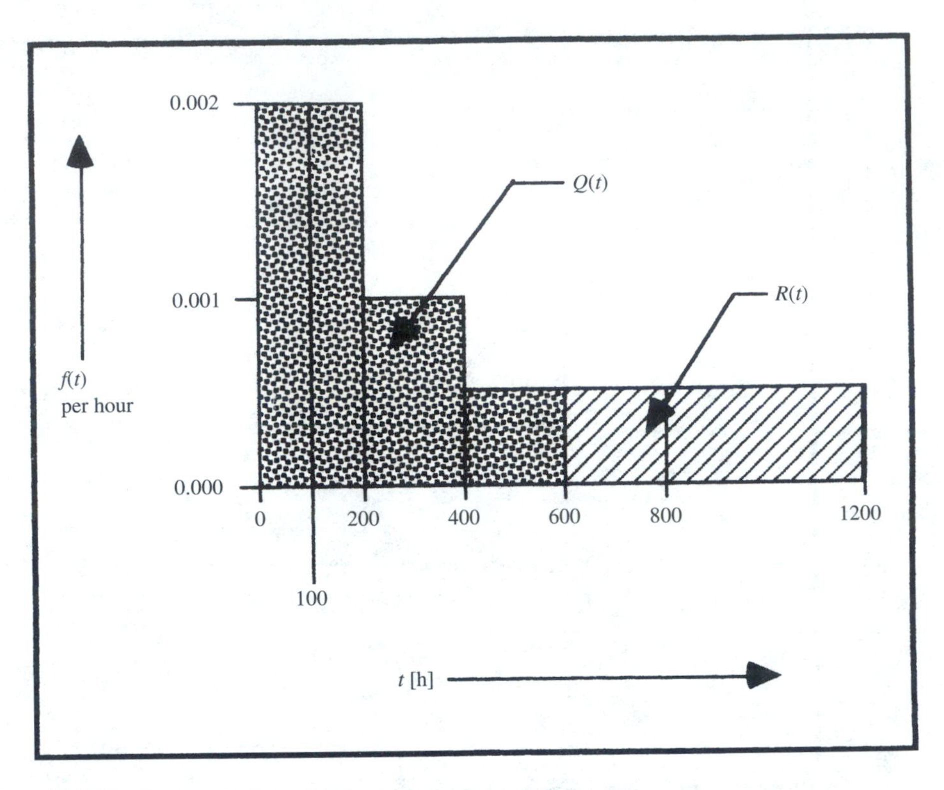

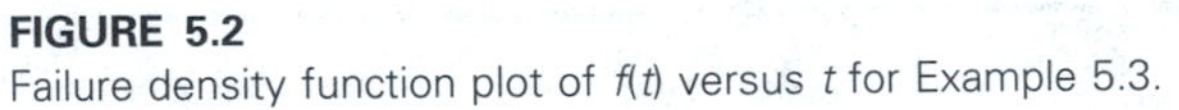

FIGURE 5.2
Failure density function plot of $f(t)$ versus t for Example 5.3.

From Equation 3.10, the reliability for a 600-h mission is the area under the $f(t)$ curve from t = 600 h to t = infinity (see Figure 3.5 again). So, the reliability is

$$R(600\text{ h}) = (200)(0.0005) + (400)(0.0005) = 0.30$$

*5.4 FAILURE RATE VERSUS OPERATING TIME

Equation 3.15 defines failure rate:

$$\lambda(t) = \frac{1}{N_s(t)} \times \frac{dN_f(t)}{dt} \tag{5.3}$$

Hence, we can estimate the time-dependent failure rate $\lambda(t)$ by the difference equation

$$\lambda(t) = \frac{1}{N_s(t)} \times \frac{\Delta N_f(t)}{\Delta t} \tag{5.4}$$

EXAMPLE 5.4 Plot $\lambda(t)$ versus t for the data of Examples 5.1 and 5.2. Use this failure rate plot to estimate the reliability for a 600-h mission.

Interval	Δt	$N_s(t)$ in	$N_f(t)$ in	$N_f(t)$ out	$\Delta N_f(t)$	$\lambda(t)$
0–100 h	100 h	5	0	1	$1 - 0 = 1$	$1/(5 \times 100 \text{ h}) = 0.002/\text{h}$
100–200 h	100 h	4	1	2	$2 - 1 = 1$	$1/(4 \times 100 \text{ h}) = 0.0025/\text{h}$
200–400 h	200 h	3	2	3	$3 - 2 = 1$	$1/(3 \times 200 \text{ h}) = 0.0017/\text{h}$
400–800 h	400 h	2	3	4	$4 - 3 = 1$	$1/(2 \times 400 \text{ h}) = 0.00125/\text{h}$
800–1200 h	400 h	1	4	5	$5 - 4 = 1$	$1/(1 \times 400 \text{ h}) = 0.0025/\text{h}$

The failure rate $\lambda(t)$ is either increasing, decreasing, or constant with operating time. The plot shown in Figure 5.3 does not exhibit evidence of an increasing or decreasing trend. The variation seen on this plot may very well be simply a natural variation expected with any type of sampling. So, there is every reason to assume that we have a constant failure rate, and the best estimate from the plotted data (the average) indicates an approximate failure rate of $\lambda = 0.002$ failures per hour (the dotted line on the plot).

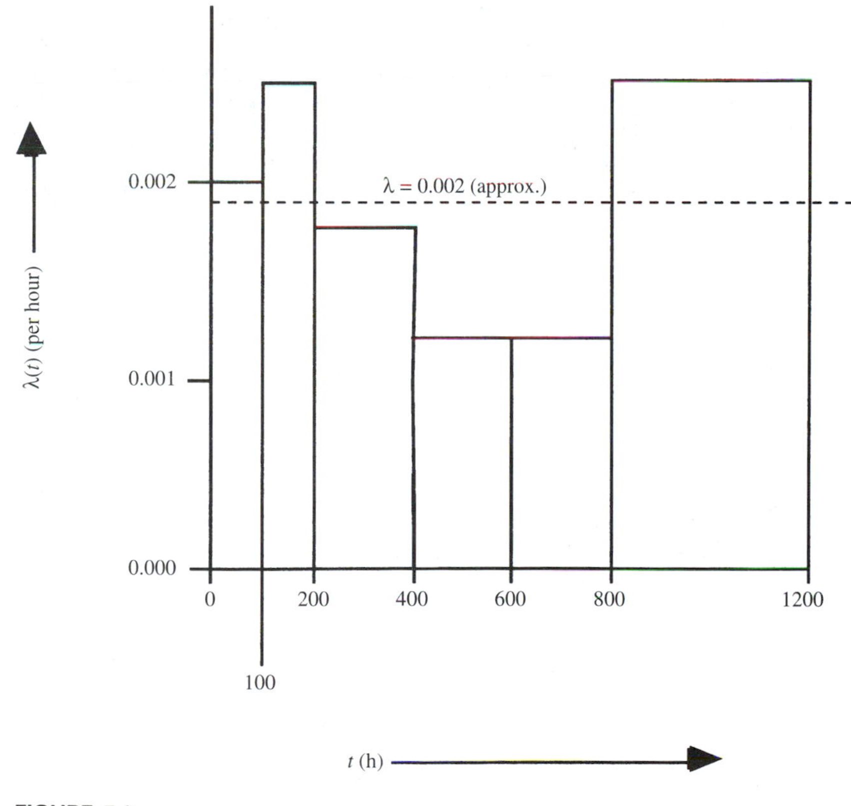

FIGURE 5.3
Plot of failure rate versus time for Example 5.4.

If failure rate is constant, we can compute a reliability estimate from Equation 3.20:

$$R(t) = e^{-\lambda t} = e^{-(0.002/\text{h} \times 600\text{h})} = e^{-1.2} = 0.30$$

If we do not make the assumption of constant failure rate, then we know reliability $R(t)$ is, from Equation 3.18,

$$R(t) = e^{-\int_0^t \lambda(T)\, dT}$$

The best estimate of $\int_0^t \lambda(T)\, dT$ is $\Sigma[\lambda(t) \times \Delta\, t]$ for all $t < 600$ h (the shaded area in the plot) is

$$(0.002)(100) + (0.0025)(100) + (0.0017)(200) + (0.00125)(200) = 1.04$$

and reliability for a 600-h mission is estimated

$$R(t) = e^{-1.04} = 0.35$$

The value $e^{-1.04}$ can be determined from Table B.8.

Notice that when treating failure rate as a variable, because the failure rate has its lowest value at 600 h, we get a higher estimate of reliability than when we treated failure rate as a constant. As observed before, since we know that if a failure rate is not constant, it must either be increasing or decreasing; because the plot gives no clear sign of either, the constant failure rate assumption is reasonable. If the plot had indicated that the failure rate is, say, increasing with operating time, the second analysis technique before (that is, first estimating the instantaneous failure rate at 600 h and then using that estimate for the failure rate prediction, as shown) would have been appropriate. However, based on our analyses in Exercises 5.1 through 5.4, our reliability prediction for a 600-h mission is 0.30.

EXAMPLE 5.5

Assuming constant failure rate for the data, as presented in Exercise 5.1, determine the reliability for a 600-h mission from the best estimate of failure rate and then by using a 60% confidence estimate of failure rate (as learned in Chapter 4).

From the data in Exercise 5.1, the first of the items put into operation fails between 0 and 100 h. Let us assume that it happens halfway through that interval, at 50 h. Similarly, we can assume that the other four items fail at 150, 350, 750, and 1150 h (halfway through their intervals). This gives us a total of 2450 part operating hours with five failures. The best estimate of failure rate from this data is 5/2450 h = 0.0020 per hour. From this estimate of constant failure rate, we can predict the reliability for a 600-h mission to be

$$R(600\text{ h}) = e^{-\lambda t} = e^{-0.0020 \times 600} = e^{-1.20} = 0.301$$

The 60% confidence estimate (from Equation 4.14) is

$$\lambda_{0.60} = (\chi^2)_{\alpha,2r}/2T$$

where $\alpha = 0.60$, $r = c = 5$ (failure terminated), and $T = 2450$ h. Thus,

$$\lambda_{0.60} = (\chi^2)_{0.60,10}/4900\text{ h} = 10.5/4900\text{ h} = 0.0021$$

giving us a failure rate prediction of

$$R(600\text{ h}) = e^{-\lambda t} = e^{-0.0021\times 600} = e^{-1.26} = 0.284$$

Notice that, as expected, when we use a confidence estimate of the failure rate, we get a more pessimistic prediction of reliability than we do by using the best estimate or than we did in the preceding exercises.

5.5 GOODNESS-OF-FIT TESTS

Either of the goodness-of-fit techniques described in Section A.10 (Appendix A) can be used for determining analytically whether or not a failure rate is constant. This determination can be accomplished by examining the operating time in intervals, comparing the expected number of failures to the observed number of failures during the interval, and measuring the comparison for significance of their differences.

If failure rate is constant, the probability of a failure occurring in the interval between a and b hours (see Section A.10) is

$$P(a < t < b) = e^{-\lambda a} - e^{-\lambda b} \quad \textbf{(5.5)}$$

where λ is the estimated constant failure rate. The expected number of failures during the interval between a and b hours is then

$$e_{a,b} = N \times P(a < t < b) = N[e^{-\lambda a} - e^{-\lambda b}] \quad \textbf{(5.6)}$$

where N is the total number of items operated until failure. If we let $o_{a,b}$ represent the actual observed number of failures during the interval a, b, then the discrepancy for the interval is $(e_{a,b} - o_{a,b})$.

When using the *Chi-Square Test for goodness of fit,* we determine the discrepancy between expected and observed number of failures $(e_i - o_i)$ for each interval i (the intervals must be equal in size) and then determine the chi-square statistic, which is the sum of the squares of all discrepancies divided by their expected number of failures:

$$\chi^2 = \Sigma\,[(e_i - o_i)^2/e_i] \quad \textbf{(5.7)}$$

If the chi-square does not exceed the critical value taken from Table B2 for a preassigned significance level, usually 5%, then the assumption of constant failure rate is accepted. The application of the chi-square test for the determination of constant failure rate is demonstrated in Example 5.6.

EXAMPLE 5.6 Use the Chi-Square Test for goodness of fit at the 5% significance level to assess whether the data in Examples 5.1 to 5.5 support the assumption of constant failure rate.

It is good practice to set up between 5 and 12 equal time intervals for this test. In this case, 6 intervals of 200 h each will be convenient. To allow the test to work, the sixth interval will go from 1000 h to infinity. From the data presented in Example 5.1, there were two failures during the first 200-h interval, one during the second 200-h interval, none during the third, one during the fourth, none during the fifth, and one during the sixth. These are the o_i values, so

$$o_1 = 2, \quad o_2 = 1, \quad o_3 = 0, \quad o_4 = 1, \quad o_5 = 0, \quad o_6 = 1$$

Using the assumed constant failure rate $\lambda = 0.002/\text{h}$ and Equation 5.5 to determine probabilities of a failure in each interval,

$$\begin{aligned}
P_1 = P(0 < t < 200\text{ h}) &= e^{-0.002\times 0} - e^{-0.002\times 200} \\
&= 1.0000 - 0.6703 = 0.330 \\
P_2 = P(200 < t < 400\text{ h}) &= e^{-0.002\times 200} - e^{-0.002\times 400} \\
&= 0.6703 - 0.4493 = 0.221 \\
P_3 = P(400 < t < 600\text{ h}) &= e^{-0.002\times 400} - e^{-0.002\times 600} \\
&= 0.4493 - 0.3012 = 0.148 \\
P_4 = P(600 < t < 800\text{ h}) &= e^{-0.002\times 600} - e^{-0.002\times 800} \\
&= 0.3012 - 0.2019 = 0.099 \\
P_5 = P(800 < t < 1000\text{ h}) &= e^{-0.002\times 800} - e^{-0.002\times 1000} \\
&= 0.2019 - 0.1353 = 0.067 \\
P_6 = P(t > 1000\text{ h}) &= e^{-0.002\times 1000} - e^{-\infty} \\
&= 0.1353 - 0.0000 = 0.135
\end{aligned}$$

We can compute the expected number of failures in each interval from Equation 5.6, $e_i = NP_i$, where $N = 5$:

$$\begin{aligned}
e_1 &= 5 \times P_1 = 5 \times 0.330 = 1.65 \\
e_2 &= 5 \times P_2 = 5 \times 0.221 = 1.11 \\
e_3 &= 5 \times P_3 = 5 \times 0.148 = 0.74 \\
e_4 &= 5 \times P_4 = 5 \times 0.099 = 0.50 \\
e_5 &= 5 \times P_5 = 5 \times 0.067 = 0.34 \\
e_6 &= 5 \times P_6 = 5 \times 0.135 = 0.68
\end{aligned}$$

The remaining computations for the test are summarized in table form.

Interval	1	2	3	4	5	6
TTF	0–200 h	200–400 h	400–600 h	600–800 h	800–1000 h	>1000 h
P_i	0.330	0.221	0.148	0.099	0.067	0.135
$e_i = 5P_i$	1.7	1.1	0.7	0.5	0.3	0.7
o_i	2	1	0	1	0	1
$\|e_i - o_i\|$	0.3	0.1	0.7	0.5	0.3	0.3
$(e_i - o_i)^2/e_i$	0.05	0.01	0.70	0.50	0.30	0.13

The chi-square statistic is the sum of the last row in the table (Equation 5.7):

$$\chi^2 = \Sigma\,[(e_i - o_i)^2/e_i] = 0.05 + 0.01 + 0.70 + 0.50 + 0.30 + 0.13 = 1.69$$

We now look up the critical chi-square value in Table B.2: $(\chi^2)_{1-\alpha,f}$, where α is the significance level (0.05) and f is the degrees of freedom, which for this test is 2 less than the number of intervals in the test. In this example, where there are 6 intervals, $f = 6 - 2 = 4$. In Table B.2,

$$(\chi^2)_{1-\alpha,f} = (\chi^2)_{0.95,4} = 9.5$$

Since the chi-square statistic for the test (1.69) does not exceed the critical value (9.5), we can accept the assumption of constant failure rate at the 5% significance level. In other words, the test does not present significant reason to reject our assumption that we are dealing with a constant failure rate of 0.002 per hour.

Now let us test the assumption of constant failure rate using another goodness-of-fit test, the *Kolmogorov-Smirnov Test.* Analytically, this is a simpler test to perform. As with the Chi-Square Test, we look at the failures that occur during each of 5 to 12 equal time intervals, but this time we compare the cumulative expected probability of failure F_i by the end of each interval with the observed cumulative percent F_n at the end of each interval. We accept our assumption at the α significance level if

$$\max |F_n - F_i| < \text{K-S}_\alpha \tag{5.8}$$

where $\max |F_n - F_i|$ is the largest absolute difference between F_n and F_i for any interval

K-S_α is the Kolmogorov-Smirnov value in Table B.3 at the α significance level and appropriate number of items operated until failure

---

EXAMPLE 5.7 Use the Kolmogorov-Smirnov Test for goodness of fit at the 5% significance level to assess whether the data in Examples 5.1 to 5.5 support the assumption of constant failure rate.

Let us use the same intervals that we used in Example 5.6. The P_i values from Example 5.6 also apply. Then the F_i values can be computed by summing the P_i values as follows: $F_1 = P_1 \cdot F_2 = P_1 + P_2$, etc. The F_n value for each interval is the total number of failures through the end of the interval divided by the total number of items tested to failure, in this case 5:

for the first interval,	$F_n = 2/5 = 0.400$
for the second interval,	$F_n = 3/5 = 0.600$
for the third interval,	$F_n = 3/5 = 0.600$
for the fourth interval,	$F_n = 4/5 = 0.800$
for the fifth interval,	$F_n = 4/5 = 0.800$
for the sixth interval,	$F_n = 5/5 = 1.000$

The remaining computations are again summarized in table form.

TTF Interval	P_i	F_i	F_n	$\lvert F_n - F_i \rvert$
0–200 h	0.330	0.330	0.400	0.070
200–400 h	0.221	0.551	0.600	0.049
400–600 h	0.148	0.699	0.600	**0.099 ← max**
600–800 h	0.099	0.798	0.800	0.002
800–1000 h	0.067	0.865	0.800	0.065
>1000 h	0.135	1.000	1.000	0.000

Notice that the largest difference is 0.099. This has to be tested against the critical K-S value. In Table B.3, where $\alpha = 0.05$ and the number of items tested until failure is 5, the K-S value is 0.563. Since the maximum $\lvert F_n - F_i \rvert$ is less than 0.563, we can once again accept our assumption of constant failure rate.

In the examples provided we examined some failure data from parts put into operation and used the data to predict reliability for a mission. By inspection of a plot of the failure data, we surmised that the failure rate is probably constant, and we estimated that constant failure rate to be about 0.002 per hour. This led us to a reliability prediction of 0.30 for a 600-h mission. The accuracy of that prediction, however, depends on our confidence in the failure rate estimate and on the validity of the constant failure rate assumption. In Example 5.5 we calculated a 60% confidence estimate of the failure rate and found it to be reasonably close to our best estimates determined both analytically and graphically. So, we can say we have confidence in our failure rate estimate, given that the failure rate can be assumed to be constant.

Examples 5.6 and 5.7 tested how well the data fit our constant failure rate assumption by comparing observed outcomes with outcomes expected if our assumption is true. If a goodness-of-fit test indicates *significance*—that is, that there is a significant difference between observed and expected outcomes—then we are advised to abandon our constant failure rate assumption. If the goodness-of-fit test indicates *no significance,* as was the case in Examples 5.6 and 5.7, we may accept our assumption.

Either the Chi-Square or Kolmogorov-Smirnov Test for goodness of fit (whichever you prefer or whichever is more convenient) can be used for analytically gaining assurance in our assumption of constant failure rate for some component or phenomenon. Chapter 6 provides methods for doing the same thing graphically. For further insight into the Chi-Square and Kolmogorov-Smirnov Tests for goodness of fit, read Section A.10 (Appendix A), which shows how these techniques are used for a variety of purposes in testing statistical assumptions. The following examples further demonstrate their use in testing constant–failure rate assumptions.

EXAMPLE 5.8 Ten devices are put into operation and are continuously monitored until they all fail. The times to failure of the devices are as follows.

Failure No.	Time (h)	Failure No.	Time (h)
1	8	6	86
2	20	7	111
3	34	8	141
4	46	9	186
5	63	10	266

Test the assumption of constant failure rate by means of the Chi-Square Test for goodness of fit at the 5% significance level.

If the failure rate is constant, the best estimate is the number of failures, 10, divided by the total device-operating time, T:

$$T = 8 + 20 + 34 + 46 + 63 + 86 + 111 + 141 + 186 + 266 = 961 \text{ h}$$

Hence, the best estimate of the failure rate, if constant, is

$$\lambda = 10/961 \text{ h} = 0.010 \text{ per hour}$$

Let us determine the 60% confidence estimate of the failure rate:

$$r = 10 \text{ (the operation is failure-terminated)} \qquad T = 961 \text{ h}$$
$$\lambda_{0.60} = (\chi^2)_{0.60,20}/2(961 \text{ h}) = 21.0/1922 \text{ h} = 0.0109 \text{ per hour}$$

So we have enough data to be confident in our estimate of $\lambda \sim 0.01$ per hour.

Let us now divide our operating time into six equal time intervals of 50 h.

Time Interval	Observed No. of Failures	Time Interval	Observed No. of Failures
0–50 h	4	150–200 h	1
50–100 h	2	200–250 h	0
100–150 h	2	250–300 h	1

We determine the P_i values from Equation 5.5:

$$P(a < x < b) = e^{-\lambda a} - e^{-\lambda b} = e^{-0.01a} - e^{-0.01b}$$

So,

$$P_1 = P(0 < x < 50) = e^{-0} - e^{-0.5} = 1.000 - 0.607 = 0.393$$
$$P_2 = P(50 < x < 100) = e^{-0.5} - e^{-1.0} = 0.607 - 0.368 = 0.239$$
$$P_3 = P(100 < x < 150) = e^{-1.0} - e^{-1.5} = 0.368 - 0.223 = 0.145$$
$$P_4 = P(150 < x < 200) = e^{-1.5} - e^{-2.0} = 0.223 - 0.135 = 0.088$$
$$P_5 = P(200 < x < 250) = e^{-2.0} - e^{-2.5} = 0.135 - 0.082 = 0.053$$

For the Chi-Square Test it is customary to allow the final duration go to infinity, so we let $P_6 = P(x > 250) = 0.082$.

The remaining computations are completed in the following table.

Interval	1	2	3	4	5	6
TTF	0–50 h	50–100 h	100–150 h	150–200 h	200–250 h	>250 h
P_i	0.393	0.239	0.145	0.088	0.053	0.082
$e_i = 10P_i$	3.93	2.39	1.45	0.88	0.53	0.82
o_i	4	2	2	1	0	1
$\lvert e_i - o_i \rvert$	0.07	0.39	0.55	0.12	0.53	0.18
$(e_i - o_i)^2/e_i$	0.001	0.064	0.209	0.016	0.530	0.040

The chi-square statistic is $\chi^2 = \Sigma(e_i - o_i)^2/e_i = 0.860$.

The number of degrees of freedom for this test is always 2 less than the number of intervals (as discussed in Section A.10 of Appendix A) $f = 6 - 2 = 4$. The significance level is $\alpha = 0.05$.

Now we go to Table B.2 to look up the critical value:

$$(\chi^2)_{1-\alpha,f} = (\chi^2)_{0.95,4} = 9.5$$

Because the chi-square statistic, 0.860, is less than this critical value, we conclude that there is no significance—that is, there is no significant reason to doubt our assumption of constant failure rate. So we can comfortably use our estimate $\lambda = 0.01$ failures per hour to make our predictions.

Notice that the conclusion to a goodness-of-fit test is a rather weak statement ("there is no significant reason to doubt our assumption that . . ."). As explained in Appendix A, this type of statistical test is applied when we already have reason to believe that our assumption is true, and we shall abandon the assumption only upon overwhelming statistical evidence that it is not true.

EXAMPLE 5.9

Test the assumption of constant failure rate for the data in Example 5.8 by means of the Kolmogorov-Smirnov Test for goodness of fit at the 5% significance level.

Let us divide the operating time into the same six intervals used in Example 5.8. Therefore, the same P_i values apply.

The remaining computations are again summarized in table form.

TTF Interval	P_i	F_i	F_n	$\lvert F_n - F_i \rvert$
0–50 h	0.393	0.393	0.400	0.007
50–100 h	0.239	0.632	0.600	0.032
100–150 h	0.145	0.777	0.800	0.023
150–200 h	0.088	0.865	0.900	**0.035 ← max**
200–250 h	0.053	0.918	0.900	0.018
>250 h	0.082	1.000	1.000	0.000

The maximum $|F_n - F_i|$ is 0.035. This value is to be compared with the critical K-S value in Table B.3 for the 0.05 significance level and the 10 devices operated until failure. That value is 0.409. Since $\max|F_n - F_i| = 0.035$ does not exceed the critical value, we conclude once again that there is no significance—that is, there is no significant reason to doubt our assumption of constant failure rate.

EXAMPLE 5.10 A set of 172 components are put on test and operated until failure. They are surveyed at 1000-h intervals with the following results.

Hour	Number of Survivors
1000	113
2000	89
3000	60
4000	30
5000	13
6000	0

Test the assumption of constant failure rate by means of the chi-square test for goodness of fit at the 5% significance level.

Let us divide the operating time into six 1000-h intervals. The numbers of observed failures during each interval are as follows.

Interval	TTF	No. of Survivors	No. of Failures
1	0–1000 h	113	59
2	1000–2000 h	89	24
3	2000–3000 h	60	29
4	3000–4000 h	30	30
5	4000–5000 h	13	17
6	5000–6000 h	0	13

To calculate the total component operating hours, assume that the time to failure for each component is halfway through the interval in which it fails. So,

$$T = (59)(500) + (24)(1500) + (29)(2500) + (30)(3500) + (17)(4500) + (13)(5500)$$
$$= 391{,}000 \text{ h}$$

The best estimate of failure rate, if constant, is 172/391,000 h = 0.000450 per hour. The P_i values can once again be determined through Equation 5.5:

$$P(a < x < b) = e^{-\lambda a} - e^{-\lambda b} = e^{-0.0005a} - e^{-0.0005b}$$

So, $P_1 = P(0 < x < 1000 \text{ h}) = e^{-0} - e^{-0.5} = 1.000 - 0.607 = 0.393$
$P_2 = P(1000 < x < 2000 \text{ h}) = e^{-0.5} - e^{-1.0} = 0.607 - 0.368 = 0.239$
$P_3 = P(2000 < x < 3000 \text{ h}) = e^{-1.0} - e^{-1.5} = 0.368 - 0.223 = 0.145$
$P_4 = P(3000 < x < 4000 \text{ h}) = e^{-1.5} - e^{-2.0} = 0.223 - 0.135 = 0.088$
$P_5 = P(4000 < x < 5000 \text{ h}) = e^{-2.0} - e^{-2.5} = 0.135 - 0.082 = 0.053$
$P_6 = P(x > 5000) = e^{-2.5} = 0.082$

As before, the remaining computations are completed in the following table:

Interval	1	2	3	4	5	6
TTF	0–1000 h	1000–2000 h	2000–3000 h	3000–4000 h	4000–5000 h	>5000 h
P_i	0.393	0.239	0.145	0.088	0.053	0.082
$e_i = 172P_i$	67.60	41.11	24.94	15.14	9.12	14.10
o_i	59	24	29	30	17	13
$\lvert e_i - o_i \rvert$	8.6	17.11	4.06	14.86	7.88	1.10
$(e_i - o_i)^2/e_i$	1.09	7.12	0.66	14.59	6.81	0.09

The chi-square statistic is $\chi^2 = \Sigma\ (e_i - o_i)^2/e_i = 30.36$.

The number of degrees of freedom for this test is always 2 less than the number of intervals (as discussed in Section A.1.10 of Appendix A): $f = 6 - 2 = 4$. The significance level is $\alpha = 0.05$.

Now we go to Table B.2 to look up the critical value:

$$(\chi^2)_{1-\alpha,f} = (\chi^2)_{0.95,4} = 9.5$$

Because the chi-square statistic, 30.36, is more than this critical value, we conclude that there *is* significance—that is, there is significant reason to doubt our assumption of constant failure rate. So we reject the assumption and will have to treat the failure rate as time dependent when making our reliability predictions.

EXAMPLE 5.11 Test the assumption of constant failure rate for the data in Example 5.10 by means of the Kolmogorov-Smirnov Test for goodness of fit at the 5% significance level.

Let us divide the operating time into the same six intervals used in Example 5.10. Therefore, the same P_i values will apply.

The remaining computations are again summarized in table form.

TTF Interval	P_i	F_i	F_n	$\lvert F_n - F_i \rvert$
0–1000 h	0.393	0.393	0.343	0.050
1000–2000 h	0.239	0.632	0.483	**0.149 ← max**
2000–3000 h	0.145	0.777	0.651	0.126
3000–4000 h	0.088	0.865	0.824	0.041
4000–5000 h	0.053	0.918	0.924	0.006
>5000 h	0.082	1.000	1.000	0.000

The maximum $|F_n - F_i|$ is 0.149. This is to be compared with the critical K-S value in Table B.3 for the 0.05 significance level and the 172 devices operated until failure. Table B.3 indicates that the critical value for $\alpha = 0.05$ is computed as

$$\text{K-S} = 1.36/N^{1/2} = 1.36/(172)^{1/2} = 0.104$$

Since $\max|F_n - F_i| = 0.149$ exceeds the critical value, we conclude once again that there *is* significance—that is, there is significant reason to doubt our assumption of constant failure rate.

Although the examples presented here dealt with failure rates of hardware, the goodness-of-fit techniques can also be applied to service areas. For example, a Chi-Square or Kolmogorov-Smirnov Test can be used to test an assumption that an error frquency has a constant rather than varying rate.

BIBLIOGRAPHY

1. I. Bazovsky. 1963. *Reliability Theory and Practice.* Upper Saddle River, N.J.: Prentice Hall.
2. A. Hald. 1952. *Statistical Theory with Engineering Applications.* New York: John Wiley.
3. P. D. T. O'Connor. 1992. *Practical Reliability Engineering,* 3d ed. New York: John Wiley.
4. M. L. Shooman. 1968. *Probabilistic Reliability: An Engineering Approach.* New York: McGraw-Hill.
5. C. O. Smith. 1976. *Introduction to Reliability in Design.* New York: McGraw-Hill.

EXERCISES

5.1. A piece of equipment is put on test, and the tenth failure occurs at the end of exactly 10,000 h of testing. The test is then stopped. The true failure rate is λ. What is the best estimate of λ? What is the 60% confidence estimate of λ? Assume constant failure rate.

5.2. Find the 90% and 95% confidence estimates of the failure rate of the equipment in Exercise 5.1.

5.3. In words, what does "90% confidence estimate of the failure rate" mean?

5.4. For the equipment in Exercise 5.1, use the 60% confidence estimate of λ to compute the MTBF.

5.5. For the equipment in Exercise 5.1, use the 60% confidence estimate to compute the reliability for a 100-h mission.

5.6. A piece of equipment is composed of the following components, all of which must operate for equipment success. Component failure rate estimates are determined to be as listed.

Component	Estimated Failure Rate (per Component)	Quantity in Equipment
Integrated circuit A	5.9×10^{-8} per hour	9
Integrated circuit B	4.4×10^{-8} per hour	3
Integrated circuit C	10.3×10^{-8} per hour	1
Resistor	0.174×10^{-8} per hour	4
Capacitor A	0.168×10^{-8} per hour	1
Capacitor B	0.280×10^{-8} per hour	14
Printed wiring board	0.590×10^{-8} per hour	1
Connector	2.970×10^{-8} per hour	1

a. Determine the MTBF of the equipment.
b. Compute the reliability for a 1000-h mission.
c. Compute the reliability for a 100,000-h mission.

6

GRAPHICAL EVALUATION FOR RELIABILITY PREDICTION

6.1 INTENT

In this chapter we shall work with two types of graph paper, *exponential* and *Weibull* paper, both of which have been designed to accomplish graphically the same evaluations that we performed analytically in Chapter 5. They can be used to test the assumption of constant, versus increasing or decreasing, failure rate. They can also be used to estimate a failure rate and to predict mission reliability. Both exponential and Weibull papers are designed so that if they are applicable (that is, the data points fit the assumed distribution), the plotted points will form a straight line. So before introducing them, let us examine the technique of *linear regression analysis,* that of

1. Finding the best-fit straight line through a set of points,
2. Determining whether the linear assumption is valid, and
3. Determining that the data adequately fit the assumed straight line.

*6.2 LINEAR REGRESSION ANALYSIS

Given N sets of data $\{(x_1, y_1), (x_2, y_2), \cdots (x_N, y_N)\}$ that appear, when plotted, to display a linear relation between x and y, our objective is to find the best-fit straight line through all the plotted points. As illustrated in Figure 6.1, **linear regression analysis** determines *the best-fit line* in the *least square sense*—that is, the line that minimizes the sums of the squares of the vertical distances between it and the plotted points. Suppose that the best-fit line is

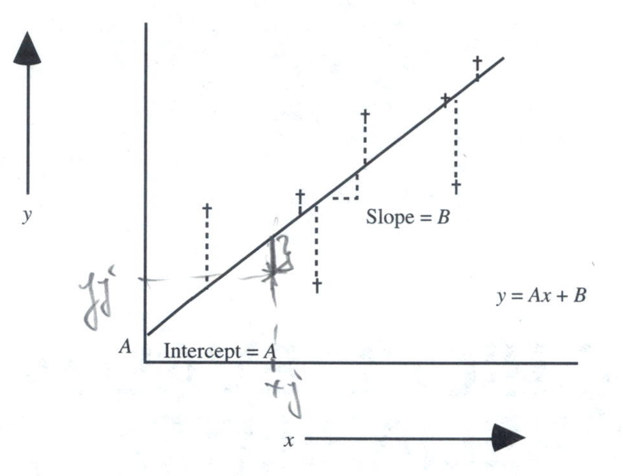

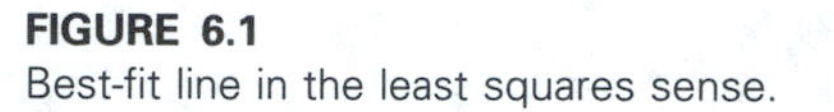

FIGURE 6.1
Best-fit line in the least squares sense.

$$y = A + Bx \tag{6.1}$$

For a given point (x_j, y_j), the vertical distance between that point and the line is

$$y_j - \{A + Bx_j\} \tag{6.2}$$

and the square of that distance is

$$(y_j - \{A + Bx_j\})^2 \tag{6.3}$$

The sum of the vertical distances between all the points and the line is

$$\sum_{j=1}^{N}(y_j - \{A + Bx_j\})^2 = \sum_{j=1}^{N}(y_j - A - Bx_j)^2 \tag{6.4}$$

By our definition, the best-fit straight line is determined by minimizing Equation 6.4. So, if we define

$$S = \sum_{j=1}^{N}(y_j - A - Bx_j)^2 \tag{6.5}$$

we have to find min S, which is that value of S such that

$$\frac{\partial S}{\partial A} + \frac{\partial S}{\partial B} = 0 \tag{6.6}$$

or of S such that

$$\frac{\partial S}{\partial A} = 0 \tag{6.7}$$

and

$$\frac{\partial S}{\partial B} = 0 \tag{6.8}$$

TABLE 6.1
Least squares summary table.

j	x_j	y_j	$x_j y_j$	x_j^2
1				
2				
3				
.				
.				
.				
N				
	$\Sigma = \square$	$\Sigma = \square$	$\Sigma = \square$	$\Sigma = \square$

Solving Equation 6.7,

$$\frac{\partial S}{\partial A} = 2\left[\sum_{j=1}^{N}(y_j - A - Bx_j)(-1)\right] = -2\sum_{j=1}^{N} y_j + 2NA + 2B\sum_{j=1}^{N} x_j = 0 \qquad \textbf{(6.9)}$$

Solving Equation 6.8,

$$\frac{\partial S}{\partial B} = 2\left[\sum_{j=1}^{N}(y_j - A - Bx_j)(-x_j)\right] = -2\sum_{j=1}^{N} x_j y_j + 2A\sum_{j=1}^{N} x_j + 2B\sum_{j=1}^{N} x_j^2 = 0 \qquad \textbf{(6.10)}$$

Hence the best-fit straight line in the least squares sense can be determined by solving Equations 6.9 and 6.10 for A and B and by inserting those values of A and B into Equation 6.1. The format of Table 6.1 will be useful in solving for A and B.

In the sections that follow we plot an $f(t)$-versus-t graph, determine a best-fit straight line through the points, and test the assumption of linearity. The best-fit straight line is

$$f(x) = A + Bx$$

where A and B are solutions of the equations

$$NA + (\Sigma x)B = \Sigma f(x) \qquad \textbf{(6.11)}$$

$$(\Sigma x)A + (\Sigma x^2)B = [\Sigma x f(x)] \qquad \textbf{(6.12)}$$

The test for the assumption of linearity is as follows.

6.3 LEAST SQUARES TEST FOR GOODNESS OF FIT

The least squares test can be used wherever there is a presumed linear relationship between x and $f(x)$. It is performed via the following steps.

1. Plot x versus $f(x)$.

2. Use the method of linear regression to determine the best-fit line, in the least square sense, through the x-versus-$f(x)$ points.

3. Compute the **correlation coefficient**

$$r_{x,f(x)} = \frac{\overline{x \cdot f(x)} - (\bar{x}) \cdot [\overline{f(x)}]}{\sigma_x \cdot \sigma_{f(x)}} = \frac{\Sigma[xf(x)] - \dfrac{(\Sigma x)[\Sigma f(x)]}{N}}{\sqrt{\left[\Sigma(x^2) - \dfrac{(\Sigma x)^2}{N}\right]\left[\Sigma[f(x]^2 - \dfrac{[\Sigma f(x)]^2}{N}\right]}} \quad \textbf{(6.13)}$$

where N is the number of data points plotted.

4. From a theorem in probability, we know that the correlation coefficient must be between -1.0 and $+1.0$:

$$|r| \leq 1.0$$

If $|r| = 1.0$, there is definite correlation.

If $r = 0$, there is definitely no correlation.

The closer that r is to $+1$ or -1, the more likely x and $f(x)$ are to correlate.

If r is positive and we accept the assumption of correlation, the correlation is positive. (As x increases, $f(x)$ also increases.)

If r is negative and we accept the assumption of correlation, the correlation is negative. (As x increases, $f(x)$ decreases.)

5. Select a significance level, α, for the test. As with other goodness-of-fit tests (See Section A.1.10), a significance level of 5% is most commonly used. A significance level of 10% is used for a stricter test and 1% is used for a more lenient test.

6. Determine the degrees of freedom, f. The number of degrees of freedom is the number of samples N minus the number of variables, which in this case is 2 (x and $f(x)$):

$$f = N - 2$$

7. From Table B.10, look up the critical correlation coefficient for the α confidence level and f degrees of freedom.

Table B.10 can be regarded as a distribution of values we would get for $r_{x,f(x)}$ if we simply matched pairs of random numbers together. That is, 10% of the time we would get an $r_{x,f(x)}$ greater than the table value corresponding to $\alpha = 0.10$, 5% of the time we would get an $r_{x,f(x)}$ greater than the table value corresponding to $\alpha = 0.05$, etc.

8. If $|r|$ is less than the critical Table B.10 value, there is no significance at the α significance level, meaning that the matchups of pairs of data points are insignificantly different than a random matching of numbers. Hence, we *reject* the assumption of correlation.

If $|r|$ is greater than the critical Table B.10 value, there *is* significance at the α significance level, meaning that there is probably something significant about the way the pairs of numbers are matched up. Hence, we *accept* the assumption of correlation.

The format of Table 6.2 will be useful in determining the statistics to solve the linear regression and determine the correlation coefficient.

TABLE 6.2
Correlation coefficient summary table.

(1) j	(2) x	(3) $f(x)$	(4) x^2	(5) $\{f(x)\}^2$	(6) $xf(x)$
1					
2					
.					
.					
.					
N					
	$\Sigma = \square$	$\Sigma = \square$	$\Sigma = \square$	$\Sigma = \square$	$\Sigma = \square$

By summing the columns in the table we determine

$$C = \Sigma x \qquad E = \Sigma x^2 \qquad G = \Sigma\{xf(x)\}$$
$$D = \Sigma f(x) \qquad F = \Sigma\{f(x)\}^2$$

Then we can determine the linear regression by solving for A and B in the following equations, which come from Equations 6.11 and 6.12:

$$NA + CB = D \tag{6.14}$$

$$CA + EB = G \tag{6.15}$$

The best-fit straight line is $f(x) = A + Bx$. **(6.16)**

The correlation coefficient can be determined from the following equation, which comes from Equation 6.13:

$$r_{x,f(x)} = \frac{G - (CD/N)}{[(E - C_2/N)(F - D^2)]^{1/2}} \tag{6.17}$$

Suppose we have two variables associated with the same process. Let us use, as an example, an engine. Two variables we may wish to study are *rotor unbalance* and *engine vibration*. We may have reason to believe that a higher unbalance of the rotor during assembly may cause higher vibrations when the engine is running. Furthermore, we may suspect that a linear relationship exists; that is, the higher the unbalance, the higher proportionally will be the magnitude of the vibrations. This suggests three questions to be answered:

1. Given the amount of unbalance, how do we predict the expected magnetude of vibration?
2. Do *unbalance* and *vibrations* actually correlate with one another? That is, can we actually say that there is a **cause-and-effect** relationship between these two variables?

3. Is the prediction technique valid? That is, if there is correlation between *unbalance* and *vibrations,* can the linear relationship we assume give us a valid prediction?

At this point it is important to point out that *correlation* between two variables does not necessarily establish a cause-and-effect relationship. Conversely, a presumed cause-and-effect relationship must be rejected if we cannot assume correlation. As an illustration, we may be able to establish correlation between the per capita rate of college educated adults in the United States over the past century and the rate of violent crime. This does not necessarily say that college attendance makes us more violent or that violence makes us more studious. Also, should we be able to assume *correlation* between *unbalance* and *vibration* and have reason to believe that there is a *cause-and-effect* relationship, those facts alone are not enough to tell us which variable is the cause and which is the effect or, indeed, whether both variables are the effects of some untested cause.

The purpose of the techniques presented here is to test a hunch about cause and effect that we already have concerning two variables x and y. Let us say that we suspect that x causes y. Mathematically we express that as $y = f(x)$: that is, y is a function of x, or whatever happens to x causes something predictable to happen to y. Now we can run an experiment with N trials, where we measure both variables x and $f(x)$ from each trial. We then use Table 6.2 to do some computations.

For each trial we enter the x and $f(x)$ measurements in columns 2 and 3, respectively. We then compute the squares, x^2 and $f(x)^2$, of columns 2 and 3 and enter the results in columns 4 and 5, respectively. For each trial, the product of the column 2 and 3 entries, $xf(x)$, is entered in column 6. Upon completing the table, we sum columns 2–6. We then use the sums of columns 2, 3, 4, and 6 to solve for linear regression parameters A and B using Equations 6.14 and 6.15. The best-fit straight line through the $x, f(x)$ points is $f(x) = A + Bx$, where A is the **intercept** and B is the **slope,** as illustrated in Figure 6.1. Finally, we use the sums of columns 2–6 to solve for the correlation coefficient $r_{x,f(x)}$ using Equation 6.13 (or 6.17).

We now turn to Table B.10 (Appendix B) and look up the critical value for the appropriate degrees of freedom, f, and selected significance level. The number of degrees of freedom is determined by subtracting 2 from the number of plotted data points: $f = N - 2$. The significance level is customarily 5%, but a lower level, such as 1%, is used when we want a stricter test, and a higher level, such as 10%, is used when we desire a more lenient test.

If the absolute value of the correlation coefficient, $|r_{x,f(x)}|$, is more than the critical value we look up in Table B.10, we can accept the assumption of linear correlation. However, if $r_{x,f(x)}$ is less than the value in Table B.10, we reject the correlation assumption. If we accept correlation and $r_{x,f(x)}$ is a positive number, we can assume positive linear correlation (as x goes up, $f(x)$ goes up). If we accept correlation and $r_{x,f(x)}$ is a negative number, we can assume negative linear correlation (as x goes up, $f(x)$ goes down).

EXAMPLE 6.1 Suppose engine rotor unbalance data are taken on 19 engines during assembly. After they are built, maximum vibration data are taken from the same engines while they are being tested at a given RPM. The recorded results are as follows.

Engine Serial Number	Rotor Unbalance (g)	Engine Vibration (mils)	Engine Serial Number	Rotor Unbalance (g)	Engine Vibration (mils)
201	0.0	2.5	211	5.0	6.0
202	1.0	1.2	212	6.0	7.2
203	1.0	4.6	213	6.0	5.3
204	2.0	4.0	214	7.0	7.0
205	2.0	2.9	215	7.0	6.5
206	3.0	4.0	216	8.0	8.1
207	3.0	5.8	217	9.0	7.0
208	4.0	4.4	218	9.0	7.5
209	4.0	5.5	219	10.0	6.8
210	5.0	6.7			

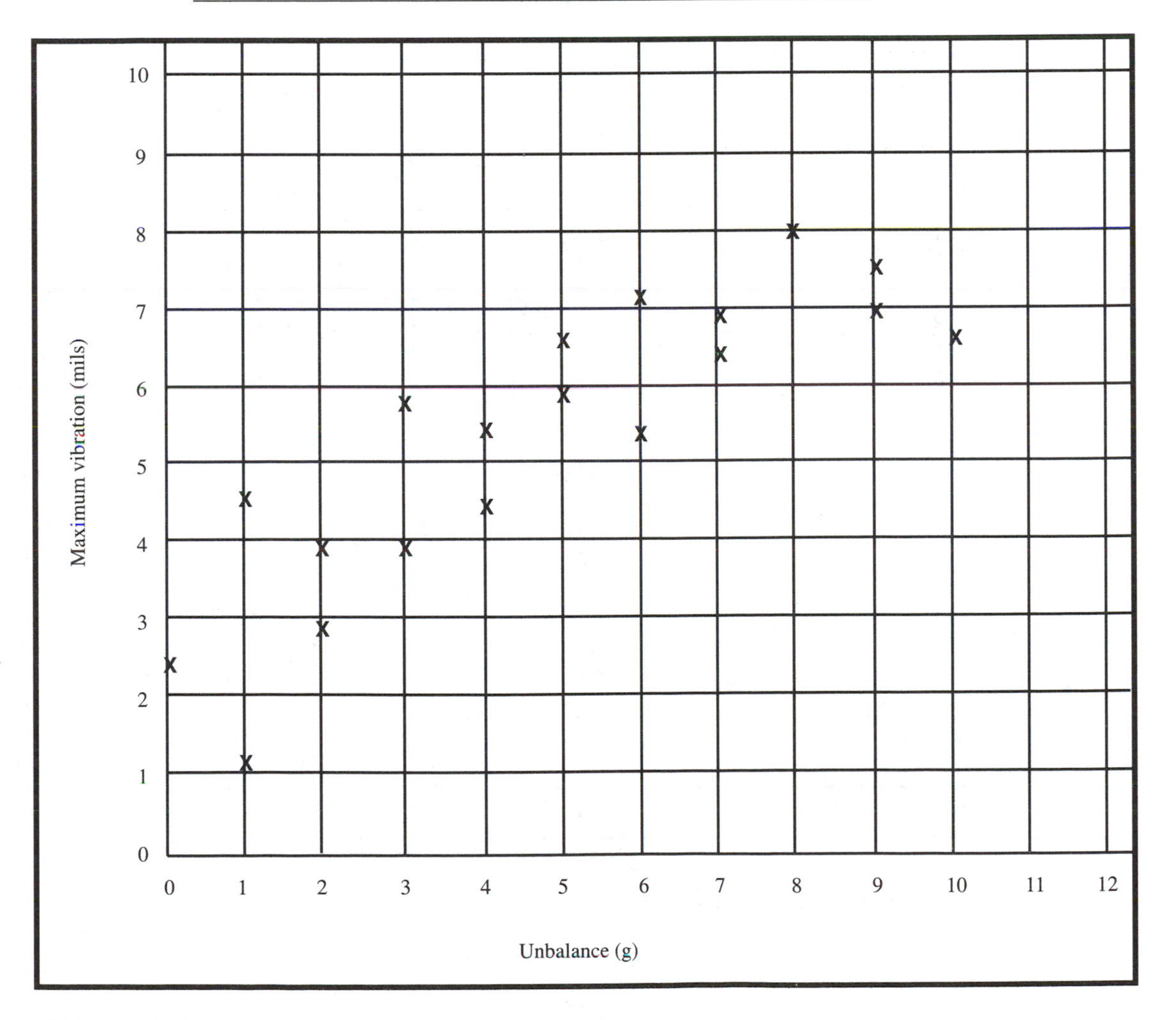

FIGURE 6.2
Vibration-versus-unbalance plot for Example 6.1.

We suspect that unbalance causes vibration and that there is a positive linear correlation. Using the linear assumption, develop a mathematical model to predict vibration from rotor unbalance, and test at the 5% significance level whether we can accept the assumption of correlation.

First of all, let us make a plot (Figure 6.2) of *vibration versus unbalance* from the recorded data on the 19 engines.

Assuming that unbalance causes vibration, let x be the unbalance (in grams) and $f(x)$ be the maximum vibration (in mils). If our suspicion regarding the correlation between these two variables holds, then we can predict vibration from unbalance by using Equation 6.1:

$$\text{vibration (in mil)} = A + Bx$$

where x is the unbalance in grams and A and B can be determined from Equations 6.14 and 6.15 by using the computational format prescribed in Table 6.2.

j	x	$f(x)$	x^2	$[f(x)]^2$	$xf(x)$
1	0	2.5	0	6.25	0.00
2	1	1.2	1	1.44	1.20
3	1	4.6	1	21.16	4.60
4	2	4.0	4	16.00	8.00
5	2	2.9	4	8.41	5.80
6	3	4.0	9	16.00	12.00
7	3	5.8	9	33.64	17.40
8	4	4.4	16	19.36	17.60
9	4	5.5	16	30.25	22.00
10	5	6.7	25	44.89	33.50
11	5	6.0	25	36.00	30.00
12	6	7.2	36	51.84	43.20
13	6	5.3	36	28.09	31.80
14	7	7.0	49	49.00	49.00
15	7	6.5	49	42.25	45.50
16	8	8.1	64	65.61	64.80
17	9	7.0	81	49.00	63.00
18	9	7.5	81	56.25	67.50
19	10	6.8	100	46.24	68.00
Totals	92	103.0	606	621.68	584.90

Applying Equations 6.14 and 6.15,

$$NA + (\Sigma x)B = \Sigma f(x) \quad \text{and} \quad (\Sigma x)A + (\Sigma x^2)B = [\Sigma xf(x)]$$

where

$$N = 19 \qquad \Sigma x^2 = 606.00$$

$$\Sigma x = 92 \qquad \Sigma xf(x) = 584.90$$

$$\Sigma f(x) = 103$$

Therefore,

$$19A + 92B = 103 \quad \text{and} \quad 92A + 606B = 584.9$$

from which we compute

$$A = 2.82 \quad \text{and} \quad B = 0.537$$

So the best-fit straight line through the 19 data points gives us the linear relationship

$$f(x) = 2.82 + 0.537x$$

or

$$\text{vibration, in mils} = 2.82 + 0.537\ (\text{unbalance, ingrams})$$

From this linear solution, we can construct the following prediction table.

Rotor Unbalance (g)	Expected Vibrations (mils)	Rotor Unbalance (g)	Expected Vibrations (mils)
0	2.82	5	5.51
1	3.36	6	6.04
2	3.89	7	6.58
3	4.43	8	7.12
4	4.97	9	7.65
		10	8.19

Now we can superimpose the best-fit straight line onto the plotted graph. See Figure 6.3.

We now calculate the correlation coefficient from Equation 6.13,

$$r_{x,f(x)} = \frac{\Sigma[xf(x)] - \dfrac{(\Sigma x)[\Sigma f(x)]}{N}}{\sqrt{\left[\Sigma(x^2) - \dfrac{(\Sigma x)^2}{N}\right]\left[\Sigma[f(x)]^2 - \dfrac{[\Sigma f(x)]^2}{N}\right]}}$$

using the table totals:

$$N = 19 \qquad \Sigma x^2 = 606.00$$
$$\Sigma x = 92 \qquad \Sigma[f(x)]^2 = 621.68$$
$$\Sigma f(x) = 103 \qquad \Sigma xf(x) = 584.90$$

$$r_{x,f(x)} = \frac{584.90 - \dfrac{(92)(103)}{19}}{\sqrt{\left[606.00 - \dfrac{(92)^2}{19}\right]\left[621.68 - \dfrac{(103)^2}{19}\right]}} = \frac{86.163}{100.813} = 0.855$$

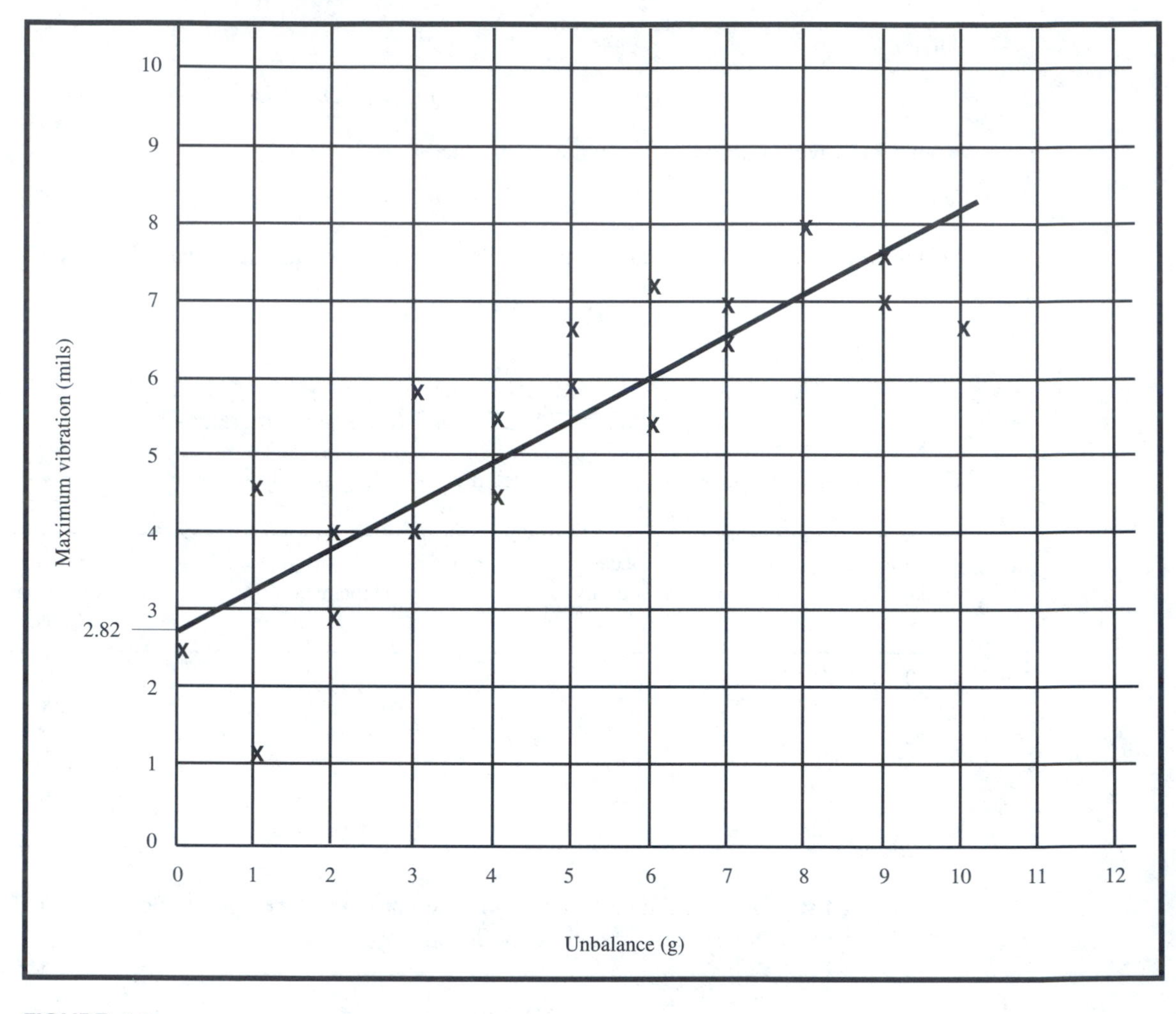

FIGURE 6.3
Best-fit straight line through Example 6.1 data.

We next extract the critical value from Table B.10 for the 0.05 significance level and for $N - 2 = 17$ degrees of freedom. The critical value is 0.456. Since our computed value exceeds the table value, we can accept that there *is* correlation at the 5% significance level.

---

*6.4 PROBABILITY GRAPH PAPERS

Linear regression can be used for related variables that are not linearly related. This can be done if the axes of the plots are **linearized.** For example, the relationship between the random variable x and the cumulative distribution function $f(x)$ *for the Normal Probability Law is not linear,* as illustrated in Figure 6.4.

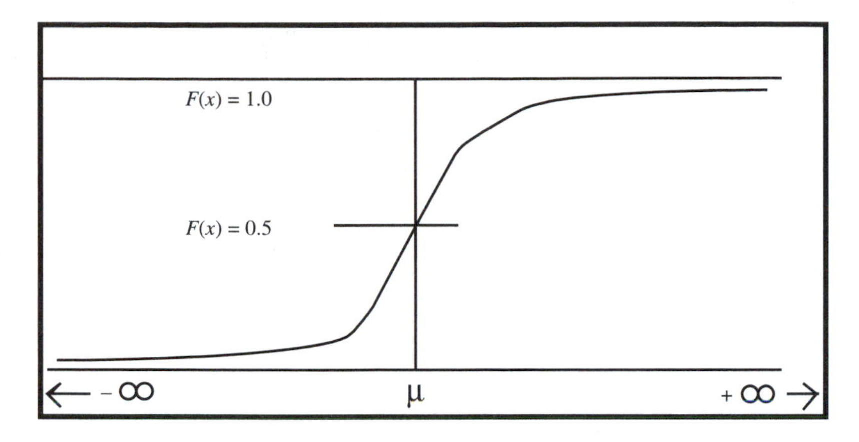

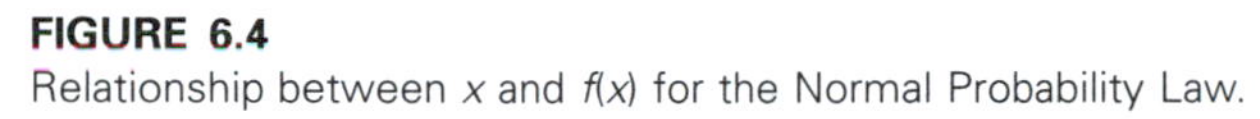
FIGURE 6.4
Relationship between x and $f(x)$ for the Normal Probability Law.

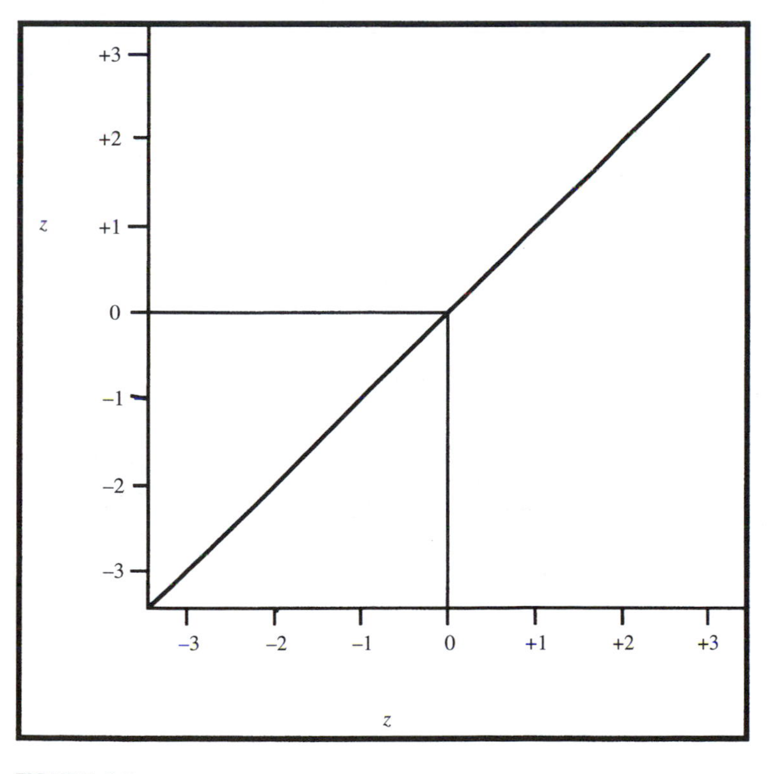

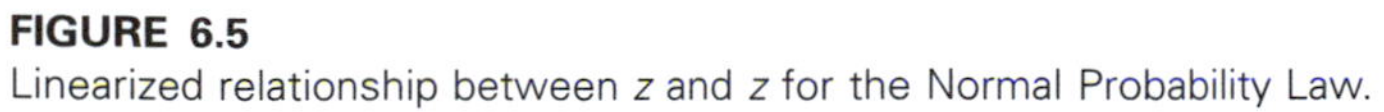
FIGURE 6.5
Linearized relationship between z and z for the Normal Probability Law.

But, we can linearize the graph through a two-step process. First, we set up two axes that are identical using a $(\mu + z\sigma)$ versus $(\mu + z\sigma)$ scale, as in Figure 6.5. This is obviously a linear relation, a 45° straight line. The second step is transforming the horizontal axis by substituting $F(\mu + z\sigma)$ for $(\mu + z\sigma)$. We end up with a horizontal axis that no longer has a linear scale, but we now have an $F(\mu + z\sigma)$-versus-$(\mu + z\sigma)$ graph (Figure 6.6), where the relationship *is* linear. Notice that the $F(\mu + z\sigma)$ values all come from Table B.1, so that the following are true:

Where we originally had $z = -3$, we now have 0.0014.

Where we originally had $z = -2$, we now have 0.0227.

Where we originally had $z = -1$, we now have 0.1587.

Where we originally had $z = 0$, we now have 0.5000.

Where we originally had $z = +1$, we now have 0.8413.

Where we originally had $z = +2$, we now have 0.9773.

Where we originally had $z = +3$, we now have 0.9987.

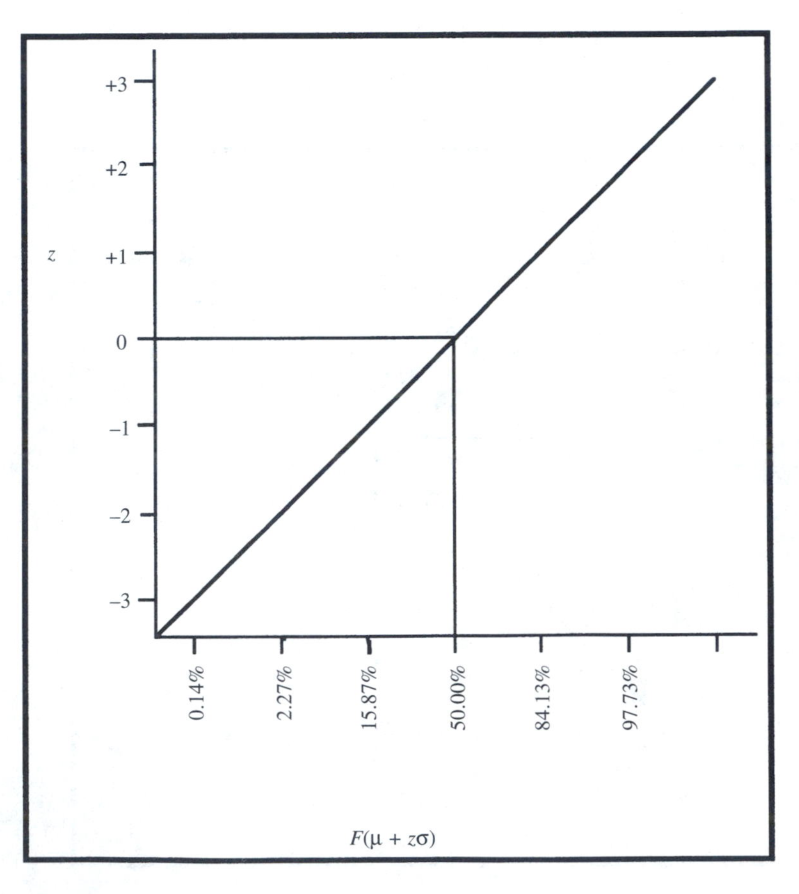

FIGURE 6.6

Linearized relationship between z and $f(z)$ for the Normal Probability Law.

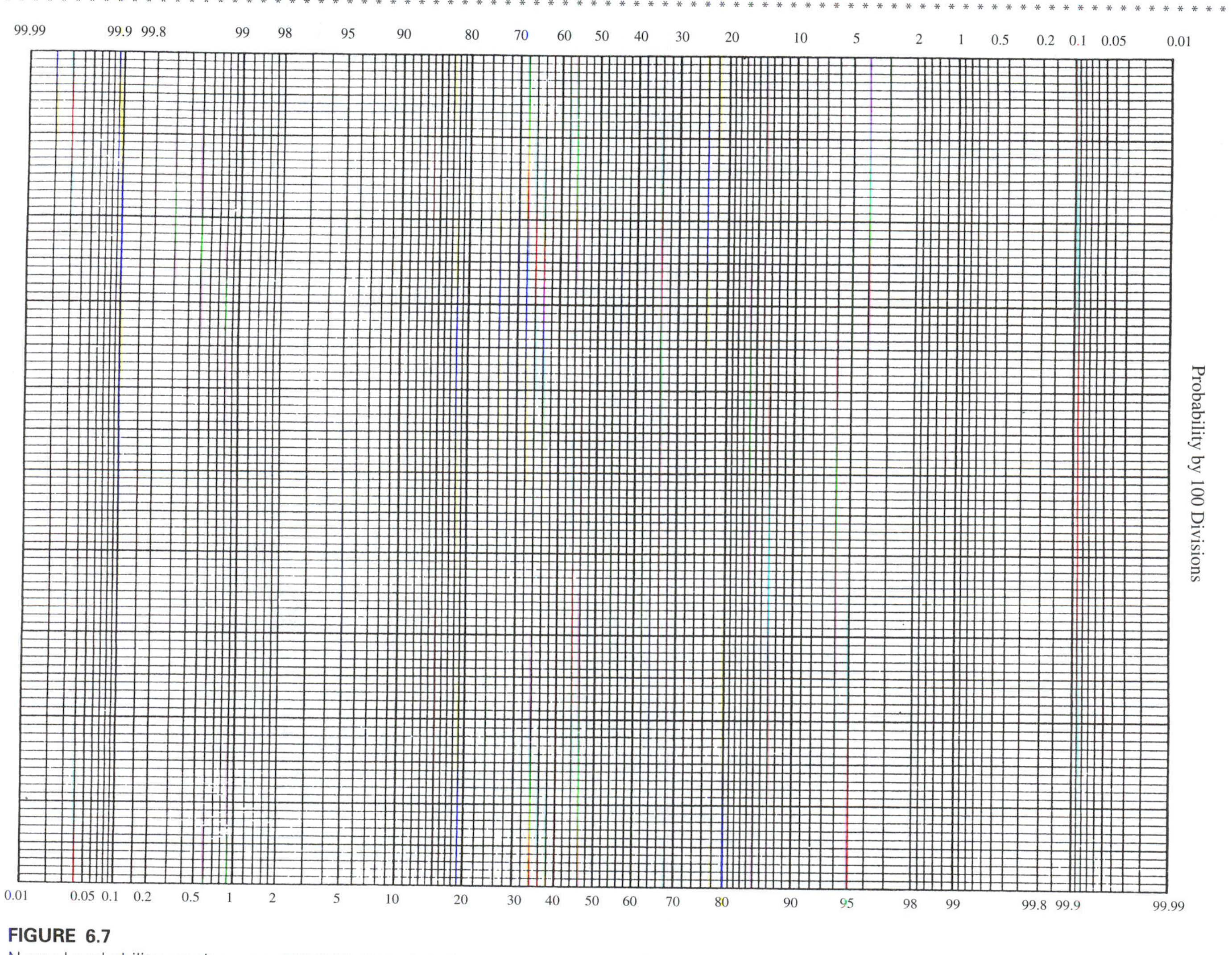

FIGURE 6.7
Normal probability graph paper. (SOURCE: From J. S. Craver, *Graph Paper from Your Computer or Copier.* Tucson: Fisher Books, 1996, p. 227.)

This graph, when fully developed, becomes the *normal probability graph* in Figure 6.7, which has z-versus-$F(\mu + z\sigma)$ axes. The vertical axis is linear, and the horizontal axis is not. The horizontal axis has been transformed by the normal cumulative distribution function. Consequently, if the distribution of data points plotted in Figure 6.8 is governed by the normal probability law, the points should fit a straight line.

When using *normal probability graph paper,* we assign values of the variable x to the vertical (linear) axis. We then determine the cumulative probabilities $F(x)$ and plot x versus $F(x)$ directly on the graph. If the plotted points form a straight line, we can assume that they are governed by the *Normal Probability Law.* Otherwise, we cannot. When the normal assumption is considered to be valid, we can determine the values of the two parameters directly from the best-fit straight line through the plotted points. As illustrated in Figure 6.8, where the line crosses the 50th percentile from the horizontal axis, we can determine an estimate of the mean μ from the vertical axis. Where the line crosses the 84th percentile from the horizontal axis, we can determine an estimate of $\mu + \sigma$ from the vertical axis.

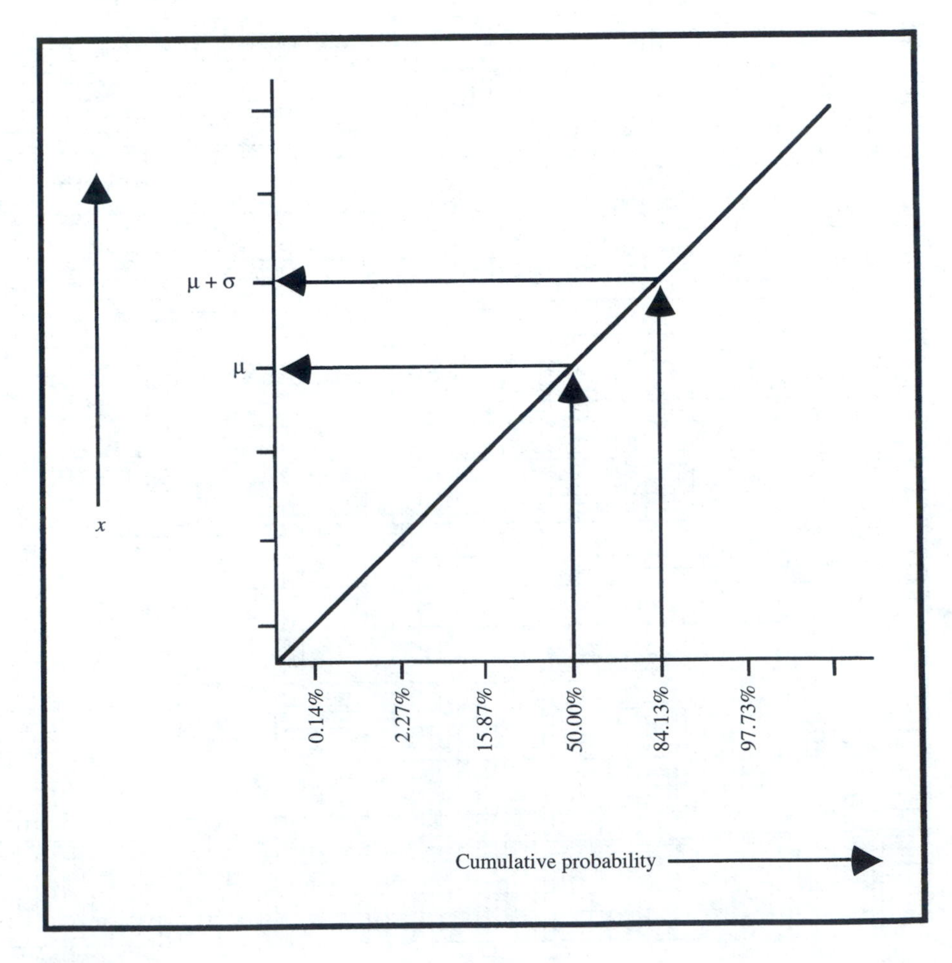

FIGURE 6.8
Using normal probability paper to estimate the parameters μ and σ.

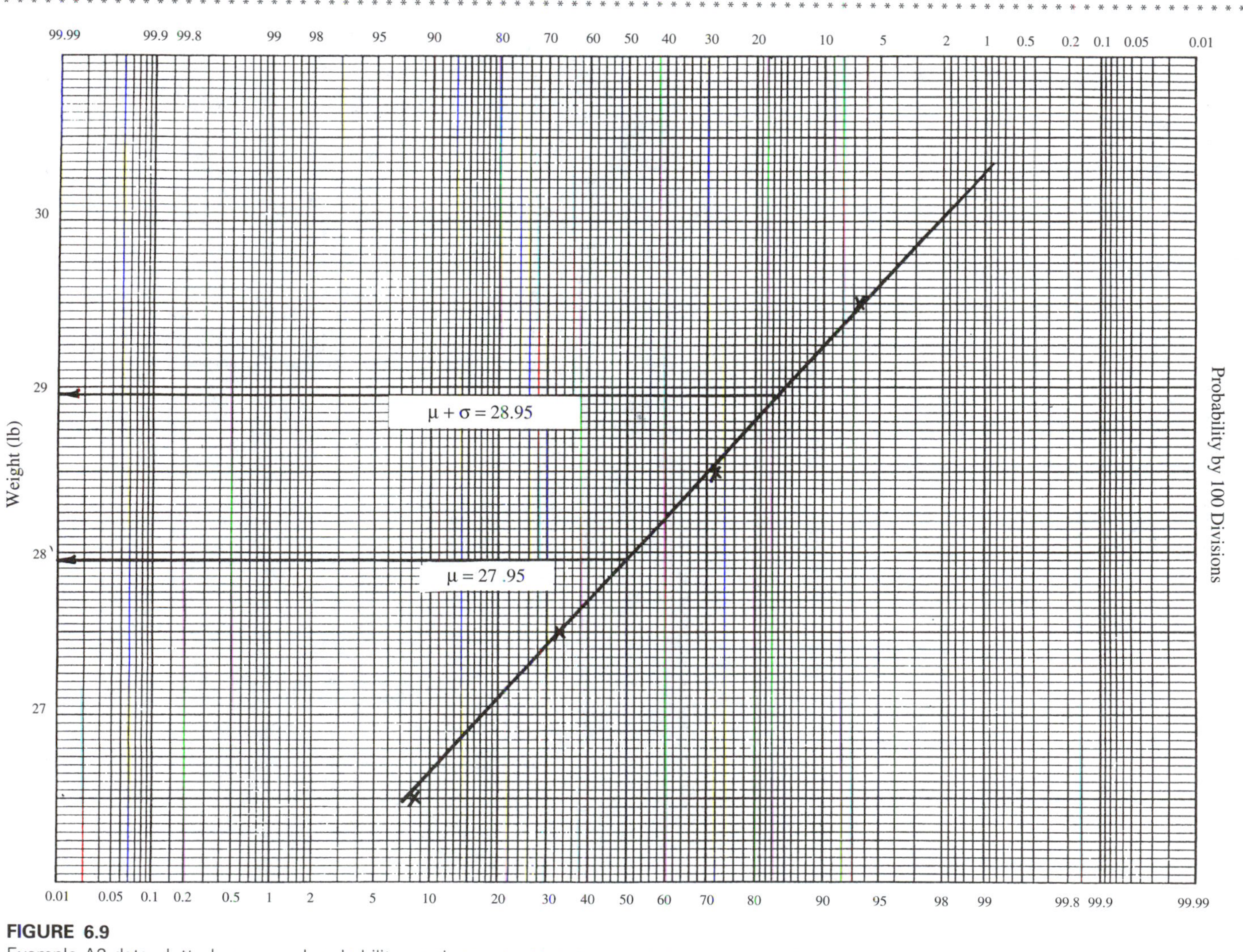

FIGURE 6.9
Example A2 data plotted on normal probability graph paper. (SOURCE: From J. S. Craver, *Graph Paper from Your Computer or Copier.* Tucson: Fisher Books, 1996, p. 227.)

In Figure 6.9 the data from Example A.2 (Appendix A) have been plotted on normal probability paper. We usually expect to have more data points to plot, because the first and last points usually are less meaningful than the others and are ignored. The remaining points in Figure 6.9 are reasonably well fit by a straight line, and from that line we can estimate $\mu = 27.95$ and $(\mu + \sigma) = 28.95$. This gives an estimate of $\sigma = 1.00$. These estimates are reasonably close to the ones we determined analytically in Example A.2.

*6.5 EXPONENTIAL GRAPH PAPER

In Chapter 3, we saw that when—and *only* when—failure rate is constant, the time-to-failure distribution follows the Exponential Probability Law (see Equation 3.21). Equation 3.20 provides the relation between **mission reliability,** $R(t)$, and failure rate, λ, when the failure rate is constant:

$$R(t) = e^{-\lambda\tau} \tag{6.18}$$

When plotted on a graph with linear axes (Figure 6.10), the resulting curve is, of course, not linear. From Equation 6.18,

$$\ln R(t) = -\lambda t \tag{6.19}$$

Applying Equation 3.2 from Chapter 3,

$$\ln [N_s(t)/N_0] = (-\lambda)t \tag{6.20}$$

If we transform the vertical axis into a logarithmic scale, so that the graph becomes $\ln R(t)$ versus t, we then plot Equation 6.19 (equivalent to Equation 6.20), which is linear. This is done in Figure 6.11, where the variable of the horizontal axis is t, the variable of the vertical axis is $\ln R(t) = \ln [N_s(t)/N_0]$, the slope is $-\lambda$, and the vertical axis intercept is 0 (= ln 1.0).

If we turn the vertical axis upside down, Figure 6.12 results. We now have a linear relationship with a positive slope, which is the constant failure rate λ. Notice that the vertical axis is

FIGURE 6.10
Reliability as a function of constant failure rate λ.

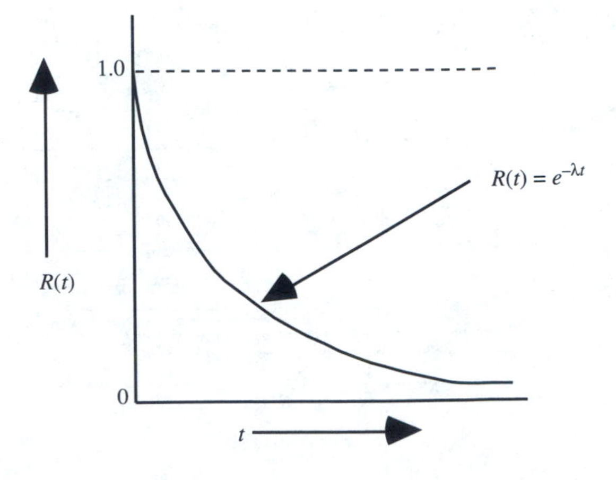

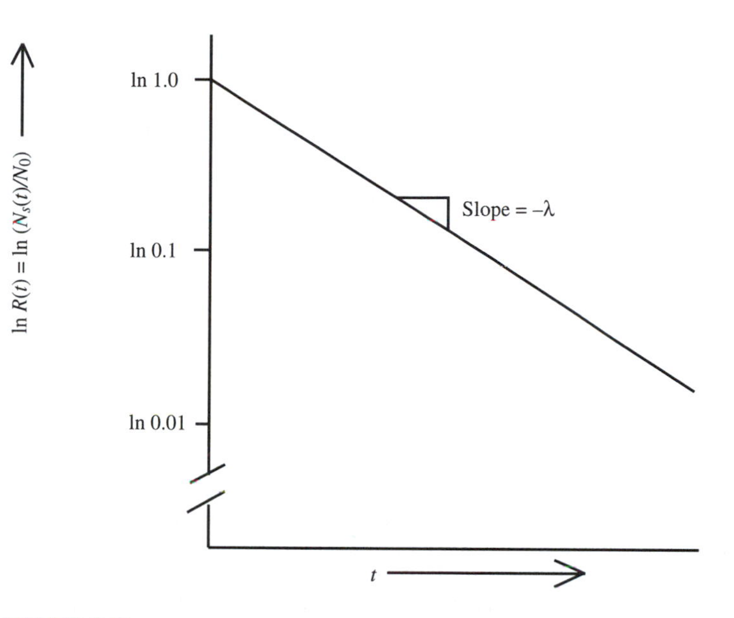

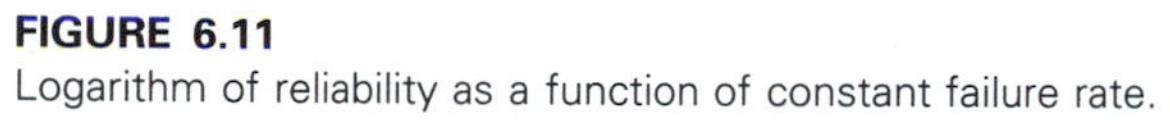

FIGURE 6.11
Logarithm of reliability as a function of constant failure rate.

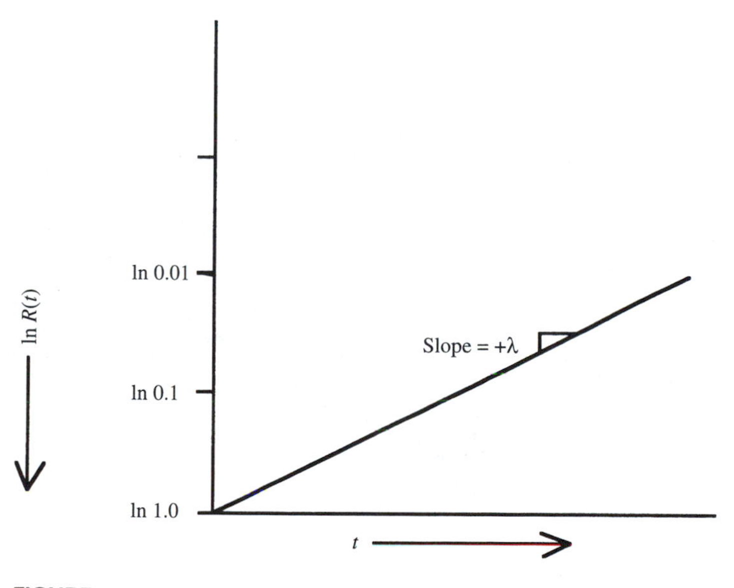

FIGURE 6.12
Logarithm of reliability as a function of constant failure rate, reversed vertical axis.

also an upside-down logarithmic scale. If we maintain that scale and relabel by removing the *ln*, the vertical axis (in Figure 6.13) becomes $N_s(t)/N_0$, which is the percent success after the operating time t. The vertical axis is still running upside down.

Let us now relabel the vertical axis as $1 - R(t)$, which is the **unreliability,** $Q(t)$, or percent failure. The values of the vertical axis are then subtracted from 1.0, resulting in Figure 6.14. The plot is still linear, with the constant failure rate λ as its slope. As this graph is applicable to constant failure rate, we can refer to it as **exponential probability paper,** the Exponential Probability Law being appropriate to constant failure rate.

We learned in Chapter 3 that if failure rate is constant—and only if failure rate is constant—the time-to-failure distribution follows the exponential probability law. Hence, if the points plotted on the graph in Figure 6.14 fit a straight line, we can consider that we are dealing with constant failure rate. Otherwise, we can consider the failure rate in question to be either *increasing* or *decreasing.* In fact, as illustrated in Figure 6.15, if the points tend to curve upward with time, we are most likely dealing with an increasing failure rate, and if the points tend to curve in the other direction, we are probably dealing with a decreasing failure rate. In doing our graphical analysis, however, we shall use the exponential probability paper to determine whether or not we can assume constant failure rate. If we cannot assume constant failure rate we shall then plot our points on **Weibull probability paper** (see Section 6.7) to determine whether increasing or decreasing failure rate fits.

If there is a linear fit through the plotted points on the exponential probability paper, we can assume constant failure rate and use the best-fit straight line to determine mission reliability. We can also use it to determine the mean time between failures (MTBF) and its reciprocal, the constant failure rate λ. The mission reliability, $R(t)$, is 1.0 minus the percent failure, which we read from the vertical axis. The MTBF is determined by recognizing that, for constant failure rate, where mission time t equals the MTBF, the mission reliability, $R(t)$, is

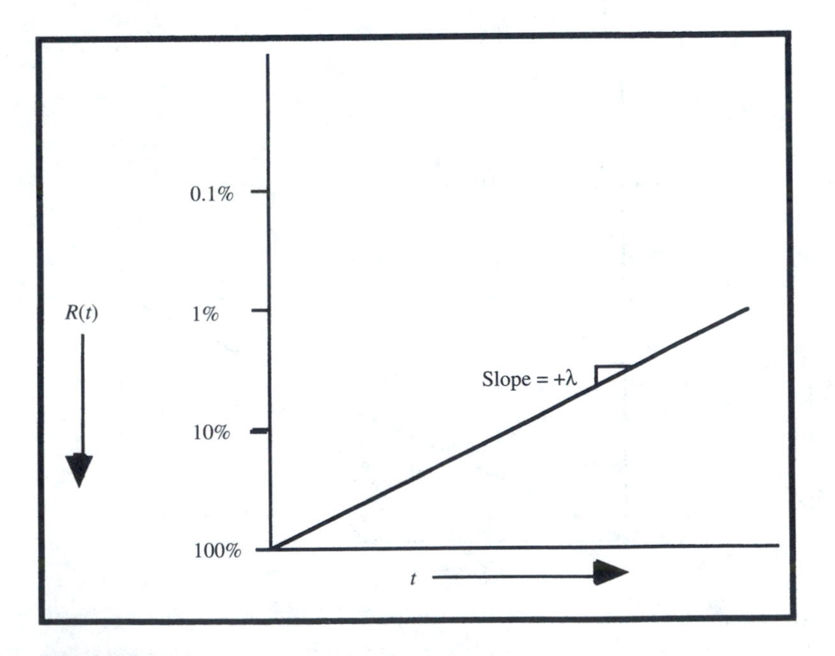

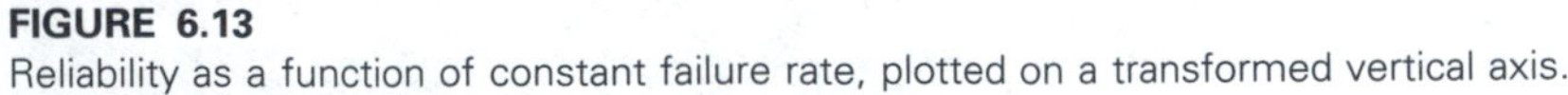
FIGURE 6.13
Reliability as a function of constant failure rate, plotted on a transformed vertical axis.

$$R(t) = R(\text{MTBF}) = e^{-\lambda t} = e^{-t/\text{MTBF}} \\ = e^{-\text{MTBF}/\text{MTBF}} = e^{-1} = 0.368 \quad \textbf{(6.21)}$$

At that point, percent failures = 1.0 − $R(t)$ = 1.0 − 0.368 = 0.632 = 63.2%.

So, we can find the MTBF from the exponential probability paper's linear plot at the point where percent failures = 63.2% (see Figure 6.16).

Notice that the exponential probability law is *not* a symmetrical distribution. It is, in fact, positively skewed. The mean is not at the median (50th percentile); it is at the 6.32 percentile

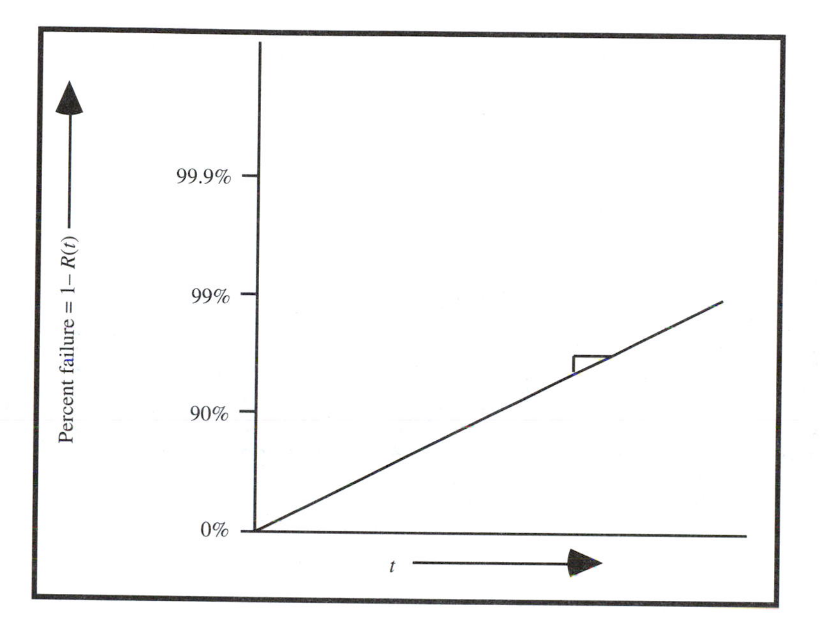

FIGURE 6.14
Percent failure as a function of time, where failure rate is constant, plotted on exponential probability paper.

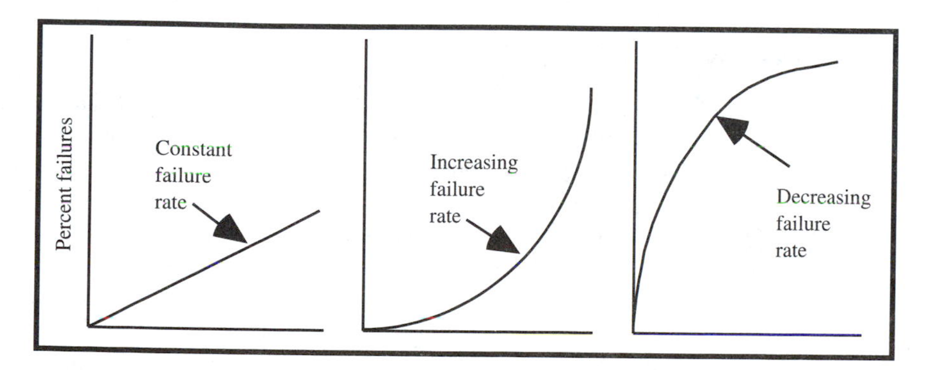

FIGURE 6.15
Shape of % failure trends when plotted on exponential probability paper.

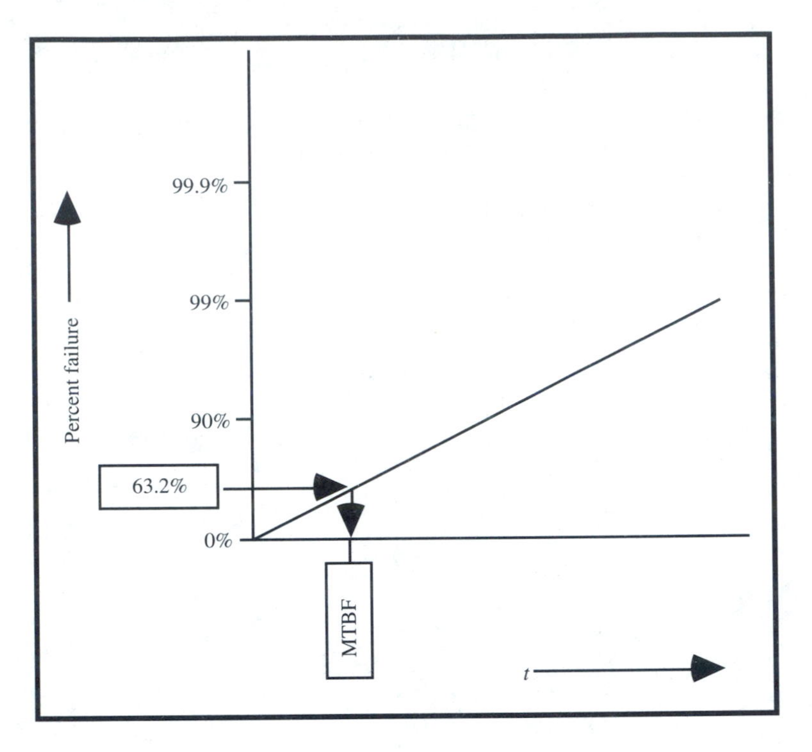

FIGURE 6.16
Location of MTBF on exponential probability plot of % failure for constant failure rate.

point. Significantly, almost two-thirds of failures are expected to occur prior to the MTBF. In other words, when we are dealing with constant failure rate, there is a probability of only 0.368 that an item will survive beyond its MTBF. We shall see in Section 6.7 that this is also the case when we are dealing with varying failure rate.

6.6 THE EXPONENTIAL PLOT

Figure 6.17 shows an exponential probability paper, which we can use for estimating *failure rate, MTBF,* and *mission reliability* whenever we are dealing with constant failure rate. It is also used for graphical *goodness of fit* testing to assess whether a constant failure rate assumption is valid.

To use exponential paper we need a set of field or test data from a quantity of similar items operated until failure. We then compute and plot on exponential paper the percent of survivors versus the operating time. If the points fit a straight line emanating from the origin (lower left-hand corner) on the exponential plot, we can assume that we have contant failure rate. In doing this we can often ignore the very first and last points plotted if the rest of the points fit the appropriate straight line.

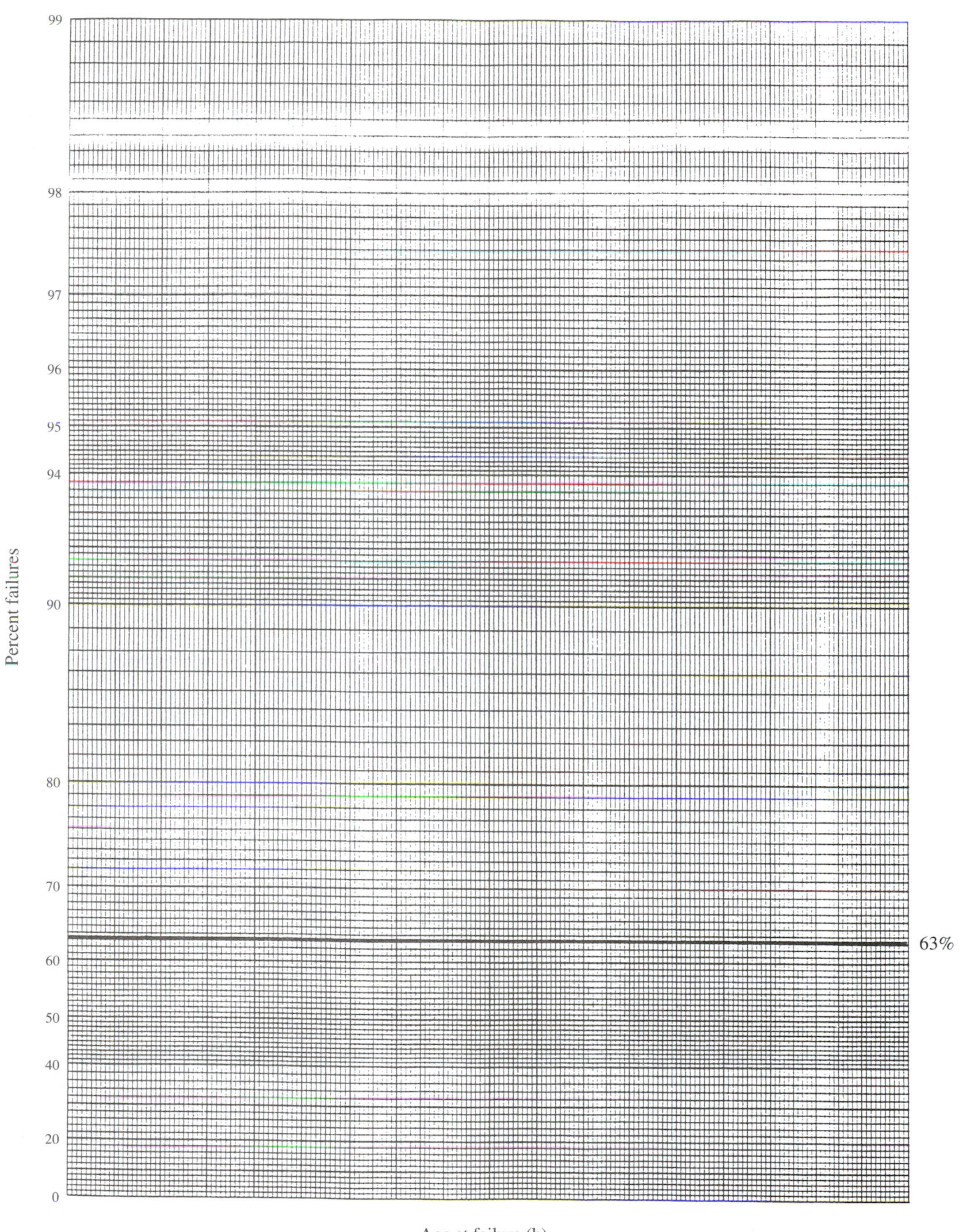

FIGURE 6.17
Exponential probability paper. (SOURCE: From J. S. Craver, *Graph Paper from Your Computer or Copier*. Tucson: Fisher Books, 1996, p. 172.)

If the constant failure rate assumption can be accepted, the plotted straight line can then be used to estimate mission reliability $R(t)$ for any mission time t. This is done by reading the *percent failures* from the vertical axis that corresponds to the *mission time* in the horizontal axis and subtracting the percent failures from 1.0:

$$R(t) = 1.0 - \%\ \text{failures} \tag{6.22}$$

Where the plotted straight line intersects the bold horizontal line emerging from the 63% *failures* point on the vertical axis, we determine the corresponding t value on the horizontal axis. That value of t is the MTBF prediction. The constant failure rate λ is then the reciprocal of the MTBF, as we learned in Chapter 3 (Equation 3.34):

$$\lambda = 1/\text{MTBF} \tag{6.23}$$

EXAMPLE 6.2

Use exponential paper to determine whether the devices of Example 5.8 can be assumed to have a constant failure rate. If so, use the plot to estimate the failure rate, the MTBF, and the reliability for a 100-h mission.

Referring to Example 5.8, 10 devices were put into operation, and the failures occurred at 8, 20, 34, 46, 63, 86, 111, 141, 186, and 266 h. Accordingly, we can plot the following % failures-versus-t points in Figure 6.17.

Operating Time t (h)	Cumulative Number of Failures	Cumulative Percent Failures
8	1	10%
20	2	20%
34	3	30%
46	4	40%
63	5	50%
86	6	60%
111	7	70%
141	8	80%
186	9	90%
266	10	100%

The points on the plot (Figure 6.18) do fit reasonably well a straight line originating at the lower left-hand corner. As discussed before, the last point is ignored in fitting our line. So we can accept the assumption of constant failure rate, and we can use this plot to make our predictions.

Our best-fit straight line intersects the bold horizontal line at $t = 90$ h. That is our MTBF estimate. Our failure rate estimate is then

$$\lambda = 1/\text{MTBF} = 1/90\ \text{h} = 0.011\ \text{failures per hour}$$

This result is in close agreement to that of Example 5.8.

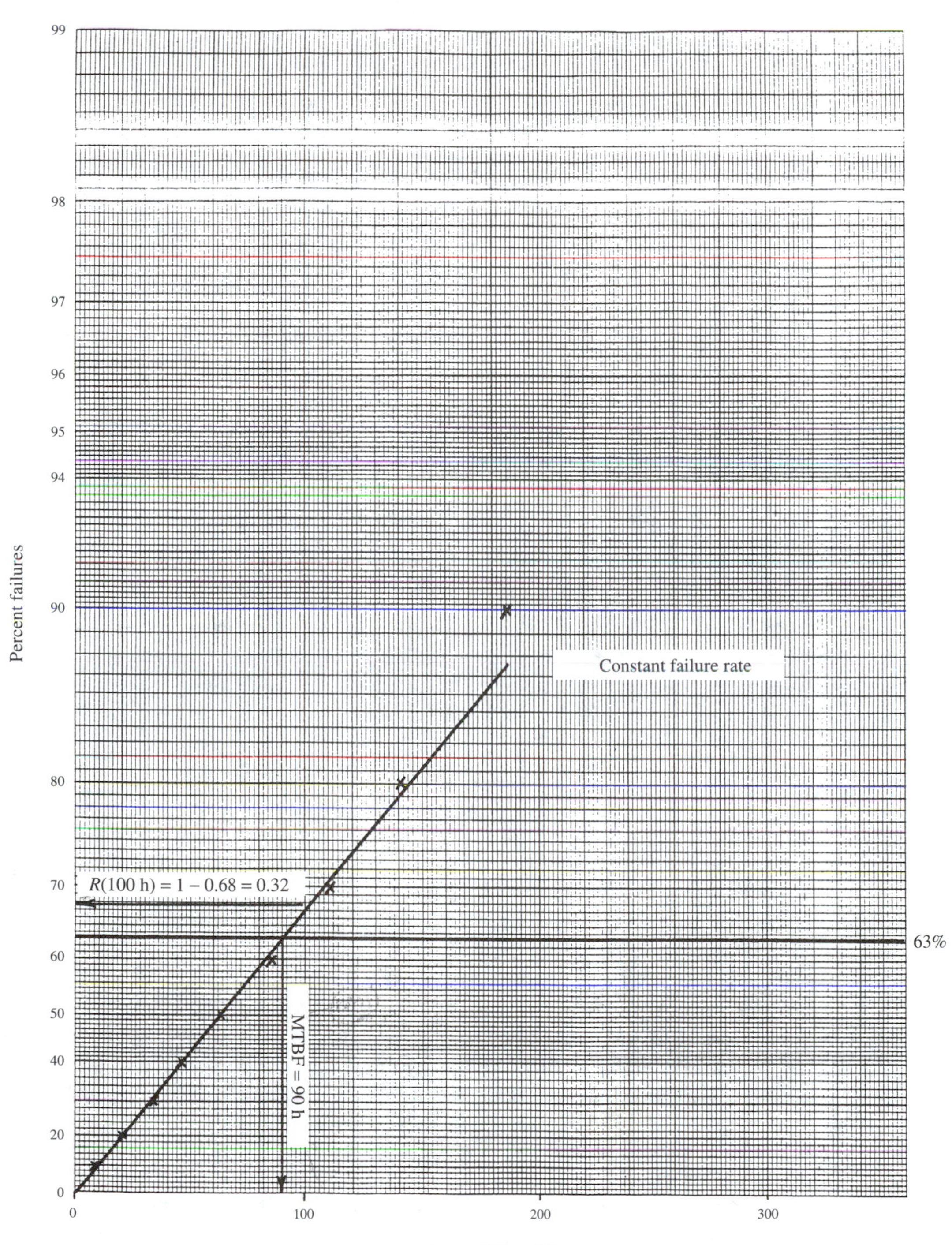

FIGURE 6.18
Plot of Example 6.2 data on exponential probability paper. SOURCE: From J. S. Craver, *Graph Paper from Your Computer or Copier*. Tucson: Fisher Books, 1996, p. 172.)

Using our plotted best-fit straight line, where $t = 100$ h, % failures = 68%. Hence, the mission reliability for a 100-h mission is

$$R(t) = 1.0 - 0.68 = 0.32$$

EXAMPLE 6.3 Use exponential paper to determine whether the components of Example 5.10 can be assumed to have a constant failure rate. If so, use the plot to estimate the failure rate, the MTBF, and the reliability for a 100-h mission.

Referring to Example 5.10, 172 components were operated until failure. They were monitored every 1000 h, with the following results.

Hour	Number of Survivors	Hour	Number of Survivors
1000	113	4000	30
2000	89	5000	13
3000	60	6000	0

Since the components were monitored only once every 1000 h. on the thousandth hour let us assume that those that failed during an unmonitored interval failed, on the average, halfway through the interval. The data can then be represented as follows.

Time (h)	Number Failed	Percent Failures
500	172 − 113 = 59	59/172 = 34%
1500	172 − 89 = 83	83/172 = 48%
2500	172 − 60 = 112	112/172 = 65%
3500	172 − 30 = 142	142/172 = 83%
4500	172 − 13 = 159	159/172 = 92%
5500	172 − 0 = 172	172/172 = 100%

Figure 6.19 shows these data plotted on exponential probability paper. These points do not fit a straight line, leading us to reject the assumption of constant failure rate. Hence, the exponential plot cannot be used to make predictions. These conclusions are in agreement with those of the analytical evaluation of these same data in Example 5.10.

Notice that the plotted points in Figure 6.19 tend to turn upward with increasing t. This suggests an *increasing failure rate* (See Figure 6.15). Were the points to curve in the other direction, a *decreasing failure rate* would be suggested, and, of course, a straight-line fit to the points suggests a *constant failure rate.*

The exponential paper is to be used for predictions *only* when we can assume constant failure rate. Otherwise, as with Example 6.3, it is necessary to go to a Weibull plot.

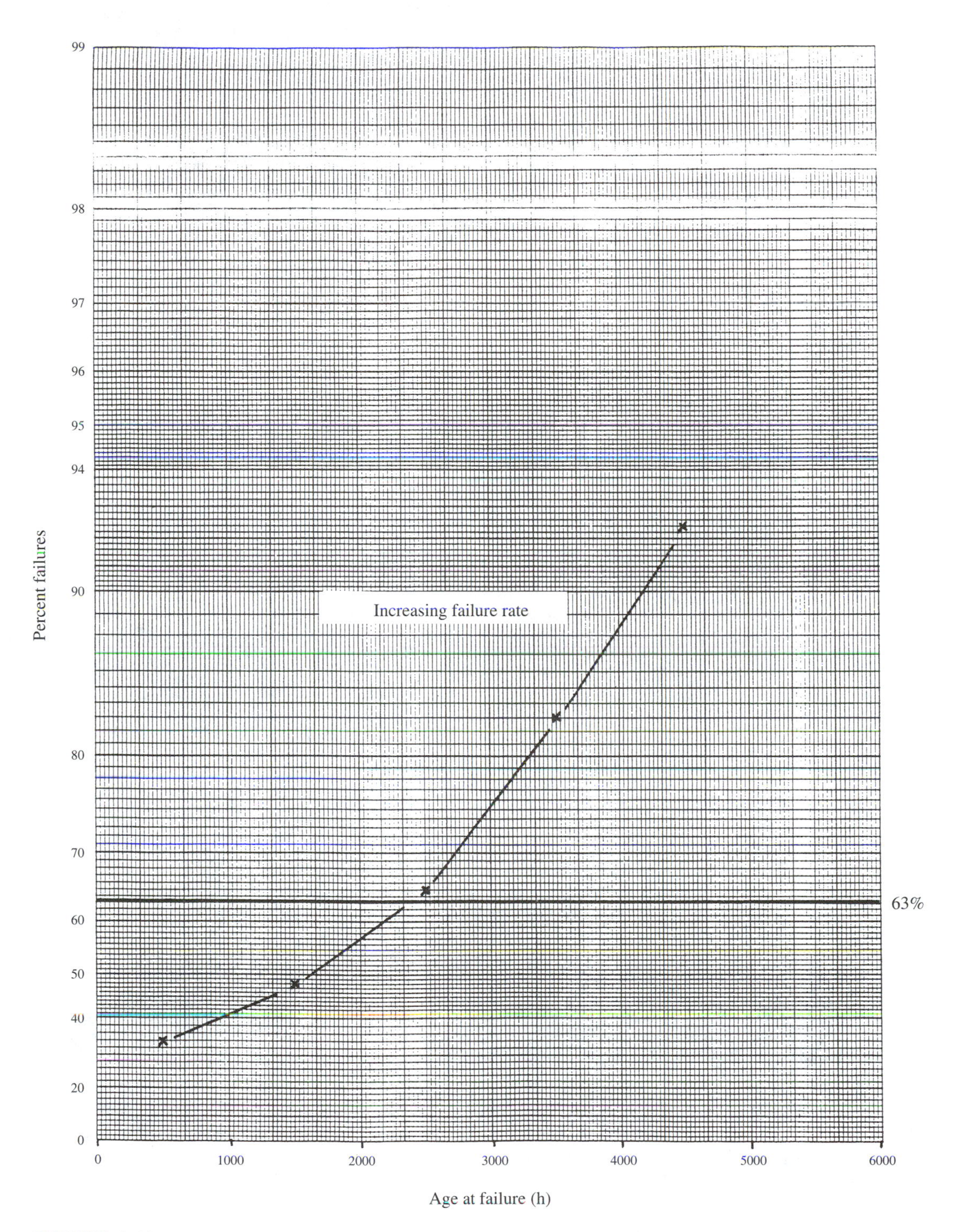

FIGURE 6.19
Plot of Example 6.3 data on exponential probability paper. (SOURCE: From J. S. Craver, *Graph Paper from Your Computer or Copier.* Tucson: Fisher Books, 1996, p. 172.)

*6.7 WEIBULL PROBABILITY GRAPH PAPER

The Weibull Probability Law, introduced by Swedish engineer and statistician Waloddi Weibull in 1951, was designed to accommodate increasing, decreasing, and constant failure rates. From Appendix A and Chapter 3, we see that the Weibull Probability Law is a more general form of the Exponential Law. In Chapter 3, Equation 3.40 states

$$f(t) = BA^{B}t^{B-1}e^{-(At)^{B}} \tag{6.24}$$

which was developed as a density function for time to failure for situations with both constant and varying failure rates $\lambda(t)$. By substituting

$$\beta = B$$
$$\mu = A$$

and allowing t to be the variable, we have the density function for the Weibull Probability Law as presented in Table A.2.

Recall from Chapter 3 that $R(t) = \int_{T}^{\infty} f(t)\, dt$. Applying Equation 6.24,

$$R(t) = \int_{T}^{\infty} f(t)\, dt = \beta\mu^{\beta}\int_{T}^{\infty} t^{\beta-1}e^{-(\mu t)^{\beta}} = e^{-(\mu t)^{\beta}} \tag{6.25}$$

From Equation 3.7, $F(t) = 1 - R(t)$. Applying Equation 6.25,

$$F(t) = 1 - e^{-(\mu t)^{\beta}} \tag{6.26}$$

from which

$$1 - F(t) = e^{-(\mu t)^{\beta}} \tag{6.27}$$

and

$$\ln [1 - F(t)] = -(\mu t)^{\beta} \tag{6.28}$$

From Equation 6.28,

$$\ln \ln [1 - F(t)] = -\beta \ln \mu - \beta \ln t \tag{6.29}$$

Equation 6.29 expresses a linear relationship between the two variables $\ln \ln [1 - F(t)]$ and $\ln t$, where $-\beta$ is the slope and $-\beta \ln \mu$ is the vertical-axis intercept.

Weibull probability paper can be constructed by adjusting the vertical axis to an upside down log log scale, on which are plotted the percent failures, and adjusting the horizontal scale to a logarithmic scale, on which is plotted the operating time. As a result of these adjustments to the axes, *% failures–versus–operating-time* plots are linear on Weibull probability paper. When the best-fit straight line is constructed through the points on Weibull paper, it can be used to directly determine mission reliability, $R(t)$, for any mission time t. It can also be used to estimate the MTBF and the variable failure rate, $\lambda(t)$, by first determining the values of the Weibull parameters $\alpha = (1/\mu)$, also known as the **scale parameter,** and β, also known as the **slope,** or **shape parameter.** The shape parameter indicates whether the failure rate is increasing, decreasing, or constant:

If $\beta < 1.0$, the failure rate is decreasing.

If $\beta = 1.0$, the failure rate is constant.

If $\beta > 1.0$, the failure rate is increasing.

The scale parameter indicates the 63.2 percentile of the time-to-failure distribution. This is true regardless of the value of β.

The Weibull probability law also has a positively skewed distribution. So, regardless of whether we are dealing with constant, increasing, or decreasing failure rate, the median time to failure is less than the mean (MTBF). Failure is always expected prior to the MTBF.

6.8 THE WEIBULL PLOT

Figure 6.20 is an example of the Weibull probability paper being used for predictions whether or not the failure rate is constant. It can be used for determining whether the failure rate $\lambda(t)$ is increasing, decreasing, or constant. It can be used to estimate the MTBF, and it can be used to predict mission reliability, $R(t)$. To use the Weibull paper, follow these instructions:

1. The horizontal axis, labeled "Age at Failure," is a two-cycle logarithmic scale. Adjust and relabel the scale to accommodate the magnitude of the TTF (time-to-failure) values to be plotted.
2. As with the exponential graph, plot the cumulative percent failures versus TTF. Then draw the best-fit straight line through the plotted points. You may find it necessary to ignore the highest and lowest points.
3. Where your plotted line intersects the "η-estimator" line (the horizontal dashed line at cumulative percent failures = 63%), read η from your adjusted TTF axis.
4. Draw a line perpendicular to your original plotted line from the "Estimation Point" in the upper-left corner of the chart to the original line.
5. Where the perpendicular line intersects the β scale at the top of the chart, read the value for β. If $\beta = 1$, assume that the failure rate is *constant.* If $\beta > 1$, assume that the failure rate is *increasing*. If $\beta < 1$, assume that the failure rate is *decreasing*.
6. Where the perpendicular line intersects the P_μ scale at the top of the chart, read the value for P_μ.
7. Find the P_μ percent value on the vertical axis, and use the originally plotted line to read the corresponding μ (= MTBF) from the horizontal axis.
8. To determine a *mission reliability,* $R(t)$, for a mission of time t, locate t on the horizontal axis, and use the originally plotted line to read the corresponding cumulative percent failure from the vertical axis.

$$\text{reliability } R(t) = 1 - \%\text{ failures}$$

EXAMPLE 6.4 Use Weibull paper to evaluate the data in Example 6.2. Determine whether the failure rate is increasing, decreasing, or constant. Estimate the MTBF. Predict the reliability for a 100-h mission.

From Example 6.2 we have the following data that we plot on the Weibull paper in Figure 6.21.

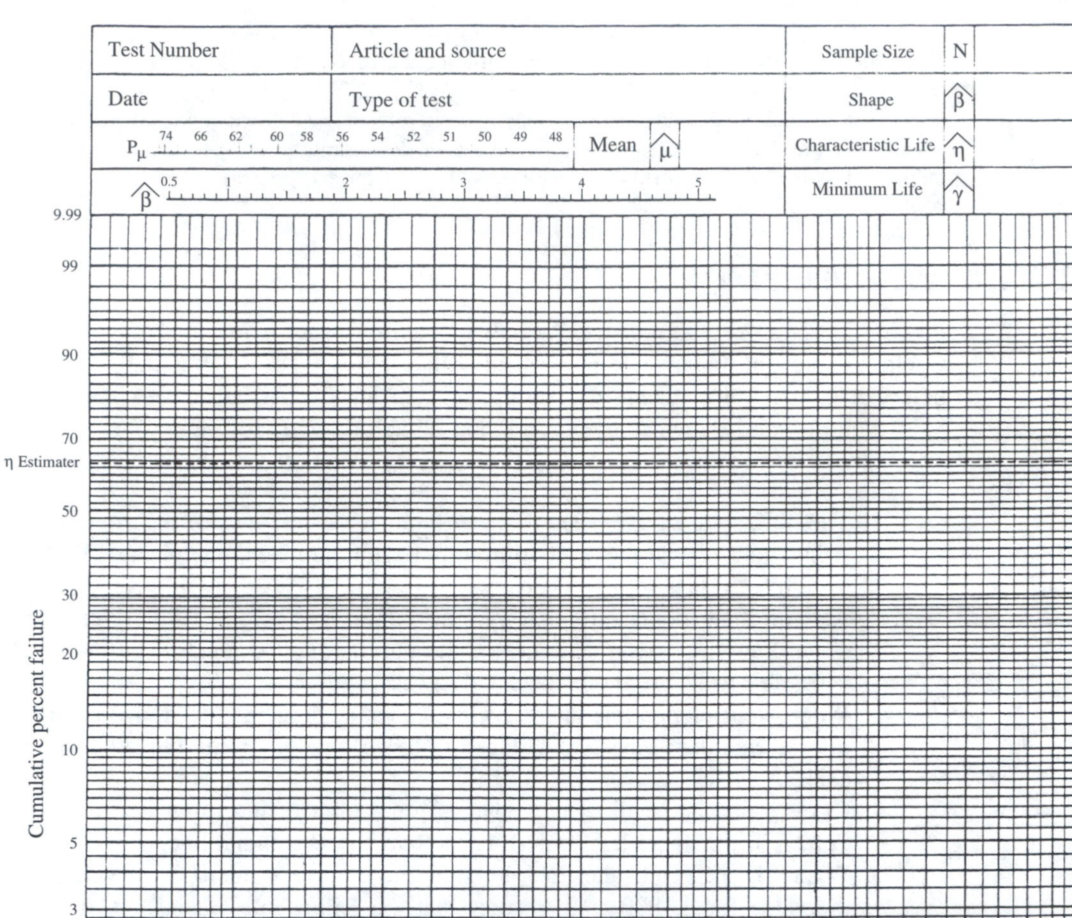

FIGURE 6.20

Weibull probability paper. (SOURCE: From J. S. Craver, *Graph Paper from Your Computer or Copier.* Tucson: Fisher Books, 1996, p. 237.)

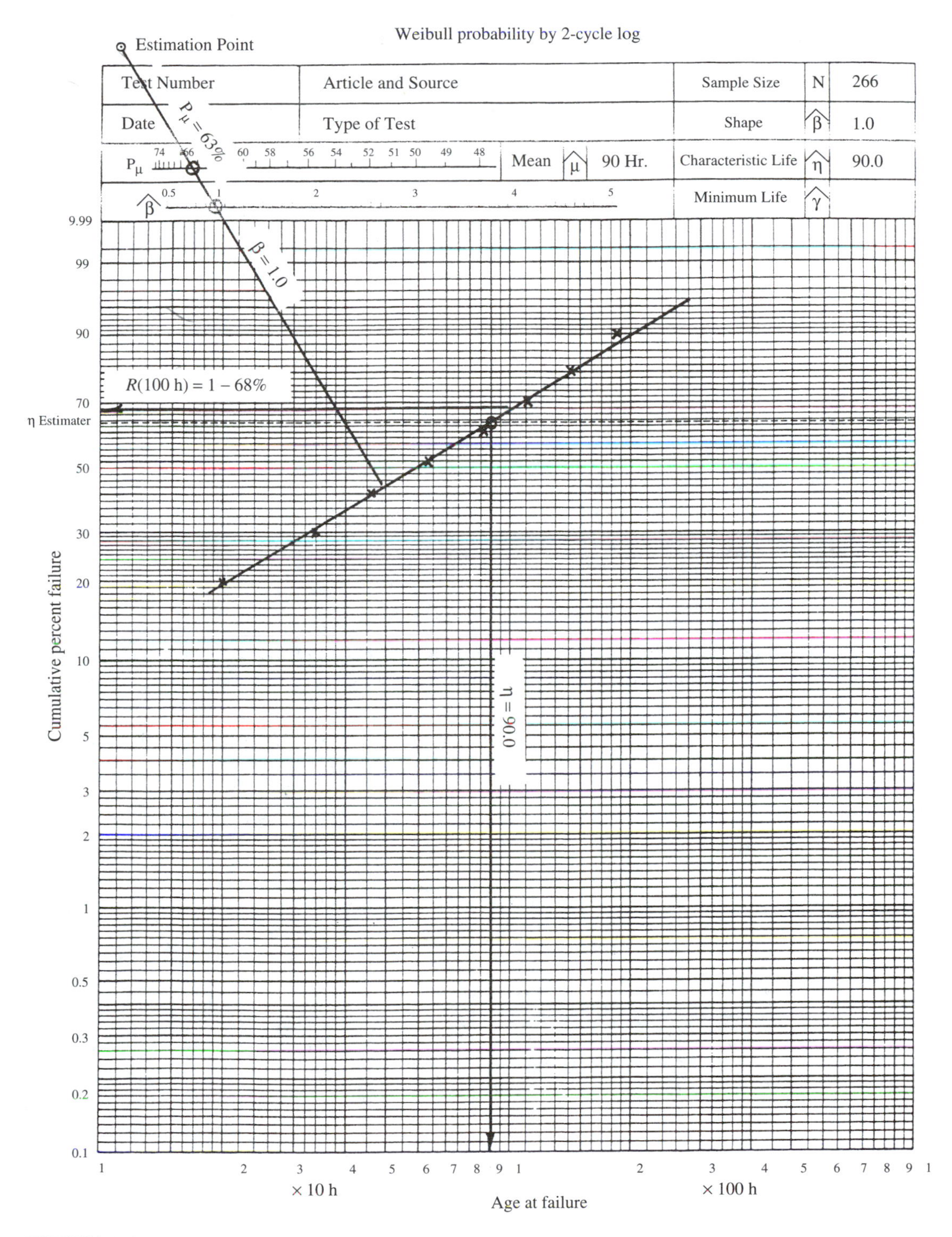

FIGURE 6.21
Plot of Example 6.4 data on Weibull probability paper. (SOURCE: From J. S. Craver, *Graph Paper from Your Computer or Copier*. Tucson: Fisher Books, 1996, p. 237.)

Operating Time t (h)	Cumulative Percent Failures
8	10%
20	20%
34	30%
46	40%
63	50%
86	60%
111	70%
141	80%
186	90%
266	100%

We then draw a best-fit line through the data points and follow the Weibull paper instructions to determine

$$\beta = 1.0$$
$$\eta = 90.0$$
$$\text{MTBF} = 90.0 \text{ h}$$
$$R(100 \text{ h}) = 1.0 - 68\% = 0.32$$

Because $\beta = 1.0$, we can assume a constant failure rate. Notice how well these results compare with those of Examples 6.2 and 5.8.

EXAMPLE 6.5 Use Weibull paper to evaluate the data in Example 6.3. Determine whether the failure rate is increasing, decreasing, or constant. Estimate the MTBF. Predict the reliability for a 100-h mission.

From Example 6.3 we have the following data that we plot on the Weibull paper in Figure 6.22.

Time (h)	Percent Failures
500	34%
1500	48%
2500	65%
3500	83%
4500	92%
5500	100%

When we draw a best-fit line through the data points and follow the Weibull paper instructions, we determine the following:

$$\beta = 1.2$$
$$\eta = 2100$$
$$\text{MTBF} = 2000 \text{ h}$$
$$R(100 \text{ h}) = 1.0 - 2.6\% = 0.974$$

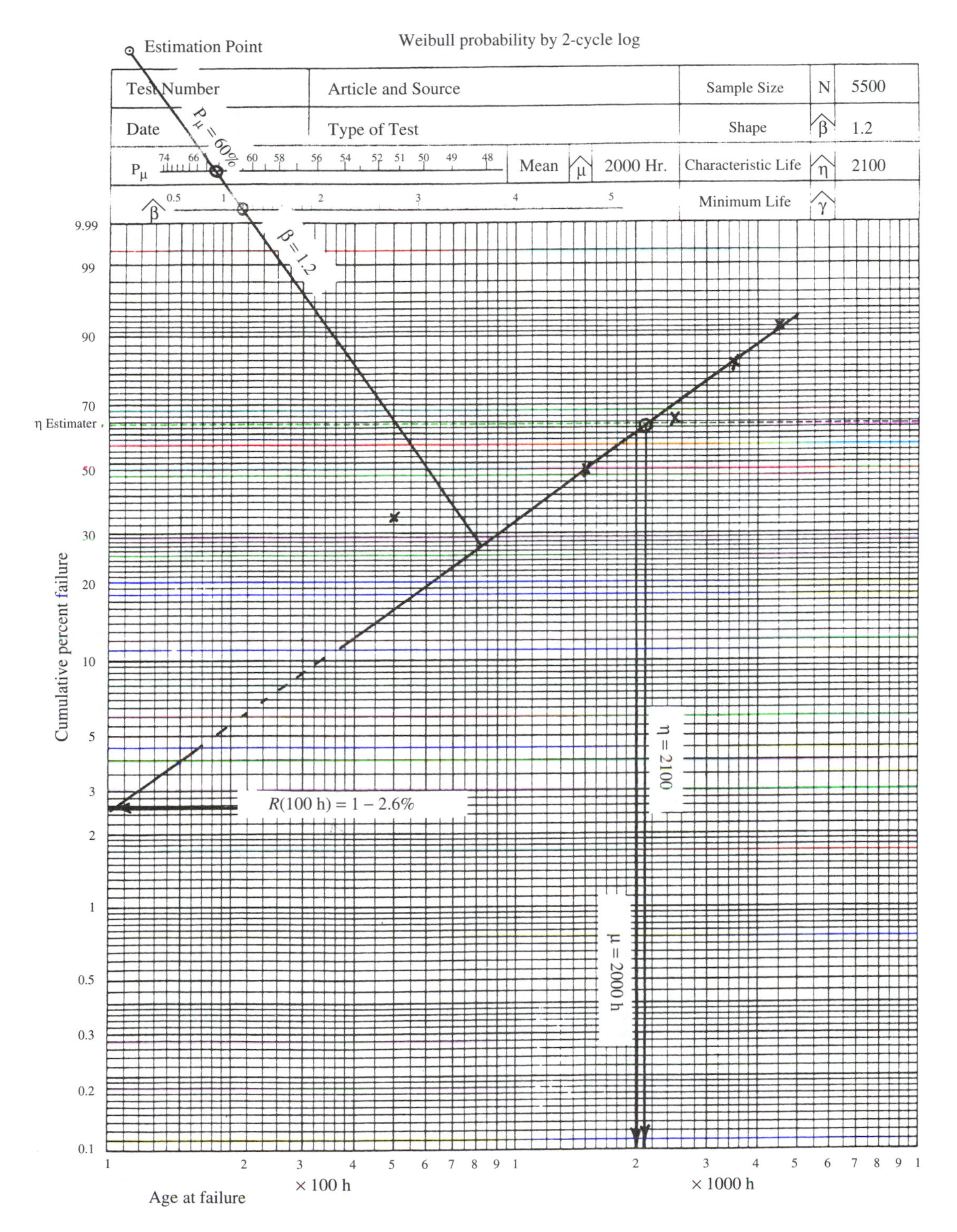

FIGURE 6.22

Plot of Example 6.5 data on Weibull probability paper. (SOURCE: From J. S. Craver, *Graph Paper from Your Computer or Copier.* Tucson: Fisher Books, 1996, p. 237.)

Because $\beta > 1.0$, we can assume an increasing failure rate, which we surmised when we analyzed the exponential plot in Example 6.3. Using the Weibull probability paper, however, allowed us to estimate MTBF and predict $R(t)$, which we were not able to do in either Example 5.10 or 6.3.

BIBLIOGRAPHY

1. J. S. Craver. 1966. *Graph Paper from Your Computer or Copier,* 3d ed. Tucson: Fisher Books.
2. A. Hald. 1952. *Statistical Theory with Engineering Applications.* New York: John Wiley.
3. P. D. T. O'Connor. 1991. *Practical Reliability Engineering,* 3d ed. New York: John Wiley.
4. M. L. Shooman. 1968. *Probabilistic Reliability: An Engineering Approach.* New York: McGraw-Hill.
5. C. O. Smith. 1976. *Introduction to Reliability in Design.* New York: McGraw-Hill.
6. P. A. Tobias and D. Trindade. 1986. *Applied Reliability.* New York: Van Nostrand Reinhold.

EXERCISES

6.1. A set of 100 devices of the same type are tested for 24 h. They are examined every 2 h, with the following results:

Time (h)	No. of Survivors	Time (h)	No. of Survivors
0	100	14	22
2	76	16	16
4	72	18	14
6	53	20	9
8	42	22	9
10	33	24	8
12	23		

a. Plot these results on both exponential and Weibull probability papers.
b. In your judgment, is the failure rate to be considered increasing, decreasing, or constant? Why?
c. Estimate the MTBF.
d. Predict the reliability for a 5-h mission.

6.2. The following set of operating times were collected from field reports on 10 devices of a given model:

Serial number 001 failed after 3214 h of operation.

Serial number 002 failed after 1984 h of operation.

Serial number 003 failed after 595 h of operation.

Serial number 004 has accumulated 4322 h and is still operating.

Serial number 005 failed after 1583 h of operation.

Serial number 006 failed after 2752 h of operation.

Serial number 007 failed after 1153 h of operation.

Serial number 008 has accumulated 3983 h and is still operating.

Serial number 009 failed after 3582 h of operation.

Serial number 010 failed after 2402 h of operation.

a. Plot these results on both exponential and Weibull probability papers.
b. Do you consider these devices to have increasing, decreasing, or constant failure rates? Why?
c. Estimate the MTBF. Predict the reliability of one of these devices for a 1000-h mission.

6.3 Fifteen items of a given model are tested and operated until failure. The times to failure (TTF) are as recorded here. Do you consider the failure rate to be increasing, decreasing, or constant? Why? Estimate the MTBF, and predict the reliability for a 200-h mission.

Item	TTF	Item	TTF	Item	TTF
1	19.8 h	6	163.4 h	11	398.6 h
2	35.2 h	7	190.7 h	12	501.9 h
3	70.5 h	8	231.3 h	13	595.8 h
4	98.0 h	9	292.1 h	14	753.8 h
5	123.6 h	10	355.0 h	15	1088.2 h

7

RESTORABILITY

7.1 INTENT

The first two chapters of this text introduced the discipline of *reliability* as a means of assuring that a product or service will function when it is required to do so by its user. These methods were presented as techniques for determining what can go wrong, how we can prevent it from going wrong, and how we can recover with minimum consequences if something does go wrong. Up to this point our discussions have related to the first two points and have dealt with predicting the frequencies of something going wrong. In this chapter we address the third point, that of recovering in the event that something does go wrong. Let us define this study of system recovery as *restorability.*

As is *reliability,* **restorability** is a feature that is inherent to the design of a system, be it a piece of equipment or a service-rendering process. It represents a potential, a capability for a system to be restored to working order within a given demand period after operability has been interrupted. It also includes measures taken to prevent or forestall an interruption during a peak demand time or to minimize the impact of such an interruption.

To summarize, restorability refers to any activities designed to prevent, delay, or correct the interruption of a system's operability or serviceability. These activities are referred to as **maintenance tasks.** When we talk about the time to perform these maintenance tasks, we refer to "spoon-fed" conditions, assuming that all materials (manuals, tools, etc.), information, and personnel necessary for performing the maintenance tasks are on hand and available. Other activities that are necessary for full restoration of an inoperable system are referred to as *administrative* or *logistics delay time* (described later). As used in this text, maintenance tasks are restricted to tasks designed into (inherent to) the system.

7.2 MAINTENANCE TASKS

Maintenance tasks can be classified as either *preventive* or *corrective* tasks. A **preventive maintenance task** is one that is performed strictly to avoid or delay a malfunction in the system. If we are dealing with a physical system, such as a piece of equipment, preventive maintenance may be a cleaning, lubricating, adjustment, or part-replacement activity. If we are dealing with the rendering of a service, preventive maintenance may be a communications, auditing, training, or calibration activity. Preventive maintenance can occur in the form of *scheduled maintenance* or *opportune maintenance*. **Scheduled maintenance,** the more common type, is usually done during some predetermined overhaul period or on some other scheduled cycle, usually designed to coincide with a minimum-demand time. **Opportune maintenance** is a preventive maintenance task that is done when system performance is interrupted for some malfunction. It may be possible to do the next scheduled maintenance tasks without further delaying the restoration of the system. Also, when the system is down for some malfunction, it may be possible to troubleshoot for other problems without extending the interruption.

A **corrective maintenance** task is one performed to rectify a system malfunction. If we are dealing with equipment, corrective maintenance may be a part adjustment, rework, or replacement. If we are dealing with the rendering of a service, corrective maintenance may be the correction of an error or revision of a procedure. Corrective maintenance can occur in the form of either *responsive* or *deferred maintenance*. **Responsive corrective maintenance** is a task performed to repair or correct some malfunction in the system. It is done in response to the malfunction and immediately upon its detection. This type of maintenance task usually occurs when the malfunction impairs the system or causes it to shut down entirely. Deferred corrective maintenance is usually in response to a malfunction that weakens or reduces the system's capability to perform. The corrective action in this case is deferred until the next scheduled maintenance or until the system is next undergoing some other corrective maintenance procedure.

7.3 LEVELS OF MAINTENANCE

The *level of maintenance* refers to the location at which certain maintenance tasks are designed to take place. Generally speaking, there are three levels of maintenance: *organizational level, intermediate level,* and *depot level.* **Organizational-level tasks** are those that take place at the user's location; **depot-level tasks** are those that must take place at the producer's facility; and the **intermediate-level tasks** are designed to take place at a specially created restoration facility.

For example, if our system is an automobile, the organizational level is the road. A typical organizational-level maintenance task is the changing of a tire. The intermediate level is an automobile repair shop, where tasks requiring trained mechanics and special diagnostic and repair equipment are performed. The depot level is for factory recalls. If our system is health care, organizational-level maintenance is any self-administered procedure, such as the taking of medication or supplements or a diet or exercise routine. Intermediate-level maintenance is any procedure designed to take place at a doctor's office, clinic, or gymnasium. Depot-level maintenance is a procedure requiring hospitalization.

7.4 DOWNTIME

Maintenance tasks may or may not impose system downtime. Some equipment can be maintained online or can have some of the maintenance actions done while the system is operating, whereas other maintenance actions require the equipment to be down. Indeed, some malfunctions may shut down the equipment. In the case of service-rendering systems, some types of malfunctions may put a halt to further activity until a full correction is made, whereas other preventive or corrective activities can be carried out while the system is in full service. As a consequence, our restorability analysis will be involved with the hours of effort put into the restorability activities as well as the system downtime imposed.

Downtime may be caused by activities associated with the maintenance tasks as well as by the tasks themselves. Often these peripheral activities may cause most of the downtime, especially with service-rendering systems. Although this text restricts itself to the restorability element of system downtime, it is appropriate to identify the other elements, which are generally controlled by user policies or regulatory laws and guidelines.

Downtime activities can be classified into three types: *restoration time, administrative time,* and *logistics delay time.* **Restoration time** is the time actually spent on corrective or preventive maintenance actions. **Administrative time** is the duration that the system is down due to any form of procedural or regulatory requirements. For example, regulatory agencies, such as the NRC, the FDA, or the NTSB, may require outage time for investigations of events that threaten public health or safety. These agencies may require removal of products from shelves, factory recalls, aircraft groundings, or power-plant shutdowns. In these extreme cases, the administrative time can far exceed the actual restoration time. More typically, administrative time consists of securing signatures of approval for restoration actions. If the restoration encompasses retraining or revisions of procedural manuals, the administrative time is time for certification or the necessary approval cycle.

Logistics delay time involves any delay in the restoration activities while awaiting replacement parts, tools, access to diagnostic equipment, or availability of personnel. It can consist of time to ship faulty items back to the factory, shipping replacement parts to a site, or, for that matter, sending experts or consultants to a site.

7.5 RESTORATION TIME

Now let us concentrate on that element of downtime that is designed into a system, the *restoration time.* Again, let us recognize that all restoration time is not necessarily *downtime.* Also, restoration tasks on the same system are expected to differ depending on whether a task is to be done at the organizational, intermediate, or depot level.

Restoration time is composed of seven possible elements:

1. Identification time
2. Diagnostic time
3. Access time
4. Interchange time
5. Reassembly time
6. Alignment time
7. Checkout time

Identification time is the time that it takes to identify that a malfunction has taken place and to identify, to the extent possible without auxiliary diagnostic equipment or outside consulting, the source of the problem. If the system is a piece of equipment, the identification element is sometimes referred to as the **localization element,** and it is the extent to which we can localize the part or component to be reworked or replaced without the need for any auxiliary test equipment. If our system is the rendering of a service, the identification element is the extent to which the problem source can be localized without the need for any diagnostic investigation or fact finding study.

Diagnostic time is the time it takes to isolate the source of the problem to the point at which the system can be restored at the applicable level of maintenance using any necessary auxiliary resources. If our system is a piece of equipment, these resources are diagnostic and test equipment. If our system is a service, these resources are the assignment of consultants or special task forces not normally employed. All restoration tasks do not necessarily require both the *identification* and *diagnostic* elements. Some require just one or the other.

Access time is the dedicated time required to reach a point at which the source of the problem or malfunction can be treated. If we are dealing with a piece of equipment, the access time is the time required to disassemble the equipment in order to access the part to be reworked or replaced. If we are dealing with a service, the access time is the time required for data gathering or accessing materials or information to treat the source of the problem. As an example, access time may be the time required to gather or create training literature or to obtain information for a revised procedure.

Interchange time is the time required for actually performing the corrective or preventive measures. If the system is the rendering of a service, the *interchange* is the actual revision of a procedure or the actual training of an operator or provider. If our system is a piece of equipment, the interchange is either the reworking of or the removal and replacement of an *LRU* **(line replaceable unit).** The designation of what is an LRU will vary depending on whether we are designing restorability tasks for an organizational-, intermediate-, or depot-level task. For a fleet of trucks, the corrective measure for a broken gear shaft may be replacement of the faulty truck at the organizational level, the replacement of the faulty gearbox at the intermediate level, or the replacement of the faulty gear at the depot level.

So the LRU designations may depend upon the capabilities of the maintenance level. But they may also be governed by expediency or by cost, space, or time trade-offs and restrictions. If the minimizaton of downtime is of paramount importance, we will want to designate LRUs at a fairly high assembly level. In the case of our trucking fleet example, if the service provided is on-time delivery, the quickest response to a truck breakdown is needed, and that response may be the replacement of the truck itself. If, on the other hand, the acquisition cost of having trucks for the fleet outweighs other considerations, we may choose to repair a disabled truck at the organizational level (thus designating truck parts as LRUs). If we are sending a ship out to sea and space restrictions are of utmost importance, then we shall want to store spare parts at low levels of assembly. This dictates that LRUs will be at that low-assembly or component level. It also means that fairly sophisticated maintenance capabilities will be necessary.

Reassembly time applies mostly to physical systems. It refers to the time required for reassembling the equipment after the LRU has been either removed or reworked. Reassembly time for a piece of equipment is usually similar in duration to the *access time.* With service-rendering systems, reassembly usually does not apply. However, if a revised procedure is to be edited into a larger service document or if updated training is to be inserted into an existing training program, such activities can be classified as reassembly.

Alignment time is the time required for any necessary alignment or adjustments of equipment after a corrective or preventive maintenance action has been accomplished. This element is not part of service maintenance and is only seldom required for equipment-maintenance tasks.

Checkout time is the time required to ascertain the effectiveness of the maintenance action and that the system is properly functioning. If the action is a corrective maintenance, we must verify the elimination of the problem. If the action is a preventive maintenance, we must assure the quality of the work. In either case we must be satisfied that new problems were not introduced by changes or maintenance actions. The checkout element usually consists of either performing some test or simply trying out the system.

All seven elements are not necessarily included in every restoration activity. Those elements that do apply do not necessarily occur in chronological order, as presented here. In fact, some elements are interrupted by others. We may start with *identification,* followed by some *diagnostic* activity, followed by some *access,* and then more *diagnostic* actions, etc.

Once we have broken down the restoration activity into its elements, we can make our restoration time predictions by evaluating each element individually. We may derive our elemental predictions from experience on similar systems, by testing and doing a time study, or from available industrial or government handbooks. Some commonly used handbooks are the GIDEP (Government-Industry Data Exchange Program) handbook on repair times and the government handbook MIL-HDBK-472 on maintainability prediction. A variety of *motion- and time-study* handbooks also provide prediction source data for estimating elements of repair times.

7.6 PREDICTION PARAMETERS

Chapter 2 defined two restorability parameters, the MTTR (mean time to restore) and MDT (mean downtime). In addition to those parameters, we shall be interested in a set called $M_{ct}(\alpha)$, which are the upper α-percentile points of the *time-to-restore* distribution for a corrective restoration task. More precisely, the restorability prediction parameters are defined as follows:

MDT The mean downtime is the expected (average) time that a system is down for a corrective or preventive restoration task. This includes administrative and logistics delay as well as restoration downtime.

MTTR The mean time to restore is the expected (average) time to perform a restoration task on a system after a malfunction, whether or not the system is down for all or part of the restoration.

$\mathbf{M_{ct}(\alpha)}$ The upper α-percentile of the time to restore distribution for a corrective restoration task is that point in time such that $\alpha\%$ of the system corrective restoration tasks will be complete. That is, $\alpha\%$ of the corrective restoration tasks are expected to be done in less time than $M_{ct}(\alpha)$. Most commonly used are $M_{ct}(90)$ and $M_{ct}(95)$. Also of particular interest is the median restoration time, $M_{ct}(50)$.

$\mathbf{M_{ct}(90)}$ 90% of the corrective restoration tasks are expected to be completed in less time than $M_{ct}(90)$.

$\mathbf{M_{ct}(95)}$ 95% of the corrective restoration tasks are expected to be completed in less time than $M_{ct}(95)$.

Median Corrective Restoration Time, $\mathbf{M_{ct}(50)}$ 50% of the corrective restoration tasks are expected to be completed in less time than the median corrective restoration time.

7.7 CORRECTIVE RESTORATION TIME DISTRIBUTION

The time-to-restore distribution for corrective restoration tasks is not symmetrical. We would expect most of the repairs to be done in less time than the MTTR. so the median, $M_{ct}(50)$, is always going to be less than the MTTR. This is true because the corrective restoration tasks that occur most frequently will be the most familiar and, consequently,

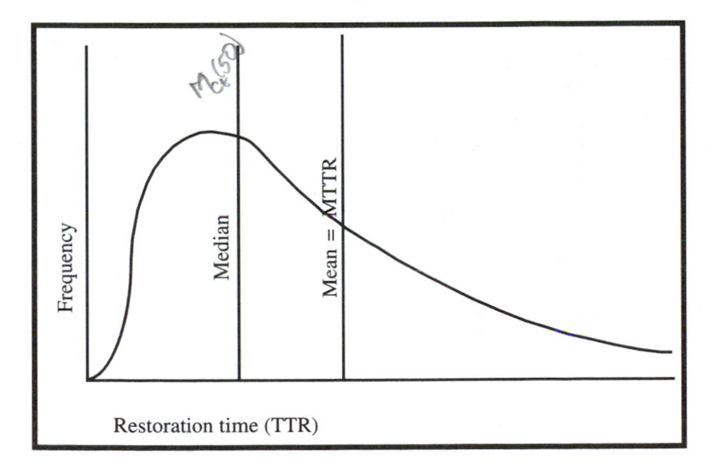

FIGURE 7.1
Typical time-to-restore distribution.

the quickest to identify and diagnose. For the same reason they will have the shortest interchange, alignment, and checkout times. Also, good restorability design practice dictates that with equipment the LRUs requiring the most frequent rework or replacement are the most accessible. This suggests shorter access and reassembly times. On the other hand, the corrective restoration tasks that occur least frequently can be expected to require the longest total restoration time. Accordingly, we can expect the time-to-restore distribution to take the shape illustrated in Figure 7.1.

*7.8 THE RESTORABILITY MATHEMATICAL MODEL

As identified in Section A.7 (Appendix A), the time-to-restore distribution of Figure 7.1 is *positively skewed.* So, as illustrated, most restoration actions will take place in less time than the MTTR.

Experience has shown that what is commonly called the **Lognormal Distribution** is a good fit to the distribution of TTR. This distribution is not actually a probability law, but a mathematical

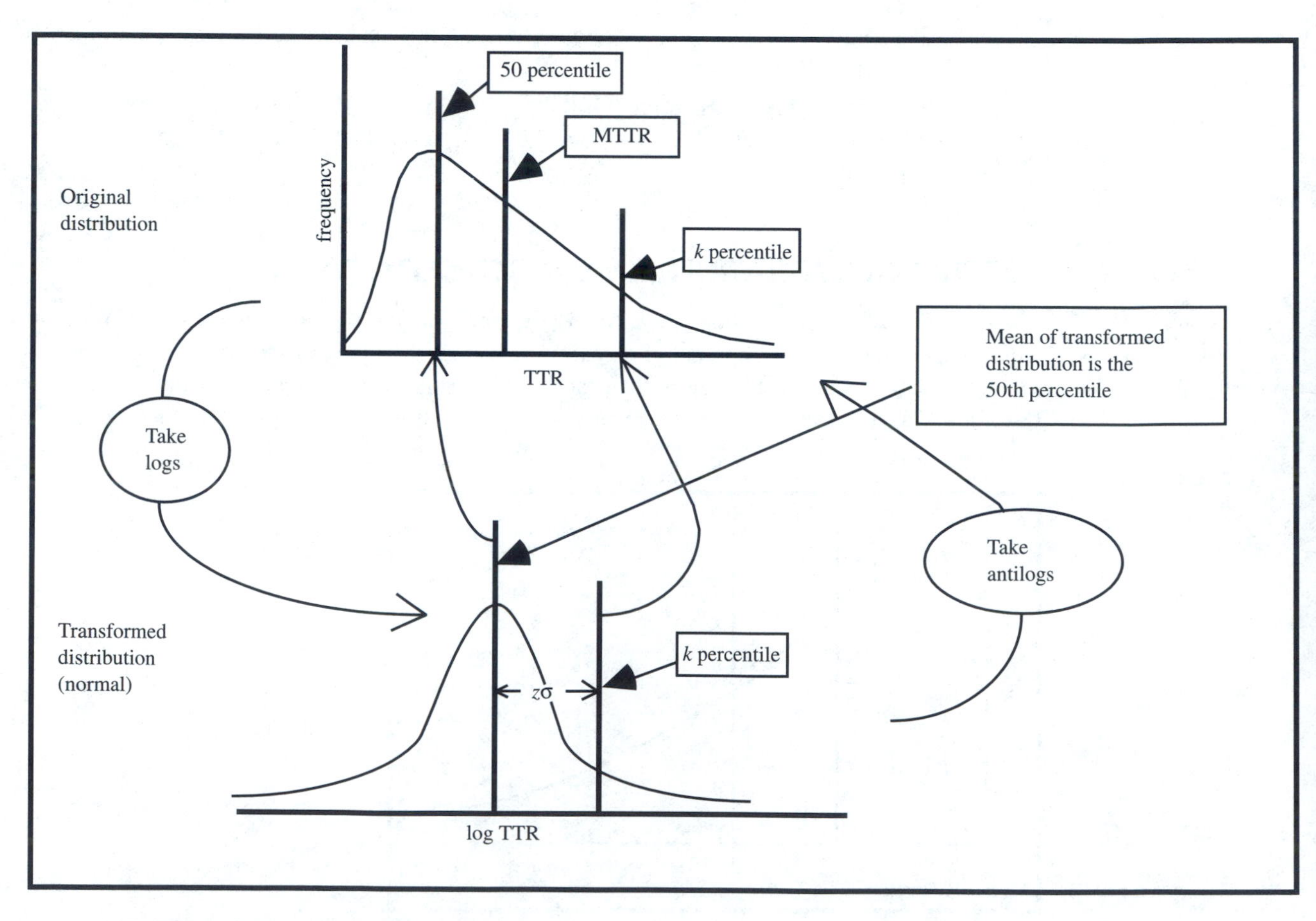

FIGURE 7.2
Transformation of the time-to-restore distribution.

transformation of TTR by taking logarithms. While the distribution of TTR is positively skewed, the distribution of log (TTR) fits the Normal Probability Law. Hence, to make percentile point predictions, we deal with the transformed distribution, which we evaluate using a Normal Probability Law assumption.

Figure 7.2 illustrates the process of using both our original and transformed distributions as prediction models. The MTTR, the mean of the TTR distribution, is estimated from the original distribution, which is positively skewed. When we transform the distribution by taking logarithms of all the TTR points, what results is the normal distribution, which is symmetrical. We can now use the transformed distribution to predict all percentile points, including the median (the 50th percentile point).

To get the median of the transformed distribution, we simply estimate its mean μ, which is the same as the median for the normal distribution. This transforms back to the median of the original distribution. So, we can get the median of the TTR distribution by taking the antilog of the mean, μ, of the transformed distribution.

Obtaining a percentile point of the transformed distribution is done the same way that percentile points are determined for any normally distributed phenomenon. The α percentile point is $\mu + z_\alpha \sigma$. This transforms back to the α percentile point of the original distribution by taking the antilog of $\mu + z_\alpha \sigma$ of the transformed distribution.

*7.9 PREDICTING MTTR FROM ESTIMATED TTR DATA

Section 7.5 identifies seven elements of restoration tasks. The estimated time for a particular restoration task i is generally accomplished by estimating times at the element level and then adding all the element time predictions to obtain a total task time, M_i. We also must estimate the frequency at which the task i is expected to occur. Assuming that corrective restoration tasks are in response to malfunctions, the applicable failure rate λ_i is the expected frequency. If a system has N possible restoration task types, the MTTR for that system is estimated by the arithmetic mean

$$\text{MTTR} = \Sigma\,(\lambda_i\,M_i)/\Sigma\,\lambda_i \qquad \textbf{(7.1)}$$

*7.10 PREDICTING MEDIAN TIME TO RESTORE FROM ESTIMATED TTR DATA

The median of the TTR distribution is its 50th percentile point, $M(50)$. This point is obtained from the transformed distribution, which, being normally distributed, has a median equal to its mean. The mean of the transformed distribution is estimated from the arithmetic mean of the logarithms of the *times to restore*. For the N possible system task types, the mean of the logarithms is

$$\mu_{\log} = \Sigma\,(\lambda_i \log M_i)/\Sigma\,\lambda_i \qquad \textbf{(7.2)}$$

The median of the original TTR distribution is obtained by transforming back—that is, by taking the antilogarithm of Equation 7.2:

$$M(50) = \log^{-1}\,[\mu_{\log}] = \log^{-1}\,[\Sigma\,(\lambda_i \log M_i)/\Sigma\,\lambda_i] \qquad \textbf{(7.3)}$$

*7.11 PREDICTING A PERCENTILE POINT FROM ESTIMATED TTR DATA

The α percentile point $M(\alpha)$ of the TTR distribution is also obtained by estimating the α percentile of the transformed distribution and transforming back. Since the transformed distribution is normally distributed, the α percentile point is that value A such that $P(x < A) = \alpha$. In this case the value is $A = \log M(\alpha)$. Following the technique of Section A.9, we set

$$\log M(\alpha) = \mu_{\log} + z_\alpha \sigma_{\log} \qquad (7.4)$$

where
$\mu_{\log}$ = the mean of the transformed distribution
$\sigma_{\log}$ = the standard deviation of the transformed distribution,
z_α = the z value from Table B.1 that corresponds to the probability $\alpha\%$

The value $\mu_{\log}$ is obtained from Equation 7.1. From Section A.7, the standard deviation is

$$\sigma_{\log} = \sqrt{\frac{\Sigma\, \lambda_i (\log M_i)^2}{\Sigma\, \lambda_i} - (\mu_{\log})^2} \qquad (7.5)$$

The value of z_α is obtained from Table B.1. Then,

$$M(\alpha) = \log^{-1}\, [\mu_{\log} + z_\alpha \sigma_{\log}] \qquad (7.6)$$

EXAMPLE 7.1 Develop an equation to compute the upper 95th percentile of the TTR distribution, assuming that TTR is lognormally distributed.

From Equation 7.6,

$$M(95) = \log^{-1}\, [\mu_{\log} + z_{0.95}\sigma_{\log}]$$

From Equation 7.2,

$$\mu_{\log} = \Sigma\, (\lambda_i \log M_i) / \Sigma\, \lambda_i$$

From Equation 7.5,

$$\sigma_{\log} = \sqrt{\frac{\Sigma\, \lambda_i (\log M_i)^2}{\Sigma\, \lambda_i} - (\mu_{\log})^2} = \sqrt{\frac{\Sigma\, \lambda_i (\log M_i)^2}{\Sigma\, \lambda_i} - \left(\frac{\Sigma\, \lambda_i \log M_i}{\Sigma\, \lambda_i}\right)^2}$$

From Table B.1, by interpolation, where $\alpha = 0.9500$, z is halfway between 1.64 and 1.65:

$$z_{0.95} = 1.645$$

Hence,

$$M(95) = \log^{-1}\left[\frac{\Sigma\, \lambda_i \log M_i}{\Sigma\, \lambda_i} + 1.645\sqrt{\frac{\Sigma\, \lambda_i (\log M_i)^2}{\Sigma\, \lambda_i} - \left(\frac{\Sigma\, \lambda_i \log M_i}{\Sigma\, \lambda_i}\right)^2}\,\right]$$

Example 7.1 derives Equation 7.10. The other equations in Section 7.12 are similarly derived.

7.12 RESTORABILITY PREDICTIONS

To estimate restorability parameters for a system, we first list all possible restorability tasks, 1 through N. For each task i we estimate the time of each of the seven elements of the restoration task. This is done by using handbook data, personal experience, or data from actual observations. The sum of all seven element times is the total restoration time estimate, M_i, for the ith task.

We next list the expected frequency of occurrence of each task i. This is usually the failure rate λ_i for the type of failure that would generate the ith corrective restoration action. For each task i we then multiply the frequency, λ_i, by the restoration time, M_i, to get what is known as the **load factor,** $\lambda_i M_i$.

We then obtain the logarithms of the restoration times, $\log M_i$, by using a calculator or looking them up in a mathematics table. We multiply frequencies by the logarithms to get $\lambda_i \log M_i$.

We then sum the frequencies, λ_i, load factors, $\lambda_i M_i$, and $\lambda_i \log M_i$ values. A table format is useful for these computations.

Task Number	Task Description	Frequency, λ_i	Restore Time, M_i	Log of Restore Time, $\log M_i$	Load Factor, $\lambda_i M_i$	$\lambda_i \log M_i$
1		λ_1	M_1	$\log M_1$	$\lambda_1 M_1$	$\lambda_1 \log M_1$
2		λ_2	M_2	$\log M_2$	$\lambda_2 M_2$	$\lambda_2 \log M_2$
.		.	.	.	.	.
.		.	.	.	.	.
.		.	.	.	.	.
N		λ_N	M_N	$\log M_N$	$\lambda_N M_N$	$\lambda_N \log M_N$
	Totals	$\Sigma\, \lambda_i$			$\Sigma\, \lambda_i M_i$	$\Sigma\, \lambda_i \log M_i$

Next, we square each of the $\log M_i$ values and multiply them by the corresponding λ_i values to get $\lambda_i(\log M_i)^2$. We then sum these products. Again a table format is useful.

Task Number	Frequency, λ_i	Log of Restore Time, $\log M_i$	Square $(\log M_i)^2$	$\lambda_i(\log M_i)^2$
1	λ_1	$\log M_1$	$(\log M_1)^2$	$\lambda_1(\log M_1)^2$
2	λ_2	$\log M_2$	$(\log M_2)^2$	$\lambda_2(\log M_2)^2$
.	.	.	.	.
.	.	.	.	.
.	.	.	.	.
N	λ_N	$\log M_N$	$(\log M_N)^2$	$\lambda_N(\log M_N)^2$
Totals	$\Sigma\, \lambda_i$			$\Sigma\, \lambda_i(\log M_i)^2$

From the summations in the tables we can now do the following computations to complete our predictions. The MTTR is determined from Equation 7.1:

$$\text{MTTR} = \frac{\Sigma\, \lambda_i M_i}{\Sigma\, \lambda_i} \tag{7.7}$$

The α percentile predictions are computed from antilogarithms, $\log^{-1}$, as derived in Equation 7.6, taken either from a calculator or from mathematics tables.

The median is determined from Equation 7.3:

$$M_{ct}(50) = \log^{-1}\left[\frac{\Sigma\, \lambda_i \log M_i}{\Sigma\, \lambda_i}\right] \tag{7.8}$$

The 90th percentile point is

$$M_{ct}(90) = \log^{-1}\left[\frac{\Sigma\, \lambda_i \log M_i}{\Sigma\, \lambda_i} + 1.28\sqrt{\frac{\Sigma\, \lambda_i (\log M_i)^2}{\Sigma\, \lambda_i} - \left(\frac{\Sigma\, \lambda_i \log M_i}{\Sigma\, \lambda_i}\right)^2}\,\right] \tag{7.9}$$

The 95th percentile is

$$M_{ct}(95) = \log^{-1}\left[\frac{\Sigma\, \lambda_i \log M_i}{\Sigma\, \lambda_i} + 1.645\sqrt{\frac{\Sigma\, \lambda_i (\log M_i)^2}{\Sigma\, \lambda_i} - \left(\frac{\Sigma\, \lambda_i \log M_i}{\Sigma\, \lambda_i}\right)^2}\,\right] \tag{7.10}$$

Finally, for any percentile α,

$$M_{ct}(\alpha) = \log^{-1}\left[\frac{\Sigma\, \lambda_i \log M_i}{\Sigma\, \lambda_i} + z_\alpha\sqrt{\frac{\Sigma\, \lambda_i (\log M_i)^2}{\Sigma\, \lambda_i} - \left(\frac{\Sigma\, \lambda_i \log M_i}{\Sigma\, \lambda_i}\right)^2}\,\right] \tag{7.11}$$

where z_α is the z value from Table B.1 that corresponds to the cumulative probability α. Notice that for a 90% probability in Table B.1, z is 1.28, and for a 95% probability, it is 1.645.

EXAMPLE 7.2 A system is known to have 10 types of malfunctions. Estimates of the failure frequency of each type and of the expected restoration time for each type are as shown in the following table.

Type	Frequency (failures per million hours)	Expected Restoration Time
1	20	25 min
2	10	1 h
3	30	20 min
4	20	35 min
5	10	50 min
6	20	12 min
7	10	1 h 45 min
8	20	30 min
9	30	25 min
10	20	40 min

Predict the mean time to restore (MTTR).

The MTTR prediction is computed via Equation 7.7, using the suggested computation table. For computational purposes, the restoration times are all converted to hours.

Type	λ (per 10^6 h)	M (h)	Load Factor, λM
1	20	0.42	8.40
2	10	1.00	10.00
3	30	0.33	10.00
4	20	0.58	11.60
5	10	0.83	8.30
6	20	0.20	4.00
7	10	1.75	17.50
8	20	0.50	10.00
9	30	0.42	12.60
10	20	0.67	13.40
Totals	190		105.80

From Equation 7.7,

$$\text{MTTR} = 105.80/190 = 0.557 \text{ h} = 33.4 \text{ min}$$

EXAMPLE 7.3 Determine the median and the upper 90th percentile of the time-to-restore distribution for the system of Example 7.2.

This computation is done using a table.

Type	λ (per 10^6 h)	M (h)	$\log M$	$(\log M)^2$	$\lambda(\log M)$	$\lambda(\log M)^2$
1	20	0.42	−0.3768	0.1419	−7.54	2.84
2	10	1.00	0.0000	0.0000	0.00	0.00
3	30	0.33	−0.4815	0.2318	−14.45	6.95
4	20	0.58	−0.2366	0.0560	−4.73	1.12
5	10	0.83	−0.0809	0.0065	−0.81	0.07
6	20	0.20	−0.6990	0.4886	−13.98	9.77
7	10	1.75	0.2430	0.0591	2.43	0.59
8	20	0.50	−0.3010	0.0906	−6.02	1.81
9	30	0.42	−0.3768	0.1419	−11.30	4.26
10	20	0.67	−0.1739	0.0302	−3.48	0.60
Totals	190				−59.88	28.01

From Equation 7.8, the median is

$$M_{ct}(50) = \log^{-1}[-59.88/190] = \log^{-1}[-0.3152]$$
$$= 0.48 \text{ h} = 29 \text{ min}$$

Notice that, as we expected, the median is less than the MTTR.

From Equation 7.9, the upper 90th percentile of the TTR distribution is

$$M_{ct}(90) = \log^{-1}\left[\frac{-59.88}{190} + 1.28\sqrt{\frac{28.01}{190} - \left(\frac{-59.88}{190}\right)^2}\right] = 0.924 \text{ h} = 55 \text{ min}$$

EXAMPLE 7.4 Determine the upper 99th percentile of the time-to-restore distribution for the system of Examples 7.2 and 7.3.

This is solved using Equation 7.11 and the summations from the computation table in Example 7.3. The z value for Equation 7.5 is the z corresponding to 99%, or a probability of 0.9900, in Table B.1:

$$z = +2.33$$

From Equation 7.11,

$$M_{ct}(99) = \log^{-1}\left[\frac{-59.88}{190} + 2.33\sqrt{\frac{28.01}{190} - \left(\frac{-59.88}{190}\right)^2}\right] = 1.570 \text{ h}$$
$$= 94 \text{ min} = 1 \text{ h } 34 \text{ min}$$

For the system in Examples 7.2, 7.3, and 7.4, the MTTR is 33 min, but 50% of the restoration tasks will be complete by 29 min, 90% of the restoration tasks will be complete by 55 min, and 99% of the restoration tasks will be complete by 94 min.

EXAMPLE 7.5 For the systems in Examples 7.2 through 7.4, what percentage of the restoration tasks will be complete in less time than the MTTR?

Setting Equation 7.11 equal to the MTTR = 0.557 h,

$$M_{ct}(\alpha) = \log^{-1}\left[\frac{-59.88}{190} + z_\alpha\sqrt{\frac{28.01}{190} - \left(\frac{-59.88}{190}\right)^2}\right] = 0.557 \text{ h},$$

from which

$$\left[\frac{-59.88}{190} + z_\alpha\sqrt{\frac{28.01}{190} - \left(\frac{-59.88}{190}\right)^2}\right] = \log 0.557 = -0.2541$$

Solving for z_α gives

$$\left[\frac{-59.88}{190} + 0.2193 \cdot z_\alpha\right] = 0.2193 \cdot z_\alpha - 0.315 = -2541$$

and

$$z_\alpha = \frac{0.061}{0.2193} = 0.278$$

From Table B.1, where $z = 0.278$, the probability is 0.6095. Therefore,

$$\alpha = 61\%$$

The significance of this result is to demonstrate that more than 60% of the restoration tasks on the system in question will be completed in less time than the MTTR.

We can always expect most of the restoration tasks to be completed in less time than the MTTR. Of equal interest is the fact, discovered in Chapter 6, that most parts put into operation will fail in less time than the MTBF.

7.13 GRAPHICAL ANALYSIS OF RESTORABILITY DATA

If the TTR distribution fits the *lognormal* assumption, the data should plot as a straight line on graph paper with a *logarithm-by–normal probability* scale. Figure 7.3 is such a plot. It shows the data from Example 7.2 plotted on *two-cycle log-by-probability* paper. Refer to Sections 6.2 through 6.4 for an understanding of the mechanism behind the *log-by-probability* plot.

The vertical axis of Figure 7.3 is a logarithm scale, and the horizontal axis is a normal probability scale. For reasons presented in Chapter 6, if we use the vertical axis to represent TTR and the horizontal axis to represent cumulative probability, then the plotted points fitting a straight line indicates a lognormal fit. The best-fit straight line can then be used to estimate the median time to repair and any percentile point of the TTR distribution from the vertical axis. This graph cannot be used to estimate the MTTR.

EXAMPLE 7.6 Plot the data from Example 7.2 in the text on log-by-probability graph paper to determine whether the lognormal assumption is a good fit. If so, use the plot to estimate the median time to repair as well as the upper 90th and 99th percentiles of the TTR distribution. Compare these results to those of Examples 7.3 and 7.4. Why can't this plot be used to estimate the MTTR?

It is first necessary to rearrange the list of task times in ascending order so that cumulative probabilities can be determined.

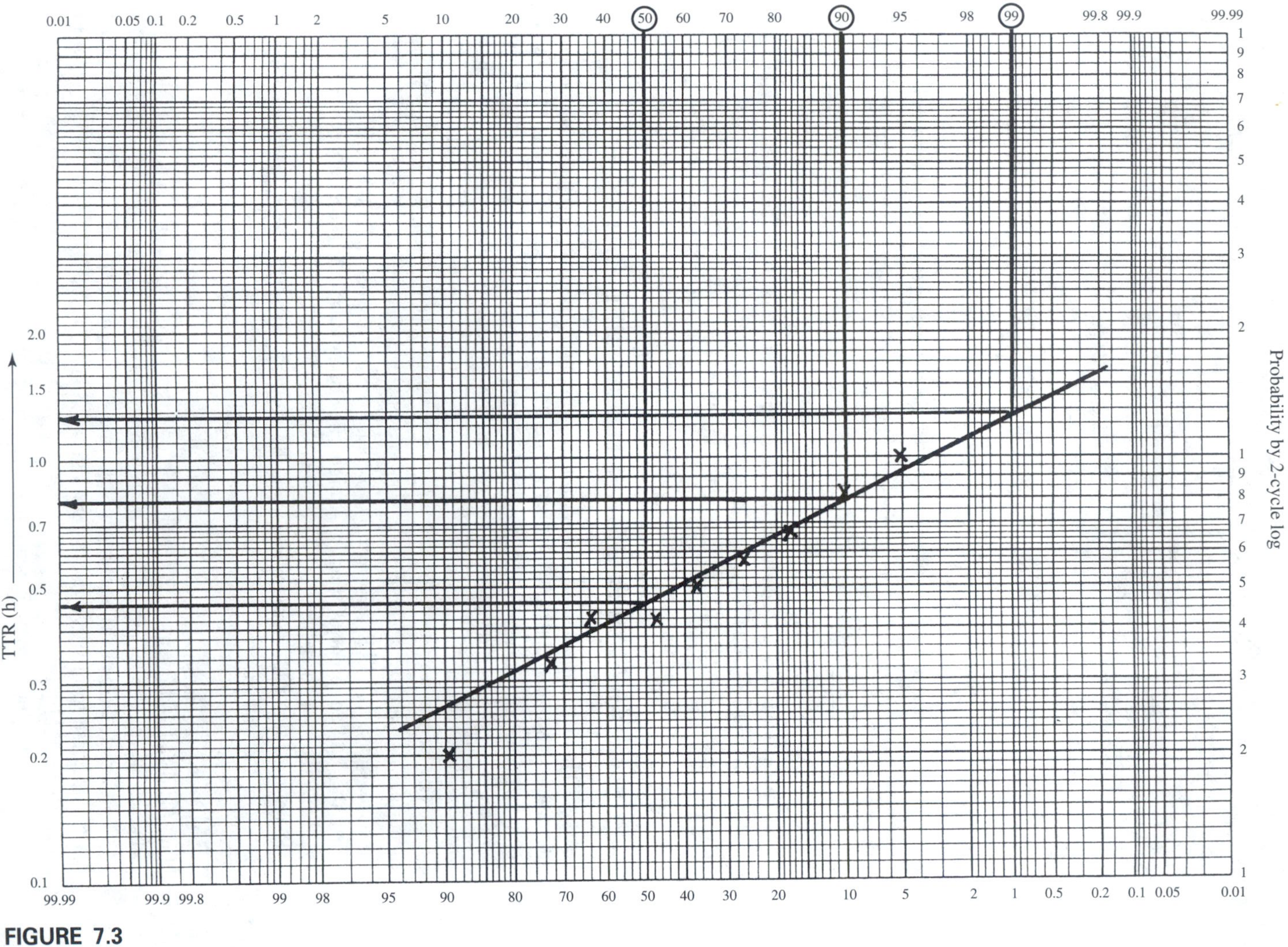

FIGURE 7.3
Solution to Example 7.6 on two-cycle log-by-probability paper. (SOURCE: From J. S. Craver, *Graph Paper from Your Computer or Copier*. Tucson: Fisher Books, 1996, p. 229.)

(1) Task Type	(2) Expected Restoration Time (h)	(3) Frequency (per million hours)	(4) Frequency as Percent of Total Frequency	(5) Cumulative Percent Frequency
6	0.20	20	10.5	10.5
3	0.33	30	15.8	26.3
1	0.42	20	10.5	36.8
9	0.42	30	15.8	52.6
8	0.50	20	10.5	63.1
4	0.58	20	10.5	73.6
10	0.67	20	10.5	84.1
5	0.83	10	5.3	89.4
2	1.00	10	5.3	94.7
7	1.75	10	5.3	100.0
		Total 190		

We next use the vertical axis of Figure 7.3 to designate the restoration times, in hours, to accommodate every column 2 value (from 0.20 to 1.75 h). Two-cycle log paper is sufficient. If the difference between the longest and shortest restoration times had spanned three magnitudes, it would have been necessary to use three-cycle log-by-probability paper.

We next plot the expected restoration time (Column 2) versus cumulative probability (Column 5) on Figure 7.3. Notice that the points do suggest a linear relationship, so we can assume a lognormal fit. From the resulting plot, the estimated median, $M(50)$, is about 0.46 h, which is reasonably close to the analytical result of Example 7.3. The plot estimates the upper 90th percentile to be about 0.8 h. This is somewhat lower than the 0.9 h estimated analytically in Example 7.3. The plot estimates the upper 99th percentile to be about 1.3 h, which is less than the 1.6 h estimated analytically in Example 7.4.

These results and the comparison point out the limitations of the graphical approach, especially at the tails of the distribution. However, the plot is a quick method for determining the adequacy of the fit of the lognormal assumption and for getting reasonable estimates of percentile points of the TTR distribution.

Because the TTR distribution is *not* symmetrical, the mean does not equal the median, and we cannot estimate the MTTR from the plot. The MTTR must be estimated analytically. We do know, however, that the MTTR is larger than the median. So we can state for this example that the MTTR is more than 0.46 h.

BIBLIOGRAPHY

1. F. L. Ankenbrandt. 1963. *Maintainability Design.* Elizabeth, N.J.: Engineering Publishers.
2. ARINC Research Corp. 1964. *Reliability Engineering.* Upper Saddle River, N.J.: Prentice Hall.

3. J. S. Craver. 1996. *Graph Paper from Your Copier,* 3d ed. Tucson: Fisher Books.
4. P. D. T. O'Connor. 1991. *Practical Reliability Engineering,* 3d ed. New York: Wiley.
5. MIL-STD-470A. 1983. *Maintainability Program Requirements.* Arlington, Va.: Department of Defense.
6. MIL-STD-472. 1966. *Maintainability Prediction.* Arlington, Va.: Department of Defense.

EXERCISES

7.1. Compute the MTTR for the equipment composed of the following LRUs.

LRU Type	Expected Restoration Time (h)	Failure Rate (per 10^6 h)
A	0.9	148
B	1.3	152
C	0.6	40
D	2.5	8
E	1.1	50
F	0.7	150
G	0.5	35
H	1.4	60
I	1.5	20
J	0.3	2
K	0.4	3
L	1.2	40
M	1.0	59
N	0.8	169
O	2.0	62

7.2. Compute the median restoration time for the equipment of Exercise 7.1.
7.3. Compute $M_{ct}(0.90)$ for the equipment of Exercise 7.1.
7.4. Compute $M_{ct}(0.95)$ for the equipment of Exercise 7.1.
7.5. Plot the Exercise 7.1 data on the log-by-probability paper in Figure 7.4. Draw a best-fit straight line through the data points. Examine the straight-line fit to evaluate whether a lognormal assumption can be made for this time-to-restore distribution.
7.6. From your plot in Exercise 7.5, estimate the *median restoration time,* $M_{ct}(0.90)$, and $M_{ct}(0.95)$.
7.7. Compare your graphical determinations of Exercise 7.6 with your analytic results of Exercises 7.2, 7.3, and 7.4. Account for any differences in these results. What are the relative merits of the analytic and graphical techniques?
7.8. In Exercise 7.6 we used the plot of Exercise 7.5 to estimate the median restoration time, $M_{ct}(0.90)$, and $M_{ct}(0.95)$. Why is it that we cannot use this plot to estimate the MTTR as well?

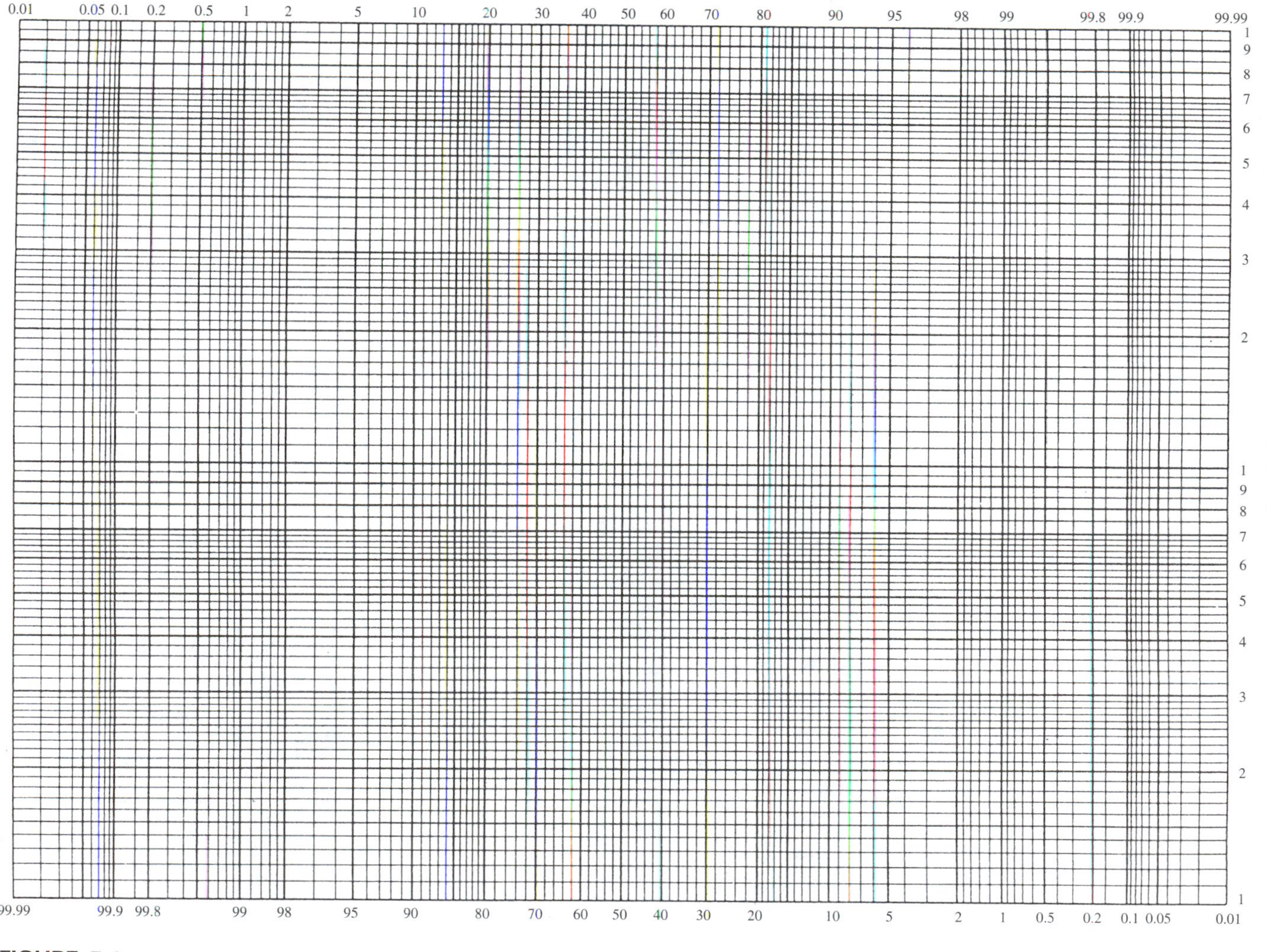

FIGURE 7.4
Two-cycle log-by-probability paper. (SOURCE: From J. S. Craver, *Graph Paper from Your Computer or Copier*. Tucson: Fisher Books, 1996, p. 229.)

8

RELIABILITY MODELING FOR SYSTEM PREDICTIONS

8.1 INTENT

In Book 1 we learned that *reliability* is an assessment of the likelihood of a manufactured product or service to perform for its user when the user demands it. For this reason we wish to be able to predict the reliability of each of our products or services. How likely is a VCR to perform or not perform for our customer throughout the time this customer plans to use it? How likely is it that a hotel's meal-delivery service will satisfy the customers whenever they want that service? How sure are we that a parcel-delivery service will deliver the right package to the right destination on time and in acceptable condition? The discipline known as *reliability* addresses such questions by making predictions for products or services.

As we learned in Chapter 3, the technique used considers our product or service to be a *system,* which, in turn, is composed of constituent elements. These elements can be components of a piece of hardware or procedural steps of a service process. Chapters 4–6 deal with means of reliability prediction at the element (or component) level. In this and the next two chapters we put these component predictions together to reach our actual objective, a prediction at the system level. The strategy for doing this is to develop a *math model*—that is, a means of relating a system to its constituent elements mathematically.

We might say that a **reliability math model** is a mathematical description of how all the components behave within their system. Using the proper reliability math model, we can now apply component-failure and component-restoration rates, determined through the techniques of Chapters 4–7, to estimate a system's **mission reliability, MTBF,** or **availability.** With this information we can predict the likelihood of our VCR to perform satisfactorily for our customer, how often a customer will be dissatisfied with hotel room service, or the likelihood of a problem with a package delivery. The reliability math model for a system is used not only for making such predictions, it can also be used to evaluate cost-effective ways of improving performance.

8.2 SYSTEMS

Throughout the next three chapters, when we refer to a **system** we shall be using the definition at the beginning of Chapter 2. Recall that under this definition a system can be a functional piece of equipment or a process for rendering a service. A system is composed of what we shall refer to as **elements,** which, in the case of equipment, may be subassemblies or components (that is, some form of hardware) or may be a software package. In the case of a service-rendering system, an element may be a procedure, a piece of support equipment, an operator, software, or training.

Reliability, as we learned in Chapter 1, is the most customer-focused portion of quality assurance. It deals with the ability of the system to function as the user requires at the time that the user demands it. And, as we learned in Chapter 7, *restorability* deals with the capability of the system to be maintained in or restored to a functioning state to meet the user's demands. Together the intention of reliability and restorability design is to assure the customer's satisfaction with functionability of a system. If a system fails to satisfy a customer's needs, the most annoying comment he or she can receive from the producer is, "Well, we did our best, but nobody is perfect."

Indeed, nobody is perfect and nothing is perfect, certainly at the element level. So, an objective of reliability and restorability design is to assure that a system is insensitive to the lack of perfection of its elements. This is accomplished by developing an in-depth understanding of how elements function, or fail to function, as part of a system and how they interact with one another. We then design our system to be able to function to meet the user's demands at the demand times, despite the fallibility of its constituent elements. In other words, although we may have to accept the existence of faults within elements, our system must be *fault-tolerant.*

As we shall see in this and the next two chapters system, fault-tolerance can be accomplished through a variety of features. These features include element redundancy, standby backup elements, and on-line restoration of elements. Chapter 13 presents methods for actively examining a system's design to search for weaknesses in its fault-tolerant ability. The topic of this and the next two chapters is the evaluation of the reliability of a system that already possesses element redundancy, standby, monitoring, and on-line restoration capabilities. Each of these features enlarges the degree of complexity for the reliability analyst. So, we shall address them one at a time, starting, in this chapter with *nonrestored* systems. By **nonrestored systems** we mean systems for which no restorations are performed unless the system is down.

8.3 SERIES AND PARALLEL SYSTEMS

A **series system** is composed of a series of elements, the failure of *any* of which will result in a system failure. A **parallel system** is composed of a group of elements, the failure of *all* of which is necessary for a system failure. A **redundant system** is composed of a group of elements, the failure of one of which will not result in a system failure. Hence, a *parallel* system is a particular type of *redundant* system; it is a redundant system that does not fail unless all of its redundant elements fail. Readers who are electrical engineers are cautioned that this text uses the terms *series* and *parallel* as defined in the *reliability* sense. It is quite possible for two components to be electrically parallel but in series in the reliability sense.

A reliability math model is generally accompanied by a **reliability block diagram,** such as those on the following pages, to illustrate the system logic.

8.4 SERIES SYSTEM

A series system (in the reliability sense) is composed of n elements, the failure of any of which will cause a system failure. Appendix A (fifth property of probability functions) shows us that if all elements are independent of one another (that is, the failure of one element does not make another of the elements more vulnerable), the system reliability is

$$R_{\text{sys}} = R_1 \times R_2 \times \cdots \times R_n = \prod_{i=1}^{n} R_i \tag{8.1}$$

where R_{sys} is the system reliability and R_i is the reliability of the *ith* element in the system. For a series system whose operation is time dependent, the reliability for a mission of specified time t is

$$R_{\text{sys}}(t) = R_1(t) \times R_2(t) \times \cdots \times R_n(t) = \prod_{i=1}^{n} R_i(t) \tag{8.2}$$

Figure 8.3 is an example of a reliability block diagram for a time-dependent series system. Here we have a communications system composed of three elements, a transmitter, a receiver, and a processor. The failure of any one of these three elements will constitute a system failure.

When we are evaluating hardware systems, we can assume that worn-out or nearly worn-out items will be on a planned replacement schedule. So elements that exhibit increasing failure rate will, because of planned replacement or other forms of scheduled preventive restoration activities, behave as though they have constant failure rates within the system. Accordingly, Equation 3.26 is appropriate, and applying it to Equation 8.2,

$$R_{\text{sys}}(t) = \prod_{i=1}^{n} R_i(t) = \prod_{i=1}^{n} e^{-\lambda_i t} = e^{-t\sum_{i=1}^{n}\lambda_i} \tag{8.3}$$

Recognizing that for a series system the system effective failure rate, sysFR (in this text the notation λ is restricted to an element-failure rate), is the sum of all the series element failure rates, and applying the constant–failure rate assumption, we can state that the system MTBF is

$$\text{MTBF} = \frac{1}{\text{sysFR}} = \frac{1}{\sum_{i=1}^{n} \lambda_i} \tag{8.4}$$

assuming that all restorations are to the $t = 0$ condition. Then we can conclude that for a series system, the reliability for a mission of time t, where wear-out items are on a planned restoration schedule and all restorations are to the $t = 0$ condition,

$$R_{\text{sys}} = e^{-t/\text{MTBF}} \tag{8.5}$$

EXAMPLE 8.1 Given a system with four independent elements in series, find the system reliability if each element has the same reliability, $R = 0.940$.

From Equation 8.1,

$$R_{\text{sys}} = \prod_{i=1}^{n} R_i = (0.940)^4 = 0.781$$

EXAMPLE 8.2 Consider a time-dependent communications system composed of a transmitting unit, a receiving unit, and a processing unit. Suppose the transmitter has an MTBF of 40 h, the receiver has an MTBF of 500 h, and the processor has an MTBF of 100 h. Find the system MTBF and the system reliability for a 5-h mission.

Unit	MTBF	λ
1. Transmitter	40 h	1/40 = 0.025/h
2. Receiver	500 h	1/500 = 0.002/h
3. Processor	100 h	1/100 = 0.010/h
		Total system FR = 0.037/h

From Equation 8.4, the system MTBF is 1/sys FR = (1/0.037) h = 27.0 h. From Equation 8.5, the reliability for a 5-h mission is

$$R_{\text{sys}} = e^{-t/\text{MTBF}} = e^{-5/27.0} = e^{-0.185} = 0.831$$

The mission reliability can also be computed from Equation 8.2 by using the following table.

Unit	λ	R(5 h)
1. Transmitter	0.025/h	$e^{-0.025\times 5} = e^{-0.125} = 0.883$
2. Receiver	0.002/h	$e^{-0.002\times 5} = e^{-0.010} = 0.990$
3. Processor	0.010/h	$e^{-0.010\times 5} = e^{-0.050} = 0.951$

From Equation 8.2,

$$R_{\text{sys}}(5\text{ h}) = \prod_{i=1}^{3} R_i(5\text{ h}) = R_1(5\text{ h}) \times R_2(5\text{ h}) \times R_3(5\text{ h})$$

$$= 0.883 \times 0.990 \times 0.951 = 0.831$$

8.5 DUTY CYCLING

In a system with time-dependent operation, all the elements within the system do not necessarily have to operate whenever the system is operating. Some elements within such a system may operate in cycles or may be one-shot items. Of the time-dependent elements within the system, some elements may be required to operate for only the first $t = T_A$ into a mission or for some specified time segment within the mission, whereas others are on specified duty cycles. When an item is required for just the first $t = T_A$ into the mission, its mission reliability is obviously $e^{-\lambda T_A}$. However, if an element within the system is required to have on-off duty cycling, in evaluating that element any of the following may apply:

1. We may consider only the active portion of the mission and neglect the dormant portion, that is, assume the dormant failure rate is $\lambda_d = 0$.
2. We may consider that a dormancy factor D applies, that is, assume that the dormant failure rate is $\lambda_d = \lambda D$.
3. If the element is an electronic component, we may take the dormant failure rate from MIL-HDBK-217E.
4. We may consider the switching function for that element to be perfect or to have a failure rate of its own. In that case subtracting that failure rate (in failures per cycle) from 1 will provide the probability p_{sw} of the switch functioning upon demand.

EXAMPLE 8.3 Consider once again the communication system of Example 8.2. Suppose it is on an 8-h mission, during which the transmitter is needed for 6 h, the receiver is needed for the entire 8-h mission, and the processor is needed for only half the mission. Consider only the active portion of the mission for each element in determining the mission reliability for the 8-h mission of the system.

Unit	Operating Time per 8 h of Communication	Failure Rate, λ	λt	$R(t) = e^{-\lambda t}$
1. Transmitter	6 h	0.025/h	0.150	0.8607
2. Receiver	8 h	0.002/h	0.016	0.9841
3. Processor	4 h	0.010/h	0.040	0.9608
			$\Sigma \lambda t = 0.206$	

The system reliability for the 8-h mission is

$$R_{sys}(8\text{ h}) = e^{-0.206} = 0.814$$

or

$$R_{sys}(8\text{ h}) = 0.8607 \times 0.9841 \times 0.9608 = 0.814$$

EXAMPLE 8.4 Determine the 8-h mission reliability for the system of Example 8.3 by using a dormancy factor of 0.10 for all units.

Unit	t	λ	Dormancy Factor D	$D\lambda t$
1. Transmitter active	6 h	0.025/h	1.0	0.150
1. Transmitter dormant	2 h	0.025/h	0.1	0.005
2. Receiver active	8 h	0.002/h	1.0	0.016
2. Receiver dormant	0 h	0.002/h	0.1	0.000
3. Processor active	4 h	0.010/h	1.0	0.040
3. Processor dormant	4 h	0.010/h	0.1	0.004
				$\Sigma D\lambda t = 0.215$

In this example there is only one way to compute mission reliability for the system:

$$R_{sys}(t) = e^{-\Sigma \lambda t} = e^{-0.215} = 0.807$$

8.6 PARALLEL SYSTEMS

A parallel system (in the reliability sense) is composed of n elements that perform identical functions, the success of any of which will lead to system success. In other words, all the components must fail in order to have a system failure.

Consider the **reliability block diagram** of Figure 8.1 to represent a parallel system composed of n identical elements, each with reliability R. Each element must then have a probability $Q = 1 - R$ of failing its specified mission. The system reliability R_{sys} is the probability that the system does not fail its specified mission: $R_{sys} = 1 - Q_{sys}$. But, by definition, a parallel system fails only if all n elements fail. Hence, $Q_{sys} = Q^n = (1 - R)^n$. So, for a parallel system of n identical elements,

FIGURE 8.1
Reliability block diagram for a parallel system.

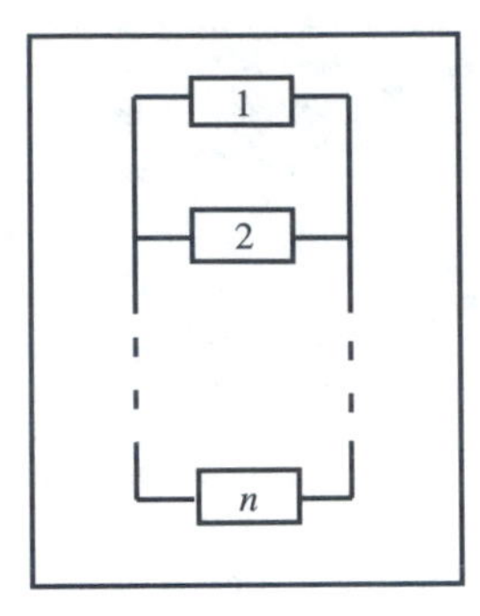

$$R_{sys} = 1 - (1 - R)^n \tag{8.6}$$

where R is the reliability for each element.

If each of the elements in Figure 8.1 has a different reliability—that is, element 1 has reliability R_1, element 2 has reliability R_2, etc.—then the system reliability becomes

$$R_{sys} = 1 - Q_{sys} = 1 - Q_1 Q_2 \cdots Q_n = 1 - (1 - R_1)(1 - R_2) \cdots (1 - R_n) \tag{8.7}$$

If the system represented by Figure 8.1 has time-dependent operation, its reliability for a mission of specified time t is

$$R_{sys}(t) = 1 - (1 - R_1(t))(1 - R_2(t)) \cdots (1 - R_n(t)). \tag{8.8}$$

Assuming constant failure rates, Equation 8.8 becomes

$$R_{sys}(t) = 1 - \prod_{i=1}^{n}(1 - e^{-\lambda_i t}) \tag{8.9}$$

If all n elements have identical failure rate λ,

$$\begin{aligned} R_{sys}(t) &= 1 - (1 - e^{-\lambda t})^n \\ &= 1 - (1 - R(t))^n \end{aligned} \tag{8.10}$$

EXAMPLE 8.5 Suppose a system is composed of two assemblies, one with a reliability of 0.95 and the other with a reliability of 0.85. What is the system reliability if system success is success of either of the two assemblies?

This is a parallel system, represented by the reliability block diagram in Figure 8.2. Its reliability is (see Equation 8.7)

$$\begin{aligned} R_{sys} = 1 - (1 - R_1)\,(1 - R_2) &= 1 - (1 - 0.95)\,(1 - 0.85) \\ &= 1 - (0.05)\,(0.15) = 0.9925 \end{aligned}$$

FIGURE 8.2
Reliability block diagram for Example 8.5.

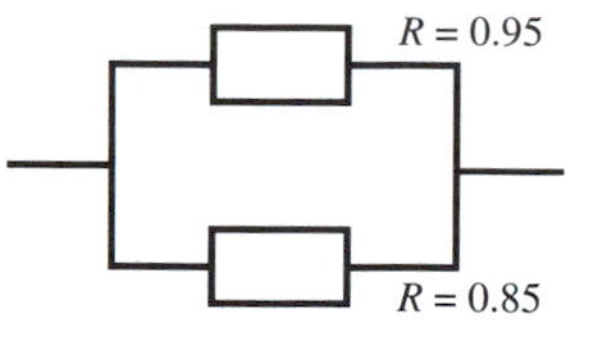

EXAMPLE 8.6

Refer to the system described in Example 8.2. Suppose we are once again interested in solving the reliability for the 5-h mission. The reliability block diagram for that system is as shown in Figure 8.3.

The solution is shown in Figure 8.4.

Suppose we wish to improve the system by adding some redundancy. Because the transmitter has the lowest reliability, let us add another transmitter. See Figure 8.5.

Now let us examine the improvement to be realized with the introduction of a redundant receiver, where each receiver is dedicated to a transmitter (Figure 8.6).

If we reconfigure this same system so that the receivers need not be uniquely dedicated to a transmitter, we can determine how the removal of this restriction can improve the system reliability. See Figure 8.7.

Let us compute the system reliability if we add a redundant processor so that each processor is dedicated to a receiver, as shown in Figure 8.8.

Then, we can compute the system reliability if we reconfigure the system so that we have independently parallel sets of each unit. See Figure 8.9.

Finally, let us compare the last result (Figure 8.10) to the system reliability result if we have two sets of the original system (Figure 8.3) configured in parallel at the system level.

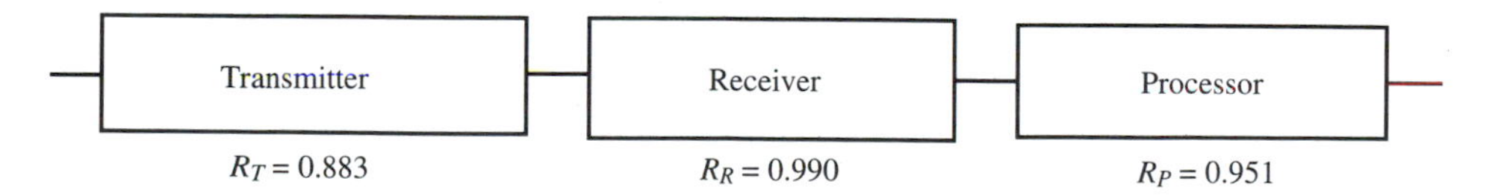

FIGURE 8.3
Reliability block diagram for Example 8.6.

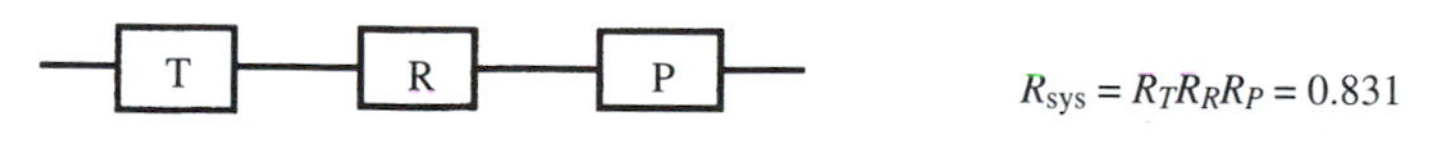

FIGURE 8.4

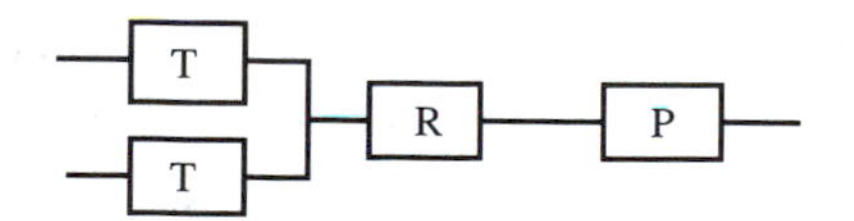

$R_{sys} = [1 - (1 - R_T)^2] R_R R_P = 0.929$

FIGURE 8.5

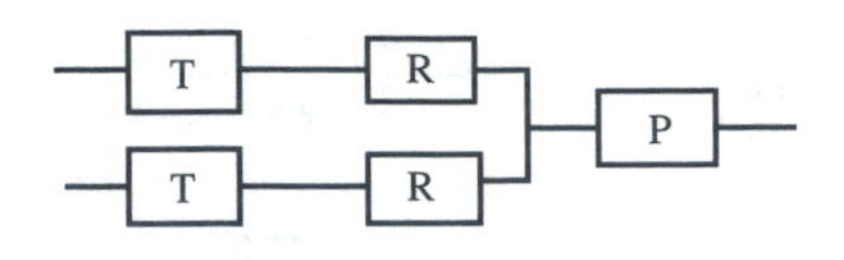

$$R_{\text{sys}} = [1 - (1 - R_T R_R)^2]\, R_P = 0.936$$

FIGURE 8.6

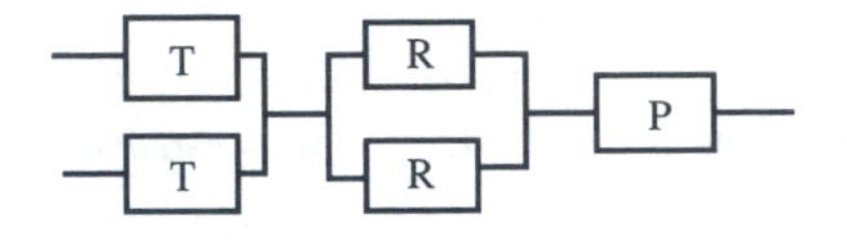

$$R_{\text{sys}} = [1 - (1 - R_T)^2]\,[1 - (1 - R_R)^2]\, R_P = 0.938$$

FIGURE 8.7

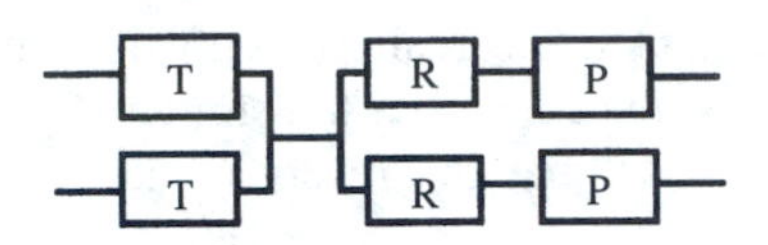

$$R_{\text{sys}} = [1 - (1 - R_T)^2]\,[1 - (1 - R_R R_P)^2] = 0.983$$

FIGURE 8.8

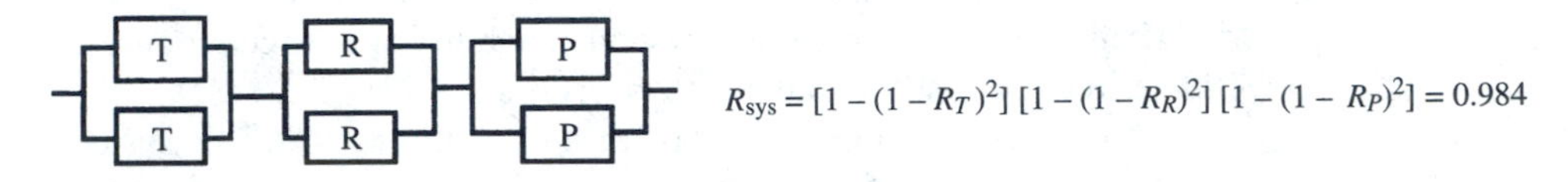

FIGURE 8.9

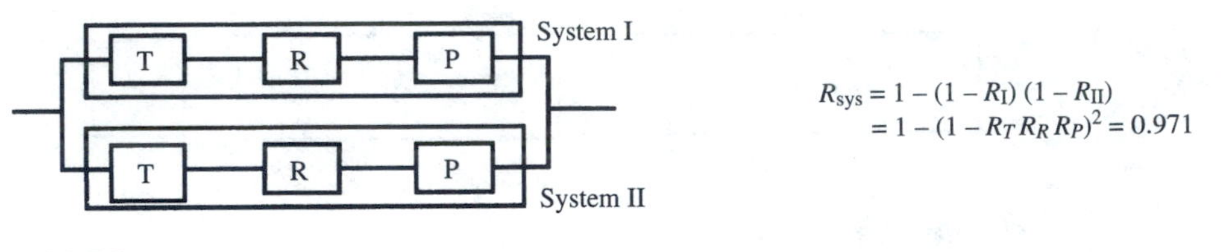

FIGURE 8.10

8.7 *K*-OUT-OF-*N* REDUNDANCY

Now let us consider a system composed of N identical elements, all active, of which any K must be successful for system success. The reliability block diagram for such a system is shown in Figure 8.11. In this system the maximum allowable number of failures is $F = N - K$. The probability of success of any of the elements is R, and the probability of failure is $Q = 1 - R$. The parallel system, discussed in the preceding section, is a special case of the K-out-of-N redundant system, where $K = N$ and the maximum number of failures is $F = 0$.

*8.7.1 Derivation of the *K*-out-of-*N* Reliability Model

The system success probability, which is the probability of at least K of N successes, is clearly governed by the *Binomial Probability Law* (refer to Table A.1)

$$R_{sys} = P(x \geq K) = \sum_{x=K}^{N} \binom{N}{x} R^x(1 - R)^{N-x} \quad \textbf{(8.11)}$$

where x is the number of successes, or

$$R_{sys} = P(x \leq F) = \sum_{x=0}^{F} \binom{N}{x} Q^x(1 - Q)^{N-x} = \sum_{x=0}^{F} \binom{N}{x} R^{N-x}(1 - R)^x \quad \textbf{(8.12)}$$

where x is the number of failures.

Notice that Equations 8.11 and 8.12 are identical to one another and are to be used whenever the maximum allowable number of failures is less than or equal to $N/2$. Otherwise, we can reduce the number of computations by recalling that

$$\sum_{x=0}^{N} \binom{N}{x} R^x(1 - R)^{N-x} = 1 \quad \textbf{(8.13)}$$

from which follows

$$R_{sys} = \sum_{x=K}^{N} \binom{N}{x} R^x(1 - R)^{N-x} = 1 - \sum_{x=0}^{K-1} \binom{N}{x} R^x(1 - R)^{N-x} \quad \textbf{(8.14)}$$

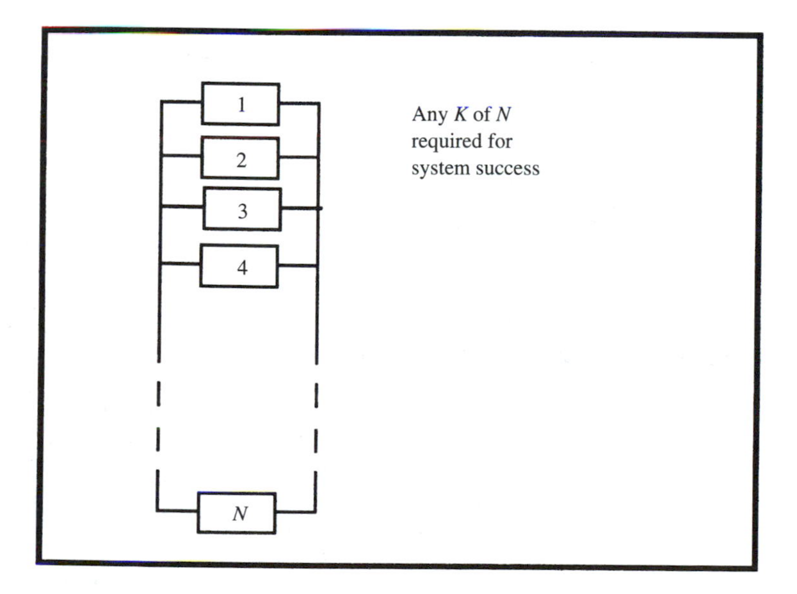

FIGURE 8.11
Reliability block diagram for *K*-out-of-*N* redundancy.

If we have a time-dependent system where K out of N identical items are required to survive a mission of time t, the mission reliability for the system is

$$R_{\text{sys}}(t) = \sum_{x=0}^{F} \binom{N}{x} \{R(t)\}^{N-x} \{1 - R(t)\}^{x} = 1 - \sum_{x=0}^{K-1} \binom{N}{x} \{R(t)\}^{x} \{1 - R(t)\}^{N-x} \tag{8.15}$$

You may use either form of Equation 8.14 or 8.15, depending on which will generate fewer computations.

From Section A.6 (the discussion on combinations),

$$\binom{N}{x} = \frac{N!}{x!(N-x)!}$$

So Equations 8.14 and 8.15 become

$$R_{\text{sys}} = \sum_{x=K}^{N} \left(\frac{N!}{x!(N-x)!} \right) R^{x}(1-R)^{N-x} = 1 - \sum_{x=0}^{K-1} \left(\frac{N!}{x!(N-x)!} \right) R^{x}(1-R)^{N-x} \tag{8.16}$$

and

$$\begin{aligned} R_{\text{sys}}(t) &= \sum_{x=K}^{N} \left(\frac{N!}{x!(N-x)!} \right) \{R(t)\}^{x} \{1 - R(t)\}^{N-x} \\ &= 1 - \sum_{x=0}^{K-1} \left(\frac{N!}{x!(N-x)!} \right) \{R(t)\}^{x} \{1 - R(t)\}^{N-x} \end{aligned} \tag{8.17}$$

Equation 8.17 can be rewritten in terms of the maximum allowable number of failures F instead of the minimum allowable number of successes:

$$\begin{aligned} R_{\text{sys}}(t) &= \sum_{x=0}^{F} \left(\frac{N!}{x!(N-x)!} \right) \{R(t)\}^{N-x} \{1 - R(t)\}^{x} \\ &= 1 - \sum_{x=F+1}^{N} \left(\frac{N!}{x!(N-x)!} \right) \{R(t)\}^{N-x} \{1 - R(t)\}^{x} \end{aligned} \tag{8.18}$$

8.7.2 Solving *K*-out-of-*N* System Reliability

If we are evaluating a cycle-dependent system, the system reliability can be computed from either of the following two equations (derived as Equation 8.16):

$$R_{\text{sys}} = \sum_{x=k}^{N} \left(\frac{N!}{x!(N-x)!} \right) R^{x}(1-R)^{N-x} \tag{8.19}$$

$$R_{\text{sys}} = 1 - \sum_{x=0}^{K-1} \left(\frac{N!}{x!(N-x)!} \right) R^{x}(1-R)^{N-x} \tag{8.20}$$

Between Equation 8.19 and 8.20 we generally choose the one that will generate the fewer computations. If $K \le (N/2)$, Equation 8.20 will generate fewer computations. If $K > (N/2)$, Equation 8.19 will generate fewer computations.

If we are evaluating a time-dependent system, the system reliability can be computed from either of the following two equations (derived as Equation 8.17):

$$R_{sys} = \sum_{x=k}^{N} \left(\frac{N!}{x!(N-x)!} \right) \{R(t)\}^{x} \{1 - R(t)\}^{N-x} \quad \textbf{(8.21)}$$

$$R_{sys} = 1 - \sum_{x=0}^{K-1} \left(\frac{N!}{x!(N-x)!} \right) \{R(t)\}^{x} \{1 - R(t)\}^{N-x} \quad \textbf{(8.22)}$$

As before, between Equation 8.21 and 8.22 we generally choose the one that will generate the fewer computations. If $K \le (N/2)$, Equation 8.22 will generate fewer computations. If $K > (N/2)$, Equation 8.21 will generate fewer computations.

EXAMPLE 8.7

Consider a system composed of four identical elements, each with a reliability $R = 0.90$. What is the system reliability if three of the four active elements are required? See Figure 8.12.

Since $K = 3 > N/2$, Equation 8.19 is more convenient to use for this application:

$$\begin{aligned} R_{sys} &= \sum_{x=k}^{N} \left(\frac{N!}{x!(N-x)!} \right) R^{x}(1 - R)^{N-x} \\ &= \sum_{x=3}^{4} \left(\frac{4!}{x!(4-x)!} \right) (0.90)^{x}(0.10)^{4-x} \\ &= \frac{4!}{3!1!} 0.90^{3} 0.10^{1} + \frac{4!}{4!0!} 0.90^{4} 0.10^{0} = 0.6561 + 0.2916 = 0.948 \end{aligned}$$

FIGURE 8.12

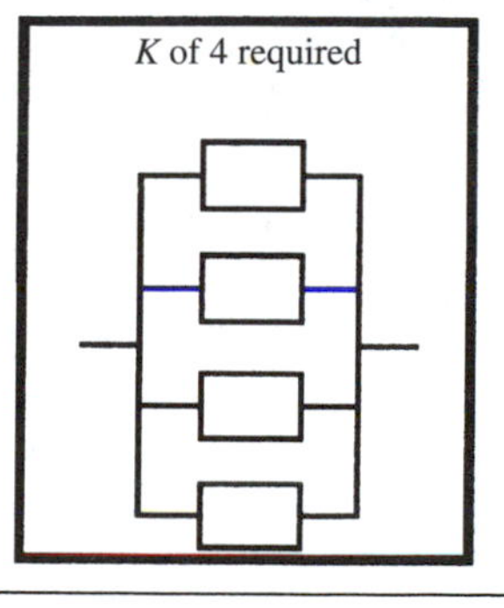

EXAMPLE 8.8 Suppose that only one of the four elements in Example 8.7 is required for system success. In that case, what is the system reliability?

In this example $K = 1 < N/2$, so Equation 8.20 is more convenient to use than Equation 8.19.

$$R_{\text{sys}} = 1 - \sum_{x=0}^{1-1}\left(\frac{4!}{x!(4-x)!}\right)0.90^{x}(0.10)^{4-x}$$
$$= 1 - \frac{4!}{0!4!}0.90^{0}0.10^{4} = 1 - (0.10)^{4} = 0.9999$$

If two out of four are required for system success, then $K = 2 = N/2$, in which case Equation 8.20 is more convenient to use:

$$R_{\text{sys}} = 1 - \sum_{x=0}^{2-1}\left(\frac{4!}{x!(4-x)!}\right)0.90^{x}(0.10)^{4-x} = 1 - \frac{4!}{0!4!}0.90^{0}0.10^{4} - \frac{4!}{1!3!}0.90^{1}0.10^{3}$$
$$= 1.0000 - 0.0001 - 0.0036 = 0.9963$$

*8.7.3. Solving *K*-out-of-*N* System MTBF

In Equation 3.30, we saw that if a system whose operation is time dependent is always restored upon failure to the $t = 0$ condition, then

$$\text{MTBF} = \int_{0}^{\infty} R(t)\, dt \tag{8.23}$$

Let us use this relationship to solve the MTBF of systems in some of the previous examples. In solving for MTBF it will usually be necessary to make use of the following solution from integral calculus:

$$\int_{0}^{\infty} t^{n} e^{-\lambda t}\, dt = \frac{n!}{\lambda^{n+1}} \tag{8.24}$$

EXAMPLE 8.9 What is the MTBF for the system in Example 8.2?

This is a system composed of three units in series.

$$R_{\text{sys}}(t) = R_1(t) \cdot R_2(t) \cdot R_3(t) = e^{-\lambda_1 t}e^{-\lambda_2 t}e^{-\lambda_3 t} = e^{-(\lambda_1+\lambda_2+\lambda_3)t}$$
$$\text{MTBF} = \int_{0}^{\infty} R_{\text{sys}}(t)\, dt = \int_{0}^{\infty} e^{-(\lambda_1+\lambda_2+\lambda_3)t}\, dt$$

Applying Equation 8.24,

$$\text{MTBF} = \frac{1}{\lambda_1 + \lambda_2 + \lambda_3} = \frac{1}{0.037} = 27 \text{ h}$$

EXAMPLE 8.10

What is the MTBF of the system from Example 8.6 whose reliability block diagram is shown in Figure 8.5? The element failure rates are

$$\lambda_T = 0.025$$
$$\lambda_R = 0.002$$
$$\lambda_P = 0.010$$

Referring to the example of Figure 8.5,

$$\begin{aligned} R_{sys}(t) &= [1 - \{1 - R_T(t)\}^2]R_R(t)R\}_P(t) = \{2R_T(t) - [R_T(t)]^2\}R_R(t)R_P(t) \\ &= (2e^{-\lambda_T t} - e^{-2\lambda_T t})e^{-(\lambda_R+\lambda_P)t} = 2e^{-(\lambda_T+\lambda_R+\lambda_P)t} - e^{-(2\lambda_T+\lambda_R+\lambda_P)t} \end{aligned}$$

Applying Equation 8.24,

$$\begin{aligned} \text{MTBF} &= 2\int_0^{\infty} e^{-(\lambda_T+\lambda_R+\lambda_P)t}\,dt - \int_0^{\infty} e^{-(2\lambda_T+\lambda_R+\lambda_P)t}\,dt \\ &= \frac{2}{\lambda_T + \lambda_R + \lambda_P} - \frac{1}{2\lambda_T + \lambda_R + \lambda_P} = \frac{2}{0.037} - \frac{1}{0.062} = 37.93 \text{ h} \end{aligned}$$

EXAMPLE 8.11

A system is composed of five identical elements, all active and each with a failure rate $\lambda = 0.001$ per hour. If any four of the five are required for system success, compute the reliability for a 100-h mission and the system MTBF. See Figure 8.13.

FIGURE 8.13

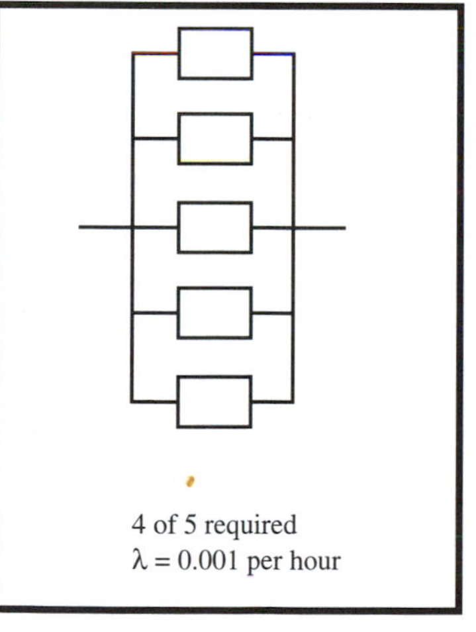

$$\begin{aligned} R_{sys}(t) &= \sum_{x=0}^{1}\binom{5}{x}(R(t))^{5-x}(1 - R(t))^x \\ &= (R(t))^5 + 5(R(t))^4(1 - R(t)) = 5(R(t))^4 - 4(R(t))^5 \\ &= 5e^{-4\lambda t} - 4e^{-5\lambda t}. \end{aligned}$$

$$R_{sys}(100\ \text{h}) = 5e^{-4\times 0.001\times 100} - 4e^{-5\times 0.001\times 100} = 0.925$$

$$\text{MTBF} = \int_0^\infty R(t)\,dt = 5\int_0^\infty e^{-4\lambda t} - 4\int_0^\infty e^{-5\lambda t}$$

$$= \frac{5}{4\lambda} - \frac{4}{5\lambda} = \frac{9}{20\lambda} = 450\ \text{h}$$

8.8 STANDBY REDUNDANCY

Now let us consider a time-dependent system composed of N identical elements, of which K must be successful for system success, as in Section 8.7. But this time the remaining $n = N - K$ are in *standby* at the start of the mission. In other words, at the time $t = 0$, K active elements are required and the $N - K$ redundant elements are in a standby mode.

Figure 8.14 shows a reliability block diagram for this system. If, at any time during the mission, one of the K active elements should fail, the system is required to detect that failure and to switch on one of the standby elements. Accordingly, the success definition for this system has two conditions:

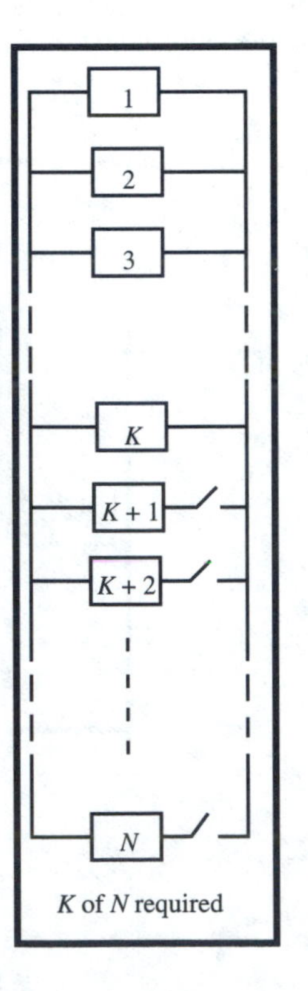

FIGURE 8.14
Reliability block diagram for a system with standby redundancy.

Condition 1 The K active elements operate throughout the mission time t.

Condition 2 Some of the active elements fail during the mission, but every time that happens the system sensing mechanism detects the failure and activates a switch, which successfully turns on one of the standby elements to replace it. Also, a standby element is available when needed to replace a failed active element, so that there are always K operating elements throughout the mission.

*8.8.1 Derivation of the Standby Redundancy Reliability Model

Consider first the simplest model for a system with standby redundancy, that illustrated by the block diagram in Figure 8.15. One of two identical items, A and B, is required by the system. When active, each has a constant rate λ, and the standby failure rate is 0. Assume perfect switching and sensing of malfunctions, so that if the active item fails, the system will automatically sense the failure and the standby item will automatically turn on. The mission time is t. The system success definitions are as follows:

Condition 1 Item A operates throughout the time t without failure.

Condition 2 Item A fails at some time t_1 ($0 \leq t_1 \leq t$). Then item B switches on and remains operating without failure until the time t.

The probability of condition 1 is simply the reliability of the item A for the mission of time t: $R_A(t) = e^{-\lambda t}$.

The probability of condition 2 is the probability that item A fails at time t_1 and that item B operates throughout the remaining time $t_2 = t - t_1$. These events are independent, so their probabilities are multiplied to derive the probability of condition 2: $Q_A(t_1)R_B(t_2)$.

Conditions 1 and 2 are mutually exclusive. Hence, the system reliability is the sum of their probabilities:

$$R_{sys}(t) = R_A(t) + Q_A(t_1)R_B(t_2) \tag{8.25}$$

Observe in Figure 8.16 that, t being a constant,

$$R_A(t) = e^{-\lambda t} \tag{8.26}$$

and, t_1 and t_2 being variables,

$$Q_A(t_1) = \int_{t_1=0}^{t} f(t_1)\, dt_1 = \int_0^t \lambda e^{-\lambda t_1}\, dt_1 \tag{8.27}$$

and

$$R_B(t_2) = \int_{t_2=t-t_1}^{\infty} f(t_2)\, dt_2 = \int_{t-t_1}^{\infty} \lambda e^{-\lambda t_2}\, dt_2 \tag{8.28}$$

FIGURE 8.15
Reliability block diagram for an active and a standby redundant item.

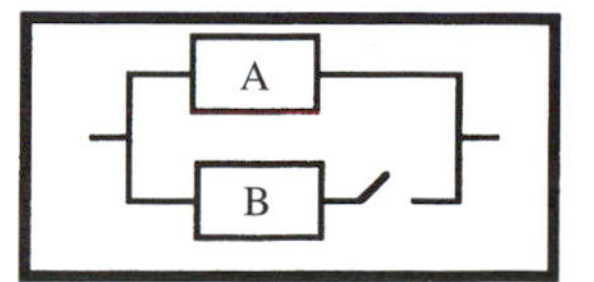

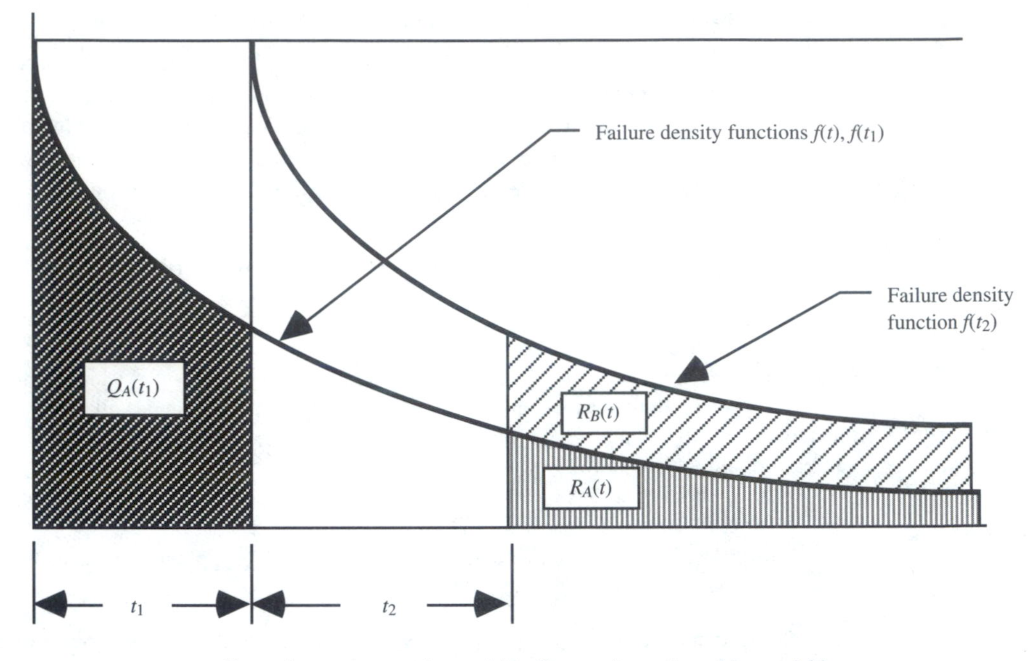

Note: t is a constant; t_1 is a variable ($0 < t_1 < t$); $t_2 = (t - t_1)$ is a variable

FIGURE 8.16
Illustration of mission reliability $R(t)$ for a system with standby redundancy.

Hence,

$$
\begin{aligned}
R_{\text{sys}}(t) &= e^{-\lambda t} + \lambda^2 \int_{t_1=0}^{t} \int_{t_2=t-t_1}^{\infty} e^{-\lambda t_1} e^{-\lambda t_2}\, dt_2\, dt_1 \\
&= e^{-\lambda t} + \lambda^2 \int_{t_1=0}^{t} e^{-\lambda t_1} \left(\frac{e^{-\lambda t_2}}{-\lambda} \right) \Bigg|_{t_2=t-t_1}^{\infty} dt_1 \\
&= e^{-\lambda t} - \lambda \int_{t_1=0}^{t} e^{-\lambda t_1} (e^{-\infty} - e^{-\lambda (t-t_1)})\, dt_1 \\
&= e^{-\lambda t} - \lambda \int_{t_1=0}^{t} e^{-\lambda t_1} (0 - e^{-\lambda t} e^{+\lambda t_1})\, dt_1 \\
&= e^{-\lambda t} - \lambda \int_{t_1=0}^{t} -e^{-\lambda t}\, dt_1 = e^{-\lambda t} + \lambda e^{-\lambda t} [t_1] \Big|_{t_1=0}^{t} = e^{-\lambda t} + \lambda t e^{-\lambda t}
\end{aligned}
\tag{8.29}
$$

*8.8.2 Model for Perfect Sensing and Switching and a Standby Failure Rate of Zero

If we decide to neglect sensing and switching malfunctions and the standby failure rate in the system just discussed (i.e., we assume 100% probability that the system will detect a malfunction of the active item and 100% probability that the standby item will be successfully turned on and that an item cannot fail while in the standby mode), then the system success definition becomes either of the following condition 1 or condition 2, while only one item at a time is operating with constant failure rate λ.

Condition 1 No failure occurs during the time t.

Condition 2 Exactly one failure occurs during time t.

This can be analyzed through the *Poisson Probability Law* (see Appendix A and Table A.1), where

$$R_{sys} = P(x \leq 1, \mu = \lambda t) = \sum_{x=0}^{1} e^{-\mu}\left(\frac{\mu^x}{x!}\right) = e^{-\mu} + \mu e^{-\mu} = e^{-\lambda t} + \lambda t e^{-\lambda t} \tag{8.30}$$

This, of course, is the same result derived via the longhand model in Equation 8.29.

The Poisson model can be used *if and only if* we assume (1) a zero failure rate for items in standby, (2) perfect sensing, and (3) perfect switching.

Let us now solve the MTBF for the same system, composed of one active item, one standby item, perfect sensing and switching, and a standby failure rate of zero. Equation 8.24 is used for solving the integrals.

$$\begin{aligned} \text{MTBF} &= \int_0^\infty R(t)\,dt = \int_0^\infty (e^{-\lambda t} + \lambda t e^{-\lambda t})\,dt \\ &= \int_0^\infty e^{-\lambda t}\,dt + \lambda \int_0^\infty t e^{-\lambda t}\,dt = \frac{1}{\lambda} + \lambda\left(\frac{1}{\lambda^2}\right) = \frac{2}{\lambda} \end{aligned} \tag{8.31}$$

*8.8.3 Model for a Multiunit Standby System with Perfect Switching, Perfect Sensing, and Zero Standby Failure Rate

The system whose block diagram is in Figure 8.17 requires one of its n identical items to operate. The others are in standby. We assume perfect sensing and switching, that the active failure rate is λ, and that the standby failure rate is 0.

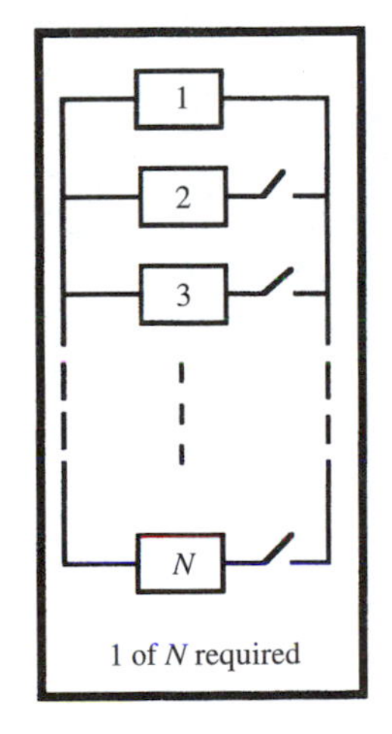

FIGURE 8.17
System with one active and (N − 1) standby items.

The Poisson Probability Law can be applied.

$$R_{sys}(t) = P(x \le n - 1, \mu = \lambda t) = e^{-\lambda t}\sum_{x=0}^{n-1}\frac{(\lambda t)^x}{x!}$$

$$= e^{-\lambda t}\left[1 + \lambda t + \frac{(\lambda t)^2}{2!} + \cdots + \frac{(\lambda t)^{n-1}}{(n-1)!}\right]$$

$$= e^{-\lambda t}\sum_{i=1}^{n}\frac{(\lambda t)^{i-1}}{(i-1)!}$$

$$\text{MTBF} = \int_0^\infty R(t)\,dt \tag{8.32}$$

$$= \int_0^\infty e^{-\lambda t}\,dt + \lambda\int_0^\infty te^{-\lambda t}\,dt + \frac{\lambda^2}{2!}\int_0^\infty t^2e^{-\lambda t}\,dt + \cdots + \frac{\lambda^{n-1}}{(n-1)!}\int_0^\infty t^{n-1}e^{-\lambda t}\,dt$$

$$= \frac{0!}{\lambda} + \lambda\left(\frac{1!}{\lambda^2}\right) + \frac{\lambda^2}{2!}\left(\frac{2!}{\lambda^3}\right) + \cdots + \frac{\lambda^{n-1}}{(n-1)!}\left(\frac{(n-1)!}{\lambda_n}\right)$$

$$= \underbrace{\frac{1}{\lambda} + \frac{1}{\lambda} + \frac{1}{\lambda} + \cdots + \frac{1}{\lambda}}_{n \text{ times}} = \frac{n}{\lambda}$$

As an exercise, review the results in Equations 8.31 and 8.33 and compare them with what you would intuitively expect.

EXAMPLE 8.12

A system is composed of three identical components, two of which are required to operate. The third is initially in standby. If we assume perfect sensing and switching, that the active failure rate of each component is λ, and that the standby failure rate is zero, find the reliability for a mission of time t and the system MTBF. See Figure 8.18.

The Poisson Probability law can be applied.

$$R_{sys}(t) = P(x \le 1, \mu = 2\lambda t) = \sum_{x=0}^{1} e^{-\mu}\left(\frac{\mu_x}{x!}\right)$$

$$= e^{-\mu} + \mu e^{-\mu} = e^{-2\lambda t} + 2\lambda te^{-2\lambda t}$$

$$\text{MTBF} = \int_0^\infty R_{sys}(t) = \int_0^\infty e^{-2\lambda t}\,dt + 2\lambda\int_0^\infty te^{-2\lambda t}$$

FIGURE 8.18
System with two active and one standby component.

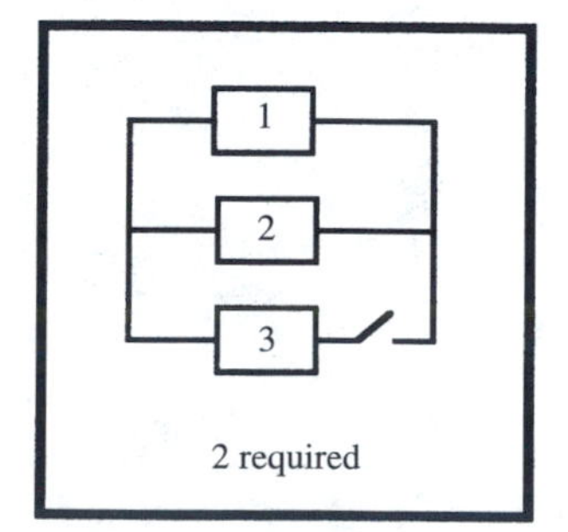

Applying Equation 8.24,

$$\text{MTBF} = \frac{1}{2\lambda} + 2\lambda\left[\frac{1}{4\lambda^2}\right] = \frac{1}{2\lambda} + \frac{1}{2\lambda} = \frac{1}{\lambda}$$

Notice that in this example the system MTBF is equal to the active MTBF of one of its components.

EXAMPLE 8.13

As a general solution to a multiunit standby system with perfect sensing and switching and a zero standby failure rate, consider the system whose reliability block diagram appears in Figure 8.14.

The system is composed of k on-line active items, all of which must operate, and n standby spares. The active failure rate for each item is λ, and the standby failure rate is assumed to be 0.

If we assume perfect sensing and switching, the system reliability for a specified mission time t is

$$R_{\text{sys}(t)} = e^{-k\lambda t}\sum_{i=1}^{n+1}\frac{(k\lambda t)^{i-1}}{(i-1)!} = e^{-k\lambda t}\sum_{i=0}^{n}\frac{(k\lambda t)^{i}}{i!} \quad \textbf{(8.34)}$$

The system MTBF is

$$\text{MTBF} = (n+1)/k\lambda \quad \textbf{(8.35)}$$

The system reliability is solved using the Poisson Probability Law, as in Example 8.12. The MTBF is then obtained by integrating the reliability between 0 and ∞, as in Example 8.12, making use of Equation 8.29.

8.8.4 Solving Reliability of Systems with Standby Redundancy, Assuming Perfect Sensing and Switching and Negligible Standby Failure Rate

Example 8.13 shows that if we make the assumptions that each of the active elements has a failure rate λ, that the standby elements and the sensing mechanism have negligible failure rates, and that the switches will always perform upon demand, the system reliability for a mission of time t is

$$\begin{aligned} R_{\text{sys}}(t) &= e^{-K\lambda t}\sum_{t=0}^{N-K}\frac{(K\lambda t)^{i}}{i!} \\ &= e^{-K\lambda t}\left[1 + K\lambda t + \frac{(K\lambda t)^2}{2!} + \frac{(K\lambda t)^3}{3!} + \cdots + \frac{(K\lambda t)^{(N-K)}}{(N-K)!}\right] \end{aligned} \quad \textbf{(8.36)}$$

where K is the required number of active elements and N is the total number of (active plus standby) elements in the system.

If we have a system (Figure 8.19) composed of one active element that is required for a mission of time t and backed up by $n = N - 1$ standby elements, all identical, the

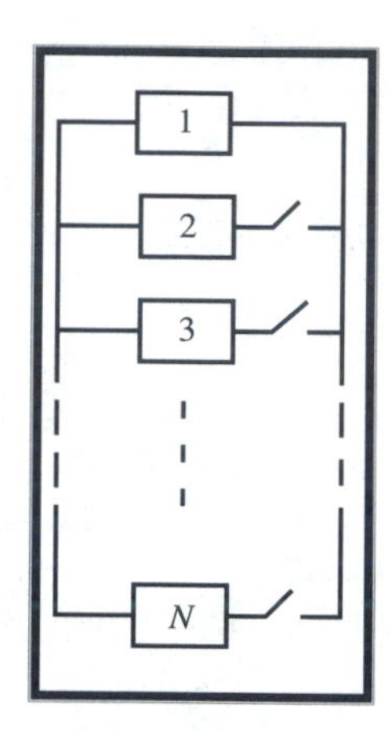

FIGURE 8.19
System with 1 active and $(N - 1)$ standby elements.

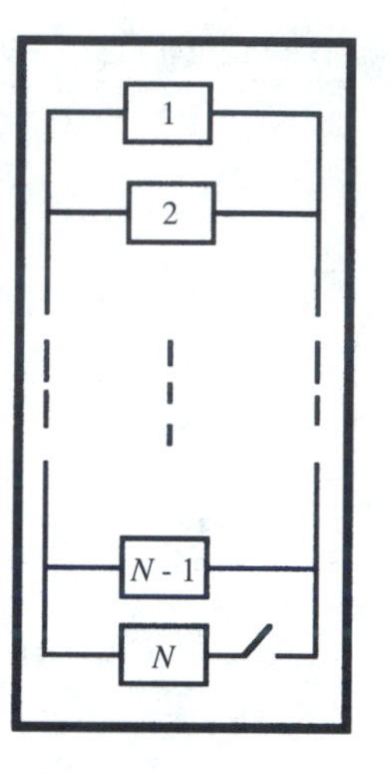

FIGURE 8.20
System with $N - 1$ active and 1 standby element.

system reliability for the mission, assuming negligible standby element and sensing function failure rates and perfect switching, is

$$R_{\text{sys}}(t) = e^{-\lambda t}\sum_{i=0}^{N-1}\frac{(\lambda t)^i}{i!} = e^{-\lambda t}\left[1 + \lambda t + \frac{(\lambda t)^2}{2!} + \frac{(\lambda t)^3}{3!} + \cdots + \frac{(\lambda t)^{(N-1)}}{(N-1)!}\right] \quad \textbf{(8.37)}$$

Equation 8.37 is Equation 8.36 with $K = 1$ and is derived as Equation 8.32 in the preceding section.

If we have a system (Figure 8.20) composed of N identical elements, of which $N - 1$ are active and required for system success and 1 is in standby redundancy, if we assume that an active element has a failure rate λ and the standby element and the sensing element failure rates are both negligible, and if we assume perfect switching, then the system failure rate for a mission of time t is

$$R_{\text{sys}}(t) = e^{-(N-1)\lambda t}\sum_{i=0}^{1}\frac{[(N-1)\lambda t)^i}{i!} = e^{-(N-1)\lambda t}[1 + (N-1)\lambda t] \quad \textbf{(8.38)}$$

This is, of course, Equation 8.36, with $K = N - 1$.

Finally, if we have a system (Figure 8.21) composed of two identical elements, one of which is active and required for system success and the other is initially in standby redundancy, if we again assume the active element failure rate to be λ and the standby element and the sensing element failure rates both to be negligible, and if we assume perfect switching, then the system failure rate for a mission of time t is

$$R_{\text{sys}}(t) = e^{-\lambda t}\sum_{i=0}^{1}\frac{(\lambda t)^i}{i!} = e^{-\lambda t}[1 + \lambda t] = e^{-\lambda t} + \lambda t e^{-\lambda t} \quad \textbf{(8.39)}$$

FIGURE 8.21
System with one active and one standby element.

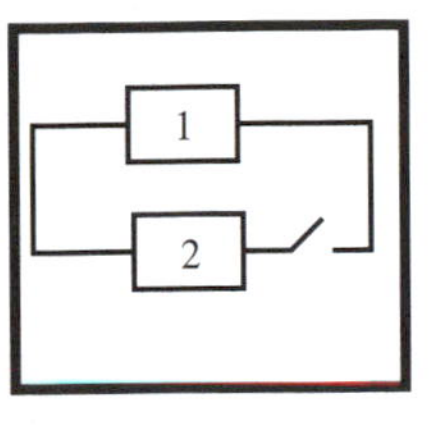

This is Equation 8.36 with $N = 2$ and $K = 1$. It is also derived in Section 8.8.2 as Equations 8.29 and 8.30.

EXAMPLE 8.14 Consider the system of Figure 8.10 in Example 8.6, but this time suppose that the redundant subsystem is in standby. If the active failure rates are as given in Example 8.2, the standby failure rates are considered to be negligible, the sensing mechanism failure rate is considered to be negligible, and perfect switching is assumed, what is the system reliability for a 5-h mission? See Figure 8.22.

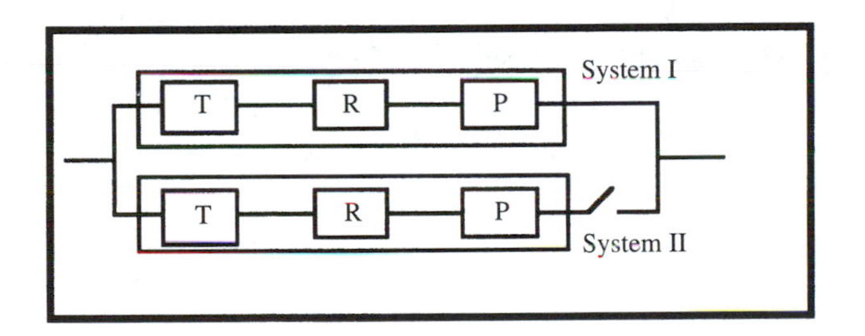

FIGURE 8.22
System composed of active and standby subsystems.

From Example 8.2, $\lambda_T = 0.025$ per hour, $\lambda_R = 0.002$ per hour, and $\lambda_P = 0.010$ per hour. So the active failure rate for one subsystem is $0.025 + 0.002 + 0.010 = 0.037$ per hour.

Equation 8.28 applies:

$$R_{\text{sys}}(t) = e^{-\lambda t} + \lambda t e^{-\lambda t} = e^{-0.037\times 5} + 0.037 \times 5e^{-0.037\times 5} = 0.985$$

Compare this result to that of Example 8.6, where both subsystems were active.

EXAMPLE 8.15 Consider a system composed of four identical elements, as in Example 8.7, where three of the four are required for system success. But this time the fourth element is in standby redundancy. Suppose the active failure rate for an element is 105.4×10^{-6} per hour and the system mission time is 1000 h, so that $e^{-\lambda t} = 0.90$. What is the system reliability

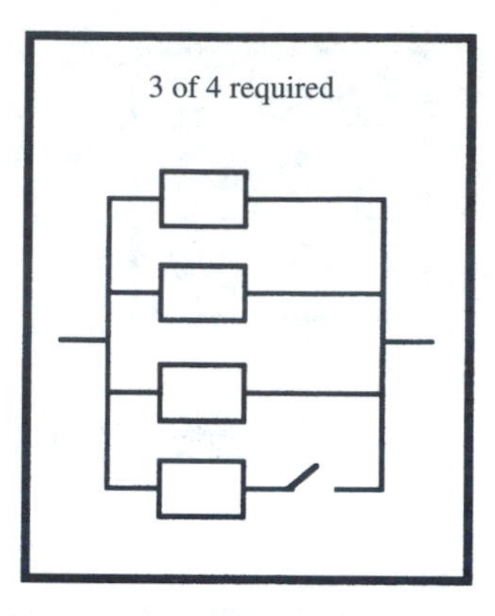

FIGURE 8.23
Three-of-four standby redundancy.

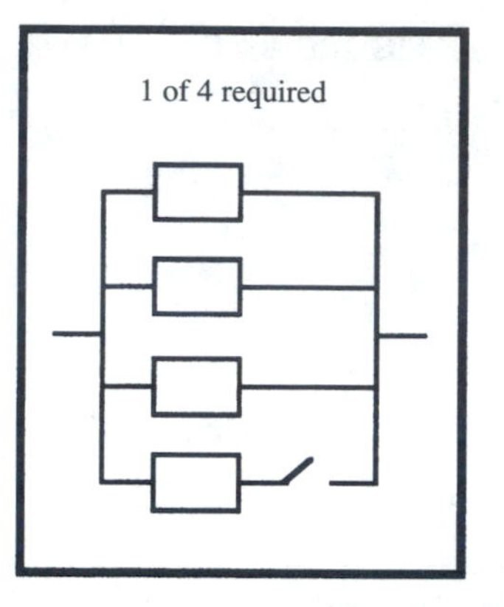

FIGURE 8.24
One-of-four standby redundancy.

for a 1000-h mission if we assume that the standby failure rate is negligible, that the sensing function failure rate is negligible, and that we have perfect switching? See Figure 8.23.

Because of our assumptions, Equation 8.38 applies, with $N = 4$ and $\lambda t = 0.1056$:

$$R_{\text{sys}}(1000\text{ h}) = e^{-(N-1)\lambda t}\sum_{i=0}^{1}\frac{[(N-1)\lambda t]_i}{i!}$$
$$= e^{-(N-1)\lambda t}[1 + (N-1)\lambda t]$$
$$= e^{-3\times 0.1056}[1 + (3 \times 0.1056)] = 0.959$$

Compare this result to that of Example 8.7.

EXAMPLE 8.16 Consider a system composed of four identical elements, as in Example 8.8, where only one of the four are required for system success. But this time the three redundant elements are in standby redundancy. Suppose the active failure rate for an element is 105.4×10^{-6} per hour and the system mission time is 1000 h, so that $e^{-\lambda t} = 0.90$. What is the system reliability for a 1000-h mission if we assume that the standby failure rate is negligible, that the sensing function failure rate is negligible, and that we have perfect switching? See Figure 8.24.

Because of our assumptions, Equation 8.37 applies, with $N = 4$ and $\lambda t = 0.1056$:

$$R_{\text{sys}}(1000\text{ h}) = e^{-\lambda t}\sum_{i=0}^{N-1}\frac{(\lambda t)_i}{i!} = e^{-0.1056}\sum_{i=0}^{3}\frac{0.1056^i}{i!}$$
$$= e^{-0.1056}\left[1 + 0.1056 + \frac{0.1056^2}{2!} + \frac{0.1056^3}{3!}\right] = 0.999995$$

Compare this result to that of Example 8.8.

---

*8.9 MODELS THAT CONSIDER STANDBY FAILURE RATES

Consider once again a system composed of two identical items A and B, as in Figure 8.15. One of the two items is required by the system, and the other, when available, is in standby. Each has an active failure rate λ_a and a standby failure rate λ_s. Let us assume perfect sensing and switching.

Because we are now considering the standby failure rate, the Poisson Probability Law is no longer valid. We must go back to the longhand technique presented at the beginning of this section.

The system success definitions are as follows:

Condition 1 Item A operates throughout the mission time t.

Condition 2 Item A fails at some time t_1 ($0 \leq t_1 \leq t$); item B in its standby mode did not fail during the period from 0 to t_1 and is available; and item B switches on and remains operating without failure until the time t.

The probability of condition 1 is $R_A = e^{-\lambda_A t}$.

The probability of condition 2 is the probability that A fails at some time t_1 ($0 \leq t_1 \leq t$), that B survived in the standby mode until that time t_1, and that B operates as an active item throughout the remaining time $t_2 = t - t_1$. These events are three independent events, so their probabilities are multiplied to derive the probability of condition 2: $Q_A(t_1)R_B{}^*(t_1)R_B(t_2)$, where R^* is the reliability in the standby mode.

Conditions 1 and 2 are mutually exclusive. Hence, the system reliability is the sum of their probabilities:

$$
\begin{aligned}
R_{\text{sys}}(t) &= R_A(t) + Q_A(t_1) \cdot R_B{}^*(t_1) \cdot R_B(t_2) \\
&= e^{-\lambda_a t} + \int_{t_1=0}^{t} \lambda_a e^{-\lambda_a t_1}\, dt_1 \cdot \int_{t_2=t_1}^{\infty} \lambda_s e^{-\lambda_s t_2}\, dt_2 \cdot \int_{t_3=t-t_1}^{\infty} \lambda_a e^{-\lambda_a t_3}\, dt_3 \\
&= e^{-\lambda_a t} + \lambda_a^2 \lambda_s \int_{t_1=0}^{t} \int_{t_2=t_1}^{\infty} \int_{t_3=t-t_1}^{\infty} e^{-\lambda_a t_1} e^{-\lambda_s t_2} e^{-\lambda_a t_3}\, dt_3\, dt_2\, dt_1 \\
&= e^{-\lambda_a t} + \lambda_a^2 \lambda_s \int_{t_1=0}^{t} \int_{t_2=t_1}^{\infty} e^{-\lambda_a t_1} e^{-\lambda_s t_2} \left[\frac{e^{-\lambda_a t_3}}{-\lambda_a}\right]\Bigg|_{t_3=t-t_1}^{\infty} dt_2\, dt_1 \\
&= e^{-\lambda_a t} - \lambda_a \lambda_s \int_{t_1=0}^{t} \int_{t_2=t_1}^{\infty} e^{-\lambda_a t_1} e^{-\lambda_s t_2} [0 - e^{-\lambda_a (t-t_1)}]\, dt_2\, dt_1 \\
&= e^{-\lambda_a t} + \lambda_a \lambda_s \int_{t_1=0}^{t} \int_{t_2=t_1}^{\infty} e^{-\lambda_a t_1} e^{-\lambda_s t_2} e^{-\lambda_a t} e^{+\lambda_a t_1}\, dt_2\, dt_1 \qquad \textbf{(8.40)} \\
&= e^{-\lambda_a t} + \lambda_a \lambda_s \int_{t_1=0}^{t} \int_{t_2=t_1}^{\infty} e^{-\lambda_s t_2} e^{-\lambda_a t}\, dt_2\, dt_1 \\
&= e^{-\lambda_a t} + \lambda_a \lambda_s \int_{t_1=0}^{t} e^{-\lambda_a t} \left[\frac{e^{-\lambda_s t_2}}{-\lambda_s}\right]\Bigg|_{t_2=t_1}^{\infty} dt_1 \\
&= e^{-\lambda_a t} - \lambda_a \int_{t_1=0}^{t} e^{-\lambda_a t} [0 - e^{-\lambda_s t_1}]\, dt_1 = e^{-\lambda_a t} + \lambda_a e^{-\lambda_a t} \int_{t_1=0}^{t} e^{-\lambda_s t_1}\, dt_1 \\
&= e^{-\lambda_a t} + \lambda_a e^{-\lambda_a t} \left[\frac{e^{-\lambda_s t_1}}{-\lambda_s}\right]\Bigg|_{t_1=0}^{t} = e^{-\lambda_a t} - \frac{\lambda_a}{\lambda_s} e^{-\lambda_a t} [e^{-\lambda_s t} - 1] \\
&= e^{-\lambda_a t} - \frac{\lambda_a}{\lambda_s} e^{-(\lambda_a + \lambda_s)t} + \frac{\lambda_a}{\lambda_s} e^{-\lambda_a t} = \left(1 + \frac{\lambda_a}{\lambda_s}\right) e^{-\lambda_a t} - \left(\frac{\lambda_a}{\lambda_s}\right) e^{-(\lambda_a + \lambda_s)t}
\end{aligned}
$$

$$
\text{MTBF} = \int_0^{\infty} R_{\text{sys}}(t)\, dt = \left(1 + \frac{\lambda_a}{\lambda_s}\right) \int_0^{\infty} e^{-\lambda_a t}\, dt - \left(\frac{\lambda_a}{\lambda_s}\right) \int_0^{\infty} e^{-(\lambda_a + \lambda_s)t}\, dt
$$

From Equation 8.29,

$$\text{MTBF} = \left(1 + \frac{\lambda_a}{\lambda_s}\right)\left(\frac{1}{\lambda_a}\right) - \left(\frac{\lambda_a}{\lambda_s}\right)\left(\frac{1}{\lambda_a + \lambda_s}\right) = \frac{\lambda_s + \lambda_a}{\lambda_s \lambda_a} - \frac{\lambda_a}{\lambda_s(\lambda_s + \lambda_a)} \tag{8.41}$$
$$= \frac{\lambda_s + 2\lambda_a}{\lambda_a(\lambda_a + \lambda_s)}$$

EXAMPLE 8.17 Suppose that in a system composed of two identical components, one in standby, the standby failure rate is 10% of the active failure rate. Find both the system reliability for a specified mission of time t and the system MTBF in terms of the active failure rate λ.

The system reliability can be found using Equation 8.29, with $\lambda_a = \lambda$ and $\lambda_s = \lambda/10$:

$$R_{\text{sys}}(t) = (1 + 10)e^{-\lambda t} - 10e^{-1.1\lambda t} = 11e^{-\lambda t} - 10e^{-1.1\lambda t}$$

The system MTBF can be computed via Equation 8.30, again with $\lambda_a = \lambda$ and $\lambda_s = \lambda/10$:

$$\text{MTBF} = (0.1 + 2)\lambda/\lambda(\lambda + 0.1\lambda) = 2.1/(1.1\lambda) = 21/(11\lambda)$$

*8.10 MODELS THAT CONSIDER SENSING AND SWITCHING FAILURE RATES

Let us again consider a system composed of two identical items (Figure 8.15), one initially in standby and one active. Let us assume that the active item has a failure rate λ, the sensing mechanism has a failure rate λ_{se}, and the switching function has a failure rate λ_{sw}. For the sake of simplicity, let us assume for now that the standby failure rate is zero.

This time the system success definitions are the following:

Condition 1 Item A operates throughout the time t without failure.

Condition 2 Item A fails at some time t_1 $(0 \le t_1 \le t)$; the failure is detected by the sensing mechanism, which sends the signal to initiate the switch; the switch functions and turns on item B, and item B remains operating without failure until the time t.

The probability of condition 1 is the reliability of item A for a mission of time t: $R_A(t) = e^{-\lambda\tau}$.

The probability of condition 2 is the probability that item A fails at some time t_1 $(0 \le t_1 \le t)$, that the sensing mechanism has not failed prior to t_1, that the switch functions, and that item B operates throughout the remaining time $(t - t_1)$. These are independent events, so the probability of condition 2 is the product of the probabilities of all these events.

The probability that item A fails at some time t_1 $(0 \le t_1 \le t)$ is

$$Q_A(t_1) = \int_{t_1=0}^{t} \lambda e^{-\lambda t_1}\, dt_1 \tag{8.42}$$

The probability that the sensing mechanism has not failed prior to t_1 is its reliability for a mission of time t_1:

$$R_{se}(t_1) = \int_{t_2=t_1}^{\infty} \lambda_{se} e^{-\lambda_{se} t_2}\, dt_2 \tag{8.43}$$

The switch operates discretely, so the probability that it functions is its reliability for a mission of one cycle, where its failure rate λ_{sw} is in failures per cycle:

$$R_{sw} = 1 - \lambda_{sw} \tag{8.44}$$

The probability that item B, when turned on, will operate for the remainder of the mission is

$$R_B(t_3) = \int_{t_3=t-t_1}^{\infty} \lambda e^{-\lambda t_3}\, dt_3 \tag{8.45}$$

Because conditions 1 and 2 are mutually exclusive, the system reliability is the sum of their probabilities:

$$\begin{aligned}
R_{sys}(t) &= R_A(t) + Q_A(t_1) \cdot R_{se}(t_1) \cdot R_{sw} \cdot R_B(t_3) \\
&= e^{-\lambda t} + \int_{t_1=0}^{t} \lambda e^{-\lambda t_1}\, dt_1 \cdot \int_{t_2=t_1}^{\infty} \lambda_{se} e^{-\lambda_{se} t_2}\, dt_2 \cdot (1-\lambda_{sw}) \cdot \int_{t_3=t-t_1}^{\infty} \lambda e^{-\lambda t_3}\, dt_3 \\
&= e^{-\lambda t} + \lambda^2 \lambda_{se}(1-\lambda_{sw}) \int_{t_1=0}^{t} \int_{t_2=t_1}^{\infty} \int_{t_3=t-t_1}^{\infty} e^{-\lambda t_1} e^{-\lambda_{se} t_2} e^{-\lambda t_3}\, dt_3\, dt_2\, dt_1 \\
&= e^{-\lambda t} + \lambda^2 \lambda_{se}(1-\lambda_{sw}) \int_{t_1=0}^{t} \int_{t_2=t_1}^{\infty} e^{-\lambda t_1} e^{-\lambda_{se} t_2} \left[\frac{e^{-\lambda t_3}}{-\lambda}\right]\Bigg|_{t_3=t-t_1}^{\infty} dt_2\, dt_1 \\
&= e^{-\lambda t} + \lambda^2 \lambda_{se}(1-\lambda_{sw}) \int_{t_1=0}^{t} \int_{t_2=t_1}^{\infty} e^{-\lambda t_1} e^{-\lambda_{se} t_2} (e^{-\lambda(t-t_1)} - 0)\, dt_2\, dt_1 \\
&= e^{-\lambda t} + \lambda \lambda_{se}(1-\lambda_{sw}) \int_{t_1=0}^{t} \int_{t_2=t_1}^{\infty} e^{-\lambda_{se} t_2} e^{-\lambda t}\, dt_2\, dt_1 \\
&= e^{-\lambda t} + \lambda \lambda_{se}(1-\lambda_{sw}) e^{-\lambda t} \int_{t_1=0}^{t} \left[\frac{e^{-\lambda_{se} t_2}}{-\lambda_{se}}\right]\Bigg|_{t_2=t_1}^{\infty} dt_1 \\
&= e^{-\lambda t} + \lambda(1-\lambda_{sw}) e^{-\lambda t} \int_{t_1=0}^{t} (e^{-\lambda_{se} t_1} - 0)\, dt_1 \\
&= e^{-\lambda t} + \lambda(1-\lambda_{sw}) e^{-\lambda t} \left[\frac{e^{-\lambda_{se} t_1}}{-\lambda_{se}}\right]\Bigg|_{t_1=0}^{t} \\
&= e^{-\lambda t} + \frac{\lambda(1-\lambda_{sw})}{\lambda_{se}} e^{-\lambda t}(1 - e^{-\lambda_{se} t}) \\
&= e^{-\lambda t}\left(1 + \frac{\lambda(1-\lambda_{sw})}{\lambda_{se}}\right) - \left(\frac{\lambda(1-\lambda_{sw})}{\lambda_{se}}\right) e^{-(\lambda+\lambda_{se})t} \\
&= e^{-\lambda t}\left(1 + \frac{\lambda R_{sw}}{\lambda_{se}}\right) - \left(\frac{\lambda R_{sw}}{\lambda_{se}}\right) e^{-(\lambda+\lambda_{se})t}
\end{aligned} \tag{8.46}$$

$$\text{MTBF} = \int_0^{\infty} R_{sys}(t)\, dt = \left(1 + \frac{\lambda R_{sw}}{\lambda_{se}}\right) \int_0^{\infty} e^{-\lambda t}\, dt - \left(\frac{\lambda R_{sw}}{\lambda_{se}}\right) \int_0^{\infty} e^{-(\lambda+\lambda_{se})t}\, dt \tag{8.47}$$

From Equation 8.24,

$$\text{MTBF} = \left(1 + \frac{\lambda R_{sw}}{\lambda_{se}}\right)\frac{1}{\lambda} - \left(\frac{\lambda R_{sw}}{\lambda_{se}}\right)\left(\frac{1}{\lambda + \lambda_{se}}\right)$$

$$= \frac{\lambda_{se} + \lambda R_{sw}}{\lambda\lambda_{se}} - \frac{\lambda R_{sw}}{\lambda_{se}(\lambda + \lambda_{se})} = \frac{(\lambda + \lambda_{se})(\lambda_{se} + \lambda R_{sw}) - \lambda^2 R_{sw}}{\lambda\lambda_{se}(\lambda + \lambda_{se})} \quad \textbf{(8.48)}$$

$$= \frac{\lambda + \lambda_{se} + \lambda R_{sw}}{\lambda(\lambda + \lambda_{se})} = \frac{\lambda(1 + R_{sw}) + \lambda_{se}}{\lambda(\lambda + \lambda_{se})} = \frac{\lambda(2 - \lambda_{sw}) + \lambda_{se}}{\lambda(\lambda + \lambda_{se})}$$

EXAMPLE 8.18

Use Equation 8.48 to solve for the MTBF if we assume perfect sensing.

With perfect sensing, $\lambda_{se} = 0$, and Equation 8.48 becomes

$$\text{MTBF} = \frac{\lambda(1 + R_{sw})}{\lambda^2} = \frac{1 + R_{sw}}{\lambda} = \frac{2 - \lambda_{sw}}{\lambda}$$

EXAMPLE 8.19

Use Equation 8.48 to solve for the MTBF if we assume perfect sensing and switching.

With perfect sensing and switching, $\lambda_{se} = \lambda_{sw} = 0$, and Equation 8.48 becomes

$$\text{MTBF} = \frac{2\lambda}{\lambda^2} = \frac{2}{\lambda}$$

Compare this result with Equation 8.31.

EXAMPLE 8.20

A system is composed of two identical components A and B. One of them is required to operate for system success. The other, when available, is in standby. The active failure rate for a component is 10^{-5} failures per hour. The standby failure rate is negligible. The failure rate for the sensing mechanism is 10^{-10} failures per hour. The switching failure rate is 0.01 failures per cycle. What is the system's reliability for a 10,000-h mission? What is the system's MTBF? See Figure 8.25.

Equation 8.46 applies.

$$R_{sys}(10{,}000\text{ h}) = e^{-\lambda t}\left(1 + \frac{\lambda(1 - \lambda_{sw})}{\lambda_{se}}\right) - \frac{\lambda(1 - \lambda_{sw})}{\lambda_{se}}e^{-(\lambda+\lambda_{se})t}$$

$$= e^{-10^{-5}\times 10{,}000}\left(1 + \frac{10^{-5}(0.99)}{10^{-10}}\right) - \frac{10^{-5}(0.99)}{10^{-10}}e^{-10^{-10}\times 10{,}000}$$

$$= e^{-0.1}(99{,}001) - e^{-0.100001}(99{,}001) = 0.99442$$

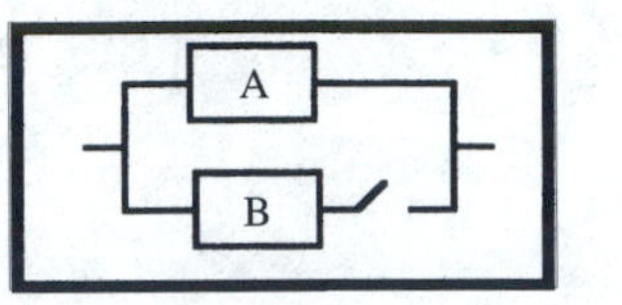

FIGURE 8.25
System of Example 8.20.

To solve the MTBF, Equation 8.48 applies:

$$\text{MTBF} = \frac{\lambda(2 - \lambda_{sw}) + \lambda_{se}}{\lambda(\lambda + \lambda_{se})} = \frac{10^{-5}(2 - 0.01) + 10^{-10}}{10^{-5}(10^{-5} + 10^{-10})}$$

$$= \frac{1.99 \times 10^{-5} + 10^{-10}}{10^{-5}(1.00001 \times 10^{-5})} = \frac{1.99001}{1.00001} \times 10^{5} = 198{,}999 \text{ h}$$

*8.11 GENERAL MODEL FOR STANDBY REDUNDANCY

This time for the system of Figure 8.15, let us consider everything: the active failure rate, the standby failure rate, the probability of switching, and the failure rate of the sensing mechanism. System success will therefore be defined by the following two conditions:

Condition 1 Item *A* operates without failure throughout its specified mission time t.

Condition 2 Item *A* fails at some time t_1 $(0 \le t_1 \le t)$, item *B* is available for operation at the time t_1, the sensing mechanism has survived until the time t_1, the switch functions, and then item *A* operates throughout the remaining time $(t - t_1)$.

As in the previous models, the probability of condition 1 is the reliability of item *A* for the mission time t: $R_A = e^{-\lambda t}$.

The probability of Condition 2 is the product of the probabilities of the five independent events that define Condition 2.

The probability of item *A* failing at some time t_1 $(0 \le t_1 \le t)$ is

$$Q_A(t_1) = \int_{t_1=0}^{t} \lambda e^{-\lambda t_1}\, dt_1 \tag{8.49}$$

where λ is the active failure rate.

The probability of item *B* being available in the standby mode at the time t_1 is

$$R_B^*(t_1) = \int_{t_2=t_1}^{\infty} \lambda_s e^{-\lambda_s t_2}\, dt_2 \tag{8.50}$$

where λ_s is the standby failure rate.

The probability that the sensing device survives until time t_1 is

$$R_{se}(t_1) = \int_{t_3=t_1}^{\infty} \lambda_{se} e^{-\lambda_{se} t_3}\, dt_3, \tag{8.51}$$

where λ_{se} is the failure rate of the sensing device.

Notice that the last two probabilities are functions of the same variable t_1 defined within the same limits. Hence, we can simplify our model by observing that the product

$$R_B^*(t_1) \cdot R_{se}(t_1) = \int_{t_2=t_1}^{\infty} (\lambda_s + \lambda_{se}) e^{-(\lambda_s + \lambda_{se})t_2}\, dt_2 \tag{8.52}$$

The probability of the switching function for one cycle is

$$R_{sw} = (1 - \lambda_{sw}) \tag{8.53}$$

where λ_{sw} is the switching failure rate in failures per cycle.

Finally, the probability of item B operating throughout the rest of the mission is

$$R_B(t_3) = \int_{t_3=t-t_1}^{\infty} \lambda e^{\lambda t_3}\, dt_3 \tag{8.54}$$

Because conditions 1 and 2 are mutually exclusive, the system reliability is the sum of the probabilities of condition 1 and condition 2:

$$\begin{aligned} R_{sys}(t) &= R_A(t) + Q_A(t_1) \cdot R_B{}^*(t_1) \cdot R_{se}\,(t_1) \cdot R_{sw} \cdot R_B(t - t_1) \\ &= e^{-\lambda t} + \int_{t_1=0}^{t} \lambda e^{-\lambda t_1}\, dt_1 \cdot \int_{t_2=t_1}^{\infty} (\lambda_{se} + \lambda_s) e^{-(\lambda_{se}+\lambda_s)t_2}\, dt_2 \cdot (1 - \lambda_{sw}) \cdot \int_{t_3=t-t_1}^{\infty} \lambda e^{-\lambda t_3}\, dt_3 \\ &= e^{-\lambda t} + \lambda^2(\lambda_s + \lambda_{se})(1 - \lambda_{sw}) \int_{t_1=0}^{t} \int_{t_2=t_1}^{\infty} \int_{t_3=t-t_1}^{\infty} e^{-\lambda t_1} e^{-(\lambda_s+\lambda_{se})t_2} e^{-\lambda t_3}\, dt_3\, dt_2\, dt_1 \end{aligned} \tag{8.55}$$

The system MTBF is

$$\begin{aligned} \text{MTBF} &= \int_0^{\infty} R_{sys}(t)\, dt \\ &= \int_{t=0}^{\infty} e^{-\lambda t} + \lambda^2(\lambda_s + \lambda_{se})(1 - \lambda_{sw}) \int_{t=0}^{\infty} \int_{t_1=0}^{t} \int_{t_2=t_1}^{\infty} \int_{t_3=t-t_1}^{\infty} e^{-\lambda t_1} e^{-(\lambda_s+\lambda_{se})t_2} e^{-\lambda t_3}\, dt_3\, dt_2\, dt_1 \end{aligned} \tag{8.56}$$

Equations 8.55 and 8.56 can be further developed into algebraic expressions similar to those of the preceding models. As you can well imagine (if you choose not to proceed on your own), to do so would lead to some rather cumbersome mathematics. In Chapters 9 and 10 we shall apply the *Markov model* technique, which will allow us to solve the reliability of this system and of even more complex systems with relative ease.

8.12 CONCLUDING REMARKS

The reliability modeling in this chapter was restricted to what we might call simple systems. We dealt only with *nonrestored* systems. Only those readers who followed the mathematical derivations of the asterisked sections were shown how to analytically determine the MTBF for these systems. Also, we had to use the advanced analytical methods of the asterisked sections to consider the effects of sensing, switching, and inactive failure rates for evaluating systems composed of standby redundancy. Readers not following the asterisked sections were restricted to making assumptions of perfect sensing and switching and had to neglect the possibility of failures of units while in the standby mode. The next chapter introduces the Markov model technique, which allows us to consider the effects of on-line restorability of parallel or redundant elements, failure rates of sensing and switching functions, both active and standby failure rates, and any variety of complexities in the logic of a system.

If you have followed the mathematics in the asterisked sections, you will appreciate that the analytical techniques used for Chapter 9 not only make it possible to solve reliability and MTBF of more complex systems, they also simplify the mathematics for the types of systems presented in this chapter.

BIBLIOGRAPHY

1. ARINC Research Corp. 1964. *Reliability Engineering.* Upper Saddle River, N.J.: Prentice Hall.
2. I. Bazovsky. 1961. *Reliability Theory and Practice.* Upper Saddle River, N.J.: Prentice Hall.
3. E. E. Lewis. 1987. *Introduction to Reliability Engineering,* New York: John Wiley.
4. D. K. Lloyd and M. Lipow. 1962. *Reliability Management, Methods, and Mathematics.* Upper Saddle River, N.J.: Prentice Hall.
5. P. D. T. O'Connor. 1991. *Practical Reliability Engineering,* 3d ed. New York: John Wiley.
6. N. H. Roberts. 1964. *Mathematical Methods in Reliability Engineering.* New York: McGraw-Hill.
7. G. H. Sandler. 1963. *System Reliability Engineering.* Upper Saddle River, N.J.: Prentice Hall.
8. M. L. Shooman. 1968. *Probabilistic Reliability: An Engineering Approach. New York: McGraw-Hill.*
9. C. O. Smith. 1976. *Introduction to Reliability in Design.* New York: McGraw-Hill.
10. P. A. Tobias and D. Trindade. 1986. *Applied Reliability.* New York: Van Nostrand Reinhold.

EXERCISES

8.1. A personal computer system is composed of a processor with a failure rate of 150 failures per million hours, a monitor with a failure rate of 50 failures per million hours, a keyboard and mouse with a failure rate of 10 failures per million hours, and a printer with a failure rate of 75 failures per million hours. A failure in any one of these elements comprises a system failure. Predict the MTBF for this system if we assume that all elements must be operable at all times.

8.2. Predict the reliability of the system in Exercise 8.1 for a 1000-h mission.

8.3. Assuming that the printer in Exercise 8.1 is necessary only 5% of the time and that all other elements are needed all the time, predict the MTBF and the reliability for a 1000-h mission. Use a 1% dormancy factor.

8.4. A workshop facilitator has a 1-h video to present to her attendees. For the presentation to be successful, she requires a video cassette recorder/player (VCR), a monitor, and a videotape cassette. Suppose the VCR has a failure rate of 500 failures per million hours, the monitor has a failure rate of 50 failures per million hours, and the cassette has a failure rate of 1000 failures per million hours. Assuming that all three are initially operable, what is the probability of her having a successful presentation? That is, what is the mission reliability for this system for a 1-h mission?

8.5. A system contains two identical elements, each with a failure rate of 1000 failures per million hours. Compute the system reliability for a 100-h mission.

8.6. A system contains five identical, redundant items, each with a reliability of 0.990. What is the system reliability if all five items are required for system success?

8.7. What is the reliability for the system of Exercise 8.6 if at least four of the five items are required for system success?

8.8. What is the reliability for the system of Exercise 8.6 if at least three of the five items are required for system success?

8.9. What is the reliability for the system of Exercise 8.6 if at least two of the five items are required for system success?

8.10. What is the reliability for the system of Exercise 8.6 if only one of the five items are required for system success?

8.11. Solve the reliability for the system whose block diagram is given in Figure 8.26, where the element reliability values are $R_A = 0.995$, $R_B = 0.950$, and $R_C = 0.900$.

8.12. Solve the reliability for the system with parallel subsystems and parallel elements shown in the block diagram of Figure 8.27, where the element reliability values are $R_A = 0.990$, $R_B = 0.850$, and $R_C = 0.700$.

8.13. A system contains two identical elements, one active and one in standby, each with an active failure rate of 1000 failures per million hours and a negligible standby failure rate. Compute the system reliability for a 100-h mission if we assume perfect sensing and switching.

8.14. Solve the reliability for a 1000-h mission for the system whose block diagram is given in Figure 8.28. Elements B an C are in standby redundancy, as shown. The active failure rates are 100 failures per million hours for element A, 500 failures per million hours for element B, and 1000 failures per million hours for element C. The standby failure rates are negligible. Assume perfect sensing and switching.

8.15. Suppose the facilitator of Exercise 8.4 has the program on a spare cassette, which she can substitute if the original cassette fails. How much improvement in reliability will be achieved by this standby cassette?

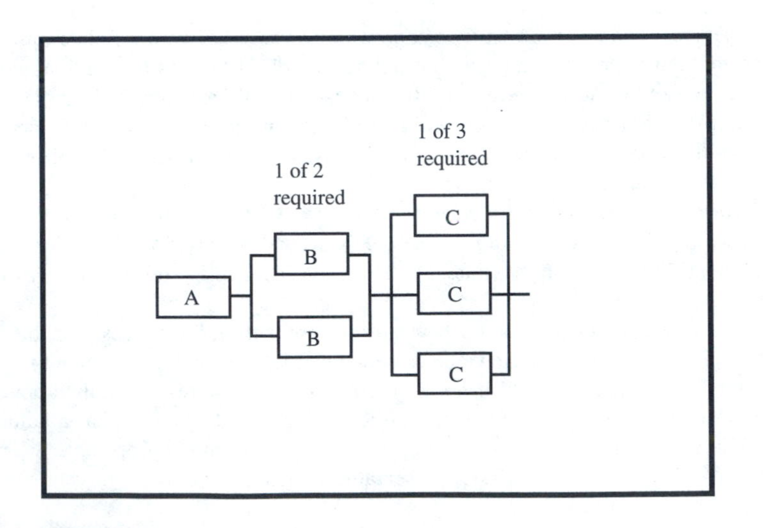

FIGURE 8.26
System of Exercise 8.11.

$R_{TOP} = [1-(1-R_B)^2]$

$R_{BOT} = [R_B(1-[1-R_C]^2)]$

$R_{sys}[1-(1-R_{TOP})(1-R_{BOT})]$

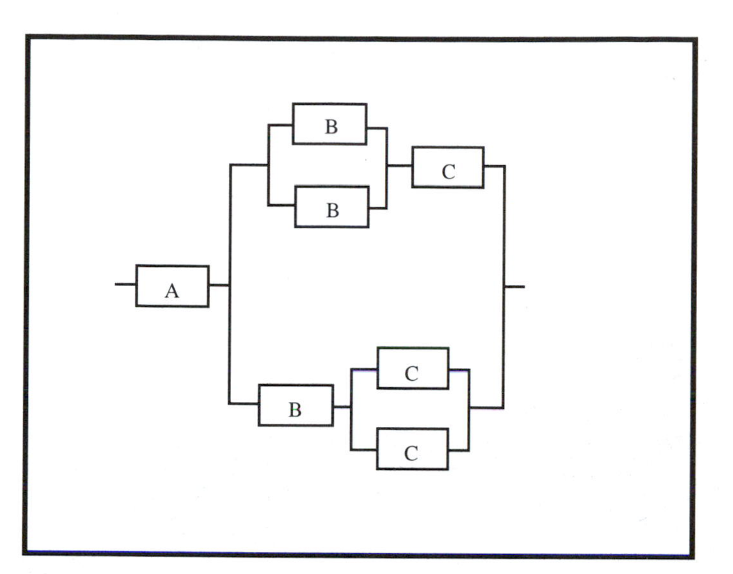

FIGURE 8.27
System of Exercise 8.12.

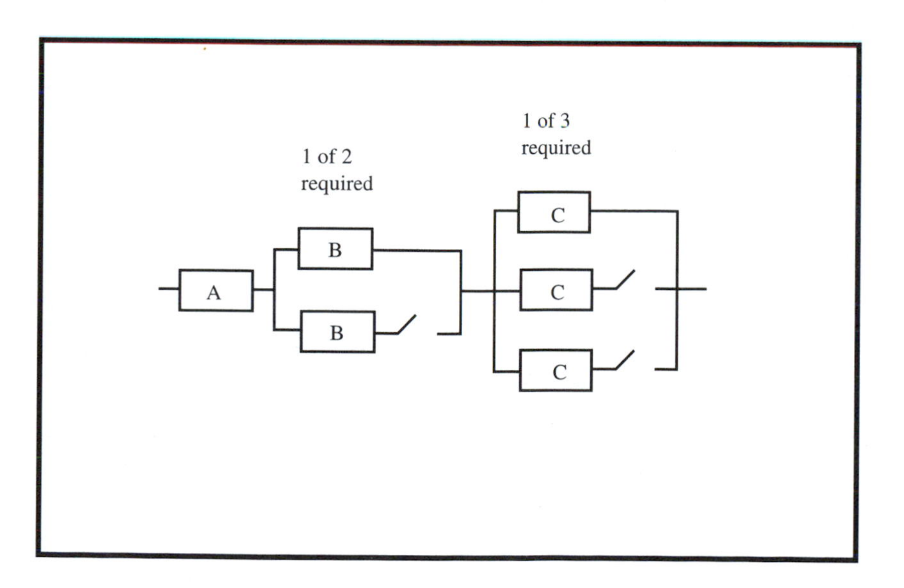

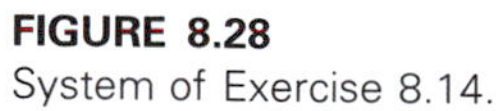

FIGURE 8.28
System of Exercise 8.14.

8.16. A hospital operating room has an emergency generator on standby in the event of a power failure. If power failures occur at an average rate of once every 5000 h and the active failure rate for the emergency generator is 2000 failures per million hours, what is the reliability for the operating room's power system for a 10-h mission? Assume perfect sensing and switching and a negligible standby failure rate for the emergency generator.

8.17. How much added protection can be achieved by adding a second emergency generator to the operating room of Exercise 8.16? In other words, if the system were composed of the primary power source and two standby generators, how much higher would the reliability be for the same 10-h mission? Once again assume perfect sensing and switching and a negligible standby failure rate for the emergency generator.

***8.18.** Review the derivations of Equations 8.31 and 8.33. Are these results the same as you would intuitively expect? Explain your answer.

***8.19.** Solve the MTBF for the system of Exercise 8.13.

***8.20.** Solve Exercise 8.16 if we assume that the probability of successful sensing and switching upon demand is 0.985 and that the generator failure rate in standby is negligible.

***8.21.** Solve Exercise 8.16 if we assume that the probability of successful sensing and switching upon demand is 0.985 and that the generator failure rate in standby is 100 failures per million hours.

***8.22.** Find the MTBF for Exercise 8.16 if we assume that the probability of successful sensing and switching upon demand is 0.985 and that the generator failure rate in standby is 100 failures per million hours.

RELIABILITY MODELING OF COMPLEX SYSTEMS

9.1 INTENT

In this chapter we shall go beyond the types of systems to which we restricted our discussion in Chapter 8, the types of systems that we classified as simple systems. The systems we evaluate in this chapter are capable of restoration during the actual system operation, they can have any degree of active or standby redundancy, they can have subsystems with monitoring functions and switching functions to activate standby elements upon demand, and the system logic can have any degree of complexity. Such system capabilities are generally features of the *continuously operating* type of system introduced in Chapter 2.

As discussed in Chapter 2 and illustrated in Figure 2.1, the reliability parameters of interest to us when analyzing continuously operating items are MTBF, MTBMA, and availability. In this chapter we shall concern ourselves with the system MTBF and MTBMA. Chapter 10 deals with the availability parameters.

9.2 SERIES SYSTEM

As presented in Chapter 8, a series system in the reliability sense is composed of elements whose failure will produce a system failure. The block diagram for such a system is as shown in Figure 9.1. Notice that each of the elements i, represented by one of the blocks,

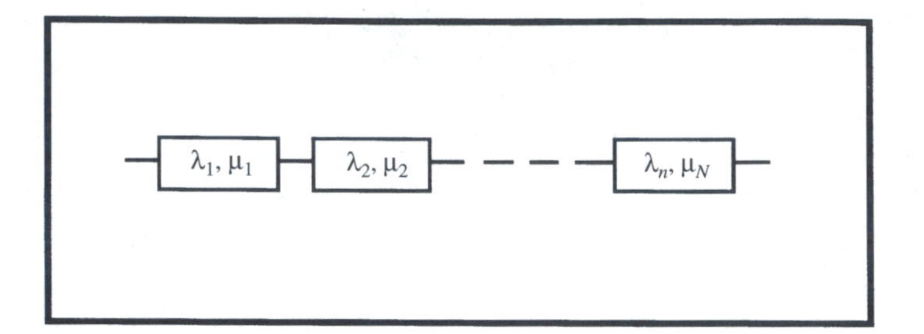

FIGURE 9.1
Block diagram for a series system with N elements.

has a failure rate λ_i and its own restoration rate μ_i. Because a failure of any of the N elements will cause the system to fail, the system failure rate is

$$\lambda_{\text{sys}} = \lambda_1 + \lambda_2 + \cdots + \lambda_N = \sum_{i=1}^{N} \lambda_i \tag{9.1}$$

Complex systems, where we generally have a planned replacement policy for elements with short wearout periods, behave as though they have constant failure rates. In fact, the more complex the system, the more likely it is to behave that way. Therefore, we can usually assume constant failure rate for a complex system and, accordingly, treat the MTBF and system failure rate as reciprocals:

$$\text{MTBF} = \frac{1}{\lambda_{\text{sys}}} = \frac{1}{\sum_{i=1}^{N} \lambda_i} = \frac{1}{\lambda_1 + \lambda_2 + \cdots + \lambda_N} \tag{9.2}$$

Because every element failure causes a system failure, there can be no restoration during operation. The restoration rates, therefore, relate to the mean downtime due to failure, or MTTR. And according to Equation 7.1,

$$\text{MTTR} = \frac{\sum_{i=1}^{N} \lambda_i M_i}{\sum_{i=1}^{N} \lambda_i} = \frac{\sum_{i=1}^{N} (\lambda_i/\mu_i)}{\sum_{i=1}^{N} \lambda_i} = \frac{(\lambda_1/\mu_1) + (\lambda_2/\mu_2) + \cdots + (\lambda_N/\mu_N)}{\lambda_1 + \lambda_2 + \cdots + \lambda_N} \tag{9.3}$$

If we include not only the failure events, but also every scheduled restoration activity (that is, planned replacements, preventive maintenance actions, logistics, or administrative interruptions), Equation 9.2 becomes the MTBMA,

$$\text{MTBMA} = \frac{1}{\sum_{i=1}^{N} \lambda_i} \tag{9.4}$$

where the values λ_i represent the rates of occurrence of all types of restoration events.

EXAMPLE 9.1 Consider a system of 50 ATM machines, 30 of which have failure rates of 1 every 150,000 h, 10 of which have failure rates of 1 every 500,000 h, and 10 of which have failure rates of 1 every 1,000,000 h. What is the MTBF of this system if all 50 machines are required to be operating all the time?

Let

$$\begin{aligned} \lambda_1 &= 1/150{,}000 \text{ h} = 6.6 \times 10^{-6} \text{ per hour} \\ \lambda_2 &= 1/500{,}000 \text{ h} = 2.0 \times 10^{-6} \text{ per hour} \\ \lambda_3 &= 1/10^6 \text{ h} \quad\;\; = 1.0 \times 10^{-6} \text{ per hour} \end{aligned}$$

The total system failure rate is

$$\begin{aligned} \lambda_{sys} = 30\lambda_1 + 10\lambda_2 + 10\lambda_3 &= (30 \times 6.6 + 10 \times 2.0 + 10 \times 1.0) \times 10^{-6} \text{ per hour} \\ &= 228.0 \times 10^{-6} \text{ per hour} \end{aligned}$$

The system MTBF is

$$\text{MTBF} = 1/\lambda_{sys} = 1/228.0 \times 10^{-6} \text{ per hour} = 4386 \text{ h}$$

EXAMPLE 9.2 If each of the ATM machines in Example 9.1 has a scheduled outage for maintenance and calibration once every 5000 operating hours and no other outages except upon failure, what is the MTBMA for the system?

The total outage rate is the rate of failure plus the rate of scheduled outages, $228.0 \times 10^{-6} + 50 \times (1/5000)$ per hour = 0.010228 per hour. Hence, the MTBMA is

$$\text{MTBMA} = 1/0.010228 \text{ per hour} = 97.8 \text{ h}$$

9.3 SYSTEMS WITH ACTIVE REDUNDANCY

Now let us consider a system (See Figure 9.2) composed of N active elements, any k of which are required for system success. Solutions for the **system effective failure rate,** λ_{sys}, and the MTBF are provided in Tables C.1 through C.6 of the Appendix C. All MTBF tables in Appendix C were developed through methods presented in Sections 9.6–9.9.

9.3.1 Identical Elements with Restoration

Table C.1 provides λ_{sys} and MTBF solutions for systems composed of N identical items, k of which are required for system success, where on-line restoration (without interruption of system functioning) is possible. Table C.1 assumes that only one restoration activity at a time is possible. Each item is considered to have failure rate λ and restoration rate μ.

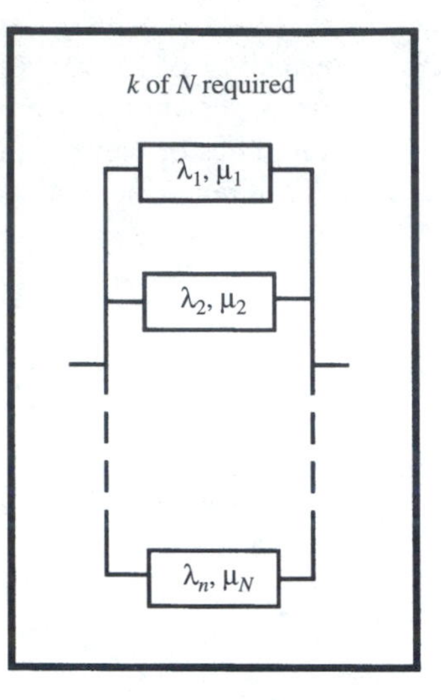

FIGURE 9.2
Block diagram for a system with N active elements, any k of which are required for system success.

Each block in Figure 9.2 may actually represent a series of components, in which case the block failure rate is the sum of the failure rates of all the series components within. The block restoration rate (in restorations per hour) is then the reciprocal of the mean restoration time for the block. The mean block restoration time is determined from Equation 9.3. Accordingly, the mean block restoration rate is

$$\mu = \frac{\sum_j \lambda_j}{\sum_j \lambda_j M_j} \tag{9.5}$$

where λ_j and M_j are, respectively, the failure rate and mean restoration time of the jth component in the block. The mean restoration time for a component is the average time that it takes to detect a failure after its occurrence plus the average time it takes to correct, repair or replace the component on-line (see Chapter 7). If a component's performance is continuously monitored and announced upon occurrence, the mean restoration time is simply the mean time to correct, repair, or replace. Suppose, on the other hand, a component is monitored once a week, and its mean repair time is 4h. When it fails, the average time to detect is 3.5 days = 84 h. Hence, the mean restoration time is 84 + 4 = 88 hours. The restoration rate is then 1/88 h = 0.011 per hour.

EXAMPLE 9.3 Suppose the system composed of 50 machines in Examples 9.1 and 9.2 requires only 48 of them to be operating at any one time. If they each have the same failure rate $\lambda = 6.6 \times 10^{-6}$ per hour and the same restoration rate $\mu = 0.20$ per hour and if on-line restoration is possible, determine the system MTBF.

The solution for $N - 2$ out of N of Table C.1 applies, with $N = 50$:

$$\text{MTBF} = \frac{\mu^2 + 2N\mu\lambda - 2\mu\lambda + 3N^2\lambda^2 - 6N\lambda^2 + 2\lambda^2}{N(N-1)(N-2)\lambda^3}$$

$$= \frac{0.04 + 0.000132 - 0.0000026 + 0.0000003 - 0.0000000 + 0.0000000}{3.38 \times 10^{-11}}$$

$$= \frac{0.04013}{3.38} \times 10^{11} \text{ h} = 11.9 \times 10^{8} \text{ h} = 1{,}190{,}000{,}000 \text{ h}$$

The MTBMA for a redundant system is determined as it is for a series system, by using Equation 9.4.

EXAMPLE 9.4 Assuming that no preventive maintenance actions are required for the system of Example 9.3, what is the system's MTBMA?

From Equation 9.4,

$$\text{MTBMA} = \frac{1}{\sum_{i=1}^{N} \lambda_i} = \frac{1}{(50 \times 6.6 \times 10^{-6}) \text{ per hour}} = 3030 \text{ h}$$

EXAMPLE 9.5 A system is composed of five identical elements, all active and each with a failure rate $\lambda = 0.001$ per hour and a restoration rate $\mu = 2.0$ per hour. If any four of the five are required for system success and on-line restoration is possible, compute the system MTBF.

The Table C.1 solution for 4 out of 5 applies:

$$\text{MTBF} = \frac{\mu + 9\lambda}{20\lambda^2} = \frac{2.0 + 0.009}{20(0.001)^2} = 100.45 \text{ h}$$

9.3.2 Identical Elements Without Restoration

Table C.4 provides λ_{sys} and MTBF solutions for systems composed of N identical items, k of which are required for system success, where on-line restoration (without interruption of system functioning) is *not* possible. Table C.4 assumes that only one restoration activity at a time is possible. Each item is considered to have a failure rate λ.

EXAMPLE 9.6 Compute the MTBF for the system of Example 9.5 if on-line restoration is not possible.

The Table C.4 solution for 4 out of 5 applies:

$$\text{MTBF} = \frac{27}{60\lambda} = \frac{27}{60(0.001)} = 450 \text{ h}$$

To fully appreciate the convenience provided by the Appendix C tables, if you are following the mathematics in the asterisked sections, compare this example with Example 8.11.

EXAMPLE 9.7 Assuming that no preventive maintenance actions are required for the systems of Examples 9.5 and 9.6, what is the system's MTBMA?

From Equation 9.4,

$$\text{MTBMA} = \frac{1}{\sum_{i=1}^{N} \lambda_i} = \frac{1}{(5 \times 0.001) \text{ per hour}} = 200 \text{ h}$$

EXAMPLE 9.8 Compute the MTBF for the system of Example 9.3 if on-line restoration is not possible.

The Table C.4 solution for $N = 50$ and $k = 48$ applies, with $\lambda = 6.6 \times 10^{-6}$ per hour:

$$\text{MTBF} = \frac{\sum_{i=k}^{N} \left(\frac{1}{i}\right)}{\lambda} = \frac{\frac{1}{48} + \frac{1}{49} + \frac{1}{50}}{6.6 \times 10^{-6}} = \frac{0.0612414}{0.0000066} = 9280 \text{ h}$$

9.3.3 Identical Elements with Failure Rates Subject to Change

Suppose we have a system composed of identical elements with constant failure rates subject to change with the additional stresses imposed by the failure of a redundant element. Tables C.2 and C.5 have been developed to cope with this situation.

EXAMPLE 9.9 A subsystem is composed of five active power supplies, any four of which are necessary for the system to function. When all five power supplies are operating, each has a failure rate $\lambda_A = 9.5 \times 10^{-6}$ per hour. If one fails, the additional load on the remaining power supplies increases their failure rates to $\lambda_B = 25.6 \times 10^{-6}$ per hour. If the mean time to restore a failed power supply is 2.0 h, determine the MTBF of this power supply subsystem.

The solution for 4 out of 5 in Table C.2 applies, with $\lambda_A = 9.5 \times 10^{-6}$ per hour, $\lambda_B = 25.6 \times 10^{-6}$ per hour, and $\mu = 1/2 \text{ h} = 0.5$ per hour:

$$\text{MTBF} = \frac{\mu + 5(\lambda_A + \lambda_B) - \lambda_B}{20\lambda_A\lambda_B} = \frac{0.5 + 5(35.1 \times 10^{-6}) - (25.6 \times 10^{-6})}{20 \times 243.2 \times 10^{-12}}$$

$$= \frac{0.5001499}{4.864 \times 10^{-9}} = 1.028 \times 10^{8} \text{ h}$$

EXAMPLE 9.10 Compute the MTBF for the subsystem of Example 9.9 if on-line restoration is not possible.

Here Table C.5 applies, with $N = 5$, $N - 1 = 4$, $\lambda_0 = 9.5 \times 10^{-6}$ per hour, and $\lambda_1 = 25.6 \times 10^{-6}$ per hour:

$$\text{MTBF} = \frac{N\lambda_0 + (N-1)\lambda_1}{N(N-1)\lambda_0\lambda_1} = \frac{5(9.5 \times 10^{-6}) + 4(25.6 \times 10^{-6})}{20 \times 9.5 \times 25.6 \times 10^{-12}} = 30{,}800 \text{ h}$$

9.3.4 Redundant Elements with Different Failure Rates

For two redundant elements that are not necessarily identical and have different failure and restoration rates, use Table C.3 for a system with on-line restoration and Table C.6 for a system where on-line restoration is not possible.

EXAMPLE 9.11 Suppose a computer has two printers, both on-line full time. One has a failure rate of 1 per 1,000,000 h and a mean restoration time of 3 h, and the other has a failure rate of 1 per 100,000 h and a mean restoration time of 30 min. What is the MTBF of this computer-printer subsystem if on-line restoration is possible?

The second equation of Table C.3 applies, with $\lambda_1 = 1 \times 10^{-6}$ per hour, $\mu_1 = 1/3$ per hour, $\lambda_2 = 10 \times 10^{-6}$ per hour, and $\mu_2 = 2$ per hour:

$$\text{MTBF} = \frac{\mu_1\mu_2 + [(\mu_1 + \mu_2)(\lambda_1 + \lambda_2)]}{\lambda_1\lambda_2[(\mu_1 + \mu_2) + (\lambda_1 + \lambda_2)]} = \frac{2/3 + [7/3 \times 11 \times 10^{-6}]}{10 \times 10^{-12}[7/3 + 11 \times 10^{-6}]}$$

$$= \frac{0.666692}{2.333} \times 10^{11} = 2.86 \times 10^{10} \text{ hours.}$$

EXAMPLE 9.12 Solve Example 9.11 if on-line restoration is not possible.

Here the equation in Table C.6 applies, with $\lambda_1 = 1 \times 10^{-6}$ per hour and $\lambda_2 = 10 \times 10^{-6}$ per hour:

$$\text{MTBF} = \frac{\lambda_1^2 + \lambda_2^2 + \lambda_1\lambda_2}{\lambda_1^2\lambda_2 + \lambda_1\lambda_2^2} = \frac{1 \times 10^{-12} + 100 \times 10^{-12} + 10 \times 10^{-12}}{10 \times 10^{-18} + 100 \times 10^{-18}}$$

$$= \frac{111}{110} \times 10^6 = 1{,}009{,}000 \text{ h}$$

There are no tables in Appendix C for systems composed of more than two redundant elements with different failure and restoration rates. For such systems it is suggested that the k-out-of-N equation from the appropriate table for systems with identical failure and restoration rates be used, assuming the highest element-failure rate and the lowest element-restoration rate. For most complex systems, this substitution will provide a valid approximation. To ascertain that the approximation is acceptable, two solutions can be computed: one using the highest failure rate and the lowest restoration rate, the other using the lowest failure rate and the highest restoration rate. The former gives a pessimistic approximation, and the latter gives an optimistic approximation. If the two results are reasonably close to each other, we can be satisfied that this approach is indeed providing a valid approximation. Otherwise it will be necessary to use the *Markov model* approach described in Sections 9.6 through 9.9.

EXAMPLE 9.13

Solve Example 9.3, 48 out of 50 machines required, where 30 of the machines have failure rates $\lambda_1 = 6.6 \times 10^{-6}$ per hour and restoration rates $\mu_1 = 0.50$ per hour, 10 of the machines have failure rates $\lambda_2 = 2.0 \times 10^{-6}$ per hour and restoration rates $\mu_2 = 0.20$ per hour, and 10 of the machines have failure rates $\lambda_3 = 1.0 \times 10^{-6}$ per hour and restoration rates $\mu_3 = 0.30$ per hour.

As with Example 9.3, the solution for $(N - 2)$ out of N of Table C.1 applies with $N = 50$:

$$\text{MTBF} = \frac{\mu^2 + 100\mu\lambda - 2\mu\lambda + 7500\lambda^2 - 300\lambda^2 + 2\lambda^2}{117{,}600\lambda^3} = \frac{\mu^2 + 98\mu\lambda + 7202\lambda^2}{117{,}600\lambda^3}$$

We get a pessimistic approximation by applying the highest failure rate and lowest restoration rate:

$$\lambda = \lambda_1 = 6.6 \times 10^{-6} \text{ per hour}$$
$$\mu = \mu_2 = 0.20 \text{ per hour}$$
$$\text{MTBF} = \frac{\mu^3 + 98\mu\lambda + 7202\lambda^2}{117{,}600\lambda^3} = \frac{0.04 + 0.0001293 + 0.0000003}{3.38 \times 10^{-11}} = 11.9 \times 10^8 \text{ h}$$

Notice that this is the same result as obtained in Example 9.3.

Now let us get an optimistic approximation by using the lowest failure rate and highest restoration rate:

$$\lambda = \lambda_3 = 1.0 \times 10^{-6} \text{ per hour}$$
$$\mu = \mu_1 = 0.50 \text{ per hour}$$
$$\text{MTBF} = \frac{\mu^2 + 98\mu\lambda + 7202\lambda^2}{117{,}600\lambda^3} = \frac{0.25 + 0.000049 + 0.0000000}{1.176 \times 10^{-13}} = 2.13 \times 10^{12} \text{ h}$$

Using this method, we obtained a pessimistic approximation of 1,190,000,000 h and an optimistic approximation of 2,130,000,000,000 hours. The solution must be somewhere in between. Because the ratio between the optimistic and pessimistic approximations is almost 1800 to 1, we might conclude that this approximation technique is inadequate and that we must use the more complicated Markov model approach of Section 9.9. However, because the magnitude of the pessimistic approximation to the MTBF is high enough to be inconsequential, we may be tempted to approximate the MTBF at 10^9 hours.

EXAMPLE 9.14

Suppose component A with failure and restoration rates of $\lambda_A = 1.0 \times 10^{-6}$ per hour and $\mu_A = 0.50$ per hour, respectively, component B with failure and restoration rates of $\lambda_B = 1.5 \times 10^{-6}$ per hour and $\mu_B = 0.50$ per hour, respectively, and component C with failure and restoration rates of $\lambda_C = 1.8 \times 10^{-6}$ per hour and $\mu_C = 1.00$ per hour, respectively, are actively redundant in a system. Any two of the components are needed for system success. What is the MTBF of this subsystem?

As with the preceding example, there is no Appendix C solution for this situation. Let us once again apply the optimistic/pessimistic approach. For the pessimistic approximation, the highest failure rate is $\lambda = \lambda_C = 1.8 \times 10^{-6}$ per hour, and the lowest restoration rate is $\mu = \mu_A = \mu_B = 0.50$ per hour. Applying the 2-out-of-3 solution of Table C.1,

$$\text{MTBF} = \frac{\mu + 5\lambda}{6\lambda^2} = \frac{0.500009}{1.944 \times 10^{-11}} = 2.57 \times 10^{10}\ \text{h}$$

For the optimistic approximation, the lowest failure rate is $\lambda = \lambda_A = 1.0 \times 10^{-6}$ per hour, and the highest restoration rate is $\mu = \mu_C = 1.00$ per hour. Applying the 2-out-of-3 solution of Table C.1,

$$\text{MTBF} = \frac{\mu + 5\lambda}{6\lambda^2} = \frac{1.000005}{6.00 \times 10^{-12}} = 16.7 \times 10^{10}\ \text{h}$$

So, we can conclude that the MTBF is somewhere between 25,700,000,000 and 167,000,000,000 h. Here there is a 6:1 ratio between the optimistic and pessimistic solutions. However, as with the previous example, the pessimistic approximation is high enough for most analysts to be tempted to report the MTBF as being more than 3×10^{10} hours.

9.4 SYSTEMS WITH STANDBY REDUNDANCY

Let us now consider a system composed of N redundant elements, k of which are active and $S = N - k$ of which are in standby, as illustrated in Figure 9.3. The k active elements are all required for system success. When one of the active elements fails, the system is required to detect that failure and to switch on one of the standby elements to replace the failed one. Solutions for the system effective failure rate λ_{sys} and the MTBF for systems with standby redundancy are provided in Tables C.7 through C.10 of Appendix C.

9.4.1 Identical Elements with Perfect Sensing and Switching

If we assume that for a system of the type illustrated in Figure 9.3, the failure rates of the standby elements and the monitoring function failure rate, as well as the probability that a switching function fails to operate when needed, are all negligible, then the solutions of Tables C.7 and C.9 apply. Table C.7 is for a system with on-line restoration capability, and Table C.9 is for a system without that capability.

EXAMPLE 9.15 Suppose a communication system has three receivers, two of which are active and are required for successful system performance and one of which is in standby redundancy. The active failure rate for a single receiver is estimated at 12×10^{-6} per hour, the standby failure rate for a receiver is assumed to be effectively zero, and the restoration rate for a receiver is assumed to be 1.0 per hour. If on-line restoration is possible and we assume perfect sensing and switching, what is the MTBF for that receiver subsystem?

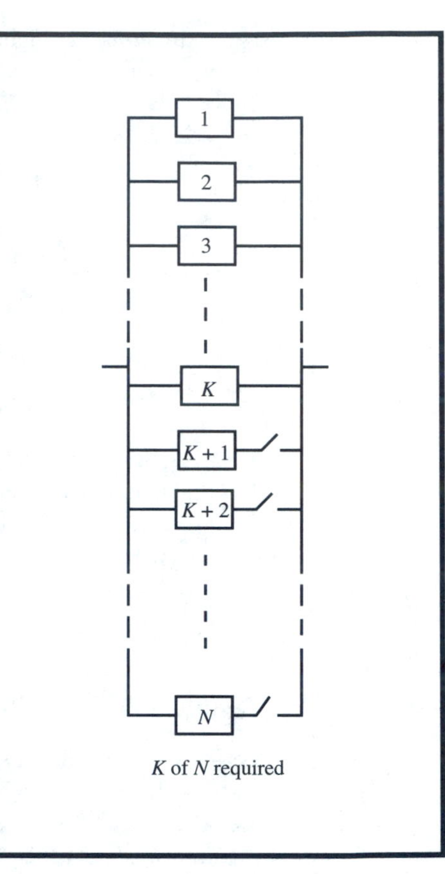

FIGURE 9.3
Block diagram for a system with k active elements and $S = (N - k)$ standby elements, where k are required for system success.

The Table C.7 solution for $N = 3$ and $k = 2$ applies:

$$\text{MTBF} = \frac{\mu + 4\lambda}{4\lambda^2} = \frac{1.0 + 4(12 \times 10^{-6})}{4 \times 144 \times 10^{-12}} = \frac{1.000048}{5.76} \times 10^{10}$$

$$= 1.736 \times 10^{9} \text{ h} = 1{,}736{,}000{,}000 \text{ h}$$

EXAMPLE 9.16 Compute the MTBF for the subsystem of Example 9.15 if on-line restoration is not possible.

The Table C.9 solution for $N = 3$ and $k = 2$ applies:

$$\text{MTBF} = \frac{1}{\lambda} = \frac{1.0}{12 \times 10^{-6}} = 83{,}300 \text{ h}$$

9.4.2 Effects of Sensing and Switching

Tables C.8 and C.10 provide system effective failure rate and MTBF solutions where the effects of monitoring and switching function failure rates are considered. The probability

p_s of successful switchover upon demand from a failed element to a standby element is a function of successful fault sensing and switching. It is computed as

$$p_s = \frac{(1 - \lambda_{sw})\mu_M}{\mu_M + \lambda_M} \tag{9.6}$$

where λ_{sw} = the failure rate of the switching function in failures per cycle
λ_M = the failure rate of the monitoring or sensing device in failures per hour
μ_M = the restoration rate of the monitoring or sensing device in restorations per hour

---

EXAMPLE 9.17 Suppose a subsystem monitoring device has a failure rate of 100×10^{-6} per hour and a mean restoration time of 100 h and its switching device has a failure rate of 1.0 per 1000 cycles. Compute the probability of a successful switchover upon demand.

From Equation 9.6,

$$p_s = \frac{(1 - \lambda_{sw})\mu_M}{\mu_M + \lambda_M} = \frac{(1.0 - 0.001)0.01}{0.01 + 0.000100} = \frac{0.009990}{0.010100} = 0.989$$

EXAMPLE 9.18 Suppose the receiver subsystem of Example 9.15 is protected by the monitoring and switching devices of Example 9.17. Compute the MTBF for the subsystem.

Table C.8 applies with $N = 3$, $k = 2$, $\lambda = 12 \times 10^{-6}$, $\mu = 1.0$, and $p_s = 0.989$:

$$\text{MTBF} = \frac{\mu + 2(1 + p_s)\lambda}{2[2\lambda + (1 - p_s)\mu]\lambda} = \frac{1.0 + 2(1.989) \times 12 \times 10^{-6}}{2[24 \times 10^{-6} + (0.011)1.0]12 \times 10^{-6}}$$

$$= \frac{1.0000477}{0.2646} \times 10^6 = 3{,}780{,}000 \text{ h}$$

EXAMPLE 9.19 Compute the MTBF for the subsystem of Example 9.18 if on-line restoration is not possible.

The Table C.10 solution for $N = 3$ and $k = 2$ applies:

$$\text{MTBF} = \frac{p_s + 1}{2\lambda} = \frac{1.989}{24 \times 10^{-6}} = 82{,}875 \text{ h}$$

9.5 USING SUBSYSTEM EFFECTIVE FAILURE RATES TO ESTIMATE COMPLEX SYSTEM MTBF

Suppose you are required to determine the MTBF of a system more complex than any of those included in the tables of Appendix C. It is sometimes possible to use the Appendix C tables if the system in question can be assumed to comprise a chain of subsystems whose system effective failure rate solutions can be determined from the tables.

Then a complex system reliability block diagram is constructed from a series of what we call **logic blocks.** Each logic block is the block diagram of a subsystem of the main system, and the system effective failure rates for all of the block diagrams can be solved using the Appendix C tables.

This technique for estimating the main system MTBF is valid under the restriction that each logic block can be assumed to have its own independent restoration capability. This restriction is, of course, not true for most systems, but the assumption is acceptable if all elements have extremely large **restoration rate–to–failure rate** (μ/λ) **ratios** of, say, greater than 1000:1. Most complex systems with on-line restoration capabilities do have elements with restoration rates many orders of magnitude larger than their failure rates. So, this technique will provide an acceptable MTBF approximation for most complex systems with on-line restoration. However, if on-line restoration is not possible, the restoration rate is $\mu = 0$, violating the large μ/λ requirement. Therefore, this technique cannot be applied unless there is on-line restoration capability throughout the system.

9.5.1 Table Approximation Technique

This technique consists of the following steps:

1. Using symbols introduced in Figure 9.4, set up a worksheet similar to that of Figure 9.5 with the following column headings:

   Logic block

   Block diagram

   Logic

   Equation

   Item failure rate λ (failures per hour)

   Item restoration rate μ (restorations per hour)

   System effective failure rate λ_{sys}
2. Construct a reliability block diagram in the second column of the worksheet that shows the system to be composed of a series chain of logic blocks. Use the third column to describe the logic.
3. Look up the applicable equations in the Appendix C tables and write them on the worksheet in the fourth column.
4. Insert the failure and restoration rates as well as applicable duty factors and switching probabilities in the fifth and sixth columns of the worksheet.
5. Applying the failure and restoration rates and the switching probabilities to the equations of column 4, compute the system effective failure rates, λ_{sys}, for each logic block.
6. The sum of the system effective failure rates for all the logic blocks approximates the system effective failure rate for the entire system. Its reciprocal approximates the system MTBF:

$$\text{MTBF} = 1/\Sigma\lambda_{sys}$$

9.5.2 Logic Blocks

A logic block within a system is described as any group of N identical items, or N items performing identical functions, any k of which must be functioning at prescribed times for system success. Any or all of the logic blocks may be on a prescribed **duty cycle** that defines when they are required to be active or dormant.

The $N - k$ items that are not required to be performing with a logic block may be in an *active* or *standby* state. A logic block may contain monitoring and switching functions to detect faults and replace faulty items with standby items, either automatically or manually. Occurrence of faults may be announced by the system or may be accomplished by a periodic troubleshooting or inspection policy.

Each logic block should be identified in the first column of the worksheet by name or by some identifying number or symbol.

9.5.3 Block Diagram Construction

The reliability block diagram is drawn on the worksheet vertically, as shown in Figure 9.5. The block diagram is composed of a series of logic blocks that can be presented in any order. Because these are reliability and not functional block diagrams, they need not be presented in functional order. Samples of symbols to be applied to the block diagrams are illustrated in Figure 9.4 (a)–(d).

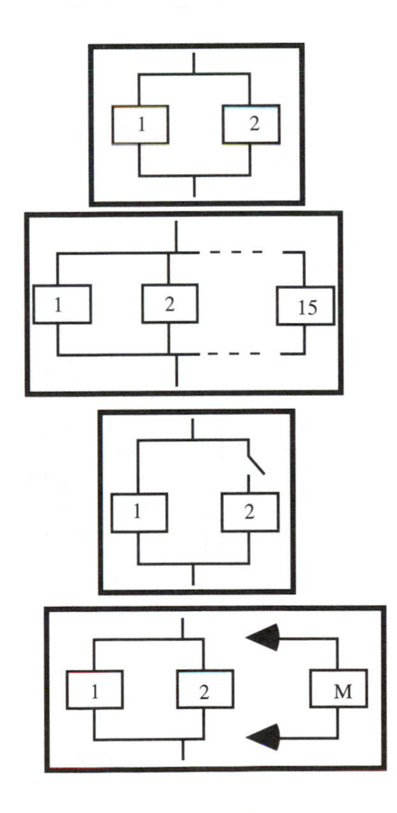

FIGURE 9.4
(a) Two on-line items with identical functions.

(b) Fifteen on-line items with identical functions.

(c) Two items with identical functions, one on-line and one in standby.

(d) Two on-line items with identical functions plus a monitoring or fault-detection function.

Logic Block	Block Diagram	Logic	Equation	Item Failure Rate	Item Restoration Rate	System Effective Failure Rate
A	 1	1 required	$\lambda_{sys} = \lambda$	$\lambda = 35 \times 10^{-6}$	------	35×10^{-6}
B	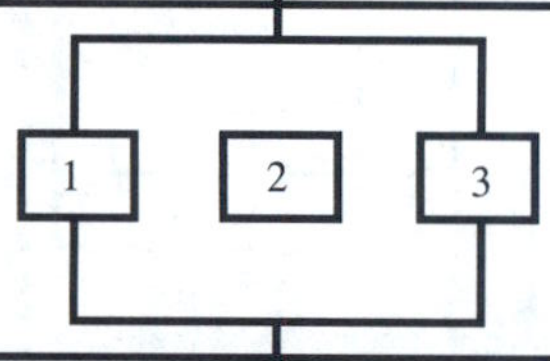 1 2 3	2 of 3 on-line restoration DF = 40%	$\lambda_{sys} = 0.4 \times \frac{6\lambda^2}{\mu + 5\lambda}$	$\lambda = 125 \times 10^{-6}$	$\mu = 0.1$	0.35×10^{-6}
C	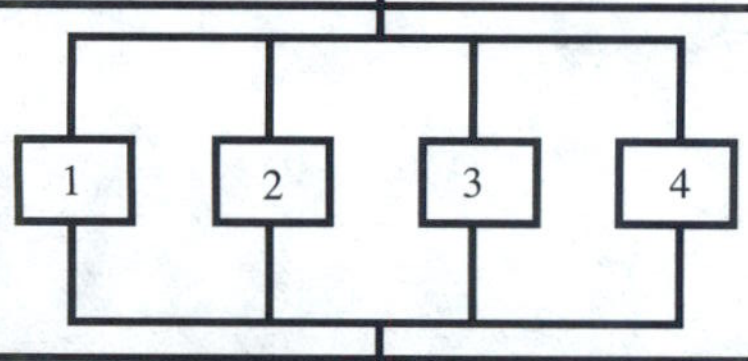 1 2 3 4	3 of 4 on-line restoration	$\lambda_{sys} = \frac{12\lambda^2}{\mu + 7\lambda}$	$\lambda = 220 \times 10^{-6}$	$\mu = 0.5$	1.20×10^{-6}
D	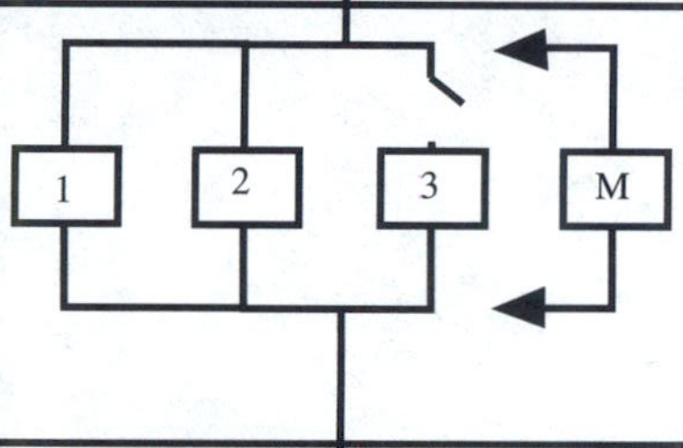 1 2 3 M	2 of 3 1 on-line 1 standby restoration sensing switch required	$\lambda_{sys} = \frac{2\lambda(2\lambda + [1 - p_s]\mu)}{\mu + 2(1 + p_s)\lambda}$ $p_s = \frac{(1 - \lambda_{sw})\mu_M}{\mu_M + \lambda_M}$	$\lambda = 25 \times 10^{-6}$ $\lambda_M = 100 \times 10^{-6}$ $\lambda_{sw} = 0.001 / cy$	$\mu = 1.0$ $\mu_M = 2.0$	0.06×10^{-6}
E	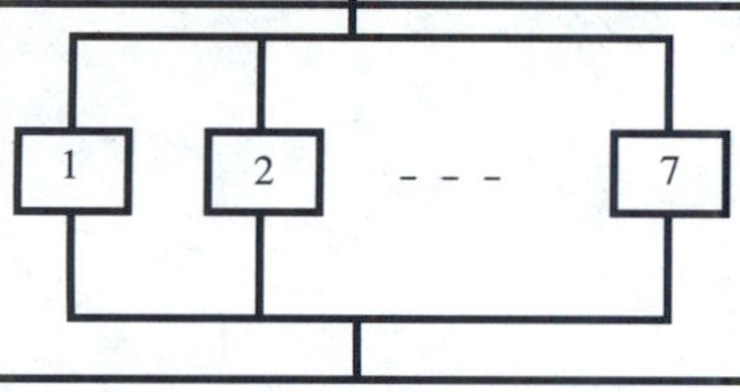 1 2 - - - 7	5 of 7 on-line restoration	$\lambda_{sys} = \frac{210\lambda^3}{\mu^2 + 12\mu\lambda + 107\lambda^2}$	$\lambda = 500 \times 10^{-6}$	$\mu = 1.0$	2.50×10^{-6}

$\Sigma\lambda_{sys} = 39.13 \times 10^{-6}$
MTBF $= 1/\Sigma\lambda_{sys} = 25{,}000$ h

FIGURE 9.5
Sample worksheet

9.5.4 Statement of Logic

The *logic* column on the worksheet is provided for including precise statements of the logic of each logic block. Table 9.1 describes possible entries for this column.

9.5.5 Equations

After the block diagrams and logic statements are entered in the worksheet, the tables in Appendix C can be used for finding the system effective failure rates for each logic block. Enter the Appendix C table that is applicable to the logic statement, and enter the system effective failure rate equation that corresponds to the block diagram. The equations are written in terms of the parameters λ, λ_k, μ, μ_k, and p_s, the values of which must be estimated and entered in the worksheet in the next two columns. Whenever a

TABLE 9.1
Description of logic statements.

Statement	Description
2 of 3 On-line Restoration	This means there are 3 on-line items, at least 2 of which must be operating. On-line restoration of a failed item is possible.
2 of 3 On-line No restoration	This is the same as the preceding example, but restoration of a failed item is not possible without interruption of the system's function. Recall that when on-line restoration is not possible this technique cannot be guaranteed to adequately estimate system MTBF.
2 of 3 On-line Restoration df = 50%	This is the same as the first example, but the logic block has a duty factor as 50%. That is, it is required to function only half the time that the system is required.
30 of 40 On-line Varying FR Restoration	This means that there are 40 on-line items, at least 30 of which must be operating. On-line restoration of a failed item is possible. As each item fails, the failure rate of the remaining items change.
3 of 5 3 on-line 2 standby Restoration Perfect sensing Perfect switching	This means that there are 5 items, 3 of which are on-line and 2 of which are in standby. 3 items must be operating. On-line restoration of a failed item is possible. Sensing of failed items and switchover to a standby replacement when needed are assumed to be perfect.
3 of 5 3 on-line 2 standby Restoration Switching, sensing required	This is the same as the preceding example, but probabilities of successful sensing and switching must be included in the calculations.

duty factor, df, applies, it should be multiplied by the table equation to derive the system effective failure rate, λ_{sys}, for the logic block.

9.5.6 System Effective Failure Rate

For each logic block on the worksheet, there should now be an equation expressing λ_{sys} as a function of

1. Duty factor, df
2. Box failure rate, λ, or rates $\lambda_1, \lambda_2, \cdots \lambda_k, \cdots$
3. Box restoration rate, μ, or rates $\mu_1, \mu_2, \cdots \mu_k, \cdots$
4. Switchover probability, p_s

There should also be values determined for each of the parameters specified by the λ_{sys} equation. So, the system effective failure rate can be calculated for each logic block. The total system effective failure rate is determined by summing the effective failure rates for all the logic blocks, and the system MTBF, as indicated previously, is the reciprocal of the total system effective failure rate:

$$\text{MTBF} = 1/\Sigma\lambda_{sys}. \qquad \textbf{(9.7)}$$

EXAMPLE 9.20 Estimate the MTBF for the system whose reliability block diagram appears in the sample worksheet of Figure 9.5.

Using the sample worksheet, the system is resolved into five logic blocks, and the block diagrams are drawn and described. From the block diagrams and logic descriptions we can select an appropriate Appendix C system effective failure rate for each logic block. The equation for logic block A comes from Table C.4; the equations for logic blocks B, C, and E are from Table C.1; and the equation for logic block D is from Table C.8. Equation 9.6 is used to determine the p_s value of logic block D. Applying the associated failure and restoration rates to the equations, the system effective failure rates are determined for each logic block. The sum of the logic block system effective failure rates estimates the effective failure rate for the entire system. (Notice that the restrictions presented at the beginning of this section are not violated here.) Its reciprocal gives an estimate for the system MTBF of 25,000 h.

9.6 THE MARKOV MODEL APPROACH TO SOLVING COMPLEX SYSTEM MTBF

The Markov model approach can be used for any system, regardless of complexity. It is usually reserved for systems too complex to be solved using the tables of Appendix C.

9.6.1 MTBF Block Diagrams

We first draw a reliability block diagram of our system and assign each block a failure and on-line restoration rate. Whenever there are standby blocks, switching probabilities and sensing function failure and restoration rates may be assigned as applicable. Then we define and describe all the success states, as illustrated in the following examples. We start with the state where no items are down, then consider all states where one item is down, then two items, etc.

EXAMPLE 9.21 Define and describe the system success states for the system whose block diagram appears in Figure 9.6.

This system has three success states.

Success State	Items Up	Items Down	Items under Repair
1	A, B, C	None	None
2	B, C	A	A
3	A, C	B	B

EXAMPLE 9.22 Define and describe the system success states for the system whose block diagram appears in Figure 9.7.

This system has four success states.

Success State	Items Up	Items Down	Items under Repair
1	A, A, B, B, C	None	None
2	A, B, B, C	A	A
3	A, A, B, C	B	B
4	A, B, C	A, B	A, B

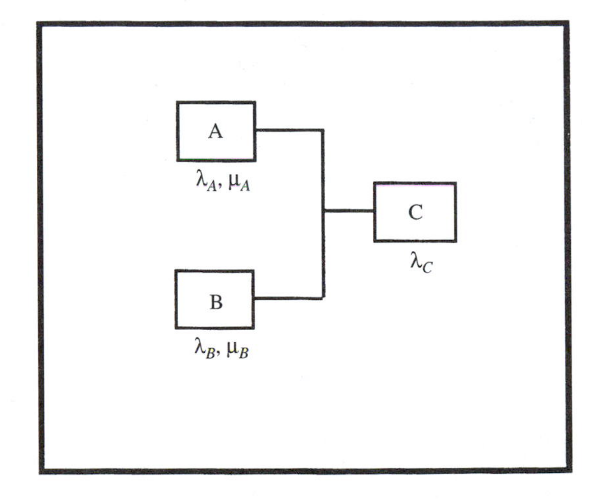

FIGURE 9.6
Block diagram for Example 9.21.

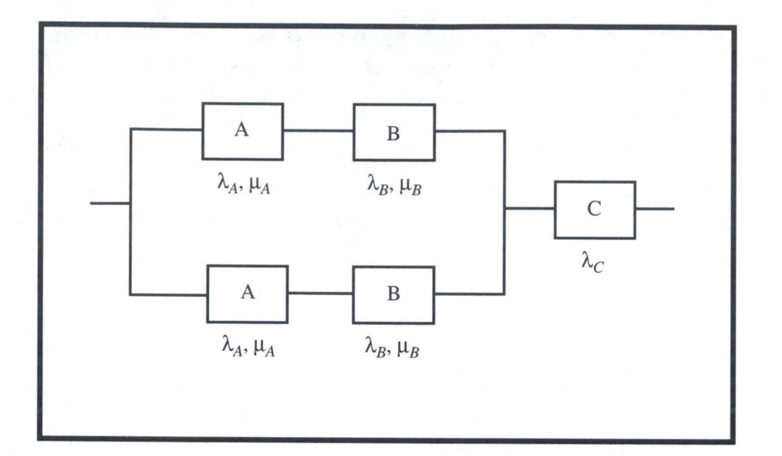

FIGURE 9.7
Block diagram for Example 9.22.

EXAMPLE 9.23 Define and describe the system success states for the system whose block diagram appears in Figure 9.8.

This system has only two success states.

Success State	Items Up	Items Down	Items in Standby	Items under Repair
1	A, A, B	None	A	None
2	A, A, B	A	None	A

FIGURE 9.8
Block diagram for Example 9.23.

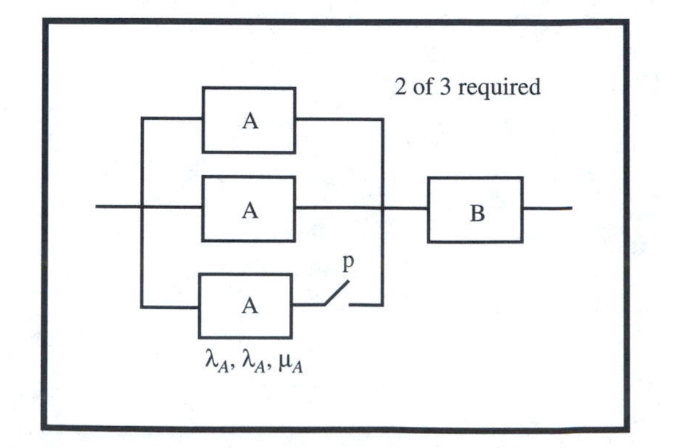

9.6.2 Transition Rate Tables

In addition to the success states defined, consider the existence of a *fail* state F. Now consider what can happen to the system upon a *single event,* that is, the failure of an up item or the restoration of a down item. Transition rate tables indicate where the system will go upon the occurrence of such a single event. It is possible for the system to transition from one state directly into another only if a single event will allow that to happen. The system cannot transition directly from a given state to another given state if no such single event is possible. Consider the three preceding examples.

EXAMPLE 9.24 Identify the state transitions that are possible for the systems of Examples 9.21 through 9.23.

First consider Example 9.21.

If the System Is in State:	And This Event Takes Place:	The System Will Go into State:
1	A fails.	2
1	B fails.	3
1	C fails.	F
2	A is restored.	1
2	B fails.	F
2	C fails.	F
3	B is restored.	1
3	A fails.	F
3	C fails.	F

Next, consider Example 9.22.

If the System Is in State:	And This Event Takes Place:	The System Will Go into State:
1	An A fails.	2
1	A B fails.	3
1	C fails.	F
2	The failed A is restored.	1
2	A B fails.	4
2	C fails.	F
2	The remaining A fails.	F
3	An A fails.	4
3	The failed B is restored.	1
3	The remaining B fails.	F
3	C fails.	F
4	The failed A is restored.	3
4	The failed B is restored.	2
4	The remaining A fails.	F
4	The remaining B fails.	F
4	C fails.	F

Finally, consider Example 9.23.

If the System Is in State:	And This Event Takes Place:	The System Will Go into State:
1	An active A fails and the switch works.	2
1	The standby A fails.	2
1	An active A fails and the switch does not work.	F
1	B fails.	F
2	A is restored.	1
2	An active A fails.	F
2	B fails.	F

Now that we have resolved the transitions possible among the states, we can establish tables to determine the transition rates.

EXAMPLE 9.25 Construct transition-rate tables for Examples 9.21, 9.22, and 9.23.

The transition rate table for Example 9.21 is as follows.

To State i	From State j	Transition Rate r_{ij}
2	1	λ_A
3	1	λ_B
F	1	λ_C
1	2	μ_A
F	2	$\lambda_B + \lambda_C$
1	3	μ_B
F	3	$\lambda_A + \lambda_C$

Notice that the system goes from state 2 to state F if either B or C fails. Hence, the transition rate from state 2 to state F is $\lambda_B + \lambda_C$.

The transition-rate table for Example 9.22 is the following:

To State i	From State j	Transition Rate r_{ij}
2	1	$2\lambda_A$
3	1	$2\lambda_B$
F	1	λ_C
1	2	μ_A
4	2	$2\lambda_B$
F	2	$\lambda_A + \lambda_C$
1	3	μ_B
4	3	$2\lambda_A$
F	3	$\lambda_B + \lambda_C$
2	4	μ_B
3	4	μ_A
F	4	$\lambda_A + \lambda_B + \lambda_C$

Notice that the system goes from state 1 to state 2 if either of the two items A fail. Hence, the transition rate from state 1 to state 2 is $2\lambda_A$.

Finally, the following is the transition-rate table for Example 9.23.

To State i	From State j	Transition Rate r_{ij}
2	1	$2p\lambda_A + \lambda'_A$
F	1	$2(1 - p)\lambda_A + \lambda_B$
1	2	μ_A
F	2	$2\lambda_A + \lambda_B$

We use the symbol r_{jj} to refer to the sum of the rates out of the state j. As examples, in Example 9.21, where there are three success states.

$$
\begin{aligned}
r_{11} &= \lambda_A + \lambda_B + \lambda_C \\
r_{22} &= \mu_A + \lambda_B + \lambda_C \\
r_{33} &= \mu_B + \lambda_A + \lambda_C
\end{aligned}
$$

In Example 9.22, where there are four success states,

$$
\begin{aligned}
r_{11} &= 2\lambda_A + 2\lambda_B + \lambda_C \\
r_{22} &= \mu_A + \lambda_A + 2\lambda_B + \lambda_C \\
r_{33} &= \mu_B + 2\lambda_A + \lambda_B + \lambda_C \\
r_{44} &= \mu_A + \mu_B + \lambda_A + \lambda_B + \lambda_C
\end{aligned}
$$

In Example 9.23, where there are two success states,

$$r_{11} = 2\lambda_A + \lambda'_A + \lambda_B$$
$$r_{22} = \mu_A + 2\lambda_A + \lambda_B$$

9.6.3 MTBF Solution

As derived mathematically in Sections 9.7–9.9 (if you are following the asterisked sections), for a system with N success states the system MTBF is

$$\text{MTBF} = x_1 + x_2 + \cdots + x_N \tag{9.8}$$

where the x_i values are the solution of the following equations:

$$\begin{aligned} -r_{11}x_1 + r_{12}x_2 + r_{13}x_3 + \cdots + r_{1N}x_N &= -1 \\ r_{21}x_1 - r_{22}x_2 + r_{23}x_3 + \cdots + r_{2N}x_N &= 0 \\ r_{31}x_1 + r_{32}x_2 - r_{33}x_3 + \cdots + r_{3N}x_N &= 0 \\ &\vdots \\ r_{N1}x_1 + r_{N2}x_2 + r_{N3}x_3 + \cdots - r_{NN}x_N &= 0 \end{aligned} \tag{9.9}$$

EXAMPLE 9.26 Solve the system MTBF for Example 9.21 if $\lambda_A = 5.0 \times 10^{-6}$ per hour, $\lambda_B = 25.0 \times 10^{-6}$ per hour, $\lambda_C = 2.0 \times 10^{-6}$ per hour, $\mu_A = 1.0$ per hour, and $\mu_B = 2.0$ per hour.

From Example 9.25,

$$\begin{aligned} r_{21} &= \lambda_A = 5.0 \times 10^{-6} \text{ per hour} \\ r_{31} &= \lambda_B = 25.0 \times 10^{-6} \text{ per hour} \\ r_{12} &= \mu_A = 1.0 \text{ per hour} \\ r_{13} &= \mu_B = 2.0 \text{ per hour} \end{aligned}$$

As before,

$$\begin{aligned} r_{11} &= \lambda_A + \lambda_B + \lambda_C = 0.000032 \text{ per hour} \\ r_{22} &= \mu_A + \lambda_B + \lambda_C = 1.000027 \text{ per hour} \\ r_{33} &= \mu_B + \lambda_A + \lambda_C = 2.000007 \text{ per hour.} \end{aligned}$$

The other rates are $r_{23} = r_{32} = 0$.

The system equations become

$$\begin{aligned} -0.000032x_1 + 1.0x_2 + 2.0x_3 &= -1 \\ 0.000005x_1 - 1.000027x_2 &= 0 \\ 0.000025x_1 - 2.000007x_3 &= 0 \end{aligned}$$

The algebraic solutions are

$$\begin{aligned} x_1 &= 500{,}000.00 \text{ h} \\ x_2 &= 2.50 \text{ h} \\ x_3 &= 6.25 \text{ h} \end{aligned}$$

From Equation 9.8,

$$\text{MTBF} = x_1 + x_2 + x_3 = 500{,}008.75 \text{ h}$$

of which an average of 500,000 h are spent in State 1, 2.5 h in State 2, and 6.25 h in State 3.

EXAMPLE 9.27 Solve the system MTBF for Example 9.22 if $\lambda_A = 0.005$ per hour, $\lambda_B = 0.025$ per hour, $\lambda_C = 0.002$ per hour, $\mu_A = 1.0$ per hour, and $\mu_B = 2.0$ per hour.

From Example 9.25,

$$\begin{aligned}
r_{21} &= 2\lambda_A = 0.010 \text{ per hour} \\
r_{31} &= 2\lambda_B = 0.050 \text{ per hour} \\
r_{12} &= \mu_A = 1.000 \text{ per hour} \\
r_{42} &= 2\lambda_B = 0.050 \text{ per hour} \\
r_{13} &= \mu_B = 2.000 \text{ per hour} \\
r_{43} &= 2\lambda_A = 0.010 \text{ per hour} \\
r_{24} &= \mu_B = 2.000 \text{ per hour} \\
r_{34} &= \mu_A = 1.000 \text{ per hour}
\end{aligned}$$

As before,

$$\begin{aligned}
r_{11} &= 2\lambda_A + 2\lambda_B + \lambda_C = 0.062 \text{ per hour} \\
r_{22} &= \mu_A + \lambda_A + 2\lambda_B + \lambda_C = 1.057 \text{ per hour} \\
r_{33} &= \mu_B + 2\lambda_A + \lambda_B + \lambda_C = 2.037 \text{ per hour} \\
r_{44} &= \mu_A + \mu_B + \lambda_A + \lambda_B + \lambda_C = 3.032 \text{ per hour}
\end{aligned}$$

The system equations become

$$\begin{aligned}
-0.062x_1 + 1.000x_2 + 2.000\,x_3 \qquad\qquad &= -1.000 \\
0.010x_1 - 1.057x_2 \qquad\qquad + 2.000x_4 &= 0.000 \\
0.050x_1 \qquad\qquad - 2.037x_3 - 1.000x_4 &= 0.000 \\
0.050x_2 + 0.010x_3 - 3.032x_4 &= 0.000
\end{aligned}$$

The algebraic solutions are

$$\begin{aligned}
x_1 &= 364.5 \text{ h} \\
x_2 &= 3.6 \text{ h} \\
x_3 &= 9.0 \text{ h} \\
x_4 &= 0.1 \text{ h}
\end{aligned}$$

From Equation 9.8,

$$\text{MTBF} = x_1 + x_2 + x_3 + x_4 = 377.2 \text{ h}$$

of which an average of 364.5 h are spent in state 1, 3.6 h in state 2, 9.0 h in state 3, and 0.1 h in state 4.

EXAMPLE 9.28 Solve the system MTBF for Example 9.23 if $\lambda_A = 0.005$ per hour, $\lambda'_A = 0.0001$ per hour, $\lambda_B = 0.0005$ per hour, $\mu_A = 1.0$ per hour, and $p = 0.90$.

From Example 9.25,

$$r_{21} = 2p\lambda_A + \lambda'_A = 0.0091 \text{ per hour}$$
$$r_{12} = \mu_A = 1.0000 \text{ per hour}$$

As before,

$$r_{11} = 2\lambda_A + \lambda'_A + \lambda_B = 0.0106 \text{ per hour}$$
$$r_{22} = \mu_A + 2\lambda_A + \lambda_B = 1.0105 \text{ per hour}$$

The system equations become

$$-0.0106x_1 + 1.0000x_2 = -1.0000$$
$$0.0091x_1 - 1.0105x_2 = 0.0000$$

The algebraic solutions are

$$x_1 = 627.12 \text{ h}$$
$$x_2 = 5.65 \text{ h}$$

From Equation 9.8,

$$\text{MTBF} = x_1 + x_2 = 632.8 \text{ h}$$

of which an average of 627.12 h are spent in state 1 and 5.65 h are spent in State 2.

EXAMPLE 9.29 Determine the MTBF for a system composed of an active element, an identical standby element, and a monitoring subsystem, as shown in Figure 9.9. The active element failure rate is $\lambda = 0.005$ per hour, the standby failure rate for the element is $\lambda' = 0.0001$ per hour, the restoration rate for the element is $\mu = 2.0$ per hour, the monitoring subsystem has a failure rate $\lambda_M = 0.01$ per hour and restoration rate $\mu_M = 0.20$ per hour, and the switching probability is $p = 0.90$.

This system has four success states, as follows:

Success State	Items Up	Items Down	Items in Standby	Items under Restoration
1	1, M	0	1	0
2	1	M	1	M
3	1, M	1	0	1
4	1	1, M	0	1
				Assume that policy is to restore the element before restoring the monitor.

If the System Is in State:	And This Event Takes Place:	The System Will Go to State:
1	Monitor fails.	2
1	Active element fails and switch works.	3
1	Standby element fails.	3
1	Active element fails and switch does not work.	F
2	Monitor is restored.	1
2	Standby element fails.	4
2	Active element fails.	F
3	Failed element is restored.	1
3	Monitor fails.	4
3	Remaining active element fails.	F
4	Failed element is restored.	2
4	Remaining active element fails.	F

The transition rates are as follows:

To State i	From State j	Transition Rate r_{ij}
2	1	$\lambda_M = 0.0100$
3	1	$p\lambda + \lambda' = 0.0046$
F	1	$(1 - p)\lambda = 0.0005$
1	2	$\mu_M = 0.2000$
4	2	$\lambda' = 0.0001$
F	2	$\lambda = 0.0050$
1	3	$\mu = 2.0000$
4	3	$\lambda_M = 0.0100$
F	3	$\lambda = 0.0050$
2	4	$\mu = 2.0000$
F	4	$\lambda = 0.0050$

The sums of the rates out of each state are

$$\lambda_{11} = 0.0151$$
$$\lambda_{22} = 0.2051$$
$$\lambda_{33} = 2.0150$$
$$\lambda_{44} = 2.0050$$

The set of equations becomes

$$\begin{aligned}
-0.0151x_1 + 0.0200x_2 + 2.0000x_3 \qquad\qquad &= -1.0000 \\
0.0100x_1 - 0.2015x_2 \qquad\qquad + 2.0000x_4 &= 0.0000 \\
0.0046x_1 \qquad\qquad - 2.0150x_3 \qquad\qquad &= 0.0000 \\
0.0001x_2 + 0.0100x_3 - 2.0050x_4 &= 0.0000
\end{aligned}$$

FIGURE 9.9
Block diagram for Example 9.29.

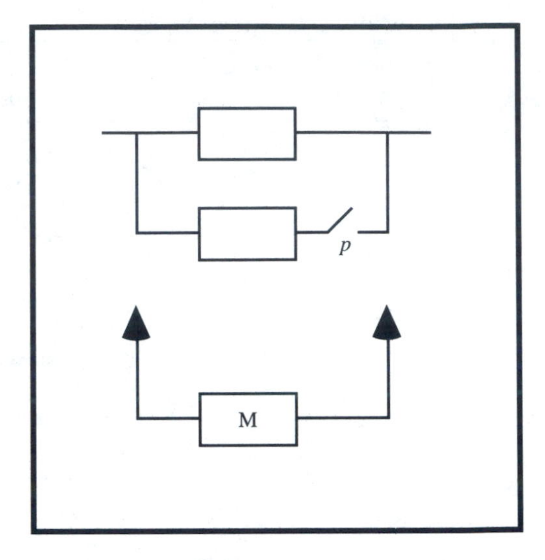

The solutions are

$$\begin{aligned} x_1 &= 1323.5 \text{ h} \\ x_2 &= 64.7 \text{ h} \\ x_3 &= 3.0 \text{ h} \\ x_4 &= 0.02 \text{ h} \end{aligned}$$

From Equation 9.8,

$$\text{MTBF} = x_1 + x_2 + x_3 + x_4 = 1391.2 \text{ h}$$

EXAMPLE 9.30 Use the tables of Appendix C to estimate the MTBF of the system of Example 9.29.

From Equation 9.6,

$$p_s = \frac{(1 - \lambda_{\text{sw}})\mu_M}{\mu_M + \lambda_M} = \frac{p \cdot \mu_M}{\mu_M + \lambda_M} = \frac{0.90 \times 0.20}{0.21} = 0.857$$

Neglect λ:

Table C.8 applies, with $N = 2$, $k = 1$, $\lambda = 0.005$ per hour, $\mu = 2.0$ per hour, and $p_s = 0.857$.

$$\text{MTBF} = \frac{\mu + (1 + p_s)\lambda}{[\lambda + (1 - p_s)\mu]\lambda} = \frac{2.000 + 1.857(0.005)}{[0.005 + (0.143)(2.000)]0.005} \text{ h} = 1381 \text{ h}$$

In the examples used, we have not seen more than four states. However, with more states there are more equations, and a computer solution for the system of equations will be necessary. Otherwise, we may choose to reduce the number of states by combining and eliminating them and working toward both an optimistic and pessimistic MTBF estimate, knowing that the actual solution is somewhere in between. This strategy can also be used with computer solutions when we want to reduce the number of equations. If both optimistic and pessimistic estimates are reasonably close to one another, an acceptable MTBF estimate will result. The following example demonstrates this technique.

EXAMPLE 9.31 Estimate the MTBF for the system of Example 9.22 by reducing the number of states. See Figure 9.10.

$$\lambda_A = 0.005 \text{ per hour}$$
$$\lambda_B = 0.025 \text{ per hour}$$
$$\lambda_C = 0.002 \text{ per hour}$$
$$\mu_A = 1.000 \text{ per hour}$$
$$\mu_B = 2.000 \text{ per hour}$$

Consider only the success states where no more than one item is down.

Success State	Items Up	Items Down	Items under Repair
1	A, A, B, B, C	0	0
2	A, B, B, C	A	A
3	A, A, B, C	B	B

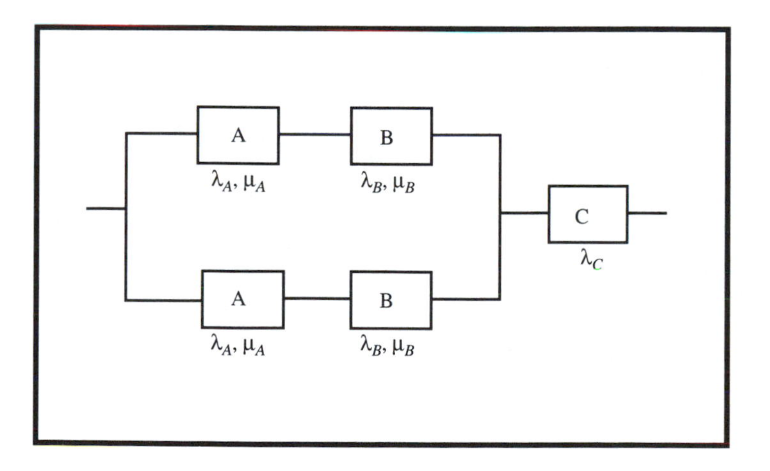

FIGURE 9.10
Block diagram for Example 9.31.

For an *optimistic* estimate, assume that no other success states exist and that the only way to enter the fail state is directly from one of the three states considered. The transition rate table is, therefore, the same as that presented in Example 9.25, with transitions involving only the first three states considered.

To State i	From State j	Transition Rate r_{ij}
2	1	$2\lambda_A = 0.010$
3	1	$2\lambda_B = 0.050$
F	1	$\lambda_C = 0.002$
1	2	$\mu_A = 1.000$
F	2	$\lambda_A + \lambda_C = 0.007$
1	3	$\mu_B = 2.000$
F	3	$\lambda_B + \lambda_C = 0.027$

The rates out of each state are

$$r_{11} = 0.062$$
$$r_{22} = 1.007$$
$$r_{33} = 2.027$$

There are three equations in the set:

$$\begin{aligned} -0.062x_1 + 1.000x_2 + 2.000x_3 &= -1.000 \\ 0.010x_1 - 1.007x_2 &= 0.000 \\ 0.050x_1 - 2.027x_3 &= 0.000 \end{aligned}$$

The algebraic solutions are

$$x_1 = 365.47 \text{ h}$$
$$x_2 = 3.63 \text{ h}$$
$$x_3 = 9.02 \text{ h}$$

From Equation 9.8,

$$\text{MTBF} = x_1 + x_2 + x_3 = 378.1 \text{ h}$$

So we can conclude that MTBF $<$ 378 h.

For a *pessimistic* estimate, assume that all states beyond the three considered are part of the fail state. Hence, the transition rate table is as before, but all rates out of states are considered, and the rates into the former state 4 (in Example 9.25) become rates into state F.

To State i	From State j	Transition Rate r_{ij}
2	1	$2\lambda_A = 0.010$
3	1	$2\lambda_B = 0.050$
F	1	$\lambda_C = 0.002$
1	2	$\mu_A = 1.000$
F	2	$2\lambda_B + \lambda_A + \lambda_C = 0.057$
1	3	$\mu_B = 2.000$
F	3	$2\lambda_A + \lambda_B + \lambda_C = 0.037$

The rates out of each state are

$$r_{11} = 0.062$$
$$r_{22} = 1.057$$
$$r_{33} = 2.037$$

There are three equations in the set:

$$\begin{aligned} -0.062x_1 + 1.000x_2 + 2.000x_3 &= -1.000 \\ 0.010x_1 - 1.057x_2 \qquad\qquad &= \quad 0.000 \\ 0.050x_1 \qquad\qquad - 2.037x_3 &= \quad 0.000 \end{aligned}$$

The algebraic solutions are

$$x_1 = 290.0 \text{ h}$$
$$x_2 = 2.7 \text{ h}$$
$$x_3 = 7.1 \text{ h}$$

From Equation 9.8,

$$\text{MTBF} = x_1 + x_2 + x_3 = 299.8 \text{ h}$$

So we can conclude that MTBF > 300 h.

Based on our optimistic and pessimistic estimates, we know that the MTBF is somewhere between 300 and 378 hours. If we are satisfied with that degree of uncertainty, we can report the MTBF as being, say, at the midrange of the interval, that is, around 340 or 350 h. Compare this with the result of Example 9.27.

---

*9.7 DERIVATION OF THE MARKOV MODEL SOLUTIONS

Before presenting this very powerful tool, let us introduce two analytic tools that are necessary for its development: moment-generating functions and Laplace transforms.

*9.7.1 Moments

In Section A.1 (Appendix A) we discuss the *statistical analysis method,* in which we use (in the second step) statistics (collected data points) to estimate parameters of an assumed probability law. Recall that a **probability law** is defined by a **density function** $f(x)$ or **mass function** $p(x)$, depending, respectively, on whether the phenomenon in question is continuous or discrete. We can use *moments about the mean* as parameters of a probability law. **Moments about the mean** are defined as follows.

Define the **expected value of the random variable** x

$$
\begin{aligned}
E[x] &= \int_{-\infty}^{\infty} x f(x)\, dx, && \text{for continuous phenomena} \\
E[x] &= \sum_{\text{all } i\text{'s}} x_i p(x_i), && \text{for discrete phenomena}
\end{aligned}
\tag{9.10}
$$

$E[x]$ is the **mean** of the probablity law governing x:

$$
E[x] = \mu \tag{9.11}
$$

The **rth moment about the mean** is defined as

$$
\begin{aligned}
E[(x - \mu)^r] &= \int_{-\infty}^{\infty} (x - \mu)^r f(x)\, dx, && \text{for continuous phenomena} \\
E[(x - \mu)^r] &= \sum_{\text{all } i\text{'s}} (x_i - \mu)^r p(x_i), && \text{for discrete phenomena}
\end{aligned}
\tag{9.12}
$$

Specific solutions for the rth moment about the mean are as follows:

- The **zero-order moment about the mean** is the entire area under the density function curve for continuous phenomena or the sum of all possible outcomes for discrete phenomena. Therefore, the zero-order moment about the mean is always unity:

$$
\begin{aligned}
E[x - \mu)^0] &= E[1.0] = 1.0 \\
E[(x - \mu)^0] &= \int_{-\infty}^{\infty} f(x)\, dx = 1.0 \\
E[(x - \mu)^0] &= \sum_{\text{all } i\text{'s}} p(x_i) = 1.0
\end{aligned}
\tag{9.13}
$$

- The **first-order moment about the mean** is zero:

$$
\begin{aligned}
E[(x - \mu)^1] &= E[x] - \mu = \mu - \mu = 0 \\
E[(x - \mu)^1] &= \int_{-\infty}^{\infty} (x - \mu) f(x)\, dx \\
&= \int_{-\infty}^{\infty} x f(x)\, dx - \mu \int_{-\infty}^{\infty} f(x)\, dx \\
&= \mu - \mu[1.0] = 0 \\
E[(x - \mu)^1] &= \sum_{\text{all } i\text{'s}} (x_i - \mu) p(x_i) \\
&= \sum_{\text{all } i\text{'s}} x_i p(x_i) - \mu \sum_{\text{all } i\text{'s}} p(x_i) \\
&= \mu - \mu[1] = 0
\end{aligned}
\tag{9.14}
$$

- The **second-order moment about the mean** is defined as the *variance:*

$$
\begin{aligned}
E[(x - \mu)^2] &= E[x^2 - 2\mu x + \mu^2] \\
&= E[x^2] - 2\mu E[x] + \mu^2 \\
&= E[x^2] - 2\mu^2 + \mu^2 \\
&= E[x^2] - \mu^2 = E[x^2] - \{E[x]\}^2 \\
&\qquad\qquad \equiv \text{var}\,(x) \equiv \sigma_x^2
\end{aligned}
\tag{9.15}
$$

$$E[(x - \mu)^2] = \int_{-\infty}^{\infty} (x - \mu)^2 f(x)\, dx \equiv \sigma_x^2, \qquad \text{for continuous phenomena}$$

$$E[(x - \mu)^2] = \sum_{\text{all } i\text{'s}} (x - \mu)^2 p(x_i) \equiv \sigma_x^2, \qquad \text{for discrete phenomena}$$

*9.7.2 Moment-Generating Functions

Moment-generating functions were invented to simplify computations of moments. They eliminate the need for complex integration operations by making use of the integral $\int e^x\, dx = e^x$ or $(d/dx)e^x = e^x$.

Define the **moment-generating function**

$$
\begin{aligned}
M_x(\theta) &\equiv E[e^{\theta x}] = \int_{-\infty}^{\infty} e^{\theta x} f(x)\, dx, &&\text{for continuous phenomena} \\
M_x(\theta) &\equiv E[e^{\theta x}] = \sum_{i=0}^{\infty} e^{\theta x_i}\, p(x_i), &&\text{for discrete phenomena}
\end{aligned}
\tag{9.16}
$$

Expanding the exponential in Equation 9.16,

$$
\begin{aligned}
M_x(\theta) \equiv E[e^{\theta x}] &= E\left[\sum_{i-0}^{\infty} \frac{(\theta x)^i}{i!}\right] = E\left[1 + (\theta x) + \frac{(\theta x)^2}{2!} + \frac{(\theta x)^3}{3!} + \cdots + \frac{(\theta x)^n}{n!} + \cdots\right] \\
&= 1 + \theta E[x] + \frac{\theta^2}{2!} E[x^2] + \frac{\theta^3}{3!} E[x^3] + \cdots + \frac{\theta^n}{n!} E[x^n] \cdots
\end{aligned}
\tag{9.17}
$$

If we take successive derivatives of the moment-generating function with respect to θ, we get the following results:

$$
\begin{aligned}
\frac{d}{d\theta} M_x(\theta) &= 0 + E[x) + \theta E[x^2] + \frac{\theta^2}{2!} E[x^3] + \cdots + \frac{\theta^{n-1}}{(n-1)!} + \cdots \\
&= E[x] + \theta E[x^2] + \frac{\theta^2}{2!} E[x^3] + \cdots
\end{aligned}
\tag{9.18}
$$

$$\frac{d^2}{d\theta} M_x(\theta) = E[x^2] + \theta E[x^3] + \frac{\theta^2}{2!} E[x^4] + \cdots \tag{9.19}$$

$$\frac{d^2}{d\theta} M_x(\theta) = E[x^3] + \theta E[x^4] + \frac{\theta^2}{2!} E[x^5] + \cdots \tag{9.20}$$

$$\vdots$$

$$
\begin{aligned}
\frac{d^n}{d\theta} M_x(\theta) &= E[x^n] + \theta E[x^{n+1}] + \frac{\theta^2}{2!} E[x^{n+2}] + \frac{\theta^3}{3!} E[x^{n+3}] + \cdots + \frac{\theta^n}{n!} E[x^{2n}] + \cdots \\
&\vdots
\end{aligned}
\tag{9.21}
$$

Now for each equation from 9.17 through 9.21, allow θ to approach zero:

$$[M_x(\theta)]|_{\theta=0} = 1.0 \tag{9.22}$$

$$[M'_x(\theta)]|_{\theta=0} = E[x] \tag{9.23}$$

$$[M'_x(\theta)]|_{\theta=0} = E[x^2] \tag{9.24}$$

$$\vdots$$

$$\left[\frac{d^n}{d\theta} M_x(\theta)\right]\Bigg|_{\theta=0} = E[x^n] \tag{9.25}$$

Notice that the mean of the probability distribution of the random variable x is

$$E[x] = \left[\frac{d}{dx} M_x(\theta)\right]\Bigg|_{\theta=0} \tag{9.26}$$

and the variance of the distribution is (see Equations 9.15, 9.23, and 9.24)

$$E[x^2] - \{E[x]\}^2 = \left[\frac{d^2}{dx} M_x(\theta) - \left\{\frac{d}{dx} M_x(\theta)\right\}^2\right]\Bigg|_{\theta=0} \tag{9.27}$$

EXAMPLE 9.32 Use moment-generating functions to find the mean and variance of the exponental probability law.

The exponential probability law governs continuous phenomena with density function $f(x) = \mu e^{-\mu x}$, $0 < x < \infty$. Then,

$$M_x(\theta) = \int_0^\infty e^{\theta x} f(x)\, dx = \int_0^\infty e^{\theta x} \cdot \mu e^{-\mu x}\, dx = \mu \int_0^\infty e^{(\theta-\mu)x}\, dx = \mu\left[\frac{-1}{\theta - \mu}\right] = \frac{\mu}{\mu - \theta}$$

Notice that Equation 8.24 was used to solve the integration.

Now we can solve the following:

$$E[x^0] = [M_x(\theta)]|_{\theta=0} = \frac{\mu}{\mu - 0} = 1, \qquad \text{which we expect}$$

$$E[x^1] = [M'_x(\theta)]|_{\theta=0} = \left[\frac{d}{d\theta}\left(\frac{\mu}{\mu - \theta}\right)\right]\Bigg|_{\theta=0} = \left[\mu\frac{d}{d\theta}(\mu - \theta)^{-1}\right]\Bigg|_{\theta=0} = \left[-\frac{\mu}{(\mu - \theta)^2}\right]\Bigg|_{\theta=0} = \frac{1}{\mu}$$

$$E[x^{12}] = [M''_x(\theta)]_{\theta=0} = \left[\frac{d}{d\theta}\left(-\frac{\mu}{(\mu - \theta)^2}\right)\right]\Bigg|_{\theta=0} = \left[-\mu\frac{d}{d\theta}(\mu - \theta)^{-2}\right]\Bigg|_{\theta=0} = \left[\frac{2\mu}{(\mu - \theta)^3}\right]\Bigg|_{\theta=0} = \frac{2}{\mu^2}$$

The mean of the exponential probability law is $E[x] = 1/\mu$. The variance is $\sigma^2 = E[x^2] - \{E[x]\}^2 = 2/\mu^2 - 1/\mu^2 = 1/\mu^2$, and the standard deviation is $\sigma = \sqrt{E[x^2] - \{E[x]\}^2} = \sqrt{1/\mu^2} = 1/\mu$.

*9.7.3 Laplace Transforms

Laplace transforms are commonly used as an aid in solving certain complex differential equations. By making use of the relation $de^x = e^x\,dx$, we can use Laplace transforms to convert differential equations into linear equations, upon which we can perform operations. We can then transform back to obtain a solution to the differential equation in the original domain. The Laplace function of a time funtion is defined as

$$\mathcal{L}\{f(t)\} \equiv F(s) \equiv \int_0^{\infty} f(t)e^{-st}\,dt \tag{9.28}$$

Compare this equation with Equation 9.16, and notice that the Laplace function is the same as the moment-generating function M_t $(\theta = -s)$, $0 < t < \infty$. We shall use this fact when solving for MTBF by means of the Markov model technique.

Table 9.2 lists the most commonly used Laplace functions $\mathcal{L}f(t)$ of various forms of $f(t)$. The transforms can be used to obtain solutions to complex differential equations by the technique illustrated in Figure 9.11:

- Start with a differential equation involving the time function $f(t)$.
- Determine the Laplace function of the differential equation. This transforms the differential equation into a linear equation involving the transformed function $F(s)$.
- Perform the algebraic manipulations to solve for $F(s)$.
- Perform the inverse transform to solve $f(t) = \mathcal{L}^{-1}F(s)$.

TABLE 9.2
Common Laplace transforms.

$f(t)$	$\mathcal{L}f(t)$
1. $f(t)$	1. $F(s) = \int_0^{\infty} f(t)e^{-st}\,dt$
2. $df(t)/dt$	2. $sF(s) - f(0)$
3. $d^2f(t)/dt$	3. $s^2F(s) - sf(0) - df(0)/dt$
4. $d^nf(t)/dt$	4. $s^nF(s) - s^{n-1}f(0) - \cdots - d^{n-1}f(0)/dt$
5. $f(at)$	5. $(1/a)F(s/a)$
6. t	6. $1/s^2$
7. e^{-at}	7. $1/(s+a)$
8. $(1/a)[1 - e^{-at}]$	8. $1/[s(s+a)]$
9. $(1/a^2)[e^{-at} + at - 1]$	9. $1/[s^2(s+a)]$
10. $(1/a^2)[(1/a) - t + (at^2/2) - (e^{-at}/a)]$	10. $1/[s^3(s+a)]$
11. $t^{n-1}e^{-at}/(n-1)!$	11. $1/(s+a)^n$
12. $[1/(b-a)][e^{-at} - e^{-bt}]$	12. $1/[(s+a)(s+b)]$
13. $(1/ab)[1 - (1/a - b)(be^{-at} - ae^{-bt})]$	13. $1/[s(s+a)(s+b)]$
14. $tf(t)$	14. $-dF(s)/ds$
15. a	15. a/s
16. 1	16. $1/s$

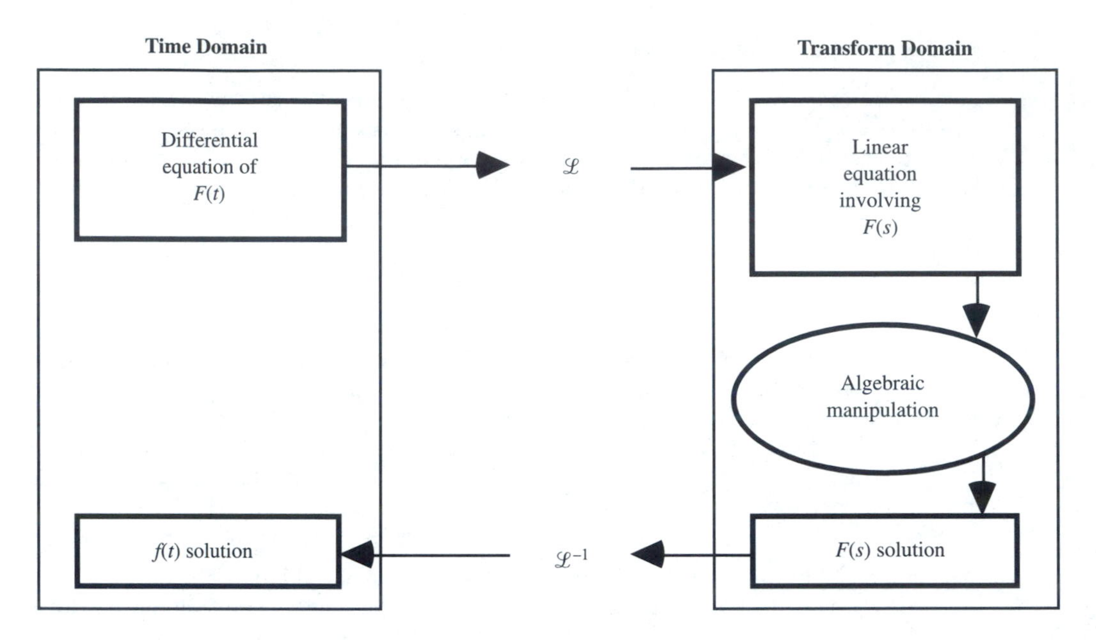

FIGURE 9.11
Laplace transform technique for solving differential equations.

EXAMPLE 9.33

Use Laplace transforms to solve the differential equation

$$\frac{df(t)}{dt} + 2f(t) = e^{-t}$$

with the boundary condition $f(0) = 0$.

From transform 2 in Table 9.2: $\mathscr{L}\frac{df(t)}{dt} = sF(s) - f(0)$

From transform 1: $\mathscr{L}f(t) = F(s)$

From transform 7: $\mathscr{L}e^{-t} = \frac{1}{s+1}$

The Laplace transform of the differential equation $df(t)/dt + 2f(t) = e^{-t}$ is

$$sF(s) - f(0) + 2F(s) = \frac{1}{s+1}$$

Applying the boundary condition $f(0) = 0$, we have $(s + 2)F(s) = 1/(s + 1)$. This has the algebraic solution

$$F(s) = \frac{1}{(s+1)(s+2)}$$

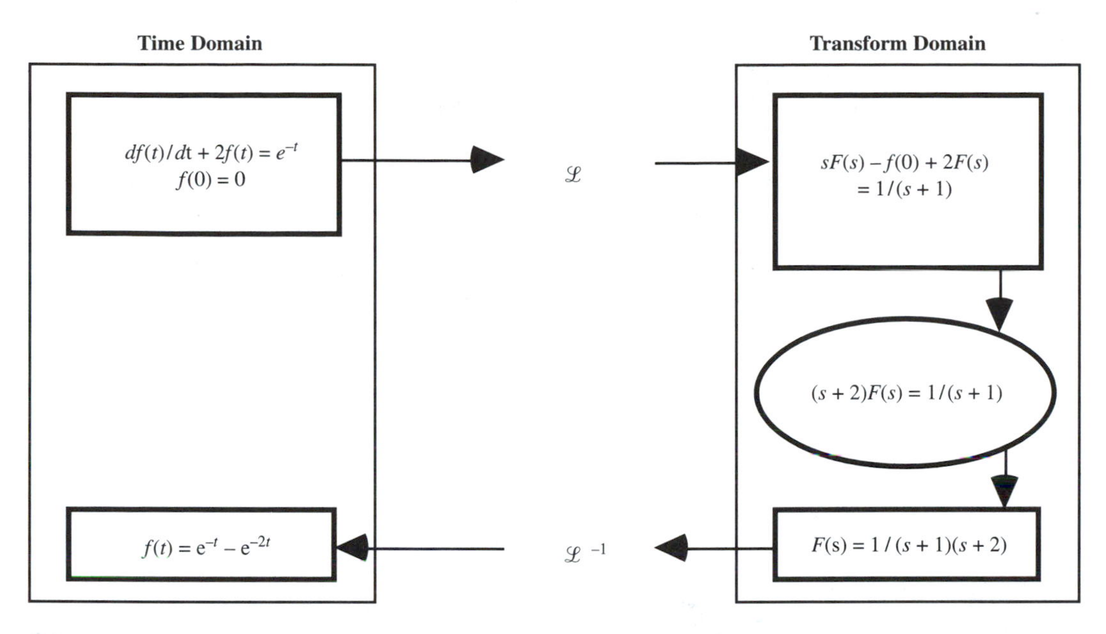

FIGURE 9.12
Laplace transform applied to Example 9.33.

Transform 12 in Table 9.2 can be used for performing the inverse transform necessary for solving $f(t)$

$$f(t) = \mathcal{L}^{-1}F(s) = \mathcal{L}^{-1}\frac{1}{(s+1)(s+2)} = \frac{1}{2-1}(e^{-t} - e^{-2t}) = e^{-t} - e^{-2t}.$$

The technique can be diagrammed as in Figure 9.12.

---

*9.7.4 Markov Models

A **stochastic process** can be defined as a time-dependent sequence of events, such as motions or changes in appearance or the state of an item as functions of time. Stochastic processes can be analyzed by means of **Markov models.**

Markov models are functions of two variables:

1. The **state,** or **location,** variable x
2. The **time** variable t

They are used for analyzing a variety of phenomena from particle motion in mechanics to assembly line simulation in operations research to our own field of interest, developing reliability models for complex systems.

Four types of models are possible:

1. The x variable is *discrete* and the t variable is *discrete*.
2. The x variable is *discrete* and the t variable is *continuous*
3. The x variable is *continuous* and the t variable is *discrete*.
4. The x variable is *continuous* and the t variable is *continuous*.

The first type, where both x and t variables are discrete, is called a **Markov chain model.** To illustrate how this works, consider an ice cube tray containing an infinite number of discrete compartments, as illustrated in Figure 9.13. Each compartment can be identified by a numbering scheme $\{m, n\}$ that refers to a row and column in this infinite tray.

Suppose that we drop a ping-pong ball into one of the compartments $\{m_1, n_1\}$ and record our starting position. Let us then strike the bottom of the tray hard enough to knock the ping-pong ball out of its compartment. The ball may then drop back into the same compartment or into another. In either event, identify this second location as $\{m_2, n_2\}$ and record it. Let us continue this experiment and collect our data so that it looks like the format in Table 9.3, where after the kth tap it is in the kth position, $\{m_k, n_k\}$.

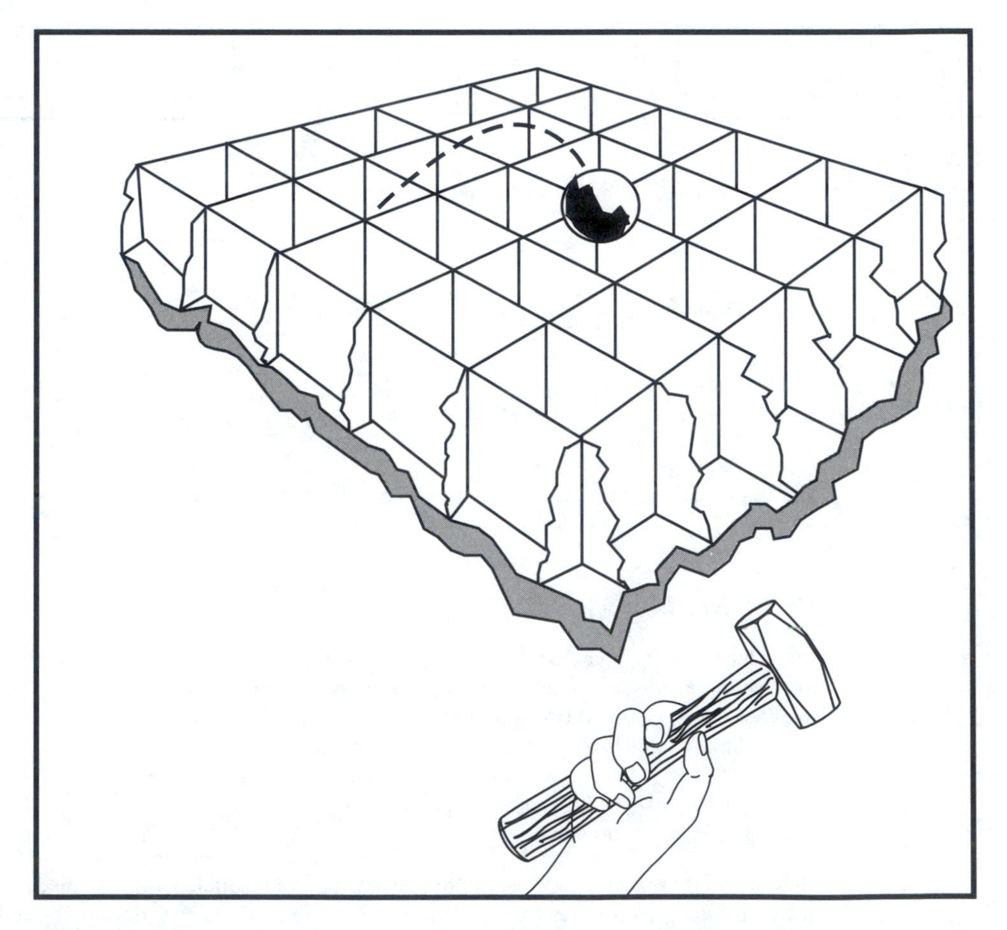

FIGURE 9.13
Illustration of a Markov model with discrete states.

TABLE 9.3
Markov chain data.

Tap Number	Sequential Position Number	Compartment
0	1	$\{m_1, n_1\}$
1	2	$\{m_2, n_2\}$
2	3	$\{m_3, n_3\}$
.	.	.
.	.	.
.	.	.
k	$k + 1$	$\{m_k, n_k\}$

This defines a **stochastic process,** because after each tap the ball must land in one of the compartments. As already pointed out, it is possible for the ball to land in the same compartment more than once; that is, $\{m_i, n_i\} = \{m_j, n_j\}$ is possible for any combination of i and j.

Most importantly, notice that the ball's location at any point in time *depends only on its immediately previous location but on no others.* That is, the probability that the ball lands in a particular compartment after the kth tap depends on where the ball was just previous to the kth tap, but it is independent of where the ball was prior to the $(k + 1)$st tap.

The second type of Markov model, where the x variable is discrete and the t variable is continuous, is called a **Markov process.** Figure 9.3 can still be used to describe this model, but instead of striking the tray to dimension the time variable in terms of a sequence of discrete taps, the tray will be vibrated to keep the ping-pong ball in continuous motion. Now the locations will have to be recorded in terms of observations made at time intervals Δt, as in Table 9.4.

Again we have defined a *stochastic process,* because at every observation the ball is in one, and only one, of the compartments. As before, it is always possible for the ball to reenter a previous compartment. And again, the ball's location at any point in time depends only on its immediately previous location. By this we mean its immediately previous *recorded* location, where the Δt is sufficiently small as to permit only one hop of the ping-pong ball.

TABLE 9.4
Markov process data.

Observation Number	Time of Observation	Compartment
1	0	$\{m_1, n_1\}$
2	Δt	$\{m_2, n_2\}$
3	$2\Delta t$	$\{m_3, n_3\}$
.	.	.
.	.	.
.	.	.
k	$(k - 1)\Delta t$	$\{m_k, n_k\}$

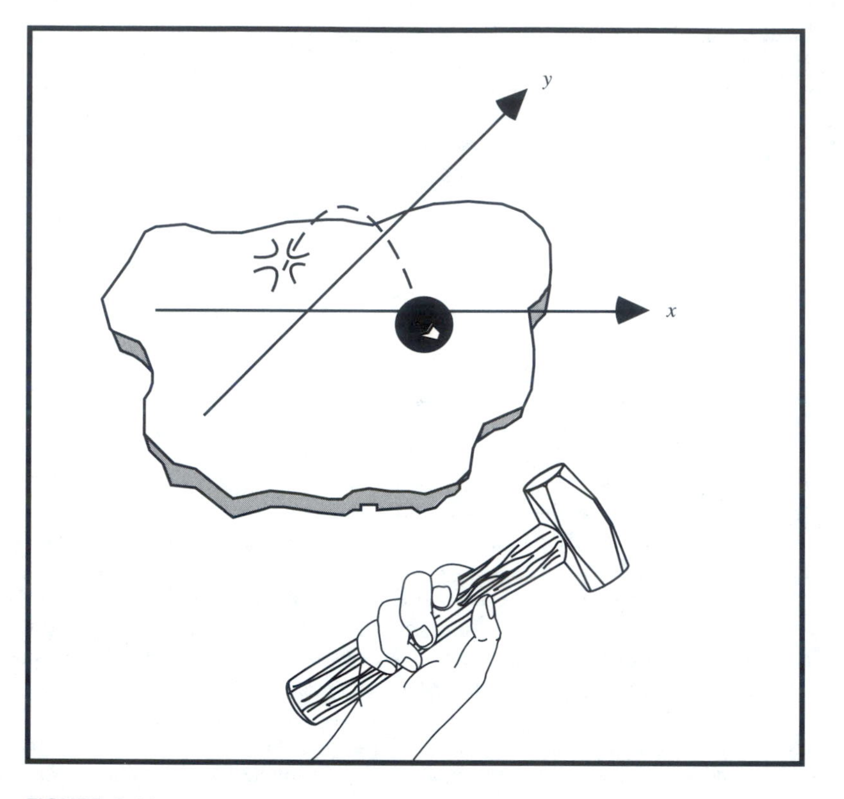

FIGURE 9.14
Illustration of a markov model with continuous states.

To demonstrate the third type of Markov model, where x is continuous and t is discrete, we use a flat magnetic tray and a steel ball, as in Figure 9.14. Now our example is analogous to that for the first type of Markov model, except that the locations are given by x and y coordinates of the continuous and infinite plane rather than by compartment numbers.

Figure 9.14 can also be used to demonstrate the fourth type of Markov model, where x and y are both continuous. However, instead of striking the tray to induce the motion of the steel ball in discrete hops, we will vibrate the tray to keep the ball in continuous motion.

*9.7.5 The Poisson Process

We are interested in the second type of Markov model, with discrete states and continuous time. Our states, instead of representing locations, will represent operational conditions of our system. The time-dependent transition amongst the states, rather than being hops of a ping-pong ball among compartments, will be *failure or restoration events* within the system. If we consider these events to be constant time-dependent failure and restoration rates, the partcular type of Markov process we are discussing is known as a **Poisson process.**

As we noted previously, arrival in a state is independent of all previous states except the last one.

*9.7.6 Applying the Markov Model to the Solution of Complex System Reliability

A complex system is composed of a given quantity of elements, any of which may be *up* or *down* or in a *standby mode* at any point in time. We may make the following general statements about any complex system:

1. At some point in time, say $t = 0$, all elements are as designated to be. That is, elements designed to be normally operating are operating, those designed to be normally in a standby mode are in standby, those designed to be performing a monitoring function are doing so, etc.
2. At any point in time, every element that is *up* (operating or operable) will eventually be down.
3. At any point in time, every element that is *down* will at some time be up (restored or replaced).
4. One-shot or cyclically operating elements can be said to be at $t = 0$ in terms of a probability statement.
5. Every combination of elements up, down, or in standby comprises a different **system state.**
6. Upon a single failure or restoration event, the system passes from one state to another.

These statements describe a Markov process, the second type of Markov model. The states are discrete, defined by what elements are *up, down,* or *in standby.* The time is continuous. In particular, if we consider failure and restoration rates to be constant we have a Poisson Process.

When determining the reliability for a mission of specified time t, we are asking the question, What is the probability that, if we are in state 1 (or any particular state) at the time $t = 0$, we never entered a *fail state* during the interval $\{0, t\}$? This is consistent with the reliability definition of a time-dependent system, that reliability for a specified mission time t is the probability that no failures occur during that mission. Hence, we can solve the reliability by contriving a Markov model where all *fail* states are combined into a single state that is an *absorbing state.*

An **absorbing state** can be defined as a state from which the system can never leave once it has entered. So by contriving our Markov model of the system in this fashion, we can equate the system reliability for a mission of time t to the probability that the system is not in the

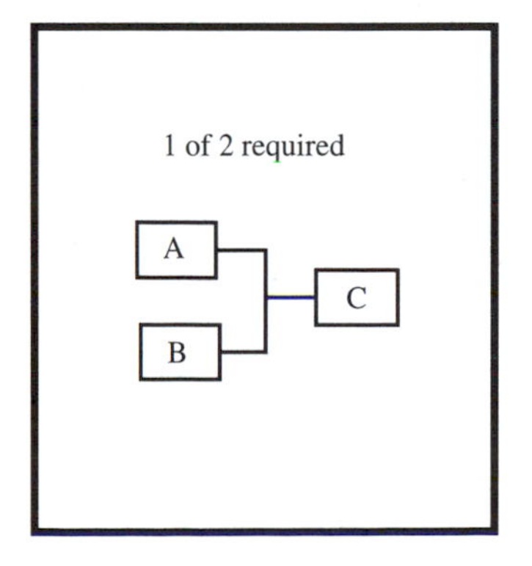

FIGURE 9.15
Redundant series block diagram.

absorbing state at the time t. If the system is not in that state at the time t, then it obviously never entered it during the interval $\{0, t\}$.

As an example, suppose we have a system as represented in the block diagram in Figure 9.15, in which there are three active elements, A, B, and C. Elements A and B are redundant to one another, and C is serial in the system. Suppose all elements have their own constant failure rates λ_A, λ_B, and λ_C, and their own restoration rates μ_A, μ_B, and μ_C. System success is defined by success of C and success of either A or B. Hence, there are three success states, defined as follows.

Success State	Elements Up	Elements Down	Elements Under Repair
1	A, B, C	None	None
2	B, C	A	A
3	A, C	B	B

All other states are *fail* states, and we shall contrive our Markov model by lumping them into a single state, F, which is an **absorbing state.** Figure 9.16 shows our ice cube tray applied to this example. It has four compartments, one for each success state and one absorbing

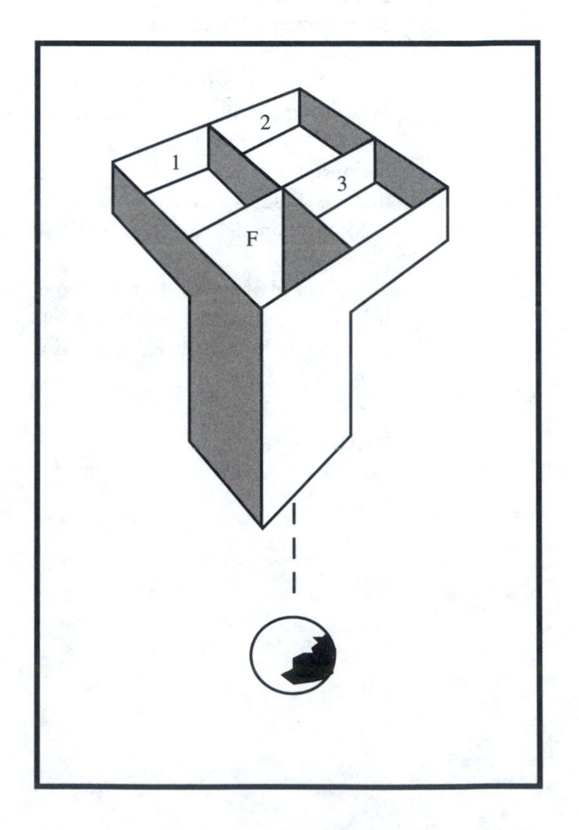

FIGURE 9.16
Illustration of a Markov model for the system of Figure 9.15.

compartment for state *F*. Once the ping-pong ball enters state F it can never return to any of the others.

Initially, the ping-pong ball will be in state 1, where all three elements are *up*. According to statement 6, the ping-pong ball can go from one state to another upon a single event. From state 1 it will go to state 2 if *element A* fails, it will go to state 3 if *element B* fails, and it will go to state F if *element C* fails. From state 2 the ping-pong ball will go to state 1 if *Element A* is restored, it will go to state F if *element B* or *element C* fails, and it cannot enter state 3 because of statement 6. (It would take more than a *single* event for the system to go from state 2 to state 3). Similarly, from state 3 the ball will go to state 1 if *Element B* is restored, it will go to state F if either *element A* or *element C* fails, and it cannot enter state 2. From state F it cannot go anywhere. If we place the ball in the state 1 compartment initially and make our first observation at time *t*, and at *t* the ball is either in State 1, 2, or 3, we can conclude that the system survived the entire interval {0, *t*} without ever having entered state F. Accordingly, the reliability *R*(*t*) is the probability that the system is in state *i* at the time *t*, so we can define reliability for a mission of specified time *t* as

$$R(t) = 1.0 - P_F(t) \tag{9.29}$$

Figure 9.17 shows a state-to-state **transition rate diagram** for this example. It shows the stresses on the ping-pong ball in all its locations. When it is in state 1, it is being drawn into state 2 at the rate λ_A (the rate at which element A is trying to fail), into state 3 at a rate λ_B, and into state F at a rate λ_C. When in state 2, it is being drawn into state 1 at a rate μ_A (the rate at which element A is being restored), into state F at a rate $\lambda_B + \lambda_C$ (the rate at which either element B or

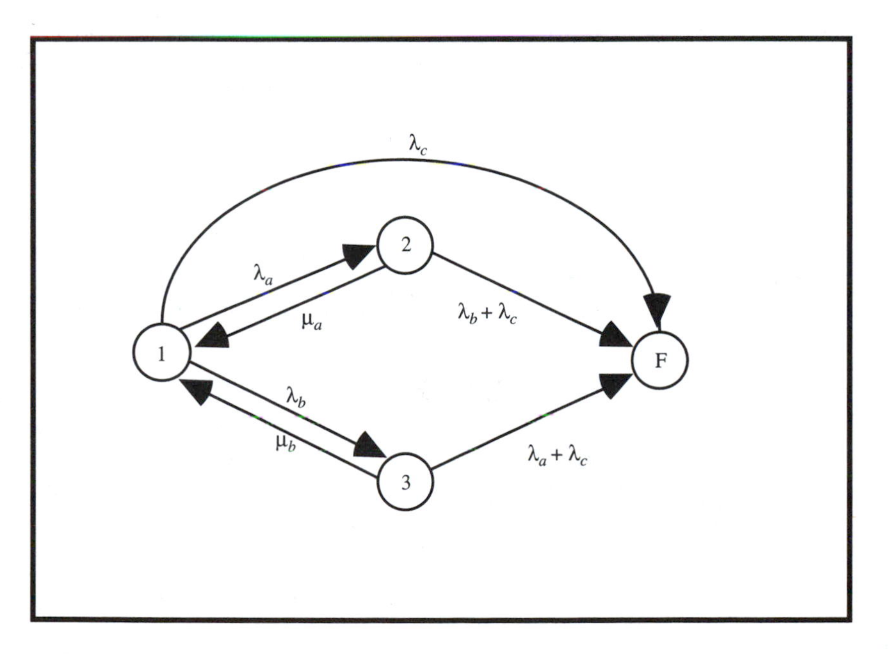

FIGURE 9.17
Transition rate diagram for the system of Figure 9.15.

C is trying to fail), and into state 3 at a rate of zero. When it is in state 3, it is being drawn into state 1 at a rate μ_B and state F at a rate $\lambda_A + \lambda_C$.

It is possible to establish a set of mutually exclusive and statistically independent Markov success states to represent any system of any degree of complexity. Once the states are defined, we can determine the transition rates among the states, as we learned in Section 9.6.2, to solve system reliability.

*9.8 THE MARKOV MODEL SOLUTION FOR SYSTEM RELIABILITY

Consider a general case (Figure 9.18) consisting of a multitude of Markov success states, with transition among the states due to failure or restoration events. Let r_{ij} represent the rate from state j to state i. (The reason for this notational eccentricity is to avoid dealing with transposing a matrix further on in the analysis.) The following assumptions are necessary in order to apply the Markov model solution.

ASSUMPTION	VALIDITY
1. Failure and restoration rates are constant, *not* time dependent.	We are operating in the "useful life" of the system, where we can consider failure rates to be approximately constant. Repair rates are, of course constant.
2. Events are independent of one another.	States can be defined so that the transitions among them are caused by independent events (malfunctions) rather than common (dependent) failure modes.
3. All equipment is as designed to be up, down, or in standby at $t = 0$.	At some point in time this is true. Define that time as $t = 0$.
4. All states are mutually exclusive.	They were defined that way.
5. When state F is entered, we cannot leave it. State F is defined as the *absorbing state.*	We have contrived our reliability model in this manner.
6. The probability of being in any state at any given point in time depends only on the immediately previous state.	This is true by definition of the Markov model.

Let us define all possible success states 1, 2, 3,..., N and the *fail state* F. Let $P_i(t)$ be the probability of being in state i ($i = 1, 2, \ldots, N, F$) at time t. Hence, $P_i(t + \Delta t)$ is the probability of being in state i at time $(t + \Delta t)$.

Because of assumption 4 and because the defined states are exhaustive,

$$\sum_{i=1}^{N} P_i(t) + P_F(t) = 1.0 \qquad \text{for all } t \tag{9.30}$$

Define a time increment Δt that is so small that the probability of more than one event taking place during Δt is negligible. The probability of being in a particular state i at the time $(t + \Delta t)$ is the probability that

1. The system is in state i at the time t and it remains in state i during Δt,

or

2. The system is in some other state at the time t, and it goes from that state into state i during Δt.

Events 1 and 2 are mutually exclusive, so $P_i(t + \Delta t)$ is the sum of the probability of event 1 plus the probability of event 2.

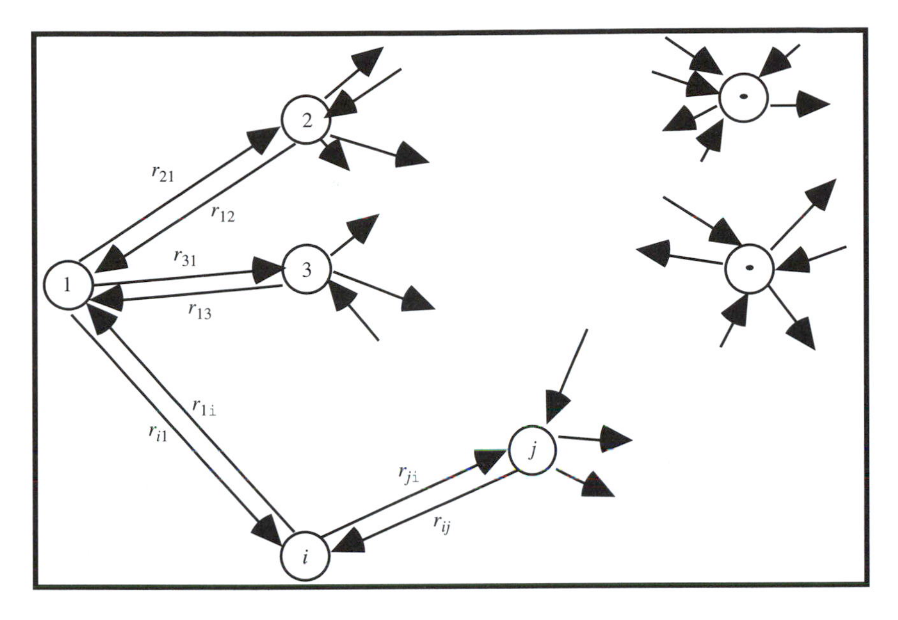

FIGURE 9.18
A general transition rate diagram.

Event 1 is the combination of two independent events, so its probability is the product of the probabilities of these two events. The probability that the system is in state i at time t is $P_i(t)$; the probability that the system remains in state i during the interval Δt is 1.0 minus the probability that it goes from the state i into another state during that interval—that is, 1.0 minus the rate at which it is drawn from state i times the interval Δt. Therefore, the probability of event 1 is $P_i(t) \cdot (1 - \sum_{\text{all } k \neq i} r_{ki}\Delta t)$.

Similarly, the probability of event (2) is the probability that the system is in some state other than i at the start of the interval times the probability that it goes from that state into state i during the interval Δt: $P_j(t)(r_{ij}\Delta t)$.

Hence, the probability of being in state i at the time $(t + \Delta t)$ is

$$P_i(t + \Delta t) = \sum_{\text{all } j \neq i} P_j(t)[r_{ij}\Delta t] + P_i(t)[1 - \sum_{\text{all } k \neq i} r_{ki}\Delta t] \quad \textbf{(9.31)}$$

Note: If there is no single (failure or restoration) event that can bring the system from a state j directly into a state i, then $r_{ij} = 0$.

For any system with N independent success states, it is possible to write a set of N independent difference equations in the form of Equation 9.31, where $i = 1, 2, 3, ..., N$. Examining the ith equation in the set, subtract $P_i(t)$ from both sides so that Equation 9.31 can be written in the form

$$P_i(t + \Delta t) - P_i(t) = \sum_{\text{all } j \neq i} P_j(t)[r_{ij}\Delta t] - P_i(t) \sum_{\text{all } k \neq i} r_{ki}\Delta t \quad \textbf{(9.32)}$$

Dividing both sides of Equation 9.32 by Δt gives

$$\frac{P_i(t + \Delta t) - P_i(t)}{\Delta t} = \sum_{\text{all } j \neq i} r_{ij} P_j(t) - P_i(t) \sum_{\text{all } k \neq i} r_{ki} \tag{9.33}$$

Allow the notation r_{ii} to represent the sum of all transition rates out of state i:

$$\left.\begin{aligned} r_{11} &= r_{21} + r_{31} + \cdots + r_{N1} \\ r_{22} &= r_{12} + r_{32} + \cdots + r_{N2} \\ &\vdots \\ r_{ii} &= \sum_{\text{all } k \neq i} r_{ki} \\ &\vdots \\ r_{NN} &= r_{1N} + r_{2N} + \cdots + r_{N-1,N} \end{aligned}\right\} \tag{9.34}$$

Applying the definition of Equation 9.34 to Equation 9.33,

$$\frac{P_i(t + \Delta t) - P_i(t)}{\Delta t} = \sum_{\text{all } j \neq i} r_{ij} P_j(t) - r_{ii} P_i(t) \tag{9.35}$$

We stated earlier that we are using the second type of Markov model, the one with discrete state and continuous time. Because time is *continuous,* let us take the limit of Equation 9.35 as Δt approaches zero:

$$\lim_{\Delta t \to 0} \frac{P_i(t + \Delta t) - P_i(t)}{\Delta t} = \frac{d}{dt} P_i(t) = \sum_{\text{all } j \neq i} r_{ij} P_j(t) - r_{ii} P_i(t) \tag{9.36}$$

The solutions of the N independent equations in the form of Equation 9.36 will give the probability of being in any success state at the time t.

From Equations 9.29 and 9.30,

$$R(t) = 1.0 - P_F(t) = P_1(t) + P_2(t) + \cdots + P_N(t) \tag{9.37}$$

Hence, the reliability at time t is the sum of the probabilities of being in each of the success states 1, 2,..., N at time t: $R(t) = \sum_{i=1}^{N} P_i(t)$. In order to find the N values of $P_i(t)$ from the N differential equations, we must have a boundary condition. We can use assumption 3, which states

$$P_1(0) = 1.0$$

and

$$P_i(0) = 0 \qquad \text{for all } i \neq 1 \tag{9.38}$$

We can also solve $R(t)$ with the system starting in any state i, not necessarily state 1; that is, the ping-pong ball in our analogy can start in any compartment in the tray. A more general boundary condition is

$$P_j(0) = 1.0$$

and

$$P_i(0) = 0 \qquad \text{for all } i \neq j \tag{9.39}$$

So the exact solution for the reliability of a complex system with N success states over the time t is

$$R(t) = \sum_{i=1}^{N} P_i(t) \tag{9.40}$$

where the $P_i(t)$ values are the solutions to the set of N differential equations

$$\begin{aligned}
\frac{d}{dt}P_1(t) &= -r_{11}P_1(t) + r_{12}P_2(t) + r_{13}P_3(t) + \cdots + r_{1N}P_N(t) \\
\frac{d}{dt}P_2(t) &= r_{21}P_1(t) - r_{22}P_2(t) + r_{23}P_3(t) + \cdots + r_{2N}P_N(t) \\
&\;\;\vdots \\
\frac{d}{dt}P_N(t) &= r_{N1}P_1(t) + r_{N2}P_2(t) + r_{N3}P_3(t) + \cdots - r_{NN}P_N(t) \\
P_1(0) &= 1.0 \quad \text{and} \quad P_i(0) = 0 \qquad \text{for all } i \neq 1
\end{aligned} \tag{9.41}$$

In matrix format the reliability of a complex system over time t can be obtained through the solutions of

$$\begin{bmatrix} P_1'(t) \\ P_2'(t) \\ \cdot \\ \cdot \\ \cdot \\ P_N'(t) \end{bmatrix} = \begin{bmatrix} -r_{11} & r_{12} \ldots r_{1N} \\ r_{21} & -r_{22} \ldots r_{2N} \\ \cdot & \\ \cdot & \\ \cdot & \\ r_{N1} & r_{N2} \ldots -r_{NN} \end{bmatrix} \begin{bmatrix} P_1(t) \\ P_2(t) \\ \cdot \\ \cdot \\ \cdot \\ P_N(t) \end{bmatrix} \tag{9.42}$$

$$P_1(0) = 1.0, \qquad P_i(0) = 0 \qquad \text{for all } i \neq 1$$

EXAMPLE 9.34 A system is composed of two identical elements that are redundant to each other. Each has a failure rate of 10^{-5} per hour. If it is a nonrestored system, what is the reliability for a 10,000-h mission?

This system has two Markov success states, as follows:

Success State	Elements Up	Elements Down
1	2	0
2	1	1

The state-to-state transition rate diagram is shown in Figure 9.19.

The system reliability is $R(t) = P_1(t) + P_2(t)$, where $P_1(t)$ and $P_2(T)$ are solutions to the following set of differential equations:

$$\frac{d}{dt}P_1(t) = -r_{11}P_1(t) + r_{12}P_2(t)$$

$$\frac{d}{dt}P_2(t) = r_{21}P_1(t) - r_{22}P_2(t)$$

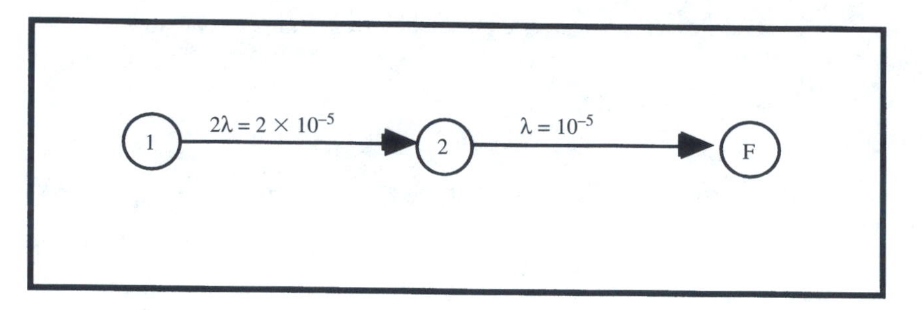

FIGURE 9.19
Transition-rate diagram for Example 9.34.

From the transition rate diagram,

$$r_{11} = 2 \times 10^{-5}, \qquad r_{12} = 0, \qquad r_{21} = 2 \times 10^{-5}, \qquad r_{22} = 10^{-5}$$

Hence, the set of differential equations becomes

$$\frac{d}{dt}P_1(t) = -2 \times 10^{-5}P_1(t)$$

$$\frac{d}{dt}P_2(t) = 2 \times 10^{-5}P_1(t) - 10^{-5}P_2(t)$$

We can solve this set of differential equations through Laplace transforms. Use Table 9.2.

$$\mathcal{L}\frac{d}{dt}P_1(t) = sF_1(s) - P_1(0) \qquad \text{(transform 2)}$$

$$\mathcal{L}\frac{d}{dt}P_2(t) = sF_2(s) - P_2(0) \qquad \text{(transform 2)}$$

But our boundary conditions are $P_1(0) = 1.0$ and $P_2(0) = 0$. So,

$$\mathcal{L}\frac{d}{dt}P_1(t) = sF_1(s) - 1 \quad \text{and} \quad \mathcal{L}\frac{d}{dt}P_2(t) = sF_2(s)$$

Also, $$\mathcal{L}P_1(t) = F_1(s) \quad \text{and} \quad \mathcal{L}P_2(t) = F_2(s) \qquad \text{(transform 1)}$$

The transformed set is as follows:

$$sF_1(s) - 1 = -2 \times 10^{-5}F_1(s)$$
$$sF_2(s) = 2 \times 10^{-5}F_1(s) - 10^{-5}F_2(s).$$

The transformed set has the algebraic solutions

$$F_1(s) = \frac{1}{(s + 2 \times 10^{-5})}$$

$$F_2(s) = \frac{2 \times 10^{-5}}{(s + 10^{-5})(s + 2 \times 10^{-5})}$$

Transforming back,

$$P_1(t) = \mathscr{L}^{-1}F_1(s) = \mathscr{L}^{-1}\frac{1}{(s + 2 \times 10^{-5})} = e^{-2\times10^{-5}t} \qquad \text{(transform 7)}$$

$$P_2(t) = \mathscr{L}^{-1}F_2(s) = \mathscr{L}^{-1}\frac{2 \times 10^{-5}}{(s + 10^{-5})(s + 2 \times 10^{-5})}$$

$$= \frac{2 \times 10^{-5}}{10^{-5}}(e^{-10^{-5}t} - e^{-2\times10^{-5}t}) \qquad \text{(transform 12)}$$

The system reliability solution is

$$R(t) = P_1(t) + P_2(t) = e^{-2\times10^{-5}t} + 2e^{-10^{-5}t} - 2e^{-2\times10^{-5}t} = 2e^{-10^{-5}t} - e^{-2\times10^{-5}t}$$

For the 10,000-h mission,

$$R(10{,}000 \text{ h}) = 2e^{-10^{-5}\times10^4} - e^{-2\times10^{-5}\times10^4} = 2e^{-0.1} - e^{-0.2} = 0.991$$

Compare this analysis with what you would have done in Chapter 8.

EXAMPLE 9.35

Suppose the system in Example 9.34 is a restored system with restoration rate $\mu = 2$ per hour for each element. Solve the system reliability for a 10,000-h mission.

This is a problem we would find impossible to solve by the techniques of Chapter 8. However, with the Markov model method, it is no more challenging than Example 9.34. There are two Markov success states, as follows.

Success State	Units Up	Units Down	Units under Repair
1	2	0	0
2	1	1	1

The state-to-state transition rate diagram is in Figure 9.20.

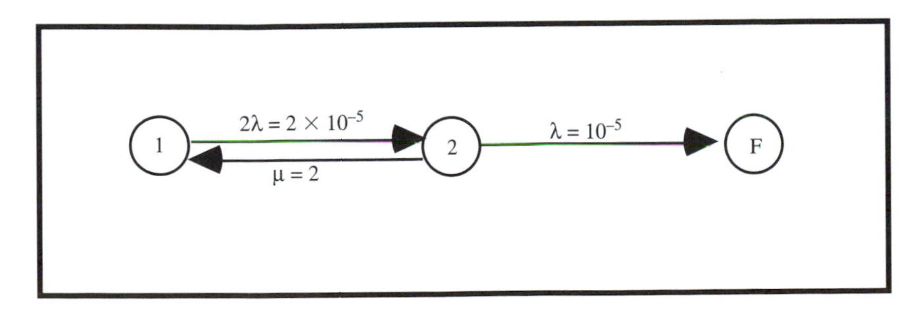

FIGURE 9.20
Transition-rate diagram for Example 9.35.

The system reliability is $R(t) = P_1(t) + P_2(t)$, where $P_1(t)$ and $P_2(T)$ are solutions to the following set of differential equations:

$$\frac{d}{dt}P_1(t) = -r_{11}P_1(t) + r_{12}P_2(t)$$

$$\frac{d}{dt}P_2(t) = r_{21}P_1(t) - r_{22}P_2(t)$$

From the transition rate diagram,

$$r_{11} = 2 \times 10^{-5}, \qquad r_{12} = 2, \qquad r_{21} = 2 \times 10^{-5}, \qquad r_{22} = 2.000010$$

Hence, the set of differential equations to be solved becomes

$$\frac{d}{dt}P_1(t) = -2 \times 10^{-5}P_1(t) + 2P_2(t)$$

$$\frac{d}{dt}P_2(t) = 2 \times 10^{-5}P_1(t) - 2.00001P_2(t)$$

where $P_1(0) = 1.0$ and $P_2(0) = 0$.

Whereas the reliability solution for this situation is expected to be almost indistinguishably close to 1.0, a more reasonable approach is to determine the unreliability and solve $R(t) = 1 - P_F(t)$. This means we have to add a third equation to the system:

$$\frac{d}{dt}P_F(t) = 10^{-5}P_2(t), \qquad \text{where } P_F(0) = 0$$

As in the previous example, we can solve this through Laplace transforms, using transforms 1 and 2 of Table 9.2:

$$\mathcal{L}P_1(t) = sF_1(s) - P_1(0) = \mathcal{L}(-2 \times 10^{-5}P_1(t) + 2P_2(t) = -2 \times 10^{-5}F_1(s) + 2F_2(s)$$
$$\mathcal{L}P_2(t) = sF_2(s) - P_2(0) = \mathcal{L}(2 \times 10^{-5}P_1(t) - 2.00001P_2(t)) = 2 \times 10^{-5}F_1(s) - 2.00001F_2(s)$$
$$\mathcal{L}P_F(t) = sF_F(s) - P_F(0) = \mathcal{L}(10^{-5}P_2(t)) = 10^{-5}F_2(s)$$

Applying the boundary conditions $P_1(0) = 1$ and $P_2(0) = P_F(0) = 0$, we have the following set of linear equations:

$$sF_1(s) - 1 = -2 \times 10^{-5}F_1(s) + 2F_2(s)$$
$$sF_2(s) = 2 \times 10^{-5}F_1(s) - 2.00001F_2(s)$$
$$sF_F(s) = 10^{-5}F_2(s)$$

The transformed set has the algebraic solution

$$F_F(s) = \frac{2 \times 10^{-10}}{s(s + 2.0000299999)(s + 10^{-10})} \approx \frac{2 \times 10^{-10}}{s^2(s + 2.00003)}$$

The system reliability is

$$R(t) = 1 - P_F(t) = 1 - \mathcal{L}^{-1}F_F(s) = 1 - 2 \times 10^{-10}\mathcal{L}^{-1}\frac{1}{s^2(s + 2.00003)}$$

Applying transform 9 of Table 9.2,

$$
\begin{aligned}
R(t) &= 1 - 2 \times 10^{-10}\left[\frac{1}{(2.00003)^2}(e^{-2.00003} + 2.00003t - 1)\right] \\
&= 1 - 0.499985 \times 10^{-10}(e^{-2.00003} + 2.00003t - 1)
\end{aligned}
$$

For $t = 10{,}000$ h,

$$R(10{,}000\text{ h}) = 1 - 0.499985 \times 10^{-10}(e^{-20.0003} + 20{,}000.3t - 1) = 0.999999$$

The advantage to operating a restored system over a nonrestored system is evident in the comparison between this result and that of Example 9.34.

In Examples 9.34 and 9.35 it was possible to solve system reliability through Laplace transforms. But the solution of mission reliability will be possible through that technique only for certain system configurations. In fact, it was possible to use the Laplace transform technique to solve Example 9.35 only after making an approximation to the $F_F(x)$ equation. Most of the time computer solutions will be used.

*9.8.1 Considerations for Computer Solution

Computer solutions will entail the solution of a set of differential equations of the form presented in Equation 9.41. Computers solve differential equations by approximating them by difference equations with sufficiently small Δt's. We want to select a Δt large enough so that a computer solution is possible and small enough so that we obtain a desired accuracy. Our strategy for testing the accuracy provided by a given Δt is to make two estimates of $R(t)$, one optimistic and one pessimistic, knowing that the actual $R(t)$ is somewhere between our two estimates. If the difference between our optimistic and pessimistic estimates is less than the error permitted by our desired accuracy, we can accept that the selected Δt is sufficiently small.

Applying Equation 9.34 to Equation 9.31,

$$P_i(t + \Delta t) = \sum_{\text{all } j \neq i} P_j(t)[r_{ij}\Delta t] + P_i(t)[1 - r_{ii}\Delta t] \tag{9.43}$$

Now we select the Δt sufficiently small so that the computer solution will be within the desired accuracy. Let

$$n = t/\Delta t \tag{9.44}$$

Then,

$$P_i(t) = P_i(n\ \Delta t) \tag{9.45}$$

and

$$P_i(t + \Delta t) = P_i([n + 1]\ \Delta t) \tag{9.46}$$

The set of N independent difference equations to be solved can be written

$$P_i([n + 1])\Delta t) = \sum_{\text{all } j \neq i} P_j(n\Delta t)[r_{ij}\Delta t] + P_i(n\Delta t)[1 - r_{ii}\Delta t] \tag{9.47}$$

where $i = 1, 2, 3,\ldots, N$.

Define the probability vector $\mathbf{\Pi}(t)$ and the matrix $[A]$ as follows:

$$\mathbf{\Pi} = \begin{bmatrix} P_1(t) \\ P_2(t) \\ \cdot \\ \cdot \\ \cdot \\ P_N(t) \end{bmatrix} \tag{9.48}$$

$$[A] = \begin{bmatrix} (1 - r_{11}\Delta t) & r_{12}\Delta t & r_{13}\Delta t & \ldots & r_{1N}\Delta t \\ r_{21}\Delta t & (1 - r_{22}\Delta t) & r_{23}\Delta t & \ldots & r_{2N}\Delta t \\ r_{31}\Delta t & r_{32}\Delta t & (1 - r_{33}\Delta t) & \ldots & r_{3N}\Delta t \\ \cdot & \cdot & \cdot & & \cdot \\ \cdot & \cdot & \cdot & & \cdot \\ \cdot & \cdot & \cdot & & \cdot \\ r_{N1}\Delta t & r_{N2}\Delta t & r_{N3}\Delta t & \ldots & (1 - r_{NN}\Delta t) \end{bmatrix} \tag{9.49}$$

In matrix form, the set of equations represented by Equation 9.47 is written

$$\mathbf{\Pi}([n + 1]\Delta t) = [A]\mathbf{\Pi}(n\Delta t) \tag{9.50}$$

from which follows

$$\mathbf{\Pi}(n\Delta t) = [A]([n - 1]\Delta t) = [A]^2\mathbf{\Pi}([n - 2]\Delta t) = \cdots = [A]^n\mathbf{\Pi}(0) \tag{9.51}$$

Hence, the solution to $\mathbf{\Pi}(t)$ is

$$\mathbf{\Pi}(t) = [A]^n\mathbf{\Pi}(0) \tag{9.52}$$

where $[A]$ is the coefficient matrix of the set of difference equations and $\mathbf{\Pi}(0)$ is known to be

$$\mathbf{\Pi}(0) = \begin{bmatrix} P_1(0) \\ P_2(0) \\ \cdot \\ \cdot \\ \cdot \\ P_N(0) \end{bmatrix} = \begin{bmatrix} 1 \\ 0 \\ \cdot \\ \cdot \\ \cdot \\ 0 \end{bmatrix} \tag{9.53}$$

if the system is in state 1 at $t = 0$. If the system is in some other state j, we define the $\mathbf{\Pi}(0)$ vector by setting $P_j(0) = 1.0$ and the other $P_i(0)$ values at zero.

So we can solve $R(t) = \sum_{i=1}^{N} P_i(t)$ by doing the following:

1. Select a sufficiently small Δt.
2. Determine $n = t/\Delta t$.

3. Determine the coefficient matrix $[A]$ from the state-to-state transition rates and raising $[A]$ to the nth power—that is, performing n successive matrix multiplications of $[A]$.
4. Finally, determine the $P_i(t)$ values through Equation 9.52.

If, for example, the system is initially in the state 1, the solution can be performed by computer as follows:

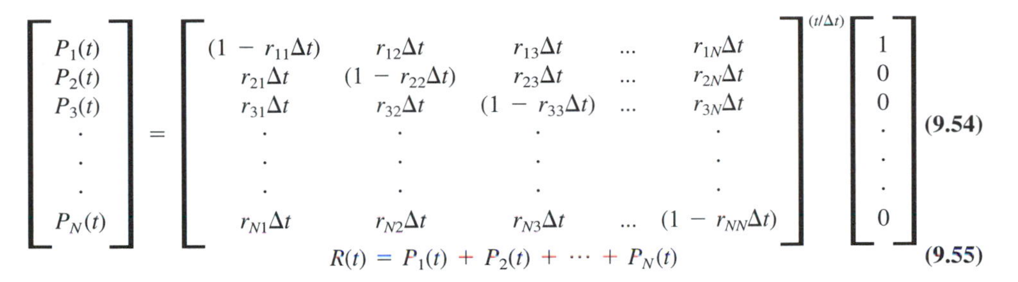

$$\begin{bmatrix} P_1(t) \\ P_2(t) \\ P_3(t) \\ \cdot \\ \cdot \\ \cdot \\ P_N(t) \end{bmatrix} = \begin{bmatrix} (1 - r_{11}\Delta t) & r_{12}\Delta t & r_{13}\Delta t & \cdots & r_{1N}\Delta t \\ r_{21}\Delta t & (1 - r_{22}\Delta t) & r_{23}\Delta t & \cdots & r_{2N}\Delta t \\ r_{31}\Delta t & r_{32}\Delta t & (1 - r_{33}\Delta t) & \cdots & r_{3N}\Delta t \\ \cdot & \cdot & \cdot & & \cdot \\ \cdot & \cdot & \cdot & & \cdot \\ \cdot & \cdot & \cdot & & \cdot \\ r_{N1}\Delta t & r_{N2}\Delta t & r_{N3}\Delta t & \cdots & (1 - r_{NN}\Delta t) \end{bmatrix}^{(t/\Delta t)} \begin{bmatrix} 1 \\ 0 \\ 0 \\ \cdot \\ \cdot \\ \cdot \\ 0 \end{bmatrix} \quad \textbf{(9.54)}$$

$$R(t) = P_1(t) + P_2(t) + \cdots + P_N(t) \quad \textbf{(9.55)}$$

*9.8.2 Combining States, Simplification, and Error Control

We now have a technique for solving the reliability of any system of any degree of complexity once we have resolved the system success logic into a Markov model. However, the degree of complexity of a system can generate an unmanageable quantity of Markov success states—that is, a quantity that is beyond the capacity of the computer. The quantity can be reduced either by breaking the system down into a series of independent subsystems or by combining states or eliminating states.

When states are combined, we shall be required to control the error introduced. This can be done by assigning rates that will generate both optimistic and pessimistic results that will envelope the true solution. Error control will be achieved by narrowing the envelope of accuracy. Suppose the five states in Figure 9.21 are reduced to the two states in Figure 9.22 by combining states i and j into state J and by combining states k, l, and m into state K. We can envelope our system

FIGURE 9.21
Five-state transition diagram.

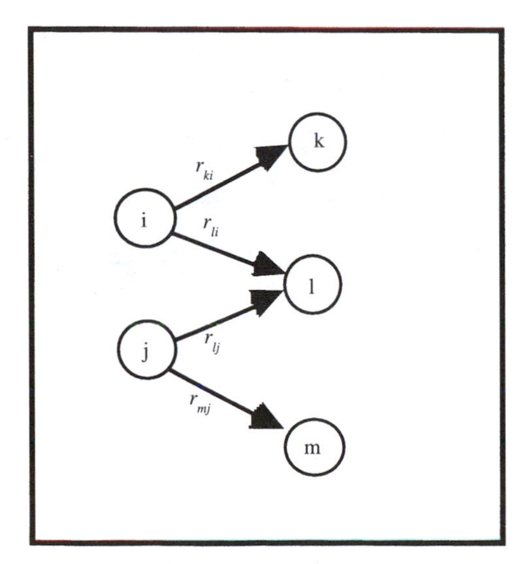

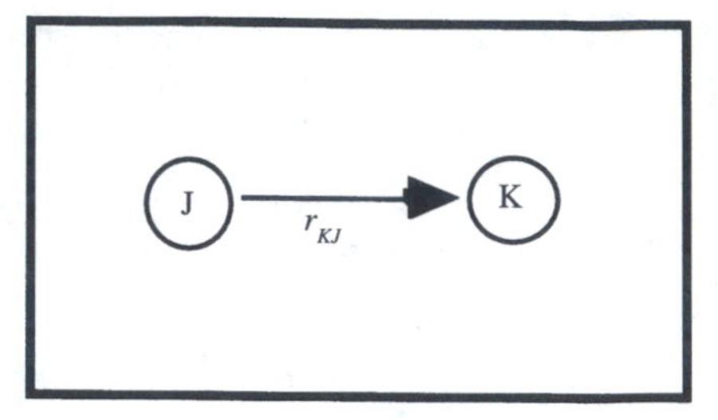

FIGURE 9.22
Reduced transition diagram.

reliability by making an optimistic and pessimistic approximation of r_{KJ}. The pessimistic assumption is

$$r_{KJ} = r_{ki} + r_{li} + r_{lj} + r_{mj} \tag{9.56}$$

and the optimistic assumption is

$$r_{KJ} = \min[r_{ki} + r_{li}, r_{lj} + r_{mj}] \tag{9.57}$$

The optimistic/pessimistic envelope is also necessary when the reduction is accomplished through elimination of states. As an example, suppose the state-to-state transition rate diagram of Figure 9.23 represents a system to be evaluated. Now, suppose that we want to reduce the complexity of the Markov model by eliminating all the success states beyond state 3. The pessimistic approximation is diagrammed in Figure 9.24, where the eliminated states are considered to be part

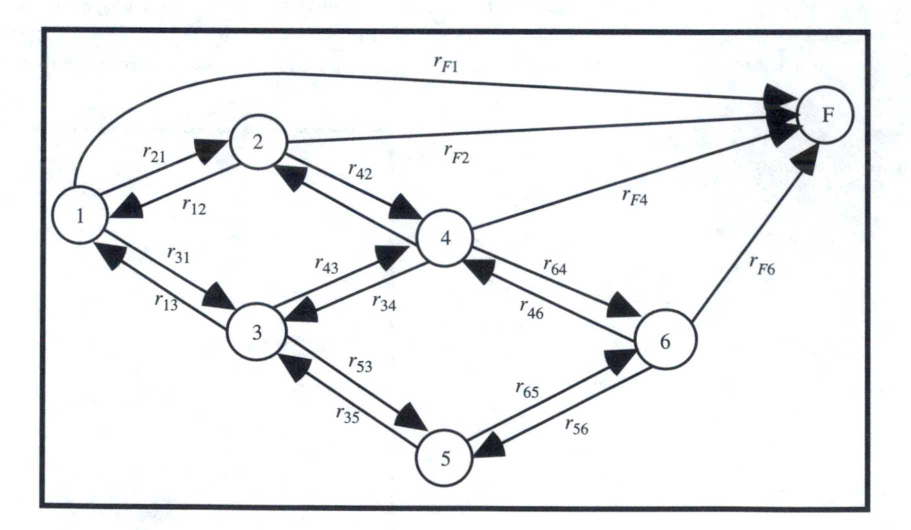

FIGURE 9.23
Complex state-to-state transition diagram.

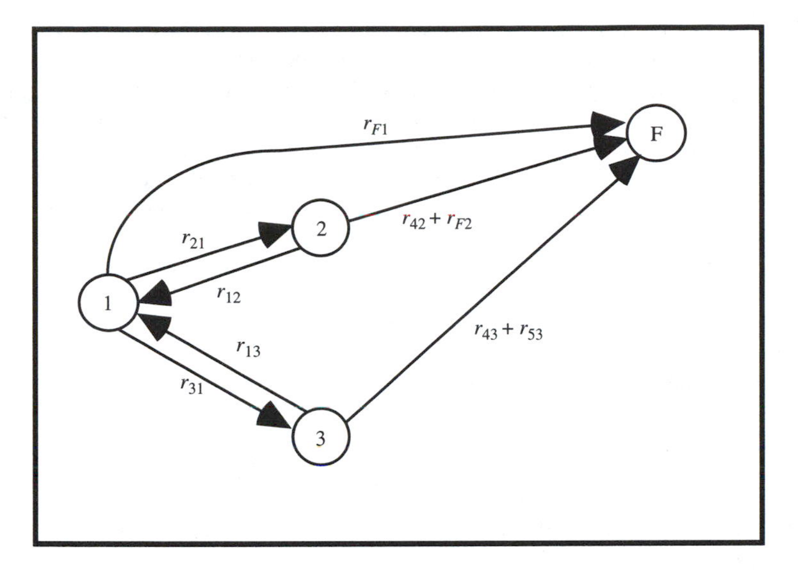

FIGURE 9.24
Pessimistic approximation.

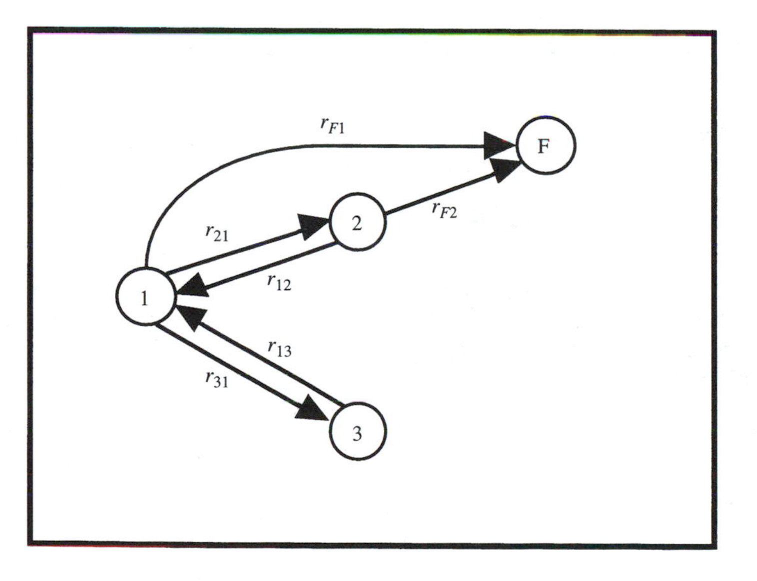

FIGURE 9.25
Optimistic approximation.

of the F state. The optimistic approximation is diagrammed in Figure 9.25, where states 4, 5, and 6 are simply ignored. We can generalize by stating that whenever we reduce the model by eliminating states, we can get an optimistic approximation by ignoring the eliminated states, and we can get a pessimistic approximation by including the eliminated states in the absorbing state.

Whenever the difference between the optimistic and pessimistic approximations is less than the required accuracy of the results, we can consider our state-reduction technique and our selection of Δt to be acceptable. Otherwise it will be necessary to adopt another state-reduction scheme or a smaller Δt or both.

The solution of mission reliability for complex systems clearly requires mathematics beyond the scope of the chapters within the text and is, accordingly, restricted to the discussion within the appendix. Fortunately, when dealing with complex systems we usually are interested in system MTBF and availability rather than the reliability of a specified mission. Although the MTBF of complex systems also requires fairly complex mathematics, the Markov model technique in conjunction with moment-generating functions and Laplace transforms makes it possible to reduce the analysis to simple algebra. Because, as we shall see, *steady-state* availability is of prime interest when evaluating complex systems, simple algebra is all that is necessary for computing complex system availability as well. The following section resolves the Markov model solutions of system MTBF into the tables of Appendix C and the algebraic methods presented in Section 9.6.

*9.9 THE MARKOV MODEL SOLUTION FOR SYSTEM MTBF

In Chapter 3 of the text and in Appendix A.2, we defined MTTF (mean time to failure) to be the mean of the time to failure (TTF) distribution, or the mean of the time required for an item to go from its initial operating state to a failed state. If we restrict our discussion to complex systems and define MTTF in Markov terms, the MTTF is the mean time required for the system to go from the initial state (presumably state 1) into a *fail* state. As wth the system reliability Markov model, it is expedient to contrive a situation where all the *fail* states are combined into a single *absorbing* state F. Then we can define MTTF as the mean time for the system to go from state 1 into state F.

If we assume that every time the system enters state F, we will always repair it back to state 1 before recommissioning it to service, we can treat MTBF (mean time between failures) as the equivalent to MTTF. With this assumption, recall our earlier results:

$$\text{MTBF} = \text{MTTF} = \int_0^{\infty} R(t)\, dt \tag{9.58}$$

Let t be the time required for the system to get from state 1 at time $t = 0$ to state F. The mean of the variable t, which is the MTTF, is the expected value of the random variable t, which, as we saw in Equation 9.26, is the first derivative of the moment-generating function $M_t(\theta)$, evaluated at $\theta = 0$:

$$\text{MTTF} = E[t] = \frac{d}{d\theta} M_t(\theta)\big|_{\theta=0} \tag{9.59}$$

where $M_t(\theta) = \int_{-\infty}^{\infty} e^{\theta t} f(t)\, dt$, the definition of the moment-generating function. Notice that $f(t)$ is the failure density function for the system.

Because t must be positive, the moment-generating function becomes

$$M_t(\theta) = \int_0^{\infty} e^{\theta t} f(t)\, dt \tag{9.60}$$

But, in Equation 9.28, we defined the Laplace transform

$$\mathscr{L}\{f(t)\} \equiv F(s) \equiv \int_0^\infty e^{-st} f(t)\, dt \tag{9.61}$$

which is the moment-generating function with $\theta = -s$. Hence,

$$\text{MTBF} = \text{MTTF} = \frac{d}{d\theta} M_t(\theta)|_{\theta=0} = -\frac{d}{ds}\mathscr{L}\{f(t)\}|_{s=0} \tag{9.62}$$

In Equation 3.14 we saw that the failure density function is

$$f(t) = -\frac{dR(t)}{dt} \tag{9.63}$$

and in Equation 9.29 we saw that

$$R(t) = 1.0 - P_F(t) \tag{9.64}$$

Hence,

$$f(t) = -\frac{dR(t)}{dt} = +\frac{dP_F(t)}{dt} \tag{9.65}$$

from which follows

$$\text{MTBF} = -\frac{d}{ds}\mathscr{L}\frac{dP_F(t)}{dt}\bigg|_{s=0} \tag{9.66}$$

From Table 9.2, transform 2,

$$\mathscr{L}\frac{d}{dt}P_i(t) = sF_i(s) - P_i(0). \tag{9.67}$$

By the boundary conditions of Equation 9.38,

$$\begin{aligned} & P_1(0) = 1.0 \\ \text{and}\quad & P_i(0) = 0 \qquad \text{for all } i \neq 1 \end{aligned} \tag{9.68}$$

Hence,

$$\begin{aligned} & \mathscr{L}\frac{d}{dt}P_1(t) = sF_1(s) - 1 \\ \text{and}\quad & \mathscr{L}\frac{d}{dt}P_i(t) = sF_i(s) \qquad \text{if } i = 2, 3, \ldots, N, F, \end{aligned} \tag{9.69}$$

from which we conclude

$$\text{MTBF} = -\frac{d}{ds}\mathscr{L}\frac{dP_F(t)}{dt}\bigg|_{s=0} = -\frac{d}{ds}sF_F(s)\bigg|_{s=0} \tag{9.70}$$

To solve the system MTBF, let us use the relations of Equation 9.69 to transform the set of differential equations of Equation 9.41:

$$\begin{aligned}
sF_1(s) - 1 &= -r_{11}F_1(s) + r_{12}F_2(s) + r_{13}F_3(s) + \cdots + r_{1N}F_N(s) \\
sF_2(s) &= r_{21}F_1(s) - r_{22}F_2(s) + r_{23}F_3(s) + \cdots + r_{2N}F_N(s) \\
sF_3(s) &= r_{31}F_1(s) + r_{32}F_2(s) - r_{33}F_3(s) + \cdots + r_NF_N(s) \\
&\vdots \\
sF_N(s) &= r_{N1}F_1(s) + r_{N2}F_2(s) + r_{N3}F_3(s) + \cdots - r_{NN}F_N(s)
\end{aligned} \tag{9.71}$$

Recall that the Markov states are mutually exclusive and exhaustive, so, as presented in Equation 9.30,

$$\sum_{i=1}^{N} P_i(t) + P_F(t) = 1.0 \tag{9.72}$$

From transform 6 of Table 9.2, $\mathcal{L}1.0 = 1/s$. Hence, the Laplace transform of Equation 9.72 is

$$\sum_{i=1}^{N} F_i(s) + F_F(s) = \frac{1}{s} \tag{9.73}$$

from which we conclude

$$sF_F(s) = 1 - s\sum_{i=1}^{N} F_i(s) \tag{9.74}$$

Applying this result to Equation 9.70,

$$\begin{aligned}
\text{MTBF} &= -\frac{d}{ds}[sF_F(s)]\bigg|_{s=0} = -\frac{d}{ds}\left[1 - s\sum_{i=1}^{N} F_i(s)\right]\bigg|_{s=0} \\
&= \frac{d}{ds}s\sum_{i=1}^{N} F_i(s)\bigg|_{s=0} = \left\{\sum_{i=1}^{N} F_i(s)\frac{ds}{ds} + s\frac{d}{ds}\sum_{i=1}^{N} F_i(s)\right\}\bigg|_{s=0} \\
&= \left\{\sum_{i=1}^{N} F_i(s) + s\frac{d}{ds}\sum_{i=1}^{N} F_i(s)\right\}\bigg|_{s=0} = \sum_{i=1}^{N} F_i(0) + 0 = \sum_{i=1}^{N} F_i(0)
\end{aligned} \tag{9.75}$$

The MTBF can be determined by summing the transforms of the $NP_i(t)$ values, namely, by summing the $F_i(s)$ values evaluated at $s = 0$. To determine the $F_i(0)$ values, we rewrite the set of Equations 9.71 for $s = 0$:

$$\begin{aligned}
-1 &= -r_{11}F_1(0) + r_{12}F_2(0) + r_{13}F_3(0) + \cdots + r_{1N}F_N(0) \\
0 &= r_{21}F_1(0) - r_{22}F_2(0) + r_{23}F_3(0) + \cdots + r_{2N}F_N(0) \\
0 &= r_{31}F_1(0) + r_{32}F_2(0) - r_{33}F_3(0) + \cdots + r_{3N}F_N(0) \\
&\vdots \\
0 &= r_{N1}F_1(0) + r_{N2}F_2(0) + r_{N3}F_3(0) + \cdots - r_{NN}F_N(0)
\end{aligned} \tag{9.76}$$

Because Equation 9.75 represents a set of N linearly independent equations with N unknowns $F_1(0)$, $F_2(0)$,..., $F_N(0)$, we have a unique solution

$$\text{MTBF} = \sum_{i=1}^{N} F_i(0) = F_1(0) + F_2(0) + \cdots + F_N(0) \qquad \textbf{(9.77)}$$

We can simplify Equations 9.76 and 9.77 by introducing the notation $x_i = F_i(0)$. Now, the MTBF of a complex system with N Markov success states is

$$\text{MTBF} = \sum_{i=1}^{N} x_i = x_1 + x_2 + \cdots + x_N \qquad \textbf{(9.78)}$$

where the x_i values are the solution of the following system of equations:

$$\begin{aligned}
-1 &= -r_{11}x_1 + r_{12}x_2 + r_{13}x_3 + \cdots + r_{1N}x_N \\
0 &= r_{21}x_1 - r_{22}x_2 + r_{23}x_3 + \cdots + r_{2N}x_N \\
0 &= r_{31}x_1 + r_{32}x_2 - r_{33}x_3 + \cdots + r_{3N}x_N \\
&\;\;\vdots \\
0 &= r_{N1}x_1 + r_{N2}x_2 + r_{N3}x_3 + \cdots - r_{NN}x_N
\end{aligned} \qquad \textbf{(9.79)}$$

When solving this set of linear equations by computer it is sometimes necessary to reduce the quantity of Markov success states to make the process more manageable. As described in the section on reliability solutions, the state reductions and error control are possible through combining or eliminating the least probable success states.

For a matrix analysis, if we define the vector

$$\mathbf{x} = \begin{bmatrix} x_1 \\ x_2 \\ \cdot \\ \cdot \\ \cdot \\ x_N \end{bmatrix} \qquad \textbf{(9.80)}$$

and the coefficient matrix

$$[B] = \begin{bmatrix} -r_{11} & r_{12} & r_{13} & \cdots & r_{1N} \\ r_{21} & -r_{22} & r_{23} & \cdots & r_{2N} \\ r_{31} & r_{32} & -r_{33} & \cdots & r_{3N} \\ \cdot & \cdot & \cdot & & \cdot \\ \cdot & \cdot & \cdot & & \cdot \\ \cdot & \cdot & \cdot & & \cdot \\ r_{N1} & r_{N2} & r_{N3} & \cdots & -r_{NN} \end{bmatrix} \qquad \textbf{(9.81)}$$

then the MTBF is computed by summing the unique x_i values of the solution to

$$[B]\mathbf{x} = \begin{bmatrix} -1 \\ 0 \\ 0 \\ \cdot \\ \cdot \\ \cdot \\ 0 \end{bmatrix} \tag{9.82}$$

$$\text{MTBF} = \sum_{i=1}^{N} x_i. \tag{9.83}$$

Recall that this solution for MTBF presumes that upon failure of the system, the system will always be restored to the $t = 0$ condition, that is, into State 1.

EXAMPLE 9.36 Solve the MTBF of a system composed of two identical, active, redundant units, each with failure rate λ. Assume that the system is nonrestored. See Figure 9.26.

There are two Markov success states.

Success State	Units Up	Units Down
1	2	0
2	1	1

The state-to-state transition-rate diagram is shown in Figure 9.27.

The transition rates are

$$r_{21} = 2\lambda, \quad r_{F1} = 0, \quad r_{12} = 0$$
$$tr_{F2} = \lambda, \quad r_{11} = 2\lambda, \quad r_{22} = \lambda$$

The coefficient matrix is

$$[B] = \begin{bmatrix} -2\lambda & 0 \\ 2\lambda & -\lambda \end{bmatrix}$$

FIGURE 9.26
Block diagram for Example 9.36.

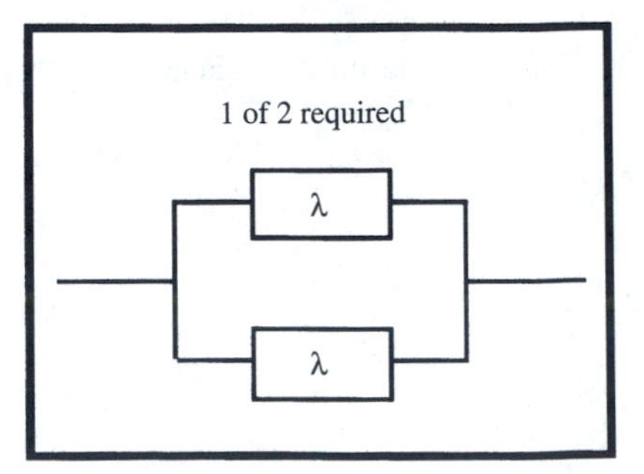

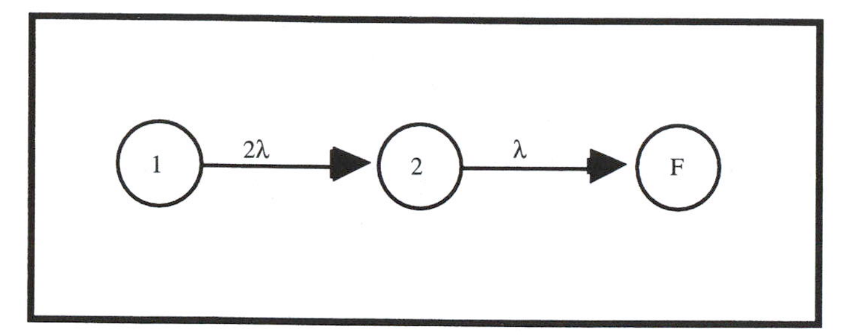

FIGURE 9.27
Transition-rate diagram for Example 9.36.

Solving for the vector **x**, $[B]\mathbf{x} = \begin{bmatrix} -1 \\ 0 \end{bmatrix}, \quad \begin{bmatrix} -2\lambda & 0 \\ 2\lambda & -\lambda \end{bmatrix} \begin{bmatrix} x_1 \\ x_2 \end{bmatrix} = \begin{bmatrix} -1 \\ 0 \end{bmatrix}$, and

$$-1 = -2\lambda x_1$$

$$0 = 2\lambda x_1 - \lambda x_2$$

The vector solution is $x_1 = 1/2\lambda$ and $x_2 = 2x_1 = 1/\lambda$. So,

$$\text{MTBF} = x_1 + x_2 = 3/2\lambda.$$

Compare this analysis with what you would do if you were following the techniques of Chapter 8.

EXAMPLE 9.37 Suppose the system in Example 9.36 is a restorable system with restoration rate μ. Solve the system MTBF. See Figure 9.28.

The MTBF solution for this example is not possible using the techniques of Chapter 8. However, by using the Markov model technique, it is no more difficult than solving Example 9.36. Again there are two Markov success states.

FIGURE 9.28
Block diagram for Example 9.37.

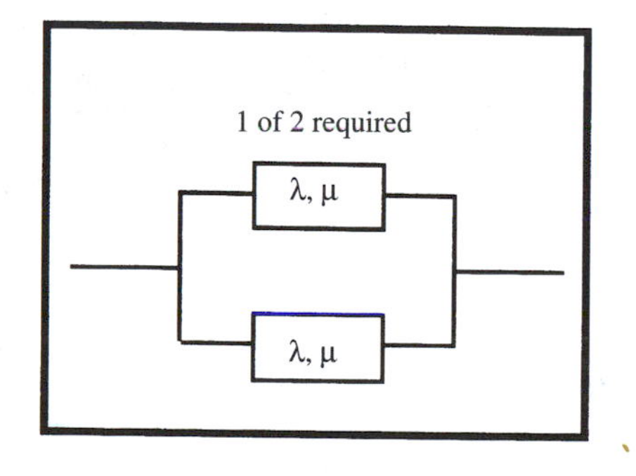

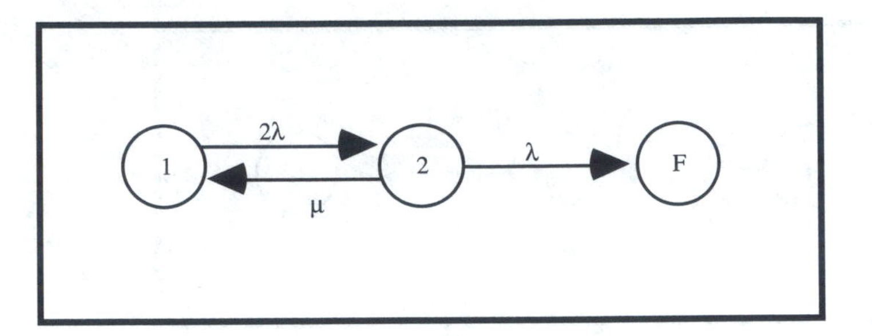

FIGURE 9.29
Transition-rate diagram for Example 9.37.

Success State	Units Up	Units Down	Units under Repair
1	2	0	0
2	1	1	1

The state-to-state transition rate is shown in Figure 9.29.
The transition rates are

$$r_{21} = 2\lambda, \quad r_{F1} = 0, \quad r_{12} = \mu$$
$$r_{F2} = \lambda, \quad r_{11} = 2\lambda, \quad r_{22} = \mu + \lambda$$

The coefficient matrix is then

$$[B] = \begin{bmatrix} -2\lambda & \mu \\ 2\lambda & -(\mu + \lambda) \end{bmatrix}$$

Solving for the vector **x**,

$$[B]\mathbf{x} = \begin{bmatrix} -1 \\ 0 \end{bmatrix}, \quad \begin{bmatrix} -2\lambda & \mu \\ 2\lambda & -(\mu + \lambda) \end{bmatrix}\begin{bmatrix} x_1 \\ x_2 \end{bmatrix} = \begin{bmatrix} -1 \\ 0 \end{bmatrix}, \quad \text{and}$$
$$-1 = -2\lambda x_1 + \mu x_2$$
$$0 = 2\lambda x_1 - (\mu + \lambda)x_2$$

The vector solution is $x_1 = (\mu + \lambda)/2\lambda^2$ and $x_2 = 1/\lambda$. So,

$$\text{MTBF} = x_1 + x_2 = \frac{(\mu + \lambda)}{2\lambda^2} + \frac{1}{\lambda} = \frac{\mu + 3\lambda}{2\lambda^2}$$

EXAMPLE 9.38 Solve the MTBF of a nonrestored system composed of a single active unit backed up by an identical standby unit if the active failure rate is λ, the standby failure rate is 0, and we assume perfect sensing and switching. See Figure 9.30.

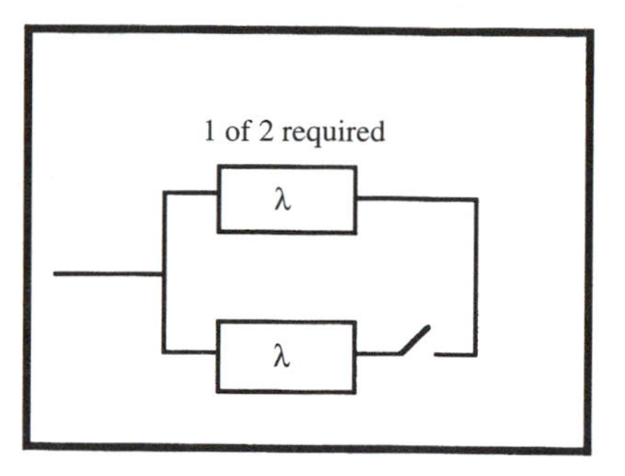

FIGURE 9.30
Block diagram for Example 9.38.

There are two Markov states as follows:

Success State	Units Up	Units Down	Units in Standby
1	1	0	1
2	1	1	0

The state-to-state transition rate diagram is shown in Figure 9.31. The transition rates are

$$r_{21} = \lambda, \quad r_{F1} = 0, \quad r_{12} = 0$$
$$r_{F2} = \lambda, \quad r_{11} = \lambda, \quad r_{22} = \lambda$$

The coefficient matrix is

$$[B] = \begin{bmatrix} -\lambda & 0 \\ \lambda & -\lambda \end{bmatrix}$$

Solving for the vector; **x**,

$$[B]\mathbf{x} = \begin{bmatrix} -1 \\ 0 \end{bmatrix}, \quad \begin{bmatrix} -\lambda & 0 \\ \lambda & -\lambda \end{bmatrix}\begin{bmatrix} x_1 \\ x_2 \end{bmatrix} = \begin{bmatrix} -1 \\ 0 \end{bmatrix}, \quad \text{and}$$
$$-1 = -\lambda x_1$$
$$0 = \lambda x_1 - \lambda x_2$$

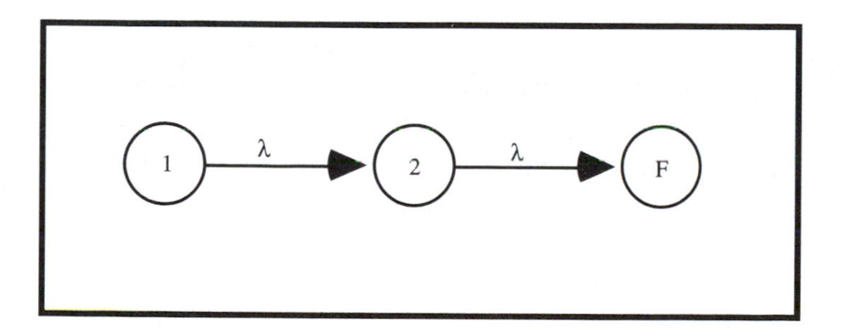

FIGURE 9.31
Transition rate diagram for Example 9.38.

The vector solution is $x_1 = 1/\lambda$ and $x_2 = x_1 = 1/\lambda$. So,

$$\text{MTBF} = x_1 + x_2 = 2/\lambda$$

Compare this approach to that of Chapter 8 leading to Equation 8.31.

In Section 9.6 we evaluated complex system MTBF through the use of the tables provided in Appendix C. The Appendix C tables were prepared to provide quick MTBF solutions for a variety of the most commonly seen system configurations. Although these tables were developed from the same Markov model approach presented in this section, their use demands no understanding of mathematics beyond the simple algebra required by the nonasterisked sections of this text. These tables can also be used by experienced reliability analysts to evaluate expeditiously the MTBF of complex systems with a pocket calculator.

As additional exercises, you are encouraged to use Equations 9.78 and 9.79 to derive the MTBF solutions provided in the tables of Appendix C.

9.10 EFFECT OF SOFTWARE

Modern systems, both manufactured products and rendered services, are typically and increasingly dependent on **software** to fulfill their functions. Designed software operates and monitors the performance of physical systems and manages the flow of information necessary to support service rendering systems. Software flaws can, therefore, impact the reliability of a system, despite the fact that reliability as defined in this text cannot be applied to software itself.

In other words, software does not have a failure rate, an MTBF, an MTTR, or availability. Software does not degrade or wear out or suffer some catastrophic failure. The software itself does not fail; rather, flaws within the software can possibly cause a failure in its dependent system. Every such flaw is the result of a programmer's error. Risks of system failure cannot be diminished through active and standby redundancy, because a fault on one software issue will exist on all of them. Also, a software fault, once detected, can be eliminated through correction of the programming error, so the fault need not cause a system failure again. Let us define the following terms relative to the affect of software upon system reliability.

Software error This refers to an error made by a programmer, such as a typographical error, an incorrect numerical value, an omission, etc.

Software fault An error leads to a fault in the software. Software faults can remain undetected for extended periods of time. Often they are not detected until they cause a system failure.

Imposed failure A software fault may or may not ultimately lead to a failure of the system. When a software-imposed system failure occurs, it is repaired by isolating the source of the fault (that is, the error) and correcting it. Then, the system should never fail in that mode again.

Software reliability This is the probability that a system will not fail during a mission due to a software fault.

Software MTBF This refers to the mean time between system failures due to software faults.

Software MTTR This is the mean time to restore a system to operation upon a failure due to a software fault. This activity usually includes the correction of the error.

Software availability This is the portion of the required operating time that a system is not down due to a software fault.

Software-reliability activities consist of fault prevention and reduction, reliability assessment and prediction, and fault elimination. Fault prevention generally consists of design practices such as modularity that are aimed at the prevention and easy detection of programming errors. Regarding assessment, there are many existing models for the prediction of faults and resulting failures. These are outside the scope of this text, and they may be found in texts that concentrate on the subject of software reliability, some of which are among the references listed at the end of this chapter. Some of the more common software-reliability models not only assess the reliability of a particular software, but reduce software faults as a byproduct of the assessment technique. To include the effects of software upon the reliability, MTBF, or availability of a complex system, we will need to estimate software reliability separately and combine results. For example,

$$R_{total}(t) = R(t) \times R_{s/w}(t), \tag{9.84}$$

where

$R(t)$ = the system mission reliability exclusive of software considerations

$R_{s/w}(t)$ = the software reliability, as defined before

$R_{total}(t)$ = the system mission reliability, inclusive of the effects of software

BIBLIOGRAPHY

1. ARINC Research Corp. 1964. *Reliability Engineering.* Upper Saddle River, N.J.: Prentice Hall.
2. I. Bazovsky. 1961. *Reliability Theory and Practice.* Upper Saddle River, N.J.: Prentice Hall.
3. E. E. Lewis. 1987. *Introduction to Reliability Engineering.* New York: John Wiley.
4. D. K. Lloyd and M. Lipow. 1962. *Reliability Management, Methods, and Mathematics.* Upper Saddle River, N.J.: Prentice Hall.
5. P. D. T. O'Connor. 1991. *Practical Reliability Engineering,* 3d ed. New York: John Wiley.
6. N. H. Roberts. 1964. *Mathematical Methods in Reliability Engineering.* New York: McGraw-Hill.
7. G. H. Sandler. 1963. *System Reliability Engineering.* Upper Saddle River, N.J.: Prentice Hall.
8. M. L. Shooman. 1968. *Probabilistic Reliability: An Engineering Approach.* New York: McGraw-Hill.
9. C. O. Smith. 1976. *Introduction to Reliability in Design.* New York: McGraw-Hill.
10. P. A. Tobias and D. Trindade. 1986. *Applied Reliability.* New York: Van Nostrand Reinhold.

EXERCISES

9.1. A system is composed of four identical, active elements, two of which are required for system success. On-line restoration is possible, with one restoration activity at a time. Each element has a failure rate of 46×10^{-6} per hour and a restoration rate of 2.0 per hour. Use Appendix C to determine the system MTBF.

9.2. Use Appendix C to find the MTBF for the system of Exercise 9.1 if three of the four elements are required for system success.

9.3. A system is composed of four identical, active elements, two of which are required for system success. On-line restoration is not possible. Each element has a failure rate of 46×10^{-6} per hour. Use Appendix C to determine the system MTBF.

9.4. Use Appendix C to find the MTBF for the system of Exercise 9.3 if three of the four elements are required for system success.

9.5. A system is composed of seven identical, active elements, four of which are required for system success. On-line restoration is not possible. Each element has a failure rate of 500×10^{-6} per hour. Use Appendix C to find the system MTBF.

9.6. A system is composed of 10 identical, active elements, 8 of which are required for system success. On-line restoration is not possible. Each element has a failure rate of 500×10^{-6} per hour. Use Appendix C to find the system MTBF.

9.7. A system is composed of two active elements, one of which is required for system success. On-line restoration is not possible. Element 1 has a failure rate of 250×10^{-6} per hour, and element 2 has a failure rate of 750×10^{-6} per hour. Use Appendix C to determine the system MTBF.

9.8. A system has two power supplies, at least one of which is required for system success. On-line restoration of these power supplies is possible (one at a time). The average restoration time is 15 min. With both power supplies operating, each has a failure rate of 25×10^{-6} per hour. If one power supply is down, the remaining one has a failure rate of 150×10^{-6} per hour. Use Appendix C to find the MTBF for this power supply subsystem.

9.9. A system contains two identical elements, one active and one in standby, each with an active failure rate of 1000 failures per million hours and a negligible standby failure rate. On-line maintenance is possible with an element restoration rate of 1.0 per hour. Compute the system MTBF if we assume perfect sensing and switching.

9.10. A hospital operating room has an emergency generator in standby in the event of a power failure. Suppose power failures occur at an average rate of once every 5000 h, and the active failure rate for the emergency generator is 200 failures per 1,000,000 h. Suppose further that it takes an average of 3 h to restore power from either of the two sources. Assuming perfect sensing and switching and a negligible standby failure rate for the emergency generator, what is the MTBF of the operating room's power system?

9.11. How much improvement in MTBF can be achieved by adding a second emergency generator to the operating room of Exercise 9.10? In other words, if the system were composed of the primary power source and two standby generators, how much higher would the MTBF be? Once again assume perfect sensing and switching and a negligible standby failure rate for the emergency generator.

9.12. Solve Exercise 9.10 if we assume that the probability of successful sensing and switching upon demand is 0.985 and that the generator failure rate in standby is negligible.

9.13. A delivery service has a fleet of five trucks, four of which must be in operation at all times. Each truck breaks down at a rate of once every 5000 h. The average restoration time is 2.0 h. What is the MTBF of this system?

9.14. A delivery service has a fleet of five trucks, three of which must be in operation at all times Each truck breaks down at a rate of once every 5000 h. The average restoration time is 2.0 h. What is the MTBF of this system?

9.15. Fifty instrumented buoys are dropped into the ocean to track the migration of an oil spill. At least 45 of these buoys must be operating for successful tracking. The failure rate of one of these buoys is 150×10^{-6} per hour. These buoys are not maintainable once in service. What is the MTBF of this system?

9.16. A customer service department has three service representatives available to take phone calls from customers. If calls come in at a steady rate of 20 per hour throughout the day and if the average length of a service telephone conversation is 10 min, how frequently would all three representatives be unavailable for an incoming call? In other words, we are looking for the MTBF of a system requiring at least one out of three representatives at all times.

9.17. A communication system is composed of a transmitter, an active receiver, and a standby receiver that can be substituted if the active one fails. Suppose the transmitter has a failure rate of 0.0025 per hour, the active receiver has a failure rate of 0.0200 per hour, and the on-line receiver restoration rate is 0.330 per hour. Assume perfect sensing and switching and a negligible standby failure rate for the receivers. Use the technique prescribed by Section 9.5 to determine this system's MTBF.

9.18. Use the method prescribed by Section 9.5 to determine the MTBF of the system whose block diagram is given in Figure 9.32. Elements B and C are in standby redundancy, as shown. The active failure rates are 100 failures per million hours for element A, 500 failures per million hours for element B, and 1000 failures per 1,000,000 h for element C. The restoration rates are 1.0 per hour for all elements. The standby failure rates are negligible. Assume perfect sensing and switching.

9.19. Use Equations 9.8 and 9.9 to solve the MTBF of a system composed of N identical, active elements, $N - 1$ of which are required for system success. On-line restoration is possible with one restoration at a time. Each element has a failure rate λ and a restoration rate μ.

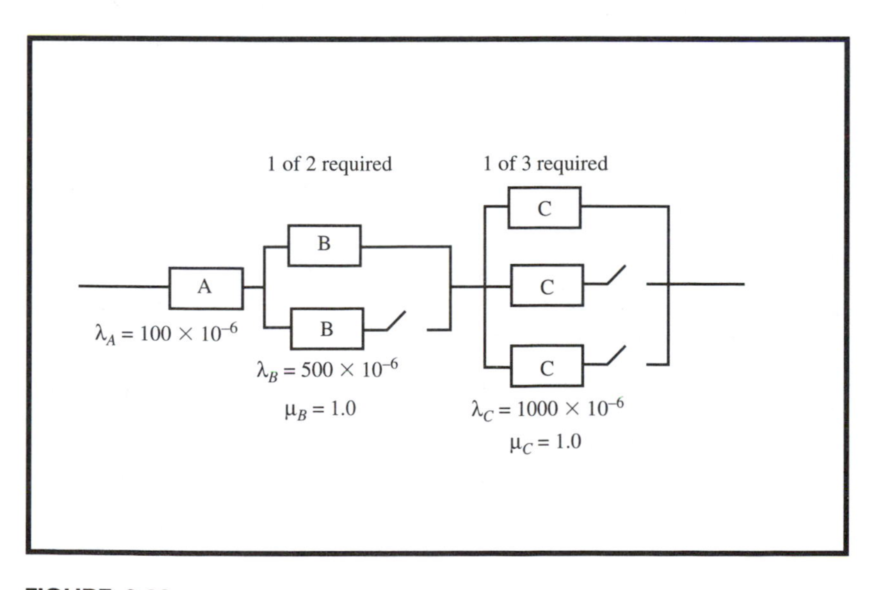

FIGURE 9.32
Block diagram for Exercise 9.18.

9.20. Use Equations 9.8 and 9.9 to solve for the MTBF of a system composed of N identical, active elements, $N - 2$ of which are required for system success. On-line restoration is possible with one restoration at a time. Each element has a failure rate λ and a restoration rate μ.

9.21. Use Equations 9.8 and 9.9 to determine the MTBF of a system composed of N identical, active elements, $N - 1$ of which are required for system success. On-line restoration is possible with one restoration at a time. With all elements up, each has a failure rate λ_A and a restoration rate μ. With one element down, each has a failure rate λ_B.

9.22. Use Equations 9.8 and 9.9 to find the MTBF of a system composed of two active elements, one of which is required for system success. On-line restoration is possible, with one restoration at a time. Element 1 has a failure rate λ_1 and a restoration rate μ_1; element 2 has a failure rate λ_2 and a restoration rate μ_2.

9.23. Use Equations 9.8 and 9.9 to find the MTBF of a system composed of N identical, active elements, $N - 1$ of which are required for system success. On-line restoration is not possible. Each element has a failure rate λ.

9.24. Use Equations 9.8 and 9.9 to determine the MTBF of a system composed of N identical, active elements, A of which are required for system success. On-line restoration is not possible. Each element has a failure rate λ.

9.25. Use Equations 9.8 and 9.9 to solve for the MTBF of a system composed of two active elements, one of which is required for system success. On-line restoration is not possible. Element 1 has a failure rate λ_1, and Element 2 has a failure rate λ_2.

9.26. Use Equations 9.8 and 9.9 to solve for the MTBF of a system composed of N identical, elements, $N - 1$ of which are active and required for system success. One element is initially in standby. On-line restoration is possible, with one restoration at a time. Each active element has a failure rate λ and a restoration rate μ. Standby failure rates are assumed negligible, and perfect sensing and switching are assumed.

9.27. Use Equations 9.8 and 9.9 to find the MTBF of a system composed of N identical, elements, $N - 1$ of which are active and required for system success. One element is initially in standby. On-line restoration is possible, with one restoration at a time. Each active element has a failure rate λ and a restoration rate μ. Standby failure rates are assumed negligible. The probability of successful switchover upon demand is p_s.

9.28. Use Equations 9.8 and 9.9 to determine the MTBF of a system composed of N identical, elements, $N - 1$ of which are active and required for system success. One element is initially in standby. On-line restoration is not possible. Each active element has a failure rate λ. Standby failure rates are assumed negligible, and perfect sensing and switching are assumed.

9.29. Use Equations 9.8 and 9.9 to find the MTBF of a system composed of N identical elements, $N - 1$ of which are active and required for system success. One element is initially in standby. On-line restoration is not possible. Each active element has a failure rate λ. Standby failure rates are assumed negligible. The probability of successful switchover upon demand is p_s.

9.30. Use Equations 9.8 and 9.9 to solve for the MTBF of a system composed of N identical, active elements, $N - 1$ of which are required for system success. On-line restoration is not possible. When all elements operate, each element has a failure rate of λ_0. When one element is down, each of the remaining ones has a failure rate λ_1. When two elements are down, each of the remaining ones has a failure rate λ_2, etc.

Compare your results in Exercises 9.19 through 9.30 with the tables in Appendix C.

10

SYSTEM AVAILABILITY AND DEPENDABILITY

10.1 INTENT

In chapter 2 we learned that for continuously operating systems the reliability parameters that concern us are MTBF and availability. The *MTBF* is a measure of how frequently the system is expected to go down, whereas the *availability* indicates that portion of the demand time that the system is operating satisfactorily. In the case of an inactive, standby system, availability indicates the portion of time that the system is operable, that is, in a state of operational readiness. Availability defined in this manner for an active or standby system is often referred to as *steady-state availability,* which is the topic of this chapter. Steady-state availability is the preferred parameter for evaluating the reliability and safety for a power plant and its systems. It is also the best parameter for dealing with computer systems, fire-protection systems, and, for that matter, any complex system that is required and depended upon on a continual basis.

10.2 SYSTEMS WITH ALL ELEMENTS ACTIVE

For a series system composed of N independent elements, with failure rates λ_1, $\lambda_2,\ldots, \lambda_N$ and with restoration rates $\mu_1, \mu_2,\ldots, \mu_N$, as illustrated in Figure 10.1, the **availability** is

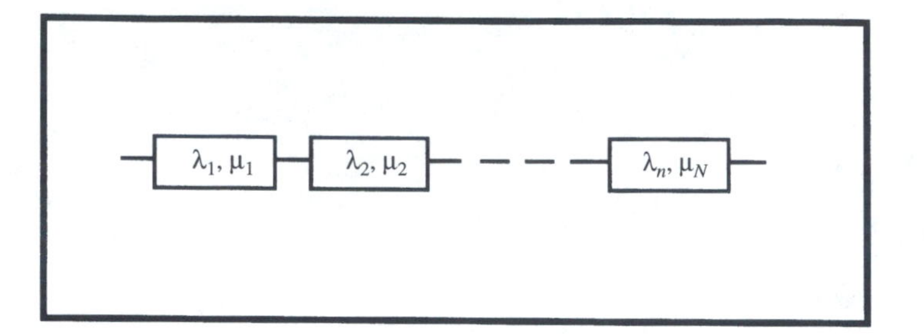

FIGURE 10.1
Block diagram for a series system with N elements.

$$A = \frac{\text{MTBF}}{\text{MTBF} + \text{MTTR}} = \frac{\mu_{sys}}{\mu_{sys} + \lambda_{sys}}$$

where
$$\lambda_{sys} = \lambda_1 + \lambda_2 + \cdots + \lambda_N \qquad \textbf{(10.1)}$$

$$\mu_{sys} = \frac{\lambda_1\mu_1 + \lambda_2\mu_2 + \cdots + \lambda_N\mu_N}{\lambda_{sys}}$$

EXAMPLE 10.1 Consider the system of ATM machines of Example 9.1, where the system effective failure rate was determined to be 228.0×10^{-6} per hour. Suppose the system restoration rate is, as given in Example 9.3, 0.20 per hour. Determine the system availability.

From Equation 10.1,

$$A = \frac{\mu_{sys}}{\mu_{sys} + \lambda_{sys}} = \frac{0.20}{0.20 + (228.0 \times 10^{-6})} = \frac{0.200000}{0.200228} = 0.99986$$

Now let us consider systems composed of N active elements, any k of which are required for system success. Availability solutions for these systems are provided in Tables C.11 through C.15 (Appendix C). All availability tables in Appendix C were derived through methods presented in Section 10.5.

10.2.1 Identical Elements with Restoration

Table C.11 provides availability solutions for systems with N identical, active elements, k of which are required for system success, where on-line restoration is possible and each element has a failure rate λ and a restoration rate μ. The tables in Appendix C assume that only one restoration activity at a time is possible. See Figure 10.2.

EXAMPLE 10.2 Suppose that at least three out of five ATM systems, identical to the one used in Example 10.1, are required at all times. What is the availability for this set of systems?

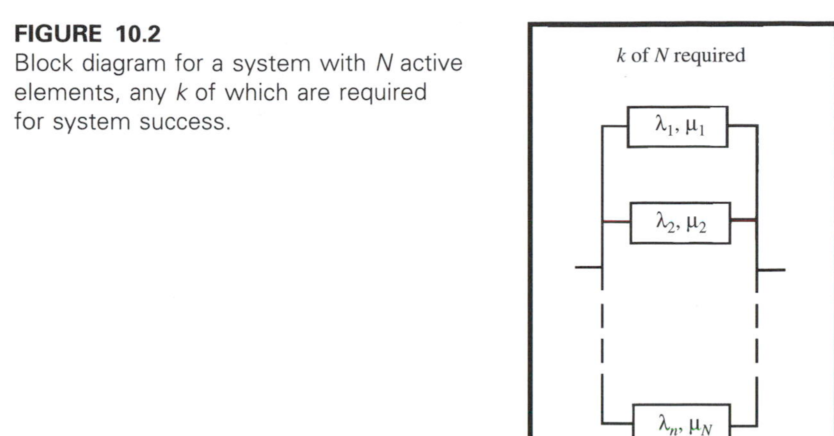

FIGURE 10.2
Block diagram for a system with *N* active elements, any *k* of which are required for system success.

The equation for three out of five from Table C.11 applies, with $\lambda = 228.0 \times 10^{-6}$ per hour and $\mu = 0.20$ per hour.

$$A = \frac{\mu^5 + 5\mu^4\lambda + 20\mu^3\lambda^2}{\mu^5 + 5\mu^4\lambda + 20\mu^3\lambda^2 + 60\mu^2\lambda^3 + 120\mu\lambda^4 + 120\lambda^5}$$

$$= \frac{0.20^5 + 5(0.20^4)(0.000228) + 20(0.20^3)(0.000228^2)}{0.20^5 + 5(0.20^4)(0.000228) + 20(0.20^3)(0.000228^2) + 60(0.20^2)(0.000228^3) + 120(0.20)(0.000228^4) + 120(0.000228^5)}$$

$$= 1 - 8.86 \times 10^{-8} = 0.99999991$$

EXAMPLE 10.3

Solve Example 10.2 if four out of five are required.

The equation for four out of five from Table C.11 applies with $\lambda = 228.0 \times 10^{-6}$ per hour and $\mu = 0.20$ per hour.

$$A = \frac{\mu^5 + 5\mu^4\lambda}{\mu^5 + 5\mu^4\lambda + 20\mu^3\lambda^2 + 60\mu^2\lambda^3 + 120\mu\lambda^4 + 120\lambda^5}$$

$$= \frac{0.20^5 + 5(0.20^4)(0.000228)}{0.20^5 + 5(0.20^4)(0.000228) + 20(0.20^3)(0.000228^2) + 60(0.20^2)(0.000228^3) + 120(0.20)(0.000228^4) + 120(0.000228^5)}$$

$$= 0.99997$$

EXAMPLE 10.4 Solve Example 10.2 if all five are required at all times.

The equation for five out of five from Table C.11 applies, with $\lambda = 228.0 \times 10^{-6}$ per hour and $\mu = 0.20$ per hour.

$$A = \frac{\mu^5}{\mu^5 + 5\mu^4\lambda + 20\mu^3\lambda^2 + 60\mu^2\lambda^3 + 120\mu\lambda^4 + 120\lambda^5}$$

$$= \frac{0.20^5}{0.20^5 + 5(0.20^4)(0.000228) + 20(0.20^3)(0.000228^2) + 60(0.20^2)(0.000228^3) + 120(0.20)(0.000228^4) + 120(0.000228^5)}$$

$$= 0.99431$$

EXAMPLE 10.5 An electronic assembly requires at least one out of two active, redundant power supplies to operate. On-line restoration is possible. If each power supply has a failure rate of 25×10^{-6} per hour and an on-line restoration rate of 2.0 per hour, determine the availability of this power supply subsystem.

The Table C.11 equation for one out of two applies:

$$A = \frac{\mu^2 + 2\mu\lambda}{\mu^2 + 2\mu\lambda + 2\lambda^2} = \frac{2.0^2 + (2 \times 2.0 \times 25 \times 10^{-6})}{2.0^2 + (2 \times 2.0 \times 25 \times 10^{-6}) + (2 \times 25^2 \times 10^{-12})}$$

$$= 1 - 3.1 \times 10^{-10} = 0.9999999997$$

10.2.2 Identical Elements without Restoration

Table C.14 provides system availability solutions for k out of N identical elements, where on-line restoration is *not* possible. Each element is assumed to have a failure rate λ, and μ is the rate at which the system is restored to service when it is down.

EXAMPLE 10.6 Solve Example 10.5 if on-line restoration is not possible.

The Table C.14 equation for one out of two applies, with $\lambda = 25 \times 10^{-6}$ per hour and $\mu = 2$ per hour.

$$A = \frac{3\mu}{3\mu + 2\lambda} = \frac{3 \times 2}{(3 \times 2) + (2 \times 25 \times 10^{-6})} = \frac{6.0}{6.0 + 0.000050} = 0.9999916$$

10.2.3 Identical Elements with Failure Rates Subject to Change

Table C.12 provides availability solutions for systems composed of identical elements with constant failure rates that can change when one or more of the elements is down.

EXAMPLE 10.7 Solve Example 10.5 if, upon the loss of one of the power supplies, the failure rate of the remaining one increases to 75×10^{-6} per hour.

Table C.12 applies, with $\lambda_A = 25 \times 10^{-6}$ per hour, $\lambda_B = 75 \times 10^{-6}$ per hour, $\mu = 2$ per hour, $N = 2$, and $k = 1$:

$$A = \frac{\dfrac{\mu^{N-1}}{(N-1)!\lambda_B^{N-1}} + \dfrac{\mu^N}{N!\lambda_A\lambda_B^{N-1}}}{\dfrac{\mu^N}{N!\lambda_A\lambda_B^{N-1}} + \sum_{i=0}^{N-1}\dfrac{\mu^i}{i!\lambda_B^i}} = \frac{\dfrac{2}{0.000075} + \dfrac{2^2}{2 \times 0.000025 \times 0.000075}}{\dfrac{2^2}{2 \times 0.000025 \times 0.000075} + \dfrac{2}{0.000075} + 1}$$

$$= 1 - 9.37 \times 10^{-10} = 0.9999999991$$

10.2.4 Redundant Elements with Different Failure Rates

For two redundant elements that are not necessarily identical and may have different failure and restoration rates, use Table C.13 to solve system availability where on-line restoration is possible, and use Table C.15 where on-line restoration is not possible.

EXAMPLE 10.8 Solve the availability of the computer subsystem of Example 9.11. There are two printers, both on-line full time. One has a failure rate of 1 per 1,000,000 h and an average restoration time of 3 h. The other has a failure rate of 1 per 100,000 h and an average restoration time of 30 min.

From Table C.13,

$$A = \frac{\mu_1\mu_2 + \mu_2\lambda_1 + \mu_1\lambda_2}{\mu_1\mu_2 + \mu_2\lambda_1 + \mu_1\lambda_2 + \lambda_1\lambda_2} = \frac{\dfrac{2}{3} + 2(1 \times 10^{-6}) + \dfrac{10 \times 10^{-6}}{3}}{\dfrac{2}{3} + 2(1 \times 10^{-6}) + \dfrac{10 \times 10^{-6}}{3} + (10 \times 10^{-12})}$$

$$= 1 - 1.5 \times 10^{-11} = 0.999999999985$$

EXAMPLE 10.9 Solve Example 10.8 if on-line restoration is *not* possible.

Assume that the average restoration time upon system failure is

$$(3 \times 10^{-6} + (0.5 \times 10^{-5})\,\frac{(3 \times 10^{-6}) + (0.5 \times 10^{-5})}{(11 \times 10^{-6})} = 0.727 \text{ h}$$

Hence, $\mu = 1.375$ per hour.

From Table C.15,

$$A = \frac{\mu(\lambda_1^2 + \lambda_2^2 + \lambda_1\lambda_2)}{\mu(\lambda_1^2 + \lambda_2^2 + \lambda_1\lambda_2) + \lambda_1^2\lambda_2 + \lambda_2^2\lambda_1} = \frac{1.375(1 + 100 + 10) \times 10^{-12}}{1.375(1 + 100 + 10) \times 10^{-12} + (10 + 100) \times 10^{-18}}$$

$$= 1 - 7 \times 10^{-7} = 0.9999993$$

10.3 SYSTEMS WITH STANDBY REDUNDANCY

Table C.16 can be used to solve the *availability* of systems with redundant elements in standby. To use this table we must assume perfect sensing and switching and that the standby elements have negligible failure rates. If we cannot make those assumptions, the Markov model technique of Section 10.5 must be used.

EXAMPLE 10.10 Solve the system availability for the receiver subsystem of Example 9.15, where two active receivers are backed up by one standby receiver. The active receiver failure rate is estimated at 12×10^{-6} per hour and the standby rate is assumed to be negligible. The restoration rate is estimated at 1 per hour, and perfect sensing and switching are assumed.

Table C.16 applies, with $N = 3$, $k = 2$, $\mu = 1$ per hour, and $\lambda = 12 \times 10^{-6}$ per hour:

$$A = \frac{\mu^3 + 2\mu^2\lambda}{\mu^3 + 2\mu^2\lambda + 4\mu\lambda^2 + 4\lambda^3}$$

$$= \frac{1 + (2 \times 1 \times 12 \times 10^{-6})}{1 + (2 \times 1 \times 12 \times 10^{-6}) + (4 \times 1 \times 144 \times 10^{-12}) + (4 \times 1728 \times 10^{-18})}$$

$$= 1 - 5.8 \times 10^{-10} = 0.99999999942$$

EXAMPLE 10.11 A company has a fleet of five buses, backed up by one spare that can be put into service upon occurrence of a breakdown of any of its active buses. Assuming a breakdown rate of 0.0003 per hour for an active bus, a restoration rate of 0.04 per hour, negligible standby failure rate, and perfect sensing and switching, estimate the availability of the fleet.

Table C.16 applies, with $N = 5$, $k = 4$, $\lambda = 0.0003$ per hour, and $\mu = 0.04$ per hour:

$$A = \frac{\mu^5 + 4\mu^4\lambda}{\mu^5 + 4\mu^4\lambda + 16\mu^3\lambda^2 + 48\mu^2\lambda^3 + 96\mu\lambda^4 + 96\lambda^5}$$

$$= \frac{0.04^5 + (4 \times 0.04^4 \times 0.003)}{0.04^5 + (4 \times 0.04^4 \times 0.003) + (16 \times 0.04^3 \times 0.003^2) + (48 \times 0.04^2 \times 0.003^3) + (96 \times 0.04 \times 0.003^4) + (96 \times 0.003^5)}$$

$$= 0.9991$$

10.4 TABLE APPROACH TO SOLVING COMPLEX SYSTEM AVAILABILITY

If you are required to determine the availability of a system more complex than those included in the tables of Appendix C, it is sometimes possible to resolve the complex system into a chain of subsystems whose availability can be solved through the Appendix C tables. Similar to the evaluation of complex system MTBF as presented in Section 9.5, the system availability can be determined from those of the subsystems.

An example of this technique appears in Figure 10.3, which analyzes the system used in Example 9.20. The same rules and restrictions that apply to the determination of

Logic	Block Diagram	Logic	Availability Equation	Item Failure Rate	Item Restoration Rate	A
A	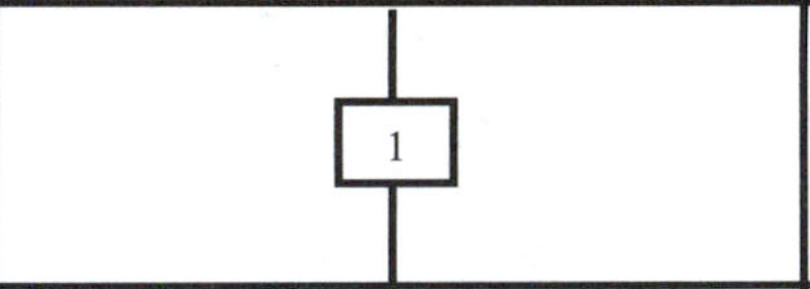 1	1 required	$\frac{\mu}{\mu + \lambda}$	$\lambda = 35 \times 10^{-6}$	$\mu = 0.5$	0.999930
B	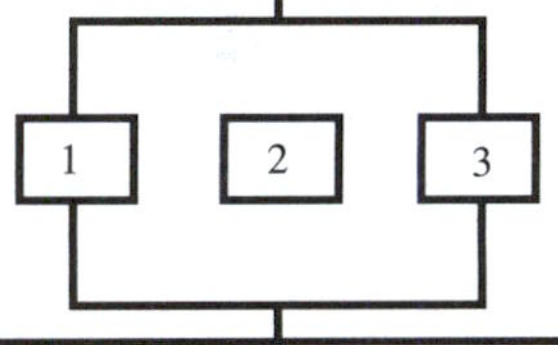 1 2 3	2 of 3 on-line restoration DF = 40%	$\frac{\mu^3 + (0.4 \times 3\mu^2\lambda)}{\mu^3 + (0.4 \times 3\mu^2\lambda) + (0.4^2 \times 6\mu\lambda^2) + (0.4^3 \times 6\lambda^3)}$	$\lambda = 125 \times 10^{-6}$	$\mu = 0.1$	0.999999
C	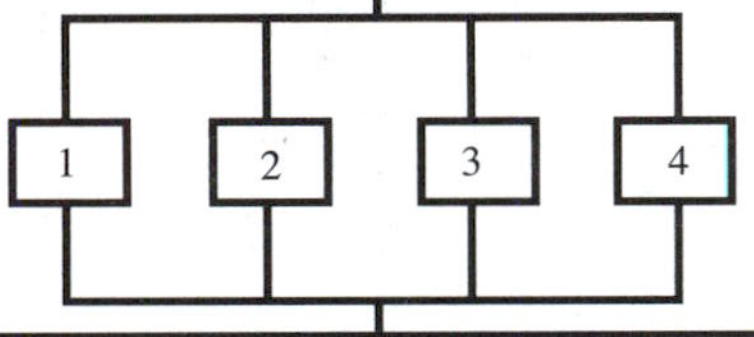 1 2 3 4	3 of 4 on-line restoration	$\frac{\mu^4 + 4\mu^3\lambda)}{\mu^4 + 4\mu^3\lambda + 12\mu^2\lambda^2 + 24\mu\lambda^3}$	$\lambda = 220 \times 10^{-6}$	$\mu = 0.5$	0.999998
D	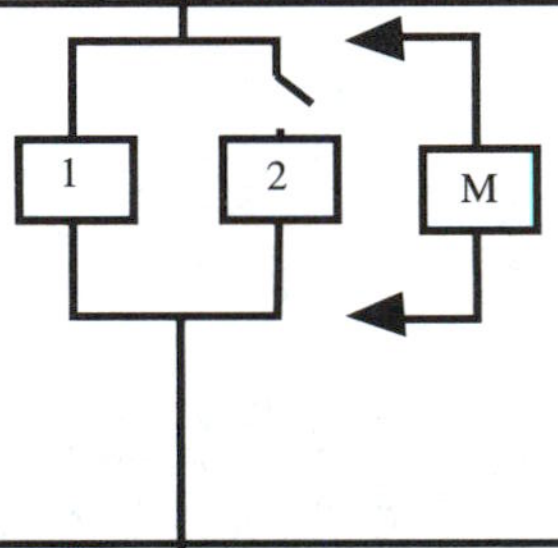 1 2 M	1 of 2 1 on-line 1 standby restoration perfect sensing and switching	$\frac{\mu^2 + \mu\lambda}{\mu^2 + \mu\lambda + \lambda^2}$	$\lambda = 25 \times 10^{-6}$	$\mu = 1.0$	1.000000
E	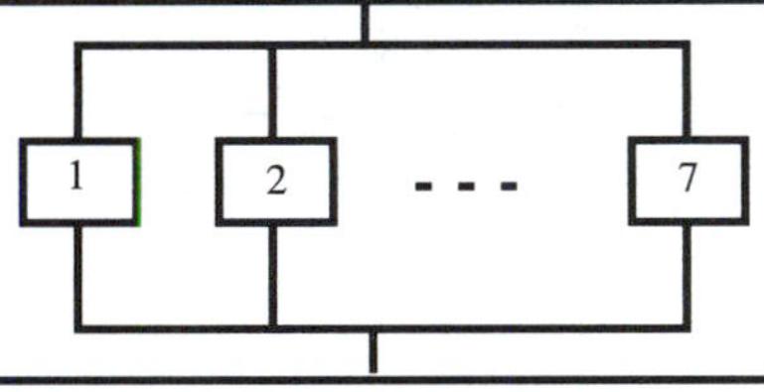 1 2 - - - 7	5 of 7 on-line restoration	$\frac{\mu^7 + 7\mu^6\lambda + 42\mu^5\lambda^2}{\mu^7 + 7\mu^6\lambda + 42\mu^5\lambda^2 + 210\mu^4\lambda^3 + 840\mu^3\lambda^4 + 2520\mu^2\lambda^5 + 5040\mu\lambda^6 +}$	$\lambda = 500 \times 10^{-6}$	$\mu = 1.0$	1.000000

FIGURE 10.3
Sample availability worksheet

system MTBF through this method also apply here. To use this method, follow all the instructions of Section 9.5, but use the Appendix C tables for availability this time. Determine the availability for each of the logic blocks, and then compute your approximation of the system availability by multiplying the availability values of all the logic blocks, as demonstrated in Example 10.12.

EXAMPLE 10.12 Estimate the availability for the system whose block diagram appears in the sample worksheet of Figure 10.3.

Using the sample worksheet, the system is resolved into five logic blocks, with the block diagrams drawn and described, as done in Example 9.20. From the block diagrams and logic descriptions, we can select an appropriate Appendix C availability equation for each logic block. The logic block A, B, C, and E equations come from Table C.11, and the logic block D equation is from Table C.16 (for which we assume perfect sensing and switching and negligible standby failure rate). Applying the associated failure and restoration rates to the equations, the availability values are determined for each logic block. The product of all logic block availability values gives an estimate of the system availability:

$$\begin{aligned} A &= A_A \times A_B \times A_C \times A_D \times A_E \\ &= 0.9999300 \times 0.9999985 \times 0.9999976 \times 1.0000000 \times 1.0000000 \\ &= 0.999926 \end{aligned}$$

10.5 MARKOV MODEL APPROACH TO SOLVING COMPLEX SYSTEM AVAILABILITY

As was the case in determining system MTBF in Chapter 9, the Markov model approach can be used for solving the availability for any system, regardless of its complexity. To use this approach, we first draw a system reliability block diagram, assigning failure and restoration rates to each block. Sensing and switching probabilities or failure and restoration rates may be assigned wherever they pertain. We then proceed as demonstrated in the examples that follow. You will notice that, in contrast to using the Markov model technique for solving MTBF, here we define *all* states, both *success* and *fail* states. Then we determine the transition rates among the defined states, as we did in Chapter 9, but this time we are concerned with the rates among the fail states as well as the success states. You may wish to review Section 9.6 for the convention used for defining the transition rates r_{ij}.

*10.5.1 Derivation of the Markov Model Solution for System Availability

The same six assumptions of Section 9.8 apply, so that at any time t the system must be in one of the N existing states. Because the states are both mutually exclusive and exhaustive, if $P_i(t)$ is the probability of being in the state i at the time t.

$$P_1(t) + P_2(t) + \cdots + P_N(t) = 1.0 \qquad \text{for all values of } t$$

At any point in time t, the system availability is the probability of the system being in one of the success states.

As before, we can find the $P_i(t)$ values through the system of equations

$$
\begin{aligned}
\frac{d}{dt}P_1(t) &= -r_{11}P_1(t) + r_{12}P_2(t) + r_{13}P_3t) + \cdots + r_{1N}P_N(t) \\
\frac{d}{dt}P_2(t) &= r_{21}P_1(t) - r_{22}P_2(t) + r_{23}P_3(t) + \cdots + r_{2N}P_N(t) \\
&\;\;\vdots \\
\frac{d}{dt}P_N(t) &= r_{N1}P_1(t) + r_{N2}P_2(t) + r_{N3}P_3(t) + \cdots - r_{NN}P_N(t)
\end{aligned}
\qquad \textbf{(10.2)}
$$

where $P_1(0) = 1.0$ and $P_i(0) = 0$ for all $i \neq 1$. The difference between the solution to this set of equations and similar ones in Chapter 9 is that some of the N states are success states, whereas others are down states.

There are three types of system availability, *instantaneous, interval,* and *steady-state* availability. The **instantaneous availability,** $A(t)$, is the probability that the system will be in a success state at a given point in time t. The **interval availability,** $A(T_1, T_2)$ is that portion of time within the interval that the system is expected to be in a success state. The **steady-state availability,** A, is the portion of time that the system is expected to be in a success state over the long haul. It is this third type of system availability with which we are generally concerned.

Mathematically, we can define the three types of system availability as follows:

$$\text{instantaneous availability} = A(t) = \sum_{\substack{\text{all success}\\ \text{states}}} P_i(t) \qquad \textbf{(10.3)}$$

$$\text{interval availability} = A(T_1, T_2) = \frac{1}{T_2 - T_1}\int_{T_1}^{T_2} A(t)\,dt \qquad \textbf{(10.4)}$$

Another form of interval availability is for the initial time T_1 into the mission and is defined as

$$A(T)_1) = \frac{1}{T_1}\int_0^{T_1} A(t)\,dt \qquad \textbf{(10.5)}$$

$$\text{steady-state availability} = A = \lim_{t \to 0} A(t) = \sum_{\substack{\text{all success}\\ \text{states}}} P_i \qquad \textbf{(10.6)}$$

where $P_i = \lim_{t \to \infty} P_i(t)$. But $\lim_{t \to \infty} \frac{d}{dt}P_i = 0$, so steady-state availability is

$$A = \sum_{\substack{\text{all success}\\ \text{states}}} P_i \qquad \textbf{(10.7)}$$

where the values of P_i are the solutions to the set of independent, linear equations

$$
\begin{aligned}
0 &= -r_{11}P_1 + r_{12}P_2 + r_{13}P_3 + \cdots + r_{1N}P_N \\
0 &= r_{21}P_1 - r_{22}P_2 + r_{23}P_3 + \cdots + r_{2N}P_N \\
0 &= r_{31}P_1 + r_{32}P_2 - r_{33}P_3 + \cdots + r_{3N}P_N \\
&\;\;\vdots \\
0 &= r_{N1}P_1 + r_{N2}P_2 + r_{N3}P_3 + \cdots - r_{NN}P_N
\end{aligned}
\qquad \textbf{(10.8)}
$$

To avoid a trivial solution, for the most trivial equation (usually the last one) in the set we can substitute

$$1 = \sum_{i=1}^{N} P_i = P_1 + P_2 + \cdots + P_N \tag{10.9}$$

Equation 10.9 is valid because all states are mutually exclusive and exhaustive.

In matrix format, if we define the vector

$$\mathbf{P} = \begin{bmatrix} P_1 \\ P_2 \\ \cdot \\ \cdot \\ \cdot \\ P_N \end{bmatrix} \tag{10.10}$$

the coefficient matrix

$$[C] = \begin{bmatrix} -r_{11} & r_{12} & r_{13} & \cdots & r_{1N} \\ r_{21} & -r_{22} & r_{23} & \cdots & r_{2N} \\ r_{31} & r_{32} & -r_{33} & \cdots & r_{3N} \\ \cdot & \cdot & \cdot & & \cdot \\ \cdot & \cdot & \cdot & & \cdot \\ \cdot & \cdot & \cdot & & \cdot \\ r_{N1} & r_{N2} & r_{N3} & \cdots & -r_{NN} \end{bmatrix} \tag{10.11}$$

and the zero vector

$$\mathbf{0} = \begin{bmatrix} 0 \\ 0 \\ \cdot \\ \cdot \\ \cdot \\ 0 \end{bmatrix} \tag{10.12}$$

then the P_i are the solution to

$$[A]\mathbf{P} = \mathbf{0} \tag{10.13}$$

The substitution of Equation 10.9 is customarily applied here.

EXAMPLE 10.13

Consider the system (of Figure 10.4) composed of three identical, active items, each of which has a failure rate λ and a restoration rate μ. Suppose that it is possible to do up to two restorations simultaneously. If at least two out of the three items are required for system success, what is the system's steady-state availability?

There are four states: one with all three items up, one with two up and one down, one with one up and two down, and one with all three items down. The first two states are system success states. The other two states, the ones with two or more items down, are both fail states. So the

FIGURE 10.4
Block diagram for Example 10.13.

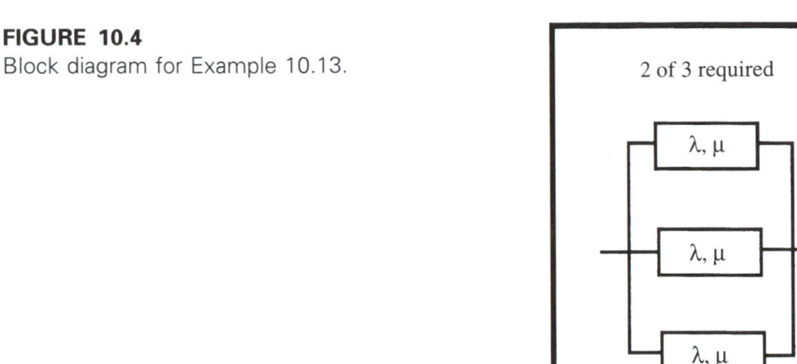

steady-state availability is the fraction of time over the long haul that the system is either in the first or the second state. The unreliability, or what we may call the *outage rate,* is the fraction of time that the system is either in the third or the fourth state. The four states are defined as follows:

Success State	Fail State	Items Up	Items Down	Items under Restoration
1		3	0	0
2		2	1	1
	3	1	2	2
	4	0	3	2

Because it is possible to do up to two restorations at a time, when the system is in state 1 there will be no restorations happening, when the system is in state 2 there will be one restoration happening, when the system is in state 3 there will be two restorations happening, and when the system is in state 4 there will still be only two restorations happening.

To evaluate the transition rates, from state 1 it is possible to go only into state 2; since a failure of any of the three items will cause that transition, $r_{21} = 3\lambda$. The other rates from state 1 are $r_{31} = r_{41} = 0$.

From State 2 it is possible to go into either state 1 (by restoring a single item) or state 3 (upon failure of either of two functioning items). Accordingly, $r_{12} = \mu$, $r_{32} = 2\lambda$, and $r_{42} = 0$.

From state 3 it is possible to go to state 2 (by restoring either of the two items that are down) or state 4 (if the remaining functioning item fails). It is impossible to go directly from state 3 to state 1. Therefore, $r_{23} = 2\mu$, $r_{43} = \lambda$, and $r_{13} = 0$.

From state 4 it is possible to go directly only into state 3 (by restoring one any two of the items that are down). So $r_{34} = 2\mu$, and $r_{14} = r_{24} = 0$.

The sums of rates out of each of the four states are $r_{11} = 3\lambda$, $r_{22} = \mu + 2\lambda$, $r_{33} = 2\mu + \lambda$, and $r_{44} = 2\mu$. The transition rate diagram is then as shown in Figure 10.5.

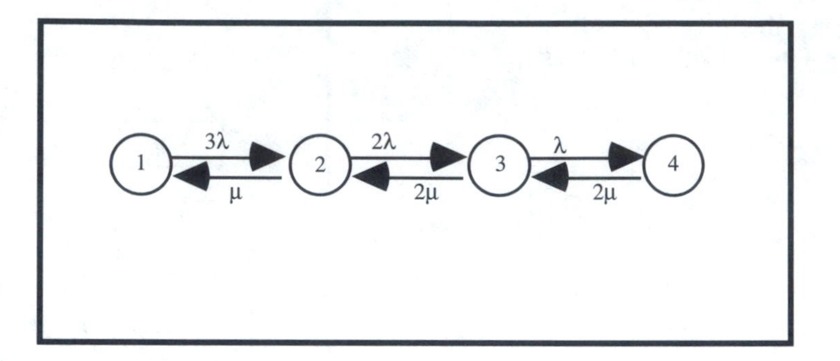

FIGURE 10.5
Transition rate diagram for Example 10.13.

From Equation 10.8, we can find the P_i values:

$$\begin{aligned}
0 &= -3\lambda P_1 + \mu P_2 \\
0 &= 3\lambda P_1 - (\mu + 2\lambda)P_2 + 2\mu P_3 \\
0 &= 2\lambda P_2 - (2\mu + \lambda)P_3 + 2\mu P_4 \\
0 &= \lambda P_3 - 2\mu P_4
\end{aligned}$$

For the fourth equation (the most trivial), we substitute Equation 10.9:

$$1 = P_1 + P_2 + P_3 + P_4.$$

The solution to the set of equations

$$\begin{aligned}
0 &= -3\lambda P_1 + \mu P_2 \\
0 &= 3\lambda P_1 - (\mu + 2\lambda)P_2 + 2\mu P_3 \\
0 &= 2\lambda P_2 - (2\mu + \lambda)P_3 + 2\mu P_4 \\
1 &= P_1 + P_2 + P_3 + P_4
\end{aligned}$$

is

$$P_1 = \frac{2\mu^3}{2\mu^3 + 6\mu^2\lambda + 6\mu\lambda^2 + 3\lambda^3}$$

$$P_2 = \frac{6\mu^2\lambda}{2\mu^3 + 6\mu^2\lambda + 6\mu\lambda^2 + 3\lambda^3}$$

$$P_3 = \frac{6\mu\lambda^2}{2\mu^3 + 6\mu^2\lambda + 6\mu\lambda^2 + 3\lambda^3}$$

$$P_4 = \frac{3\lambda^3}{2\mu^3 + 6\mu^2\lambda + 6\mu\lambda^2 + 3\lambda^3}$$

The success states are states 1 and 2. Therefore, the system steady-state availability is

$$A = P_1 + P_2 = \frac{2\mu^3 + 6\mu^2\lambda}{2\mu^3 + 6\mu^2\lambda + 6\mu\lambda^2 + 3\lambda^3}$$

EXAMPLE 10.14

Consider the same system of Example 10.13, composed of three identical, active items, each of which has a failure rate λ and a restoration rate μ. Again suppose that it is possible to do up to two restorations simultaneously. But this time consider that *at least one* out of the three items is required for system success. What is the steady-state availability for this system?

This system has the same four states as that of Example 10.13, and the steady-state probabilities of being in each of those four states are the same as calculated. This time, however, there are three success states, states 1, 2, and 3. Accordingly, the system steady-state availability is

$$A = P_1 + P_2 + P_3 = \frac{2\mu^3 + 6\mu^2\lambda + 6\mu\lambda^2}{2\mu^3 + 6\mu^2\lambda + 6\mu\lambda^2 + 3\lambda^3}$$

As additional exercises, use Equations 10.7 through 10.9 to derive the availability solutions provided in the tables of Appendix C.

10.5.2 Applying the Markov Model Solution

The derivation of Equations 10.7 through 10.9 in the preceding section demonstrates that *system availability* is

$$A = \sum_{\substack{\text{all success} \\ \text{states } i}} P_i \tag{10.14}$$

where the P_i values are the probabilities of being in the various states i and are the solutions of the system of N equations (N is the total number of states, both *success* and *fail*).

$$\begin{aligned}
0 &= -r_{11}P_1 + r_{12}P_2 + r_{13}P_3 + \cdots + r_{1N}P_N \\
0 &= r_{21}P_1 - r_{22}P_2 + r_{23}P_3 + \cdots + r_{2N}P_N \\
0 &= r_{31}P_1 + r_{32}P_2 - r_{33}P_3 + \cdots + r_{3N}P_N \\
&\ \ \vdots \\
0 &= r_{(N-1),\,1}P_1 + r_{(N-1),\,2}P_2 + \cdots + r_{(N-1),(N-1)}P_{(N-1)} + r_{(N-1),\,N}P_N \\
1 &= P_1 + P_2 + P_3 + \cdots + P_n
\end{aligned} \tag{10.15}$$

EXAMPLE 10.15

Define and describe all the system states for the system whose block diagram appears in Figure 10.6.

This system has eight states, three success states and five fail states, as follows:

Success State	Fail State	Elements Up	Elements Down	Elements under Repair
1		A, B, C	None	None
2		B, C	A	A
3		A, C	B	B
	4	A, B	C	C
	5	C	A, B	A
	6	B	A, C	C
	7	A	B, C	C
	8	None	A, B, C	C

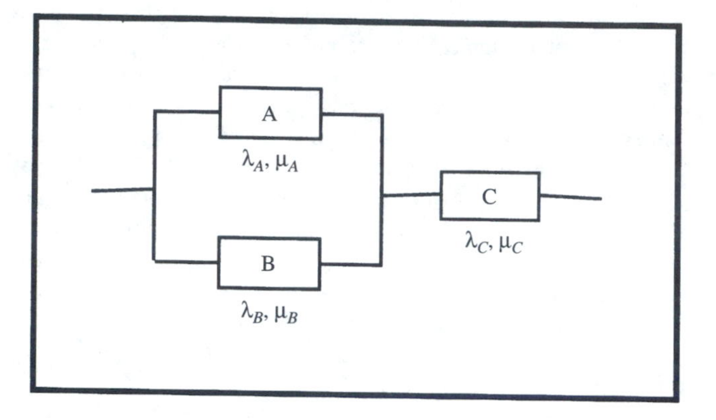

FIGURE 10.6
Block diagram for Example 10.15.

An assumption is made here that only one element at a time can be restored. We established a repair policy that creates a priority (C before A before B) to restore back to a success state as quickly as possible whenever a fail state is entered.

EXAMPLE 10.16 Identify the state transitions that are possible for the system of Example 10.15.

If the System is in State:	And This Event Takes Place:	The System Will Go into State:
1	A fails.	2
1	B fails.	3
1	C fails.	4
2	A is restored.	1
2	B fails.	5
2	C fails.	6
3	B is restored.	1
3	A fails.	5
3	C fails.	7
4	C is restored.	1
4	A fails.	6
4	B fails.	7
5	A is restored.	3
5	C fails.	8
6	C is restored.	2
6	B fails.	8
7	C is restored.	3
7	A fails.	8
8	C is restored.	5

EXAMPLE 10.17 Construct a transition-rate table for Example 10.15.

To State i	From State j	Transition Rate r_{ij}
2	1	λ_A
3	1	λ_B
4	1	λ_C
1	2	μ_A
5	2	λ_B
6	2	λ_C
1	3	μ_B
5	3	λ_A
7	3	λ_C
1	4	μ_C
6	4	λ_A
7	4	λ_B
3	5	μ_A
8	5	λ_C
2	6	μ_C
8	6	λ_B
3	7	μ_C
8	7	λ_A
5	8	μ_C

EXAMPLE 10.18 Find the availability for the system of Example 10.15 if $\lambda_A = 5.0 \times 10^{-6}$ per hour, $\lambda_B = 25.0 \times 10^{-6}$ per hour, $\lambda_C = 2.0 \times 10^{-6}$ per hour, $\mu_A = \mu_C = 1.0$ per hour, and $\mu_B = 2.0$ per hour.

From Example 10.17,

$$
\begin{aligned}
r_{21} &= \lambda_A = 5.0 \times 10^{-6} \text{ per hour} \\
r_{31} &= \lambda_B = 25.0 \times 10^{-6} \text{ per hour} \\
r_{41} &= \lambda_C = 2.0 \times 10^{-6} \text{ per hour} \\
r_{12} &= \mu_A = 1.0 \text{ per hour,} \\
r_{52} &= \lambda_B = 25.0 \times 10^{-6} \text{ per hour} \\
r_{62} &= \lambda_C = 2.0 \times 10^{-6} \text{ per hour} \\
r_{13} &= \mu_B = 2.0 \text{ per hour} \\
r_{53} &= \lambda_A = 5.0 \times 10^{-6} \text{ per hour} \\
r_{73} &= \lambda_C = 2.0 \times 10^{-6} \text{ per hour} \\
r_{14} &= \mu_C = 1.0 \text{ per hour} \\
r_{64} &= \lambda_A = 5.0 \times 10^{-6} \text{ per hour} \\
r_{72} &= \lambda_B = 25.0 \times 10^{-6} \text{ per hour} \\
r_{35} &= \mu_A = 1.0 \text{ per hour}
\end{aligned}
$$

$$
\begin{aligned}
r_{85} &= \lambda_C = 2.0 \times 10^{-6} \text{ per hour}\\
r_{26} &= \mu_C = 1.0 \text{ per hour}\\
r_{86} &= \lambda_B = 25.0 \times 10^{-6} \text{ per hour}\\
r_{37} &= \mu_C = 1.0 \text{ per hour,}\\
r_{87} &= \lambda_A = 5.0 \times 10^{-6} \text{ per hour}\\
r_{58} &= \mu_C = 1.0 \text{ per hour}
\end{aligned}
$$

As introduced in Chapter 9, the states r_{ii} represent the sum of the rates coming out of state i. Therefore,

$$
\begin{aligned}
r_{11} &= \lambda_A + \lambda_B + \lambda_C = 32.0 \times 10^{-6} \text{ per hour}\\
r_{22} &= \mu_A + \lambda_B + \lambda_C = 1.000027 \text{ per hour}\\
r_{33} &= \mu_B + \lambda_A + \lambda_C = 2.000007 \text{ per hour}\\
r_{44} &= \mu_C + \lambda_A + \lambda_B = 1.000030 \text{ per hour}\\
r_{55} &= \mu_A + \lambda_C = 1.000002 \text{ per hour}\\
r_{66} &= \mu_C + \lambda_B = 1.000025 \text{ per hour}\\
r_{77} &= \mu_C + \lambda_A = 1.000005 \text{ per hour}\\
r_{88} &= \mu_C = 1.000000 \text{ per hour}
\end{aligned}
$$

Applying the set of Equations 10.15,

$$
\begin{aligned}
0 &= -(32 \times 10^{-6})P_1 + P_2 + 2P_3 + P_4\\
0 &= (5.0 \times 10^{-6})P_1 - 1.000027P_2 + P_6\\
0 &= (25.0 \times 10^{-6})P_1 - 2.000007P_3 + P_5 + P_7\\
0 &= (2.0 \times 10^{-6})P_1 - 1.000030P_4\\
0 &= (25.0 \times 10^{-6})P_2 + (5.0 \times 10^{-6})P_3 - 1.000002P_5 + P_8\\
0 &= (2.0 \times 10^{-6})P_2 + (5.0 \times 10^{-6})P_4 - 1.000025P_6\\
0 &= (2.0 \times 10^{-6})P_3 + (25.0 \times 10^{-6})P_4 - 1.000005P_7\\
1 &= P_1 + P_2 + P_3 + P_4 + P_5 + P_6 + P_7 + P_8
\end{aligned}
$$

The solutions to the set of equations are

$$
\begin{aligned}
P_1 &= 0.9999806\\
P_2 &= 0.0000048\\
P_3 &= 0.0000125\\
P_4 &= 0.0000018\\
P_5 + P_6 + P_7 + P_8 &= 0.0000003
\end{aligned}
$$

Because the first three states are success states and all the others are fail states, the system availability is the proportion of the total time that the system is in either state 1, 2, or 3. That is,

$$A = P_1 + P_2 + P_3 = 0.999998$$

EXAMPLE 10.19 Use the Markov model technique to solve the availability of the system of Example 10.7.

Here we had two active, redundant power supplies, each with a failure rate of 0.000025 per hour and a restoration rate of 2 per hour. But upon the failure of one of the power supplies, the remaining one had a failure rate of 0.000075 per hour.

This system has three states, two of which are success states.

Success State	Fail State	Elements Up	Elements Down	Elements under Repair
1		2	0	0
2		1	1	1
	3	1	2	1

The state transitions are as follows.

If the System Is in State:	And This Event Takes Place:	The System Will Go into State:
1	A power supply fails.	2
2	The power supply is restored.	1
2	The remaining power supply fails.	3
3	A power supply is restored.	2

Then the transition rates are as follows.

To State i	From State j	Transition Rate r_{ij}
2	1	2(0.000025) = 0.000050 per hour
1	2	2 per hour
3	2	0.000075 per hour
2	3	2 per hour

The sums of the rates out of each state are

$$r_{11} = 0.000050 \text{ per hour}$$
$$r_{22} = 2.000075 \text{ per hour}$$
$$r_{33} = 2 \text{ per hour}$$

Applying these values to the set of Equations 10.15,

$$0 = -0.000050P_1 + 2P_2$$
$$0 = 0.000050P_1 + 2.000075P_2 - 2P_3$$
$$1 = P_1 + P_2 + P_3$$

The solutions to the set of equations are

$$P_1 = 0.999975$$
$$P_2 = 0.999925$$
$$P_3 = 9.37 \times 10^{-10}$$

Therefore, the system availability is

$$A = 1 - P_3 = 1 - 9.37 \times 10^{-10} = 0.9999999991$$

Compare this result to the solution of Example 10.7.

EXAMPLE 10.20 A company has four telephone lines coming into its switchboard. Suppose that during the workday, phone calls arrive at a rate of 20 per hour and that the average length of each call is 10 min. What is the availability of this system of telephones? In other words, what is the probability that somebody telephoning the company during working hours will be able to reach one of the four lines upon demand?

This system has five states:

State 1, where all four lines are available.

State 2, where one line is busy and three are available.

State 3, where two lines are busy and two are available.

State 4, where three lines are busy and one is available.

State 5, where all four lines are busy.

From state 1 the system can go only into state 2, and that happens if a call comes in. From state 2 the system can go either into state 1 or state 3; it will go to state 1 if the call is terminated, and it will go to state 3 if another call comes in. Similarly, from state 3 the system can go either to state 2 or state 4; it will go to state 2 if one of the calls is terminated, and it will go to state 4 if another call comes in. From state 4 the system can go into state 3 if one of the calls is terminated and into state 5 if another call comes in. From state 5 the system can go only into state 4, and that happens if one of the four calls is terminated. The system availability is the portion of time that the system is in one of the first four states—that is, the probability that the system is not in state 5:

$$A = 1 - P_5$$

The system is moving from each state to its successive state (state 1 to state 2, state 2 to state 3, state 3 to state 4, state 4 to state 5) at the rate at which phone calls are coming in to the switchboard, that is, 20 per hour. The rate at which the system is moving from each state to its predecessor state (state 2 to state 1, state 3 to state 2, state 4 to state 3, state 5 to state 4) is the rate at which calls are being processed, that is, 10 min (1⁄6 h) per call. So when one line is busy, the system process rate is 6 per hour, when two lines are busy the system process rate is 12 per hour, when three lines are busy it is 18 per hour, and when four lines are busy it is 24 per hour. Accordingly the transition rates among states are as illustrated in Figure 10.7.

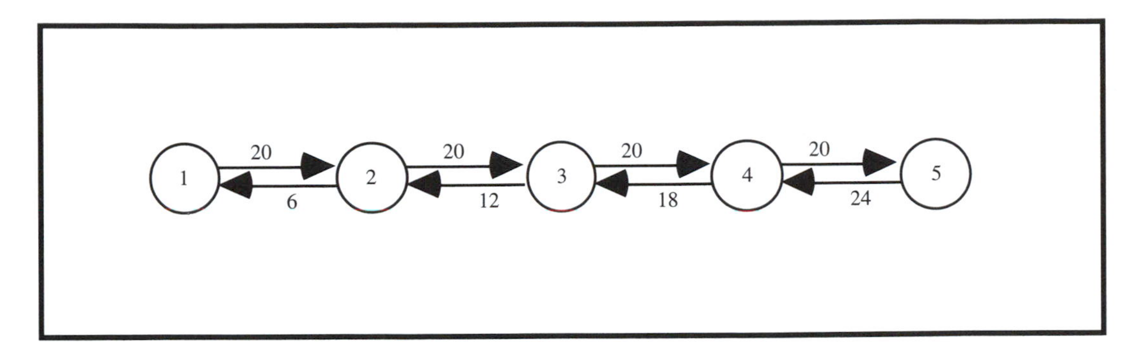

FIGURE 10.7
Transition rate diagram for Example 10.20.

The rates among the states are as follows:

To State i	From State j	Rate r_{ij}
2	1	20 per hour
1	2	6 per hour
3	2	20 per hour
2	3	12 per hour
4	3	20 per hour
3	4	18 per hour
5	4	20 per hour
4	5	24 per hour

The sums of the rates out of the states are as follows:

$$r_{11} = 20 \text{ per hour}$$
$$r_{22} = 26 \text{ per hour}$$
$$r_{33} = 32 \text{ per hour}$$
$$r_{44} = 38 \text{ per hour}$$
$$r_{44} = 24 \text{ per hour}$$

Applying the set of Equations 10.15,

$$0 = 20P_1 + 6P_2$$
$$0 = 20P_1 - 26P_2 + 12P_3$$
$$0 = 20P_2 - 32P_3 + 18P_4$$
$$0 = 20P_3 - 38P_4 + 24P_5$$
$$1 = P_1 + P_2 + P_3 + P_4 + P_5$$

The solutions to this set of equations are as follows:

$$P_1 = 0.0471$$
$$P_2 = 0.1572$$
$$P_3 = 0.2620$$
$$P_4 = 0.2911$$
$$P_5 = 0.2426$$

The availability for this system of four telephone lines is

$$A = 1 - P_5 = 1.0 - 0.2426 = 0.757$$

So, 75.7% of the time callers will be able to reach the company switchboard without delay.

Additional iterations of this analysis can be used to evaluate the potential benefits of adding phone lines to this switchboard.

EXAMPLE 10.21

An office has two copy machines. During working hours the user frequency is once every 6 min (10 per hour), for an average of 10 min. The failure rate for a copy machine is 0.0005 per hour, and its average repair time is 4 h. What is the availability of this system of office copy machines? In other words, what is the probability that, at a given time, a member of the office staff will not have to wait to use a copy machine?

This systm has six states:

State 1, where both copiers are operable and available for use.

State 2, where one copier has failed and the other is available.

State 3, where one copier is in use and the other is available.

State 4, where one copier has failed and the other is in use.

State 5, where both copiers are in use.

State 6, where both copiers have failed.

The transitions that are possible among states are as follows:

If the System Is in State:	And This Event Takes Place:	The System Will Go into State:
1	A machine fails.	2
1	A machine is used.	3
2	The failed machine is restored.	1
2	The operable machine is used.	4
2	The remaining machine fails.	6
3	Machine use is terminated.	1
3	The available machine fails.	4
3	The available machine is used.	5
4	Use of the operable machine is terminated.	2
4	The failed machine is restored.	3
5	Use of one of the machines is terminated.	3
6	One of the machines is restored.	2

When both the machines are operable, their total failure rate amounts to $2 \times 0.0005 = 0.0010$ per hour, and when one is operating its failure rate is 0.0005 per hour. The restoration rate of a failed machine is 0.25 per hour. The use rate is 10 per hour, and the recovery rates (rates of restoration from an in-use condition) are 1⁄10 min = 6 per hour when one is in use and 2⁄10 min = 12 per hour when both are in use. Consequently, the rates among the states are as follows:

To State i	From State j	Rate r_{ij}
2	1	0.0010 per hour
3	1	10 per hour
1	2	0.25 per hour
4	2	10 per hour
6	2	0.0005 per hour
1	3	6 per hour
4	3	0.0010 per hour
5	3	10 per hour
2	4	6 per hour
3	4	0.25 per hour
3	5	12 per hour
2	6	0.25 per hour

Figure 10.8 illustrates the transition rates.

The sums of the rates out of each of the states are as follows:

$$\begin{aligned} r_{11} &= 10.0010 \text{ per hour} \\ r_{22} &= 10.2505 \text{ per hour} \\ r_{33} &= 16.0010 \text{ per hour} \\ r_{44} &= 6.25 \text{ per hour} \\ r_{55} &= 12 \text{ per hour} \\ r_{66} &= 0.25 \text{ per hour} \end{aligned}$$

Applying the set of Equations 10.15,

$$\begin{aligned} 0 &= -10.0010P_1 + 0.25P_2 + 6P_3 \\ 0 &= 0.0010P_1 - 10.2505P_2 + 6P_4 + 0.25\,P_6 \\ 0 &= 10P_1 - 16.0010P_3 + 0.25P_4 + 12P_5 \\ 0 &= 10P_2 + 0.0010P_3 - 6.25P_4 \\ 0 &= 10P_3 - 12P_5 \\ 1 &= P_1 + P_2 + P_3 + P_4 + P_5 + P_6 \end{aligned}$$

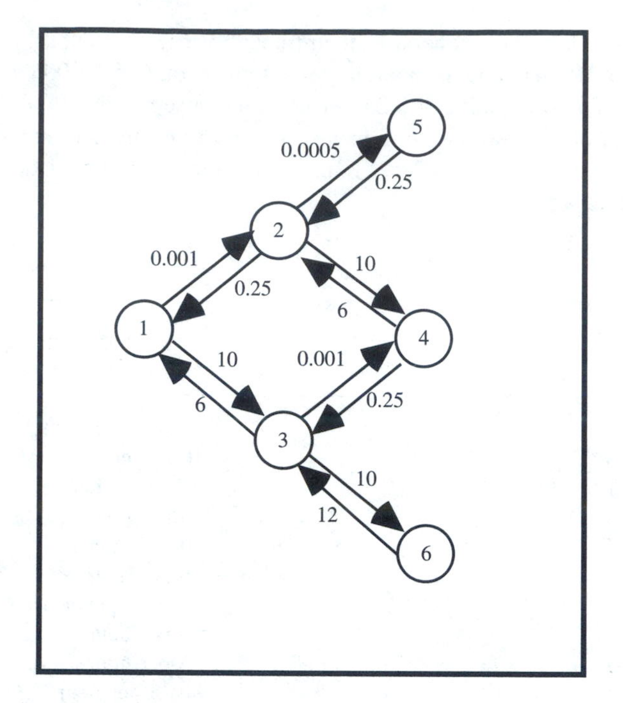

FIGURE 10.8
Transition rate diagram for Example 10.21.

The solutions to the set of equations are as follows:

$$
\begin{aligned}
P_1 &= 0.2459 \\
P_2 &= 0.0010 \\
P_3 &= 0.4099 \\
P_4 &= 0.0016 \\
P_5 &= 0.3416 \\
P_6 &= 0.0000
\end{aligned}
$$

The availability of this system of office copiers is the probability that at least one of them is available at any given time—that is, the probability that the system is in either state 1, 2, or 3. Hence,

$$A = P_1 + P_2 + P_3 = 0.2459 + 0.0010 + 0.4099 = 0.657$$

Therefore, 65.7% of the time office staff members do not have to wait to use a copier. To evaluate the benefit of adding another copy machine, repeat this analysis for three copy machines.

10.6 OUTAGE RATES

In the power industry, availability is traditionally expressed as **outage rate** in terms of **outage-hours per year.** Considering a year to be equivalent to 8760 h,

$$\text{outage rate} = (1 - A) \times 8760 \text{ h/year} \tag{10.16}$$

EXAMPLE 10.22 Determine the outage rate for the system of copiers of Example 10.21.

$$\begin{aligned}\text{outage rate} = (1 - A) \times 8760 \text{ h/year} &= (1 - 0.657) \times 8760 \text{ h/year} \\ &= 3005 \text{ h per year}\end{aligned}$$

The power industry typically classifies each outage as either a *forced* or a *planned outage*. A **forced outage** is outage time due to a failure or malfunction, and it includes all seven restoration-time elements presented in Chapter 7 as well as any logistics or administrative delay time associated with the malfunction plus additional time mandated by a regulatory agency (such as the NRC). A **planned outage** consists of time down for any scheduled, preventive maintenance or testing.

EXAMPLE 10.23 A given power plant's maintenance policy requires 500 h of planned outage per year. If its operating availability is 85%, determine its forced outage, planned outage, and total outage.

$$\text{planned outage} = 500 \text{ h per year}$$

The required operating time per year is then 8760 − 500 = 8260 h per year. Then,

$$\begin{aligned}\text{forced outage} = (1 - A) \times 8260 \text{ h/year} &= (1 - 0.85) \times 8260 \text{ h/year} \\ &= 1239 \text{ h per year}\end{aligned}$$

Hence,

$$\text{total outage} = 1239 + 500 = 1739 \text{ h per year}$$

10.7 SYSTEM DEPENDABILITY

As introduced in Chapter 2, *dependability* applies to a standby system, often one that serves as backup to an active system or a system that provides an emergency function in the event of an accident, such as a fire-protection system. Because such a system is normally in an inactive standby mode, we are concerned with its availability to serve when called upon and its ability to perform its required function throughout its demand time:

***Dependability* is the probability that a standby system will be available to operate upon demand and then continue to operate throughout the demand period:**

$$D(t) = A \times R(t) \tag{10.17}$$

A standby system's *availability* is also called its **operational readiness.** So, dependability is a standby system's operational readiness times its mission reliability.

EXAMPLE 10.24 Suppose the sprinkler system in a building is operable 99% of the time. In the event of a fire, it is required to activate and then to function for at least ½ h. Suppose its reliability for a ½-h mission is 0.999. What is the dependability of this system?

The system's operational readiness (availability in the standby mode) is $A = 0.990$. Its reliability in the active mode for a ½-h mission is $R(t) = 0.999$. Hence,

$$D(0.5 \text{ h}) = A \times R(0.5 \text{ h}) = 0.990 \times 0.999 = 0.989$$

Methods provided in this chapter and in Appendix A for determining availability and reliability for complex systems can be applied to determining the *dependability* of a complex standby system.

10.8 SPARES PROVISIONING

When concerned about the availability of a physical system within an isolated environment for an extended period—for example, aboard a ship at sea or aboard a space shuttle—it is necessary to design a spares provisioning policy. This policy involves decisions on the quantity of spare parts of each type to be carried aboard in order to support the repairs that might be necessary during the mission. Another factor of the spares provisioning policy is designating the level (i.e., component, subassembly, equipment, etc.) at which spare replacements will be supplied.

Considerations are (1) that we want to have a sufficient quantity of spare parts, (2) that we generally are dealing with volume and weight considerations, and (3) that repair capabilities are restricted by the tools, instrumentation, and the skills of the personnel aboard. If we choose to provide spares at a lower assembly level (that is, a component level), we can store more spares in less space and add less weight. However, a lower assembly level of sparing usually demands more tooling and instrumentation to perform an onboard repair, higher skill levels of onboard maintenance personnel, and a longer restoration time. If, on the other hand, we decide to supply spares at a higher assembly level, repairs can be completed in less time with less skill and fewer tools.

Obviously, trade-offs are necessary to determine the optimum level of sparing to fit each situation. Considerations of this trade-off are the criticality of consequences of running out of spares or aborting a mission, the necessity to adhere to space and weight restrictions, and the need to complete a repair quickly. This trade-off evaluation requires a means of predicting *spares sufficiency* for any assembly level of sparing being considered.

*10.8.1 Derivation

Let us suppose that a system is to be in an isolated environment for a mission of time t. Let us further suppose that the system has N such parts operating throughout the mission and that each of them has a failure rate λ. Then, the expected quantity of failures of this part during the mission is $N\lambda t$. A review of Section A.8 of Appendix A will show that the Poisson Probability Law applies and that the probability that we will have no more than c failures during the mission, and thus require no more than c spares on board, is

$$P(x \leq c) = e^{N\lambda t}\sum_{i=0}^{c}\frac{(N\lambda t)^i}{i!} \tag{10.18}$$

If we want a probability of at least P that we do not run out of spares during the mission, the spares sufficiency of the part in question is the lowest c such that $P(x \leq c) = e^{N\lambda t}\sum_{i=0}^{c}(N\lambda t)^i/i! \geq P$. Finding this value can be accomplished either analytically or through Table B.9 of Appendix B.

To use Table B.9, set $\mu = N\lambda t$ and then search the table for the lowest value of c, corresponding with μ, that will provide a probability at least as high as desired.

10.8.2 Determining Spares Sufficiency

Spares sufficiency for a given part is determined through Table B.9 of Apprendix B, as follows:

1. On the basis of the criticality of the part and the mission, select a **spares sufficiency probability.** This is the same as the probability of not running out of spares of this particular part during the mission. A spares sufficiency probability of 0.999 is commonly used for this type of analysis, although a higher or lower probability may be used as dictated by the possible consequences of running out.
2. Determine the *mission time* t.
3. Determine the *quantity* N of the part used in the system.
4. Determine the *part failure rate* λ.
5. Set $\mu = N\lambda t$.
6. In Table B.9, corresponding to μ, find the lowest value of c that provides a probability at least as high as the spares sufficiency probability selected in Step 1. The quantity c is the minimum quantity of spares required for the part in question.

EXAMPLE 10.25 A ship is about to embark on a cruise that will keep it at sea for 20 days. How many spares of a component with a failure rate of 0.0001 per hour should be onboard if 250 such components are expected to operate 24 h a day and we want a 0.995 probability of spares sufficiency?

$$\begin{aligned}\mu = N\lambda t &= 250 \times 0.0001 \text{ per hour} \times 20 \text{ days} \times 24 \text{ h per day}\\ &= 12.0\end{aligned}$$

In Table B.9, where $\mu = 12.0$, we look for the value of c that will provide a probability of at least 0.995. Notice that for $c = 21$ the probability is 0.994, and for $c = 22$ the probability is 0.997. So, $c = 22$ is the lowest quantity that will meet our spares sufficiency probability requirement. We therefore want to take 22 spares of this part for the mission.

EXAMPLE 10.26 Suppose that, in Example 10.25, we carry aboard only 15 spares of the part in question. What is the probability that we will run out of spares before the cruise is over?

In Table B.9, where $\mu = 12.0$ and $c = 15$, the probability of spares sufficiency is 0.844. Therefore, the probability that 15 spares is not sufficient for the cruise is $1 - 0.844 = 0.156$. In other words, there is a 15.6% probability of running out of spares.

BIBLIOGRAPHY

1. ARINC Research Corp. 1964. *Reliability Engineering.* Upper Saddle River, N.J.: Prentice Hall.
2. I. Bazovsky. 1961. *Reliability Theory and Practice.* Upper Saddle River, N.J.: Prentice Hall.
3. E. E. Lewis, 1987. *Introduction to Reliability Engineering.* New York: John Wiley.
4. D. K. Lloyd and M. Lipow. 1962. *Reliability Management, Methods, and Mathematics.* Upper Saddle River, N.J.: Prentice Hall.
5. P. D. T. O'Connor. 1991. *Practical Reliability Engineering,* 3d ed. New York: John Wiley.
6. N. H. Roberts. 1964. *Mathematical Methods in Reliability Engineering.* New York: McGraw-Hill.
7. G. H. Sandler. 1963. *System Reliability Engineering.* Upper Saddle River, N.J.: Prentice Hall.
8. M. L. Shooman. 1968. *Probabilistic Reliability: An Engineering Approach.* New York: McGraw-Hill.
9. C. O. Smith. 1976. *Introduction to Reliability in Design.* New York: McGraw-Hill.
10. P. A. Tobias and D. Trindade. 1986. *Applied Reliability.* New York: Van Nostrand Reinhold.

EXERCISES

10.1. A system is composed of four identical, active elements, two of which are required for system success. On-line restoration is possible, with one restoration activity at a time. Each element has a failure rate of 46×10^{-6} per hour and a restoration rate of 2.0 per hour. Use Appendix C to find the system availability.

10.2. Use Appendix C to determine the availability for the system of Exercise 10.1 if three of the four elements are required for system success.

10.3. A system is composed of four identical, active elements, two of which are required for system success. On-line restoration is not possible. Each element has a failure rate of 46×10^{-6} per hour. The system restoration rate is 2.0 per hour. Use Appendix C to find the system availability.

10.4. Use Appendix C to solve the availability for the system of Exercise 10.3 if three of the four elements are required for system success.

10.5. A system is composed of seven identical, active elements, four of which are required for system success. On-line restoration is not possible. Each element has a failure rate of 500×10^{-6} per hour. The system restoration rate is 1.0 per hour. Use Appendix C to find the system availability.

10.6. A system is composed of 10 identical, active elements, 8 of which are required for system success. On-line restoration is not possible. Each element has a failure rate of 500×10^{-6} per hour. The system restoration rate is 1.0 per hour. Use Appendix C to find the system availability.

10.7. A system is composed of two active elements, one of which is required for system success. On-line restoration is not possible. Element 1 has a failure rate of 250×10^{-6} per hour, and element 2 has a failure rate of 750×10^{-6} per hour. The system restoration rate is 1.0 per hour. Use Appendix C to determine the system availability.

10.8. A system has two power supplies, at least one of which is required for system success. On-line restoration of these power supplies is possible (one at a time). The average restoration time is 15 min. With both power supplies operating, each has a failure rate of 25×10^{-6} per hour. If one power supply is down, the remaining one has a failure rate of 150×10^{-6} per hour. Use Appendix C to solve for the availability for this power supply subsystem.

10.9. A system contains two identical elements, one active and one in standby, each with an active failure rate of 1000 failures per 1,000,000 h and a negligible standby failure rate. On-line maintenance is possible with an element restoration rate of 1.0 per hour. Compute the system availability if we assume perfect sensing and switching.

10.10. A hospital operating room has an emergency generator in standby in the event of a power failure. Suppose power failures occur at an average rate of once every 5000 h and the active failure rate for the emergency generator is also once every 5000 h. Suppose further that it takes an average of 3 h to restore power from either of the two sources. Assuming perfect sensing and switching and a negligible standby failure rate for the emergency generator, what is the availability of the operating room's power system?

10.11. How much improvement in availability can be achieved by adding a second emergency generator to the operating room of Exercise 10.10? In other words, if the system were composed of the primary power source and two standby generators, how much higher would the availability be? Once again assume perfect sensing and switching and a negligible standby failure rate for the emergency generator.

10.12. A delivery service has a fleet of 5 trucks, 4 of which must be in operation at all times. Each truck breaks down at a rate of once every 5000 h. The average restoration time is 2.0 h. What is the availability of this system?

10.13. A delivery service has a fleet of five trucks, three of which must be in operation at all times. Each truck breaks down at a rate of once every 5000 h. The average restoration time is 2.0 h. What is the availability of this system?

10.14. A customer service department has three service representatives available to take phone calls from customers. If calls come in at a steady rate of 20 per hour throughout the day and if the average length of a service telephone conversation is 10 min, what is the likelihood that all three representatives are unavailable for an incoming call? In other words, we are looking for the availability of a system requiring at least one out of three representatives at all times.

10.15. A communication system is composed of a transmitter, an active receiver, and a standby receiver that can be substituted if the active one fails. Suppose the transmitter has a failure rate of 0.0025 per hour, the active receiver has a failure rate of 0.0200 per hour, and the restoration rate for any restoration is 0.330 per hour. Assume perfect sensing and switching and a negligible standby failure rate for the receivers. Use the technique prescribed by Section 10.4 to determine this system's availability.

10.16. Use the method prescribed by Section 10.4 to solve the availability of the system whose block diagram is given in Figure 10.9. Elements B and C are in standby redundancy as shown. The active failure rates are 100 failures per 1,000,000 hours for element A, 500 failures per 1,000,000 hours for element B, and 1000 failures per 1,000,000 hours for element C. The restoration rates are 1.0 per hour for all elements. The standby failure rates are negligible. Assume perfect sensing and switching.

10.17. What is the outage rate for the communication system of Exercise 10.15?

10.18. A hotel's smoke detection is composed of two independent, redundant circuits, each capable of sounding an alarm in the event of fire within the building. If the failure rate of each circuit is 100×10^{-6} per hour, the system is checked once every week (once per 168 h), and the mean restoration time is 30 min, what is its operational readiness?

10.19. In the event that the hotel in Exercise 10.18 has a fire and an alarm sounds, it is required to continue sounding for at least 1 h. What is the likelihood that this happens? In other words, what is the alarm's reliability for a 1-h mission?

10.20. What is the dependability of the smoke detection alarm system of Exercise 10.18?

10.21. Use Equations 10.14 and 10.15 to find the availability of a system composed of N identical, active elements, $N - 1$ of which are required for system success. On-line restoration is possible, with one restoration at a time. Each element has a failure rate λ and a restoration rate μ.

10.22. Use Equations 10.14 and 10.15 to find the availability of a system composed of N identical, active elements, $N - 2$ of which are required for system success. On-line restoration is possible with one restoration at a time. Each element has a failure rate λ and a restoration rate μ.

10.23. Use Equations 10.14 and 10.15 to solve for the availability of a system composed of N identical, active elements, $N - 1$ of which are required for system success. On-line resto-

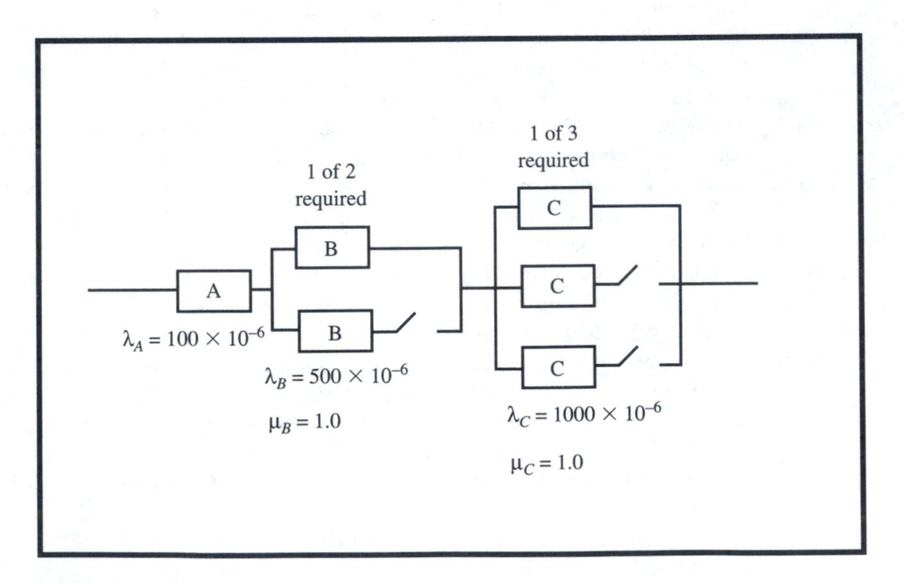

FIGURE 10.9
Block diagram for Exercise 10.16.

ration is possible, with one restoration at a time. With all elements up, each has a failure rate λ_A and a restoration rate μ. With one element down, each has a failure rate λ_B.

10.24. Use Equations 10.14 and 10.15 to solve for the availability of a system composed of two active elements, one of which is required for system success. On-line restoration is possible. Element 1 has a failure rate λ_1 and a restoration rate μ_1; element 2 has a failure rate λ_2 and a restoration rate μ_2.

10.25. Use Equations 10.14 and 10.15 to solve for the availability of a system composed of N identical elements, $N - 1$ of which are active and required for system success. One element is initially in standby. On-line restoration is possible with one restoration at a time. Each active element has a failure rate λ and a restoration rate μ. Standby failure rates are assumed negligible, and perfect sensing and switching are assumed.

Compare your results in Exercises 10.21 through 10.25 with the tables in Appendix C.

BOOK THREE Reliability Assurance and Improvement

11

RELIABILITY AND RESTORABILITY DEMONSTRATION TESTING

11.1 INTENT

We learned in the first two books of this text that *reliability* and *restorability* are properties that are inherent to a system. Be that system a service or a manufactured product, the levels of reliability and restorability that can be attained are established at the time of the system design. Whether or not a conscious effort is made to design reliability and restorability features into our system, the potential of the system to withstand failures in performance and to recover quickly when failures do occur is restricted by the design activities. Clearly, Book 2 is the most important of the text, because it provides methods for considering *reliability* and *maintainability* at the time of the system design. Through these methods, we can evaluate the reliability and restorability potential of alternative design approaches, perform design trade-offs, and make informed decisions. The methods of Book 2 also allow us to make predictions that are useful tools for writing warranties and instructions for operating and maintaining our system.

Once we have passed our design phase, we have three concerns:

1. Are the system reliability and restorability as predicted?
2. Can we enhance the reliability and restorability?
3. What happens if something goes wrong?

The three chapters of Book 3 deal with these issues. This chapter presents methods for testing our system to demonstrate that predicted *reliability, MTBF,* and *MTTR* have been achieved. The next chapter deals with the concept of *reliability growth and improvement.* From our discussions thus far, we can conclude that reliability improvement implies some form of system redesign. The final chapter investigates what can go wrong and produce failures within our system.

Demonstration testing can be regarded as a form of inspection. It is not, however, *inspection* as we usually think of it, because we are not examining a component or how a particular procedure is working out. Here we are inspecting the system in its entirety with the objective of demonstrating an already accepted conclusion about the system's reliability, MTBF, or MTTR. We have already acknowledged that demonstration testing is an activity of less importance than those presented in Book 2. As with other forms of inspection, we are not doing anything that adds value to our system or enhances its reliability in any way when we carry out a demonstration test. For this reason many reliability programs have in recent years diminished or reduced this activity altogether.

No less an authority than the late Dr. W. Edwards Deming included among his legendary *14 points* a warning against basing a quality effort upon *inspection.* It is useful to note here that Dr. Deming did not say to avoid inspection altogether. He merely stated that inspection is not to be the sole quality assurance activity. He recognized that *quality* is designed in and not inspected into a system; that by the time inspection takes place it is too late for it to impact true quality. Furthermore, *quality* cannot be achieved or improved by inspecting to separate bad parts from good. However, once an effective quality assurance effort is in place, inspection can be a useful quality tool, albeit one of lesser importance.

Similarly we would not want to use *demonstration testing* as our sole reliability effort. But it, too, is a useful tool when applied only to demonstrate physically what we have already achieved analytically. Its value is only as a supplement to our overall reliability program. The demonstration of *mission reliability* or of *MTBF* generally consists of exercising our system or a set of systems in a fashion typical of in-use operation and observing the number of failures occurring offer a period of time or quantity of cycles. As you can imagine, the demonstration of reliability or MTBF can potentially lead to a prohibitively long and expensive test. But because we are not demanding that our test results be capable of doing anything more than demonstrating a prediction or compliance with a specification, the *test-of-hypothesis* technique can be used.

11.2 THE TEST-OF-HYPOTHESIS TECHNIQUE

The **test-of-hypothesis technique** is a method used for the type of decision making where we are trying to resolve which of two paths we ought to take. The advantage of the test-of-hypothesis technique is that it helps us make this *either-or* type of decision relatively quickly and inexpensively. Its weakness, however, is that it does not permit us to make predictions or to draw a strong conclusion about the phenomenon we are studying. Its value is restricted to allowing us to decide between, say, path A and path B.

The first step in applying the test-of-hypothesis technique is to state a hypothesis, referred to as the **null hypothesis,** H_0. A null hypothesis may be that the MTBF of a system meets its specified requirement.

As a second step we may also state an **alternative hypothesis,** H_1. An example of this may be that the MTBF of the system is some value that is less than the specified requirement. As illustrated in Table 11.1, the null hypothesis may, in actuality, be either true or false. Our intention is to design a test that will direct our decision based on the actual truth or falsity of the null hypothesis. We hope that when the test indicates that the null hypothesis is true, it actually is true, and that when the test indicates that the null hypothesis is false, it actually is false. However, that will not always be the case. There will be times when the test will lead us to an erroneous conclusion, considered a test error. The probability of a test error for a particularly designed test is called the **risk** of the test.

As indicated in Table 11.1, if the test tells us that the null hypothesis is *false* when it is not, a **Type I error** has been made. If the test tells us that the null hypothesis is *true* when it is not, a **Type II error** has been made. The probability of a Type I error for a test is called the α risk; the probability of a type II error is called the β risk. In other words, α is the probability of rejecting the null hypothesis when it is true, and β is the probability of accepting the null hypothesis when it is false. One of our objectives in designing a test of hypothesis is to minimize each of these risks.

Finally, we perform the test, assure its validity, and draw a conclusion. In the case of a reliability or MTBF demonstration test, the conclusion is either that our system can meet its required reliability or MTBF or it cannot. If the test erroneously rejects the hypothesis that the requirement can be met, a Type I error has been made, and the probability of that happening is α, which we call the producer's risk. If the test erroneously accepts the hypothesis that the requirement can be met, a Type II error has been made, and the probability of that happening is β, which we call the consumer's risk.

TABLE 11.1
The test-of-hypothesis technique.

In Actuality:	The Test Says:	Then We Have:	Test Risk:
H_0 is true.	H_0 is true.	No error	—
	H_1 is true.	Type I error	α
H_1 is true.	H_0 is true.	Type II error	β
	H_1 is true.	No error	—

The test-of-hypothesis technique, despite its limitations, is quite adequate for demonstrating reliability or MTBF at this stage. Assuming that good reliability design practices have been followed throughout the design and development of the system and that reliability or MTBF predictions have already been made, all we want to do is to show that our system is capable of a reliability or MTBF level that we are already convinced it can achieve. Hence, we are satisfied with a test that is capable of only indicating to us either *yes, it can* or *no, it cannot.*

11.3 RELIABILITY DEMONSTRATION TESTING OF SYSTEMS WITH CYCLE-DEPENDENT PERFORMANCE

Here we are interested in demonstrating a *success probability.* This test consists of operating a sample of systems through an established quantity of cycles and accepting if no more than a predetermined quantity of failures are included among those cycles. The government standard MIL-STD-105D contains a variety of test plans for demonstrating success probability for any type of situation. The test plans presented in this section are taken from MIL-STD-105D.

11.3.1 Single-Test Plans

The following test parameters pertain to a **success probability demonstration test:**

- R_0 *Success probability to be demonstrated.* This is usually a specified reliability value.
- **DR** *Discrimination ratio.* In applying the test-of-hypothesis technique, it is necessary to set an alternative hypothesis. Here we usually select a *success probability* value R_1 that is less than R_0. The discrimination ratio is the ratio between R_0 and its alternative, R_1.
- R_1 *Alternative success probability.* $R_1 = R_0/\text{DR}$.
- N *Number of system cycles in the test.*
- α *Producer's risk.*
- β *Consumer's risk.*
- *AQL* *Specified maximum unreliability:* $AQL = 1.0 - R_0$.
- c *Success criterion.* The test is accepted if there are c or fewer failures in N cycles. Otherwise, the test is rejected.

These parameters relate to each other as shown in Table 11.2, which applies the Table 11.1 test of hypothesis strategy to the design of a success probability demonstration test. In designing a success probability demonstration test, we would like to keep the discrimination ratio DR as close to 1.0 as possible while keeping the required number of cycles N and the test risks α and β as low as possible. However, as the discrimination ratio gets closer to 1.0, the other parameters, α, β, and N, tend to increase. Also, as N decreases, it causes the discrimination ratio and the test risks to increase. Similarly, as either test risk decreases, it causes an increase in the other risk, the required number of cycles, and the discrimination ratio. Hence, the design of a success probability test requires balancing all the test parameters to develop a plan that will best fit the situation at hand—for example, the cost and the number of test items at hand, the success probability to be demonstrated, and whether or not the test is destructive.

For most situations a plan can be chosen from among those presented in Table 11.3, which was derived from MIL-STD-105. To use this table, find an R_0 value that is equal to or close to the success probability to be demonstrated. Then select the test plan,

TABLE 11.2
The test-of-hypothesis technique applied to success probability demonstration tests. (This tests the hypothesis that the true reliability is R_0.)

In Actuality:	The Test Says:	Then We Have:	Test Risk:
Reliability = R_0	Accept	No error	—
	Reject	Type I error	α
Reliability = R_1	Accept	Type II error	β
	Reject	No error	—

TABLE 11.3
Success probability demonstration test plans.

Plan	N	R_0	R_1	DR	α	β	c	$c + 1$
1	1250	0.9999	0.9982	1.002	10%	10%	0	1
2	125	0.999	0.982	1.02	10%	10%	0	1
3	500	0.999	0.992	1.01	10%	10%	1	2
4	32	0.996	0.940	1.06	10%	10%	0	1
5	125	0.996	0.969	1.03	10%	10%	1	2
6	200	0.996	0.968	1.03	5%	5%	2	3
7	13	0.990	0.838	1.18	10%	10%	0	1
8	50	0.990	0.924	1.07	10%	10%	1	2
9	80	0.990	0.923	1.07	5%	5%	2	3
10	125	0.990	0.938	1.05	5%	5%	3	4
11	200	0.990	0.940	1.05	2%	2%	5	6
12	20	0.96	0.72	1.33	5%	5%	2	3
13	50	0.96	0.80	1.20	1%	5%	5	6
14	80	0.96	0.84	1.14	1%	5%	7	8
15	5	0.90	0.42	2.14	10%	10%	1	2
16	20	0.90	0.54	1.67	1%	5%	5	6
17	50	0.90	0.68	1.32	1%	5%	10	11

from among those with the appropriate R_0, that has the highest number of cycles N that you can tolerate. The table provides all the test parameters for the test you have selected, as well as the accept and reject criteria. To conduct the test, operate your system through N cycles and observe the number of successes and failures. If the number of failures is c or fewer, *accept* the hypothesis that the reliability (success probability) is equal to R_0. If the number of failures is $c + 1$ or more, *reject* that hypothesis, and assume that your system reliability is not as high as R_0.

Your test risks are as indicated by the plan selected. If the reliability is actually equal to R_0, the probability is α that the test will, nevertheless, be rejected (that is, have more than c failures in N cycles). If, on the other hand, the true reliability equals R_1, the probability is β that the test will be accepted (have c or fewer failures).

EXAMPLE 11.1

Design a reliability demonstration test to demonstrate a system success probability of 0.990 in fewer than 100 cycles.

This can be accomplished through Test Plan 9 of Table 11.3. This plan requires 80 cycles. If the quantity of failures among the 80 cycles is 2 or fewer, the test is accepted, and if there are 3 or more failures among the 80 cycles, the test is rejected. This test features a discrimination ratio of 1.07 and decision risks of 5%. That is, if the true system

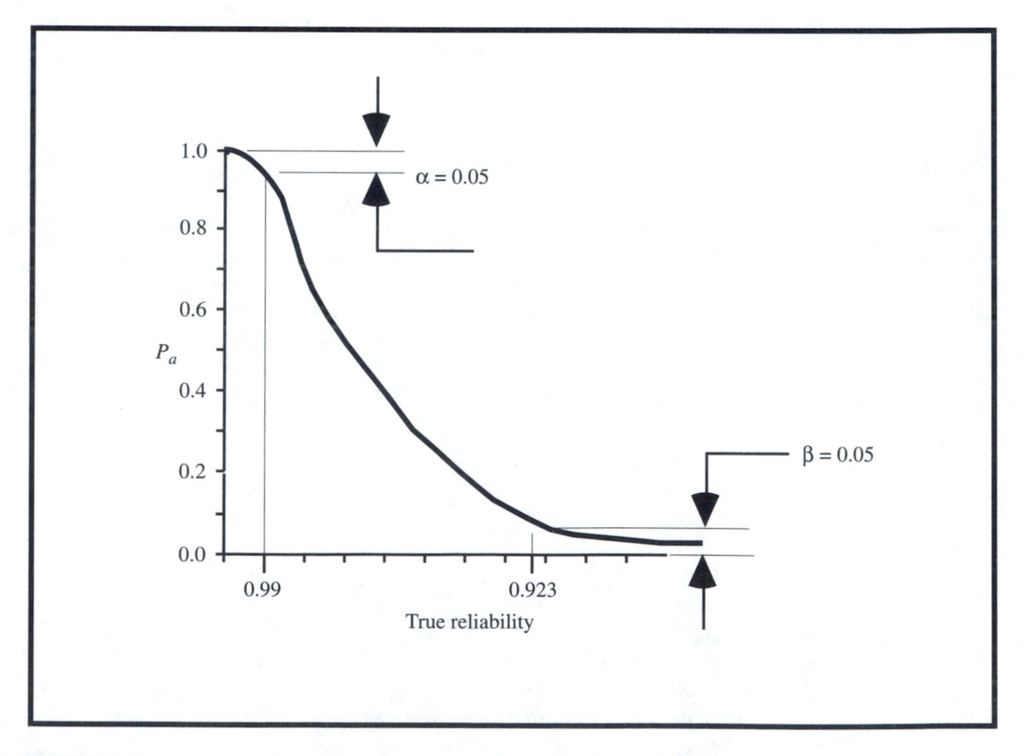

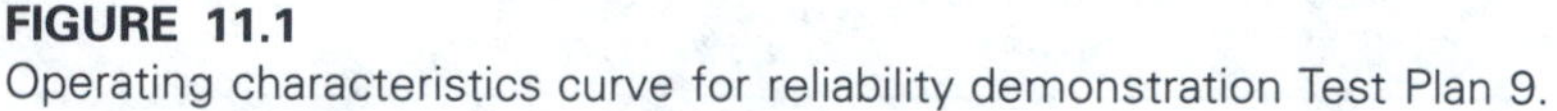
FIGURE 11.1
Operating characteristics curve for reliability demonstration Test Plan 9.

success probability is 0.990, there is a 5% probability that the test will be rejected (with more than 2 failures among the 80 cycles), and if the true success probability is 0.924, there is a 5% probability that the test will be accepted.

In determining the adequacy of a test plan, producer and consumer will want to examine what is known as an **operating characteristics curve** (or *O-C curve*). The O-C curve for a test plan indicates the likelihood of a system to be accepted or rejected at various reliability levels. The O-C curve for the test plan of Example 11.1 is illustrated in Figure 11.1. Here you can see that if the system submitted to this test has an actual reliability of 0.990, there is a 95% probability of acceptance, P_a, and a 5% probability of rejection. If the actual reliability of the system is higher than 0.990, the acceptance probability is higher. For example, if the system reliability is 0.995, the probability of acceptance is around 99%. If the true reliability of the system is as low as 0.980, for instance, the probability of acceptance is only about 78%. At a 0.970 system reliability, the probability of acceptance is only slightly more than 50%. Finally, if the system reliability is as low as the alternative reliability $R_1 = 0.923$, the success probability is 5%. Figure 11.1 also illustrates the producer's risk, α, the probability of a system with a 0.990 reliability being rejected, and the consumer's risk, β, the probability of a system with a 0.923 reliability being accepted. The following section shows how to compute test acceptance probabilities as well as how to design success probability demonstration tests should the selection of Table 11.3 be inadequate.

*11.3.2 Construction of an O-C Curve for a Success Probability Demonstration Test

The operating characteristics (O-C) curve plots the probability of test acceptance P_a versus the actual success probability R, as illustrated in Figure 11.1. The probability of test acceptance is the probability that out of N cycles there will be no more than c failures:

$$P_a = P(x \leq c) \tag{11.1}$$

where x is the number of failures and c is the maximum allowable number of failures for test acceptance. Referring to Section A.8, you will see that this is governed by the Binomial Probability Law, and

$$P_a = P(x \leq c) = \sum_{x=0}^{c} \binom{N}{x} (1 - R)^x R^{N-x} \tag{11.2}$$

Using Example 11.1 to illustrate the construction of an O-C curve, the test plan calls for testing through $N = 80$ cycles, and the test is accepted if there are $c = 2$ or fewer failures. Hence, the probability of acceptance is

$$\begin{aligned} P_a &= \sum_{x=0}^{c} \binom{N}{x} (1 - R)^x R^{N-x} = \sum_{x=0}^{2} \binom{80}{x} (1 - R)^x R^{80-x} \\ &= R^{80} + 80(1 - R)R^{79} + 3160(1 - R)^2 R^{78} = (3081R^2 - 6240R + 3160)R^{78} \end{aligned} \tag{11.3}$$

If the actual success probability is $R = 0.990$, the probability of acceptance is

$$P_a = (3081 \cdot 0.990^2 - 6240 \cdot 0.990 + 3160)0.990^{78} = 0.953 \quad \textbf{(11.4)}$$

If the actual success probability is $R = 0.970$, the probability of acceptance is

$$P_a = (3081 \cdot 0.970^2 - 6240 \cdot 0.970 + 3160)0.970^{78} = 0.568 \quad \textbf{(11.5)}$$

If the actual success probability is $R = 0.950$, the probability of acceptance is

$$P_a = (3081 \cdot 0.950^2 - 6240 \cdot 0.950 + 3160)0.950^{78} = 0.231 \quad \textbf{(11.6)}$$

If the actual success probability is $R = 0.930$, the probability of acceptance is

$$P_a = (3081 \cdot 0.930^2 - 6240 \cdot 0.930 + 3160)0.930^{78} = 0.075 \quad \textbf{(11.7)}$$

If the actual success probability is $R = 0.923$, the probability of acceptance is

$$P_a = (3081 \cdot 0.923^2 - 6240 \cdot 0.923 + 3160)0.923^{78} = 0.049 \quad \textbf{(11.8)}$$

The results of computations 11.4 through 11.8 can be used to construct the O-C curve of Figure 11.1. Notice that the producer's risk α is the probability of rejection if $R = 0.990$, which is $1.0 - 0.953 = 5\%$. Notice also that the consumer's risk β is the probability of acceptance if $R = 0.930$, which is 95%.

11.3.3 Multiple Test Plans

The use of **multiple test plans** is a strategy for reducing the expected quantity of cycles without paying the price of increasing the decision risks or discrimination ratio. However, a multiple plan is somewhat complicated. In this type of plan you observe an initial quantity of cycles, and on the basis of the number of failures you reach either an *accept, reject,* or *continue testing* decision. With an *accept* or *reject* decision, no further testing is necessary. But with a *continue* decision, an additional set of observations is required. The results of these additional observed cycles lead to an *accept, reject,* or *continue* decision. This process is continued until an *accept* or *reject* decision is ultimately reached. Most of the time a multiple plan will lead to an *accept* or *reject* decision with fewer cycles than are required for the equivalent single plan. It is possible, but unlikely, that a long series of *continue* decisions will result in more cycles than the equivalent single plan.

Because the appeal of multiple plans is restricted to situations where a larger-than-desired quantity of cycles N is specified by a single sampling plan, only five such plans are presented in Table 11.4. There is no multiple test plan equivalent to Plan 1 of Table 11.3. You may refer to MIL-STD-105D, from which Table 11.4 is derived, for a wider variety of multiple test plans.

As an example of the application of Table 11.4, Test Plan 3A is the multiple plan equivalent to Test Plan 3 of Table 11.3. Notice that the decision risks and discrimination ratio are the same. To follow Plan 3A, an initial 125 cycles are observed. An *accept*

TABLE 11.4
Success probability multiple demonstration test plans.

Plan	R_0	R_1	DR	α	β	Cumulative N	Accept with No. of Failures ≤:	Reject with No. of Failures ≥:
3A	0.999	0.992	1.01	10%	10%	125	—	2
						250	—	2
						375	0	2
						500	0	3
						625	1	3
						750	1	3
						875	2	3
5A	0.996	0.969	1.03	10%	10%	32	—	2
						64	—	2
						96	0	2
						128	0	3
						160	1	3
						192	1	3
						224	2	3
6A	0.996	0.968	1.03	5%	5%	50	—	2
						100	0	3
						150	0	3
						200	1	4
						250	2	4
						300	3	5
						350	4	5
9A	0.990	0.923	1.07	5%	5%	20	—	2
						40	0	3
						60	0	3
						80	1	4
						100	2	4
						120	3	5
						140	4	5
12A	0.960	0.720	1.33	5%	5%	5	—	2
						10	0	3
						15	0	3
						20	1	4
						25	2	4
						30	3	5
						35	4	5

decision cannot be reached on the basis of these 125 cycles, but if there are 2 or more failures among them, a *reject* decision is reached. Otherwise, an additional 125 cycles must be observed, for a cumulative total of 250 cycles. At this point if there are 2 or more failures, a *reject* decision is reached; otherwise, an additional 125 cycles must be observed, for a cumulative total of 375 cycles. If there are no failures among the 375 cycles, an *accept* decision is reached; if there are 2 or more failures among the 375 cycles, a *reject* decision is reached; if there is one failure among the 375 cycles an additional 125 cycles must be observed, for a cumulative total of 500 cycles. As described before, this process is continued until the *accept* or *reject* decision is reached. The decision will most likely be reached with fewer than the 500 cycles required by Test Plan 3.

The O-C curve for the test plans of Table 11.4 are similar to those of the corresponding plans of Table 11.3. Consequently, whenever a corresponding Table 11.4 plan is available, it can be used in place of the Table 11.3 plan if you want to take advantage of the potential advantages of a multiple test plan.

11.4 MTBF DEMONSTRATION TESTING OF SYSTEMS WITH TIME-DEPENDENT PERFORMANCE

Here we are interested in demonstrating a system *mean time between failures (MTBF).* This test consists of operating a sample of systems through a given operating time, measured in system operating hours, and accepting if no more than a predetermined quantity of failures take place among all the systems on test during the test time. An MTBF demonstration test can take the form of either a *fixed-length* or a *sequential* test plan. Both types of test plans, as presented in this section, are taken from the government standard MIL-HDBK-781D, which contains a variety of plans for many types of time-dependent performance situations.

The MTBF demonstration tests the hypothesis that the true system MTBF equals a specified value θ_0 against an alternative hypothesis that the true system MTBF equals θ_1, a value distinctly less than θ_0. The Table 11.1 test-of-hypothesis strategy applied to the design of an MTBF demonstration test is shown in Table 11.5.

TABLE 11.5
The test-of-hypothesis technique applied to MTBF demonstration tests. (This tests the hypothesis that the true MTBF is θ_0.)

In Actuality:	The Test Says:	Then We Have:	Test Risk:
MTBF = θ_0	Accept	No error	—
	Reject	Type I error	α
MTBF = θ_1	Accept	Type II error	β
	Reject	No error	—

11.4.1 Fixed-Length Tests

A **fixed-length MTBF demonstration test** is one requiring the test systems to be operated until an established amount of operating time has been accumulated. Whenever a failure occurs, the system will be restored or replaced, and operation will be resumed as soon as possible. If during that fixed operating time, the total number of failures occurring is equal to some quantity c or fewer, the test is *accepted*. However, if the total numbr of failures is equal to or greater than $c + 1$, the test is *rejected*. The following test parameters pertain to a fixed-length MTBF demonstration test:

θ_0 *MTBF to be demonstrated.* This is usually a specified *MTBF* value.

DR *Discrimination ratio.* In applying the test-of-hypothesis technique, it is necessary to set an alternative hypothesis. Here we usually select an MTBF value θ_1 that is less than θ_0. The discrimination ratio is the ratio between θ_0 and its alternative, θ_1.

θ_1 *Alternative MTBF.* $\theta_1 = \theta_0/\text{DR}$.

T *The total test time required.* This is the operating time to be accumulated among all the systems on test.

α *Producer's risk.* This is the probability of rejecting a test when the true $MTBF = \theta_0$.

β *Consumer's risk.* This is the probability of accepting a test when the true $MTBF = \theta_1$.

c *Success criterion.* The test is accepted if there are c or fewer failures in N cycles. Otherwise, the test is rejected.

As with the success probability demonstration test discussed in Section 11.3.1, in designing an MTBF demonstration test we wish to keep the discrimination ratio, DR, as close to 1.0 as possible, while keeping the total test time, T, and the decision risks, α and β, as low as possible. However, we are again faced with the dilemma that as the discrimination ratio, DR, gets closer to 1.0, the other parameters, α, β, and T, tend to increase; as T decreases, the other parameters, DR, α, and β, tend to increase; and as either decision risk decreases, it causes an increase in the other risk, the test time, and the discrimination ratio. As with the success probability demonstration test, the design of an MTBF demonstration test requires us to balance all the test parameters to derive a test plan that will best fit the situation at hand. In developing an optimum plan, we must take into consideration the magnitude of the MTBF to be demonstrated, the cost and the number of test systems and test facilities available to us, the amount of time available to us in drawing a timely conclusion to the test, and whether or not the test is destructive.

For most situations a plan can be chosen from among those presented in Table 11.6, which comes from MIL-HDBK-781D. To use the table, start with the desired quantity of test systems to be used and the amount of calendar time available for completion of the test. The amount of calendar time available is determined by how quickly you need a conclusion to the test, the cost and availability of test facilities, personnel, and test

TABLE 11.6
MTBF demonstration test plans.

Plan	α	β	DR	T	c	$c + 1$
1	10%	10%	1.5	$30.00\theta_0$	36	37
2	20%	20%	1.5	$14.07\theta_0$	17	18
3	10%	10%	2.0	$9.40\theta_0$	13	14
4	20%	20%	2.0	$3.90\theta_0$	5	6
5	10%	10%	3.0	$3.10\theta_0$	5	6
6	20%	20%	3.0	$1.43\theta_0$	2	3
7	30%	30%	1.5	$5.33\theta_0$	6	7
8	30%	30%	2.0	$1.85\theta_0$	2	3
9	30%	30%	3.0	$0.37\theta_0$	0	1

systems, and whether the test can be conducted on a 24-h-a-day schedule, including weekends and holidays. The total available test time, T, can be estimated as the quantity of test systems times the available calendar time for the test. Divide T by θ_0, and use that quotient to select a test plan or a set of plans from Table 11.6 that will satisfy your initial conditions. From among the selected plans, pick one that has the most desirable combination of total required test time, decision risks, and discrimination ratio. If you cannot find such a plan, some compromises will be necessary among the originally desired quantity of test systems, operating time, test scheduling, discrimination ratio, or decision risks. Section 11.4.2, which shows how to compute test acceptance probabilities for an o-c curve, also shows how to design MTBF demonstration tests should the Table 11.6 plans not offer a test plan to satisfy your constraints.

EXAMPLE 11.2

We are interested in demonstrating that the design of a given system has satisfied a specified MTBF of 1000 h. We have five prototype systems available for this test and have available five test facilities for a 1-month period. We wish to test on a two-shift basis (a maximum of 16 h a day), 5 days a week. We also wish to have a discrimination ratio and decision risks as low as possible. Select a suitable test plan from Table 11.6.

Operating five test systems for 16 h a day, 5 days a week, for 4 weeks provides a total test time of

$$T = 5 \times 16 \times 5 \times 4 = 1600 \text{ h}$$

Setting θ_0 equal to the specified MTBF of 1000 h, the test time can be

$$T = 1600 \text{ h} \div 1000 \text{ h} = 1.6\theta_0$$

Of the test plans in Table 11.6, the one with the best combination of discrimination ratio and decision risks within the maximum allowable T is Test Plan 6, with the following test parameters:

$$\begin{aligned}\alpha &= 20\% \\ \beta &= 20\% \\ \theta_0 &= 1000 \text{ h} \\ \text{DR} &= 3.0 \\ \theta_1 &= \theta_0 \div \text{DR} = 333 \text{ h} \\ c &= 2 \\ c + 1 &= 3\end{aligned}$$

According to this plan, the five test systems will be operated 16 h a day, 5 days a week until $1.43\theta_0 = 1430$ operating hours are accumulated among the systems. Whenever a failure occurs on one of the systems, it will be restored, and the system operation will resume. When the 1430 operating hours are accumulated, if the total number of failures accumulated among the five test systems equals 2 or fewer, the test will be accepted. If 3 or more failures occur among the five test systems during the test, the test will be rejected. If the true MTBF equals 1000 h, the probability of test rejection is approximately 20%. However, if the true MTBF equals 333 h, the probability of test acceptance is 20%.

Suppose we are dissatisfied with decision risks as high as 20%. Let us examine Test Plan 5, which offers risks of just 10%, which is more to our liking. The discrimination ratio and, consequently, θ_0 and θ_1 would be the same as for Test Plan 6, but the price we have to pay for the lower decision risks is a longer total test time T. According to Test Plan 5 the operating time to be accumulated among the five test systems will have to be

$$3.10\ \theta_0 = 3100 \text{ h}$$

With each of the five systems accumulating the same amount of operating time, the test will take $3100 \div 5 = 620$ h. To complete the test within the same 4-week period, the systems will have to be operated $620 \div 4 = 155$ h a week. This can be accomplished if the test is conducted continuously 24 h a day, 7 days a week. Or, if we were able to extend the time available for the test by a second month, we could maintain the 16-h-a-day, 5 day-a-week schedule. The parameters for this test plan are

$$\begin{aligned}T &= 3100 \text{ h} \\ \alpha &= 10\% \\ \beta &= 10\% \\ \theta_0 &= 1000 \text{ h} \\ \text{DR} &= 3.0 \\ \theta_1 &= \theta_0 \div \text{DR} = 333 \text{ h} \\ C &= 5 \\ c + 1 &= 6\end{aligned}$$

According to this plan, the five test systems will be operated 24 h a day, 7 days a week until $3.10\theta_0 = 3100$ operating hours are accumulated among the systems. Whenever a failure occurs on one of the systems, it will be restored, and the system operation will resume. When the 3100 operating hours are accumulated, if the total number of failures accumulated among the five test systems equals 5 or fewer, the test will be accepted. If 6 or more failures occur among the five test systems during the test, the test will be rejected. If the true MTBF equals 1000 h, the probability of test rejection is approximately 10%. However, if the true MTBF equals 333 hours, the probability of test acceptance is 10%.

As with the success probability demonstration test plans, the adequacy of an MTBF demonstration test plan can be evaluated through its o-c curve, which indicates the likelihood of an accept or reject decision at various MTBF levels. The o-c curve for Test Plan 5 of Table 11.6, with the values of Example 11.2 superimposed, is illustrated in Figure 11.2. Notice that if the system submitted for test has an inherent MTBF of 1000 h, the probability of test acceptance is 90%. If its MTBF is as low as 333 h, the

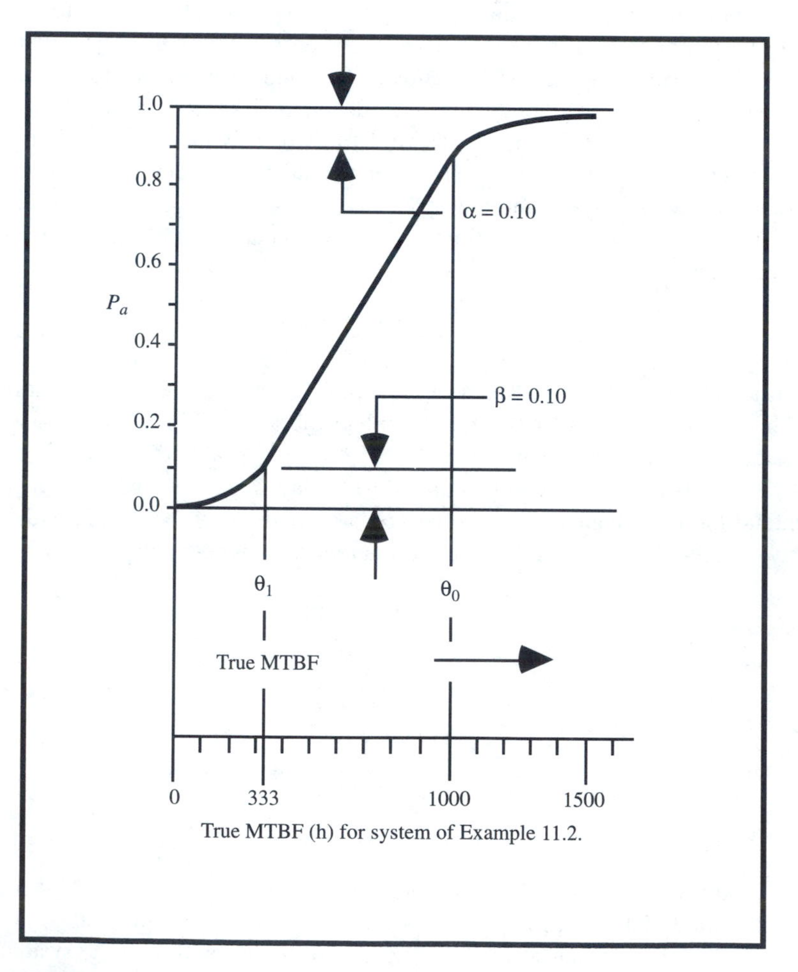

FIGURE 11.2
Operating characteristics curve for MTBF demonstration Test 5.

probability of test acceptance is only 10%. If the system is designed to have an MTBF 20% higher than its specified value—that is, 1200 h—the probability of test acceptance is around 93%.

*11.4.2 Derivation of Fixed-Length MTBF Demonstration Tests

Fixed-length MTBF demonstration tests assume an exponential time-to-failure distribution for the equipment on test. This assumption is valid if a planned replacement policy precludes wearout failure modes, so that we are dealing with only catastrophic failure modes within the equipment being tested. The fixed-length MTBF test consists of testing until we accumulate a total equipment operating time T. The test is accepted if there are c or fewer scorable failures during the time T. The test is rejected if there are $r = c + 1$ or more scorable failures during the time T.

From Equation 4.12, we can compute the operating characteristics—that is, the probability of acceptance, P_a, versus actual equipment MTBF. The equipment MTBF, say, θ_i, that will provide a P_a of γ can be computed

$$\theta_i = \frac{2T}{\chi^2_{1-\gamma,2r}} \quad \textbf{(11.9)}$$

where $\chi^2_{1-\gamma,2r}$ is the chi-square value at the $1 - \gamma$ level and at $2r$ degrees of freedom. This, of course, is the probability that there are c or fewer failures in time T if the MTBF is θ_i.

The specified MTBF θ_0 is then

$$\theta_0 = \frac{2T}{\chi^2_{\alpha,2r}} \quad \textbf{(11.10)}$$

where α is the producer's risk, that is, the probability of having r or more failures during time T if the true MTBF is θ_0. Hence, the total test operating time, T, is

$$T = \frac{\theta_0 \chi^2_{\alpha,2r}}{2} \quad \textbf{(11.11)}$$

Similarly, the alternative MTBF θ_1 is

$$\theta_1 = \frac{2T}{\chi^2_{1-\beta,2r}} \quad \textbf{(11.12)}$$

where β is the consumer's risk, that is, the probability of having c or fewer failures during the time T if the true MTBF is θ_1. Hence, the total test operating time, T, is also

$$T = \frac{\theta_1 \chi^2_{1-\beta,2r}}{2} \quad \textbf{(11.13)}$$

From Equations 11.11 and 11.13 we can derive the discrimination ratio DR $= (\theta_0/\theta_1)$:

$$\left(\frac{\theta_0}{\theta_1}\right) = \frac{\chi^2_{1-\beta,2r}}{\chi^2_{\alpha,2r}} \quad \textbf{(11.14)}$$

In designing a test we first select a desirable discrimination ratio DR $= (\theta_0/\theta_1)$ and set of risks θ_0 and θ_1. We next determine the smallest value of r that will satisfy Equation 11.14. This can be accomplished through a table. See Table 11.7.

TABLE 11.7
Determination of test reject criterion.

r	$\chi^2_{1-\beta,2r}/\chi^2_{\alpha,2r}$	$\leq (\theta_0/\theta_1)$
1		No
2		No
3		.
.		.
.		.
.		
$(r-1)$		No
r	$\longleftarrow$	Yes

From Equation 11.11, we can compute the fixed-length test time in multiples of θ_0:

$$\frac{T}{\theta_0} = \frac{\chi^2_{\alpha,2r}}{2} \tag{11.15}$$

EXAMPLE 11.3

Design a fixed-length test plan to demonstrate an MTBF of 1000 h, with decision risks $\alpha = \beta =$ 20% and a discrimination ratio of 2.0.

Applying Table 11.7 to determine the reject criterion r gives Table 11.8.

For this test, $r = 6$, $c = 5$, and from Equation 11.11, the required fixed-length test time is

$$T = \frac{\theta_0 \chi^2_{\alpha,2r}}{2} = \frac{7.81 \times 1000 \text{ h}}{2} = 3900 \text{ h}$$

So we are required to test until 3900 equipment operating hours are accumulated, and the test is accepted if there are 5 or fewer scorable failures and rejected if there are 6 or more scorable failures. The null hypothesis is that MTBF $= \theta_0 =$ 1000 h, and the alternative hypothesis is MTBF $= \theta_1 =$ 500 h. The actual decision risks are determined as follows.

The producer's risk is

$$\begin{aligned}\alpha &= P(x \geq 6 \mid \text{MTBF} = 1000 \text{ h}) \\ &= 1 - P(x \leq 5 \mid \text{MTBF} = 1000 \text{ h})\end{aligned}$$

This can be determined through the Poisson Probability Law with

$$\mu = \text{T/MTBF} = 3900/1000 = 3.90$$

From Table B.9, $P(x \leq 5 \mid \mu = 3.9) = 0.801$. Hence, $\alpha = 1 - 0.801 = 19.9\%$.

The consumer's risk is $\beta = P(x \leq 5 \mid \text{MTBF} = 500 \text{ h})$, which can be determined through the Poisson Probability Law with

$$\mu = \text{T/MTBF} = 3900/500 = 7.80$$

From Table B.9, $P(x \leq 5 \mid \mu = 7.8) = 0.210$. Hence, $\beta = 21.0\%$.

TABLE 11.8
Determination of reject criterion for Example 11.3.

r	$\chi^2_{1-\beta,2r}/\chi^2_{\alpha,2r} = \chi^2_{0.80,2r}/\chi^2_{0.20,2r}$	$\leq (\theta_0/\theta_1)$
1	3.22/0.446 = 7.2	No
2	5.99/1.65 = 3.6	No
3	8.56/3.07 = 2.8	No
4	11.0/4.59 = 2.4	No
5	13.4/6.18 = 2.2	No
6	15.8/7.81 = 2.0	Yes

11.4.3 Sequential Tests

Suppose that in Example 11.2 we wish to have the lower decision risks without paying the price of the additional operating time. We may decide to use the strategy of **sequential demonstration testing.** Sequential test plans, commonly referred to as **PRST** (or **probability ratio sequential test**) plans, for demonstrating system MTBF serve the same purpose as multiple test plans for demonstrating system success probability: They provide lower decision risks with the potential for less testing. Test Plans 1 through 8 of Table 11.6 have alternative plans, Test Plans 1A through 8A of Table 11.9, whereby on the basis of the accumulated quantity of failures versus accumulated operating time, we may reach *accept, reject,* or *continue testing* decisions at any point during the test.

EXAMPLE 11.4 Use the equivalent sequential MTBF demonstration test, providing 10% decision risks, to demonstrate the specified 100-h MTBF on the system of Example 11.2 through the same five prototypes. Discuss the potential advantages of using a sequential test plan in this circumstance.

Test Plan 5A features test parameters $\alpha = 10\%$, $\beta = 10\%$, DR = 3.0, $\theta_0 = 1000$ h, and $\theta_1 = 333$ h, as did Test Plan 5 in Example 11.2. The accept/reject criteria and test times, from Table 11.9, are determined to be the following.

Cumulative Number of Failures	Reject If Total Equipment Operating Hours ≤:	Accept If Total Equipment Operating Hours ≥:
0	N/A	1250
1	N/A	1800
2	190	2350
3	740	2900
4	1290	3450
5	1840	3450
6	3390	3450
7	3450	N/A

TABLE 11.9
PRST plans for demonstrating system MTBF.

Plan	α	β	DR	Number of Failures	Reject if Total Test Time ≤:	Accept if Total Test Time ≥:
1A	10%	10%	1.5	0	N/A	$4.40\theta_0$
				1	N/A	$5.21\theta_0$
				2	N/A	$6.20\theta_0$
				3	N/A	$6.83\theta_0$
				4	N/A	$7.64\theta_0$
				5	N/A	$8.45\theta_0$
				6	$0.45\theta_0$	$9.27\theta_0$
				7	$1.26\theta_0$	$10.08\theta_0$
				8	$2.07\theta_0$	$10.89\theta_0$
				9	$2.89\theta_0$	$11.70\theta_0$
				10	$3.69\theta_0$	$12.51\theta_0$
				11	$4.50\theta_0$	$13.32\theta_0$
				12	$5.31\theta_0$	$14.13\theta_0$
				13	$6.12\theta_0$	$14.94\theta_0$
				14	$6.93\theta_0$	$15.75\theta_0$
				15	$7.74\theta_0$	$16.56\theta_0$
				16	$8.55\theta_0$	$17.37\theta_0$
				17	$9.37\theta_0$	$18.19\theta_0$
				18	$10.18\theta_0$	$19.00\theta_0$
				19	$10.99\theta_0$	$19.81\theta_0$
				20	$11.80\theta_0$	$20.62\theta_0$
				21	$12.61\theta_0$	$21.43\theta_0$
				22	$13.42\theta_0$	$22.24\theta_0$
				23	$14.23\theta_0$	$23.05\theta_0$
				24	$15.04\theta_0$	$23.86\theta_0$
				25	$15.85\theta_0$	$24.67\theta_0$
				26	$16.66\theta_0$	$25.48\theta_0$
				27	$17.47\theta_0$	$26.29\theta_0$
				28	$18.29\theta_0$	$27.11\theta_0$
				29	$19.10\theta_0$	$27.92\theta_0$
				30	$19.90\theta_0$	$28.73\theta_0$
				31	$20.72\theta_0$	$29.54\theta_0$
				32	$21.53\theta_0$	$30.35\theta_0$
				33	$22.34\theta_0$	$31.16\theta_0$
				34	$23.15\theta_0$	$31.97\theta_0$
				35	$23.96\theta_0$	$32.78\theta_0$
				36	$24.77\theta_0$	$33.00\theta_0$
				37	$25.58\theta_0$	$33.00\theta_0$
				38	$26.39\theta_0$	$33.00\theta_0$

TABLE 11.9
Continued

Plan	α	β	DR	Number of Failures	Reject if Total Test Time ≤:	Accept if Total Test Time ≥:
2A	20%	20%	1.5	0	N/A	$2.79\theta_0$
				1	N/A	$3.60\theta_0$
				2	N/A	$4.41\theta_0$
				3	$0.16\theta_0$	$5.22\theta_0$
				4	$0.97\theta_0$	$6.03\theta_0$
				5	$1.78\theta_0$	$6.84\theta_0$
				6	$2.60\theta_0$	$7.66\theta_0$
				7	$3.41\theta_0$	$8.47\theta_0$
				8	$4.22\theta_0$	$9.28\theta_0$
				9	$5.03\theta_0$	$10.09\theta_0$
				10	$5.84\theta_0$	$10.90\theta_0$
				11	$6.65\theta_0$	$11.71\theta_0$
				12	$7.46\theta_0$	$12.49\theta_0$
				13	$8.27\theta_0$	$13.33\theta_0$
				14	$9.10\theta_0$	$14.14\theta_0$
				15	$9.89\theta_0$	$14.60\theta_0$
				16	$10.70\theta_0$	$14.60\theta_0$
				17	$11.52\theta_0$	$14.60\theta_0$
				18	$12.55\theta_0$	$14.60\theta_0$
				19	$14.60\theta_0$	N/A
3A	10%	10%	2.0	0	N/A	$2.20\theta_0$
				1	N/A	$2.90\theta_0$
				2	N/A	$3.59\theta_0$
				3	$0.35\theta_0$	$4.28\theta_0$
				4	$1.04\theta_0$	$4.97\theta_0$
				5	$1.74\theta_0$	$5.67\theta_0$
				6	$2.43\theta_0$	$6.38\theta_0$
				7	$3.12\theta_0$	$7.05\theta_0$
				8	$3.82\theta_0$	$7.75\theta_0$
				9	$4.51\theta_0$	$8.44\theta_0$
				10	$5.20\theta_0$	$9.13\theta_0$
				11	$5.90\theta_0$	$9.83\theta_0$
				12	$6.59\theta_0$	$10.30\theta_0$
				13	$7.28\theta_0$	$10.30\theta_0$
				14	$7.97\theta_0$	$10.30\theta_0$
				15	$8.67\theta_0$	$10.30\theta_0$
				16	$10.30\theta_0$	N/A

TABLE 11.9
Continued

Plan	α	β	DR	Number of Failures	Reject if Total Test Time ≤:	Accept if Total Test Time ≥:
4A	20%	20%	2.0	0	N/A	$1.400\theta_0$
				1	N/A	$2.090\theta_0$
				2	$0.350\theta_0$	$2.790\theta_0$
				3	$1.040\theta_0$	$3.480\theta_0$
				4	$1.730\theta_0$	$4.170\theta_0$
				5	$2.430\theta_0$	$4.870\theta_0$
				6	$3.120\theta_0$	$4.870\theta_0$
				7	$3.810\theta_0$	$4.870\theta_0$
				8	$4.870\theta_0$	N/A
5A	10%	10%	3.0	0	N/A	$1.250\theta_0$
				1	N/A	$1.800\theta_0$
				2	$0.190\theta_0$	$2.350\theta_0$
				3	$0.740\theta_0$	$2.900\theta_0$
				4	$1.290\theta_0$	$3.450\theta_0$
				5	$1.840\theta_0$	$3.450\theta_0$
				6	$3.390\theta_0$	$3.450\theta_0$
				7	$3.450\theta_0$	N/A
6A	20%	20%	3.0	0	N/A	$0.890\theta_0$
				1	N/A	$1.440\theta_0$
				2	$0.120\theta_0$	$1.500\theta_0$
				3	$1.500\theta_0$	N/A
7A	30%	30%	1.5	0	N/A	$2.100\theta_0$
				1	N/A	$2.910\theta_0$
				2	N/A	$3.720\theta_0$
				3	$0.810\theta_0$	$4.530\theta_0$
				4	$1.620\theta_0$	$4.530\theta_0$
				5	$4.430\theta_0$	$4.530\theta_0$
				6	$4.530\theta_0$	N/A
8A	30%	30%	2.0	0	N/A	$0.860\theta_0$
				1	N/A	$1.550\theta_0$
				2	N/A	$2.250\theta_0$
				3	$2.250\theta_0$	N/A

According to this plan, an accept decision can be reached with no failures in 1250 equipment operating hours or no more than 1 failure in 1800 equipment operating hours. In either of these events the test can be concluded within the requisite 4-week period on the two-shift, 5-day basis. Only if the test is accepted with between 2 and 6 failures will additional test time be necessary. If the test is to be rejected, it will most likely be completed within the same 4-week period. The worst-case scenario requires 3450 equipment operating hours, and that can be accomplished on a 5-day, two-shift schedule within 2 months.

Hence, by specifying the sequential Test Plan 5A, we have a very good chance of accomplishing all our objectives, namely, completing the test within the desired period on the desired workday schedule and with no more than the desired 10% decision risks. There is, however, the uncertainty that the test can run for more than 4 weeks, up to a possible maximum of 9 weeks.

*11.4.4 Derivation of PRST Method for Demonstrating MTBF

In these tests the accept/reject criteria are defined through three regions, defined in Figure 11.3. At any time during the sequential test that the *accept* region is entered on the operating time–versus–number of failures plot, the test is accepted. At any time that the *reject* region is entered on the operating time–versus–number of failures plot, the test is rejected.

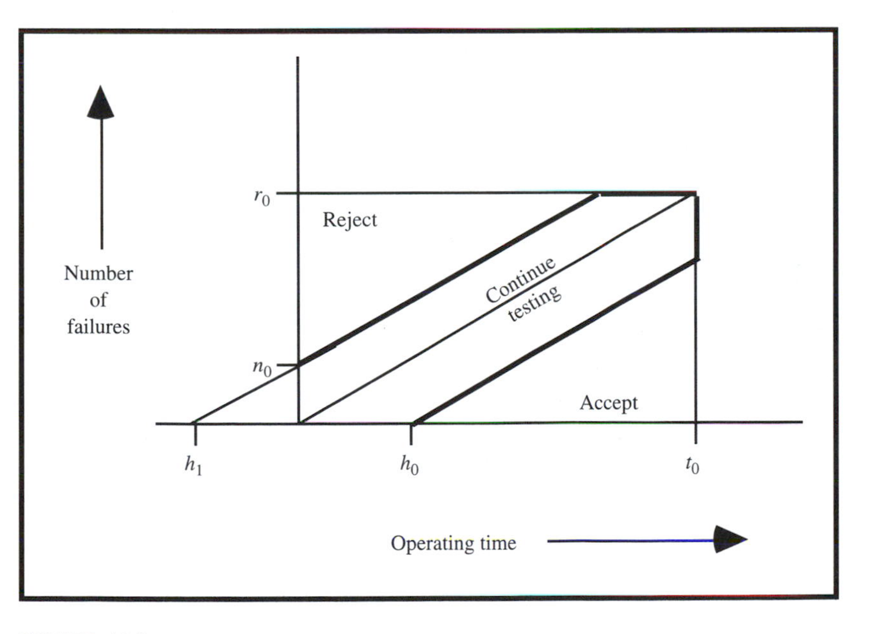

FIGURE 11.3
PRST accept reject criteria.

To design a PRST, first determine the accept and reject ratios, as follows:

$$\text{reject ratio} = A = \frac{\text{probability of rejecting ``rejectable'' system}}{\text{probability of rejecting ``acceptable'' system}} = \frac{1-\beta}{\alpha}$$

$$\text{accept ratio} = B = \frac{\text{probability of accepting ``rejectable'' system}}{\text{probability of accepting ``acceptable'' system}} = \frac{\beta}{1-\alpha}$$

Let h_0 = the minimum accept time. At that point,

$$B = \frac{e^{-\lambda_1 h_0}}{e^{-\lambda_0 h_0}} = e^{-(\lambda_1 - \lambda_0)h_0} = e^{-(1/\theta_1 - 1/\theta_0)h_0} \tag{11.16}$$

Therefore,

$$h_0 = \frac{-\ln B}{\left(\frac{1}{\theta_1} - \frac{1}{\theta_0}\right)}$$

Similarly,

$$h_1 = \frac{-\ln A}{\left(\frac{1}{\theta_1} - \frac{1}{\theta_0}\right)} \tag{11.17}$$

Solving for the slope S in Figure 11.3, illustrated in Figure 11.4,

$$S = \frac{n_0}{-h_1} = \frac{n_0}{\left[\frac{+\ln A}{1/\theta_1 - 1/\theta_0}\right]} = \frac{n_0\left(\frac{1}{\theta_1} - \frac{1}{\theta_0}\right)}{\ln A} \tag{11.18}$$

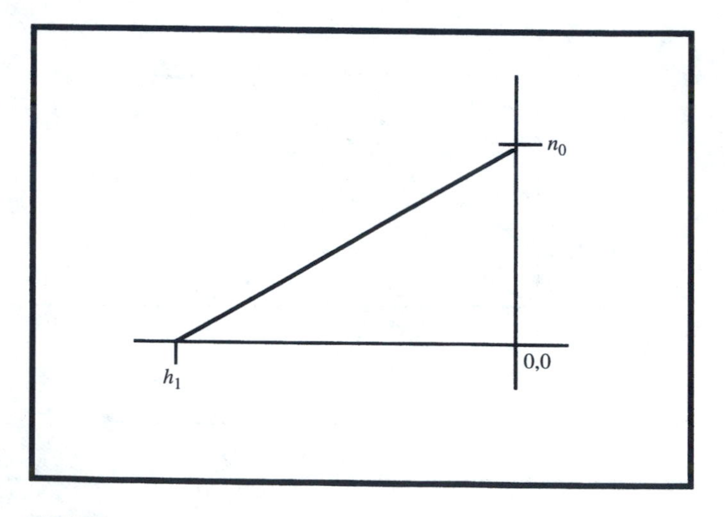

FIGURE 11.4
Slope of Figure 11.3.

The accept ratio, A, is

$$A = \frac{\text{probability of test rejection if MTBF} = \theta_1}{\text{probability of test rejection if MTBF} = \theta_0}$$

$$= \frac{\dfrac{(T/\theta_1)^{n_T} e^{-T/\theta_1}}{n_T!}}{\dfrac{(T/\theta_0)^{n_T} e^{-T/\theta_0}}{n_T!}} = \left(\frac{\theta_0}{\theta_1}\right)^{n_T} e^{-T(1/\theta_1 - 1/\theta_0)} \tag{11.19}$$

where n_T is the reject number at the time T. At time $T = 0$,

$$A = \left(\frac{\theta_0}{\theta_1}\right)^{n_0} e^0 = \left(\frac{\theta_0}{\theta_1}\right)^{n_0} \tag{11.20}$$

Therefore,

$$\ln A = n_0 \ln\left(\frac{\theta_0}{\theta_1}\right), \quad \text{and} \quad S = \frac{n_0\left(\dfrac{1}{\theta_1} - \dfrac{1}{\theta_0}\right)}{\ln A} = \frac{\left(\dfrac{1}{\theta_1} - \dfrac{1}{\theta_0}\right)}{\ln(\theta_0/\theta_1)} \tag{11.21}$$

Let the truncation point r_0 in Figure 11.3 be $3c$, where c is the maximum number of failures allowed for an accept decision for a fixed-length test with the same decision risks a and b and the same discrimination ratio (θ_0/θ_1). The truncation time, as illustrated in Figure 11.3, is

$$t_0 = S\,r_0 = \frac{\left(\dfrac{1}{\theta_1} - \dfrac{1}{\theta_0}\right) r_0}{\ln(\theta_0/\theta_1)} \tag{11.22}$$

S is the slope of the region boundaries in Figure 11.3.

11.5 RESTORABILITY DEMONSTRATION TESTS

Restorability demonstration tests generally consist of simulating faults on a system in question and timing the restoration activities, with the objective of assessing system MTTR (mean time to restore) and an upper α-percentile point $M_{ct}(\alpha)$. Although this is typically done on equipment, a restorability demonstration test can also be performed on a service-rendering system (for example, to evaluate the time to restore interrupted service).

To demonstrate system restorability, it is usually possible to demonstrate the MTTR and $M_{ct}(\alpha)$ at fairly high confidence levels, usually at the 90% confidence level, so the test-of-hypothesis technique is not necessary here. Hence, contrary to reliability and MTBF demonstration testing, restorability demonstration test results can be used for making predictions. The technique described here very closely follows that of the government standard MIL-STD-471A.

11.5.1 Restorability Specifications

Restorability of equipment is usually specified in terms of two requirements:

1. System MTTR
2. System $M_{ct}(0.90)$, the upper 90th percentile of the time-to-restore distribution

or

1. System MTTR
2. System $M_{ct}(0.95)$, the upper 95th percentile of the time-to-restore distribution.

This technique demonstrates achievement of the specified rquirements at the 90% confidence levels.

11.5.2 LRU Task Allocation

Restoration time on a system is expected to vary depending on the level of maintenance (See Chapter 7). Different maintenance actions are performed at organizational, intermediate, or depot levels. So, the first step in demonstrating restorability parameters is to identify the LRUs *(line replaceable units)* for the pertinent maintenance level. Next, the failure rate predictions are listed for all the LRUs comprising the system. Recognizing that various modes of failure within a single LRU can require different restoration times, it is necessary to allocate the LRU failure rates down to each of the failure modes. The *failure-modes-and-effects (FMEA)* technique, presented in Chapter 13, is useful here.

Because we expect, as we learned in Chapter 7, the system restoration times to fit a *lognormal* distribution, estimating the MTTR and $M_{ct}(\alpha)$ will require the demonstration of at least 50 restoration tasks. It is customary to demonstrate either 50, 100, or 150 tasks, depending on the anticipated length of these restoration times and the time and resources available for testing. The frequency of each of the tasks types will be proportional to demand frequencies, that is, failure rates.

11.5.3 Demonstration Tasks

To design a demonstration test for a system, we first make a list of all task types. A **task type** is the corrective maintenance activity necessary to restore the system from a particular failure mode of a particular LRU. So, each failure mode of each LRU will represent one task type. If we then list the allocated failure rate for each failure mode of each LRU and add up these failure rates, the number of tasks to be performed for each task type is determined from

$$N_i = \frac{\lambda_i}{\lambda_T} N \tag{11.23}$$

where

N_i = the number of tasks required on the ith task type
λ_i = the allocated failure rate for the ith task type
λ_T = the total of the failure rates for all task types
N = the total number of tasks to be demonstrated

EXAMPLE 11.5 Suppose a system contains 10 LRUs, the first of which has 5 failure modes, another 6 of which have 3 failure modes each, 2 of which have 2 failure modes each, and the last of which has only 1 failure mode. This amounts to 28 tasks. Suppose the failure rates are allocated as follows.

Task Type	LRU	Failure Mode	Failure Rate per 10^6 h
1	1	1	42.6
2	1	2	13.2
3	1	3	6.0
4	1	4	2.1
5	1	5	1.0
6	2	1	55.9
7	2	2	36.4
8	2	3	11.9
9	3	1	10.2
10	3	2	6.3
11	3	3	22.4
12	4	1	19.6
13	4	2	16.6
14	4	3	10.3
15	5	1	4.4
16	5	2	7.7
17	5	3	6.2
18	6	1	5.7
19	6	2	5.5
20	6	3	2.0
21	7	1	111.0
22	7	2	99.6
23	7	3	84.3
24	8	1	9.7
25	8	2	3.2
26	9	1	24.8
27	9	2	15.1
28	10	1	109.0

Allocate the number of tasks of each type to be performed for a restorability demonstration test to be composed of 100 tasks.

The total failure rate for all LRU's is computed to be 742.7×10^{-6} per hour. The number of each type of task, N_i, is determined through Equation 11.23 to be as follows.

Task Type	LRU	Failure Mode	Failure Rate per 10^6 h	N_i
1	1	1	42.6	5.7
2	1	2	13.2	1.8
3	1	3	6.0	0.8
4	1	4	2.1	0.3
5	1	5	1.0	0.1
6	2	1	55.9	7.5
7	2	2	36.4	4.9
8	2	3	11.9	1.6
9	3	1	10.2	1.4
10	3	2	6.3	0.8
11	3	3	22.4	3.0
12	4	1	19.6	2.6
13	4	2	16.6	2.2
14	4	3	10.3	1.4
15	5	1	4.4	0.6
16	5	2	7.7	1.0
17	5	3	6.2	0.8
18	6	1	5.7	0.8
19	6	2	5.5	0.7
20	6	3	2.0	0.3
21	7	1	111.0	14.9
22	7	2	99.6	13.4
23	7	3	84.3	11.4
24	8	1	9.7	1.3
25	8	2	3.2	0.4
26	9	1	24.8	3.3
27	9	2	15.1	2.0
28	10	1	109.0	14.7

If we round off the N_i values to the nearest integer, four of the task types (Type 4, 5, 20, and 25) would require no task demonstrations. However, if we pool them together, their total, N_i, adds up to 1.1, requiring one demonstration task from a set composed of all four types. This leads to the following results.

Task Type	LRU	Failure Mode	Failure Rate per 10^6 h	N_i	Number of Tasks to be Performed
1	1	1	42.6	5.7	6
2	1	2	13.2	1.8	2
3	1	3	6.0	0.8	1
6	2	1	55.9	7.5	7
7	2	2	36.4	4.9	5
8	2	3	11.9	1.6	2
9	3	1	10.2	1.4	1
10	3	2	6.3	0.8	1
11	3	3	22.4	3.0	3
12	4	1	19.6	2.6	3
13	4	2	16.6	2.2	2
14	4	3	10.3	1.4	1
15	5	1	4.4	0.6	1
16	5	2	7.7	1.0	1
17	5	3	6.2	0.8	1
18	6	1	5.7	0.8	1
19	6	2	5.5	0.7	1
21	7	1	111.0	14.9	15
22	7	2	99.6	13.4	13
23	7	3	84.3	11.4	11
24	8	1	9.7	1.3	1
26	9	1	24.8	3.3	3
27	9	2	15.1	2.0	2
28	10	1	109.0	14.7	15
4, 5, 20,	1	4, 5			
25	6	3	8.3	1.1	1
	8	2			
				Total	100

It is customary to develop a list of candidate tasks containing two to three times the required number of tasks. In other words, for each task of a given type to be actually used, two or three demonstration tasks will be designed. In the case of Example 11.5, we may wish to design 300 tasks, of which 18 will be of Type 1, 6 will be of Type 2, etc., and 3 will be distributed from among Types 4, 5, 20, and 25. Of the 300 designed, candidate tasks, of course only 100 will actually be demonstrated.

11.5.4 Means of Introducing Faults

For each of the designed tasks, it will be necessary to devise a means of inducing into the system a fault that will require the maintenance task in question. Prior to performing the test, each of these faults will individually be induced into the test system to verify that they produce the expected symptoms.

11.5.5 Task-Selection Process

The actual test is to be conducted in an atmosphere that simulates the in-use conditions and environment for the pertinent level of maintenance. All tools, equipment, and documentation expected to be on hand at the intended restoration site will be available for use during the test. The tasks will be performed by personnel of the same skill levels and provided the same training as the on-site maintenance personnel.

The test will start with test maintenance personnel *not* present. The first task to be performed will be selected at random from the list of candidate tasks. This random selection can be computer-generated or by physically drawing a task card at random from a "hat" containing cards representing all the candidates. The fault requiring the performance of the selected task will be simulated on the test system, and the symptoms will be verified.

The test maintenance personnel will then be summoned, and symptoms will be reported. The restoration task will then be performed and timed by an observer, who will record the times as described in the next section. Upon completion, the test maintenance personnel will again leave, while the next task is randomly selected and the next fault is induced. This process will continue until all tasks are completed.

Upon completion of the test, the quantity of each task type is to be as required by the test plan (that is, as determined by Equation 11.1). To assure that this is the case, when the required number of tasks of a particular task type (or set of task types) have been conducted, all remaining tasks of that type will be removed from the candidate list. Another way of assuring this condition is that when a task of a given type is selected and the required quantity of tasks of that type have already been completed, the selection will be ignored, and another random selection will be made.

11.5.6 Timing of Repairs

The designated observer/timekeeper will maintain a test log, upon which will be recorded the observed times for each restoration time element (*identification, diagnostic, access, interchange, reassembly, alignment, and checkout times,* as defined in Chapter 7) of each task performed. The observer/timekeeper will have to observe carefully which element is being worked upon and record the time of each. The sum of all element times for each task is the total restoration time for that task. A task restoration time represents the time from the report of symptoms to the final checkout and verification of a successful restoration. If we were working to the test of Example 11.1, by the end of the test we would have a recorded total restoration time for each of the 100 tasks.

11.5.7 Confidence Estimates of MTTR and $M_{ct}(\alpha)$

At the conclusion of the demonstration test we shall have the task restoration time Rp_i for each of the demonstrated tasks i. We shall then compute and list next to each of the task restoration times the squares of the restoration times $(Rp_i)^2$, the logarithms ($\log Rp_i$), and the squares of the logarithms ($(\log Rp_i)^2$). We shall need the sum of each of these, which we can determine through a data sheet set up as illustrated in Figure 11.5.

Task	Rp	$(Rp)^2$	$\log Rp$	$(\log Rp)^2$
Totals	$\sum Rp$	$\sum (Rp)^2$	$\sum \log Rp$	$\sum (\log Rp)^2$

FIGURE 11.5
Restorability demonstration test data form.

From the $\sum Rp$, $\sum (Rp)^2$, $\sum \log Rp$, and $\sum (\log Rp)^2$ values, we compute the following parameters from relationships derived in Section 11.5.9:

$$\mathrm{MTTR}_{0.90} = \frac{\sum Rp}{N} + 1.28\sqrt{\frac{N\sum Rp^2 - (\sum Rp)^2}{N^2(N-1)}} \tag{11.24}$$

$$Rp_{0.90}(0.50) = \log^{-1}\left[\frac{\sum \log Rp}{N} + 1.28\sqrt{\frac{N\sum (\log Rp)^2 - (\sum \log Rp)^2}{N^2(N-1)}}\right] \tag{11.25}$$

$$Rp_{0.90}(0.90) = \log^{-1}\left[\frac{\sum \log Rp}{N} + \left(\frac{1.28}{\sqrt{N}} + 1.4\right)\sqrt{\frac{N\sum (\log Rp)^2 - (\sum \log Rp)^2}{N(N-1)}}\right] \tag{11.26}$$

$$Rp_{0.90}(0.95) = \log^{-1}\left[\frac{\sum \log Rp}{N} + \left(\frac{1.28}{\sqrt{N}} + 1.8\right)\sqrt{\frac{N\sum (\log Rp)^2 - (\sum \log Rp)^2}{N(N-1)}}\right] \tag{11.27}$$

where $MTTR_{0.90}$ is the 90% confidence estimate of the $MTTR$

$Rp_{0.90}(0.50)$ is the 90% confidence estimate of the median of the restoration time distribution

$Rp_{0.90}(0.90)$ is the 90% confidence estimate of the upper 90th percentile of the restoration time distribution

$Rp_{0.90}(0.95)$ is the 90% confidence estimate of the upper 95th percentile of the restoration time distribution,

N is the number of restoration tasks performed for the demonstration test

11.5.8 Accept-Reject Criteria

The test is accepted if the $MTTR_{0.90}$ and the $Rp_{0.90}(0.90)$ do not exceed the specified MTTR and $Rp(0.90)$, respectively, or if the $MTTR_{0.90}$ and the $Rp_{0.90}(0.95)$ do not exceed the specified MTTR and $Rp(0.95)$, respectively. If the median repair time is specified instead of the MTTR, then test acceptance will depend on the $Rp_{0.90}(0.50)$ not exceeding that specified median. Derivations of this test plan and Equations 11.24 through 11.27 are in Sections 11.5.9 and 11.5.10.

EXAMPLE 11.6 A system has a specified MTTR of 75 min and a specified maximum restoration time at the 95th percentile of 3 h. A restorability demonstration composed of 50 tasks is conducted according to the practices of this section. The restoration task times are recorded as shown in Table 11.10. Using 90% confidence estimates, as suggested in this section, has compliance with the equipment restorability specifications been demonstrated?

Applying these data to the computation form of Figure 11.5 yields the results shown in Table 11.11.

TABLE 11.10
Task times for Example 11.6.

Task	Time (h)	Task	Time (h)	Task	Time (h)
1	0.60	21	0.96	36	1.40
2	1.04	22	0.68	37	0.65
3	1.18	23	0.75	38	0.86
4	0.80	24	1.86	39	0.48
5	1.35	25	0.99	40	1.28
6	0.73	26	0.95	41	1.60
7	0.52	27	0.85	42	0.57
8	1.70	28	0.46	43	0.94
9	0.40	29	0.93	44	0.84
10	1.50	30	0.83	45	0.58
11	1.25	31	1.08	46	0.78
12	0.78	32	0.60	47	2.12
13	2.25	33	0.98	48	1.00
14	0.92	34	0.56	49	0.70
15	1.20	35	1.30	50	1.18
16	0.85				
17	0.65				
18	1.55				
19	0.70				
20	1.16				

TABLE 11.11
Example 11.6 computations.

Task	*Rp*	$(Rp)^2$	log *Rp*	$(\log Rp)^2$
1	0.60	0.3600	−0.2218	0.0492
2	1.04	1.0816	+0.0170	0.0003
3	1.18	1.3924	+0.0719	0.0052
4	0.80	0.6400	−0.0969	0.0094
5	1.35	1.8225	+0.1303	0.0170
6	0.73	0.5329	−0.1367	0.0187
7	0.52	0.2704	−0.2840	0.0807
8	1.70	2.8900	+0.2304	0.0531
9	0.40	0.1600	−0.3979	0.1584
10	1.50	2.2500	+0.1761	0.0310
11	1.25	1.5625	+0.0969	0.0094
12	0.78	0.6084	−0.1079	0.0116
13	2.25	5.0625	+0.3522	0.1240
14	0.92	0.8464	−0.0362	0.0013
15	1.20	1.4400	+0.0792	0.0063
16	0.85	0.7225	−0.0706	0.0050
17	0.65	0.4225	−0.1871	0.0350
18	1.55	2.4025	+0.1903	0.0362
19	0.70	0.4900	−0.1549	0.0240
20	1.16	1.3456	+0.0645	0.0042
21	0.96	0.9216	−0.0177	0.0003
22	0.68	0.4624	−0.1675	0.0281
23	0.75	0.5625	−0.1249	0.0156
24	1.86	3.4596	+0.2695	0.0726
25	0.99	0.9801	−0.0044	0.0000
26	0.95	0.9025	−0.0223	0.0005
27	0.85	0.7225	−0.0706	0.0050
28	0.46	0.2116	−0.3372	0.1137
29	0.93	0.8649	−0.0315	0.0010
30	0.83	0.6889	−0.0809	0.0065
31	1.08	1.1664	+0.0034	0.0011
32	0.60	0.3600	−0.2218	0.0492
33	0.98	0.9604	−0.0088	0.0001
34	0.56	0.3136	−0.2518	0.0634
35	1.30	1.6900	+0.1139	0.0130
36	1.40	1.9600	+0.1461	0.0214
37	0.65	0.4225	−0.1871	0.0350
38	0.86	0.7396	−0.0655	0.0043
39	0.48	0.2304	−0.3188	0.1016
40	1.28	1.6384	+0.1072	0.0115
41	1.60	2.5600	+0.2041	0.0417
42	0.57	0.3249	−0.2441	0.0596
43	0.94	0.8836	−0.0269	0.0007
44	0.84	0.7056	−0.0757	0.0057
45	0.58	0.3364	−0.2366	0.0560
46	0.78	0.6084	−0.1079	0.0116
47	2.12	4.4944	+0.3263	0.1065
48	1.00	1.0000	0.0000	0.0000
49	0.70	0.4900	−0.1549	0.0240
50	1.18	1.3924	+0.0719	0.0052
Totals:	49.89	58.3563	−1.7697	1.5349

From Equation 11.24,

$$\text{MTTR}_{0.90} = \frac{\sum Rp}{N} + 1.28\sqrt{\frac{N\sum Rp^2 - (\sum Rp)^2}{N^2(N-1)}}$$

$$= \frac{49.89}{50} + 1.28\sqrt{\frac{(50)(58.3563) - (49.89)^2}{(50)^2(49)}} = 1.07 \text{ h}$$

$$= 1 \text{ h } 4 \text{ min}$$

This does not exceed the specified 75 min.

From Equation 11.27,

$$Rp_{0.90}(0.95) = \log^{-1}\left[\frac{\sum \log Rp}{N} + \left(\frac{1.28}{\sqrt{N}} + 1.8\right)\sqrt{\frac{N\sum (\log Rp)^2 - (\sum \log Rp)^2}{N(N-1)}}\right]$$

$$= \log^{-1}\left[\frac{-1.7697}{50} + 1.98\sqrt{\frac{(50)(1.5349) - (1.7697)^2}{(50)(49)}}\right] = \log^{-1}[0.3078]$$

$$= 2 \text{ h } 2 \text{ min}$$

This also does not exceed its specification of 3 h.

An accept decision is reached.

---

*11.5.9 Predicting Restorability Parameters from Demonstration Test Data, using Confidence Estimates.

As we learned earlier, the confidence in our prediction is restricted by the amount of data. In a demonstration test we estimate the MTTR and the parameters $\mu_{\log}$ and $\sigma_{\log}$ of the transformed distribution based upon a quantity N tasks. These tasks are selected at random but are stratified so that the distribution of task types will be proportional to their expected frequencies.

Let T_i be the measured time it took to perform the *ith* task. Then the best estimate of the MTTR is

$$\text{MTTR} = \sum T_i/N \tag{11.28}$$

By taking the logarithms of all N tasks times T_i, we can plot a histogram and examine the transformed distribution. We can also perform an analytical goodness-of-fit test from Section A.10 to test our lognormal distribution assumption.

The best estimate of the mean of the transformed distribution is

$$\mu_{\text{log-est}} = \sum (\log T_i)/N, \tag{11.29}$$

and the best estimate of the standard deviation is

$$\sigma_{\text{log-est}} = \sqrt{\frac{N}{(N-1)}\left[\frac{\sum (\log T_i)^2}{N} - (\mu_{\text{log-est}})^2\right]} \tag{11.30}$$

Now we can estimate the median as

$$M(50) = \log^{-1} \mu_{\text{log-est}} = \log^{-1} [\textstyle\sum T_i/N] \tag{11.31}$$

the 90th percentile as

$$M(90) = \log^{-1} [\mu_{\text{log-est}} + 1.28\sigma_{\text{log-est}}], \tag{11.32}$$

the 95th percentile as

$$M(95) = \log^{-1} [\mu_{\text{log-est}} + 1.645\sigma_{\text{log-est}}] \tag{11.33}$$

and the α percentile as

$$M(\alpha) = \log^{-1} [\mu_{\text{log-est}} + z_\alpha \sigma_{\text{log-est}}]. \tag{11.34}$$

It is generally recommended that a restorability test use no fewer than 50 demonstration tasks. The larger the quantity of tasks the more confidence we have in our estimates of *MTTR*, $\mu_{\log}$, and $\sigma_{\log}$. However, we may wish to apply a confidence estimate to our predictions. Given that we shall always have at least 50 tasks from which to make our estimates, we can use the **Central Limit Theorem** from probability theory. The Central Limit Theorem states that a distribution of means is governed by the Normal Probability Law, regardless of the distribution of the original measurements, and that the mean of the distribution of means is the same as the mean of the original distribution of measurements,

$$\mu_\mu = \mu \tag{11.35}$$

Also, the standard deviation of the distribution of means is

$$\sigma_\mu = \sigma/\sqrt{N} \tag{11.36}$$

where σ is the standard deviation of the original distribution of measurements
N is the quantity of measurements (tasks)

Hence, if we are interested in the upper β confidence estimate of the MTTR, we compute first the standard deviation estimate for the original distribution of task times:

$$s = \sqrt{\frac{N}{(N-1)}\left[\frac{\sum T^2}{N} - (\text{MTTR})^2\right]} = \sqrt{\frac{\sum T^2 - N(\text{MTTR})^2}{(N-1)}} \tag{11.37}$$

Note that s estimates the standard deviation of a distribution that is positively skewed and, therefore, *not* normal. The standard deviation of the distribution of means, which *is* normal, is

$$\sigma_\mu = \frac{s}{\sqrt{N}} = \sqrt{\frac{1}{(N-1)}\left[\frac{\sum T^2}{N} - (\text{MTTR})^2\right]} = \sqrt{\frac{\sum T^2 - N(\text{MTTR})^2}{N(N-1)}} \tag{11.38}$$

Therefore the upper β confidence estimate of the MTTR is

$$\begin{aligned} \text{MTTR}_\beta &= \text{MTTR} + z_\beta\,(\sigma_\mu) \\ &= \text{MTTR} + z_\beta\sqrt{\frac{\sum T^2 - N(\text{MTTR})^2}{N(N-1)}} \end{aligned} \tag{11.39}$$

where z_β is the z value corresponding to the probability α in Table B.1. If we are interested in a 90% confidence estimate, from Table B.1, $z_{0.90} = 1.28$, and

$$\text{MTTR}_{0.90} = \text{MTTR} + 1.28\sqrt{\frac{\sum T^2 - N(\text{MTTR})^2}{N(N-1)}} \tag{11.40}$$

If we are interested in the upper β confidence estimate of the median or the α percentile point, we must make a confidence estimate of the mean of the transformed distribution—that is, the distribution of the logarithms. Again, using the Central Limit Theorem, the standard deviation of the mean of the logarithms is

$$\sigma_{\mu\text{-log}} = \frac{\sigma_{\log}}{\sqrt{N}} = \sqrt{\frac{1}{(N-1)}\left[\frac{\sum (\log T)^2}{N} - (\mu_{\text{log-est}})^2\right]} \tag{11.41}$$

So the upper β confidence estimate of $\mu_{\text{log-est}}$ is

$$\begin{aligned} \mu_{\text{log-0.90}} &= \mu_{\text{log-est}} + z_\beta \sigma_{\mu\text{-log}} \\ &= \frac{\sum (\log T_i)}{N} + z_\beta \sqrt{\frac{1}{(N-1)}\left[\frac{\sum (\log T_i)^2}{N} - (\mu_{\text{log-est}})^2\right]} \end{aligned} \tag{11.42}$$

Therefore, from Equations 11.34 and 11.42, the β confidence estimate of the median restoration time and the α percentile of the TTR distribution are, respectively,

$$\begin{aligned} M(50)_\beta &= \log^{-1}[\mu_{\text{log-est}} + z_\beta \sigma_{\mu\text{-log}}] \\ &= \log^{-1}\left(\frac{\sigma (\log T_i)}{N} + z_\beta \sqrt{\frac{1}{(N-1)}\left[\frac{\sum (\log T_i)^2}{N} - (\mu_{\text{log-est}})^2\right]}\right) \end{aligned} \tag{11.43}$$

$$\begin{aligned} M(\alpha)_\beta &= \log^{-1}[\mu_{\text{log-est}} + z_\beta \sigma_{\mu\text{-log}} + z_\alpha\, \sigma_{\text{log-est}}] \\ &= \log^{-1}\left\{\frac{\sum (\log T_i)}{N} + z_\beta \sqrt{\frac{1}{(N-1)}\left[\frac{\sum (\log T_i)^2}{N} - (\mu_{\text{log-est}})^2\right]}\right. \\ &\quad \left. + z_\alpha \sqrt{\frac{N}{(N-1)}\left[\frac{\sum (\log T_i)^2}{N} - (\mu_{\text{log-est}})^2\right]}\right\} \end{aligned} \tag{11.44}$$

EXAMPLE 11.7

A restorability demonstration test on a system, was composed of 100 restoration tasks, with the following results:

$$\sum T_i = 205 \text{ h} \qquad \sum (\log T_i) = 24.7$$
$$\sum (T_i)^2 = 4520 \qquad \sum (\log T_i)^2 = 11.1$$

Compute an 80% confidence estimate of the MTTR, the median time to restore, and the upper 95th percentile of the TTR distribution. Assume a lognormal fit.

From Table B.1, the z value that corresponds to a probability of 0.800 is

$$z_{0.80} = 0.842$$

The best estimate of the MTTR is

$$\text{MTTR} = 205/100 = 2.05 \text{ h}$$

From Equation 11.39, the 80% confidence estimate of the MTTR is

$$\text{MTTR}_{0.80} = \text{MTTR} + z_\beta \sqrt{\frac{\sum T^2 - N(\text{MTTR})^2}{N(N-1)}}$$

$$= 2.05 + 0.842\sqrt{\frac{4520 - 100(2.05)^2}{100(99)}} = 2.59 \text{ h}$$

From Equation 11.43, the 80% confidence estimate in the median time to restore is

$$M(50)_{0.80} = \log^{-1}\left\{\frac{24.7}{100} + 0.842\sqrt{\frac{1}{(99)}\left[\frac{11.1}{100} - (0.247)^2\right]}\right\} = 1.84 \text{ h}$$

From Table B.1, the z value that corresponds to a probability of 0.9500 is

$$z_{0.95} = 1.645$$

From Equation 11.44, the 80% confidence estimate in the upper 95th percentile of the TTR distribution is

$$M(95)_{0.80} = \log^{-1}\left\{\frac{24.7}{100} + 0.842\sqrt{\frac{1}{99}\left[\frac{11.1}{100} - (0.247)^2\right]} + 1.645\sqrt{\frac{100}{999}\left[\frac{11.1}{100} - (0.247)^2\right]}\right\}$$
$$= 4.32 \text{ h}$$

The results indicate that based on the test data, we are 80% confident that the mean time to restore is no more than 2.59 h, that the median time to restore is no more than 1.84 h, and that 95% of the restoration tasks can be accomplished in less than 4.32 h.

*11.5.10 Derivation of Restorability Demonstration Test Plans

The restorability demonstration test plan technique and Equations 11.24 through 11.27 were derived under the assumption that the times to restore a system are lognormally distributed, as discussed in Chapter 7. If the task frequency allocation is proportional to the estimated failure rates, as prescribed in Section 11.5.3, we can then expect the resulting distribution of demonstrated task times to be lognormal as well. This is an assumption that can be tested through a statistical goodness-of-fit test (See Chapter 5) or by plotting the results on logarithm-versus-probability paper, as shown in Example 7.6 of Chapter 7.

The 90% confidence estimate of the MTTR is determined through Equation 11.40. Substituting the notation R_P for the demonstrated task times T,

$$\text{MTTR} = \sum R_P / N \tag{11.45}$$

we obtain Equation 11.24,

$$\text{MTTR}_{0.90} = \frac{\sum Rp}{N} + 1.28\sqrt{\frac{N\sum Rp^2 - (\Sigma\, Rp)^2}{N^2(N-1)}} \tag{11.46}$$

The percentile points of the time-to-restore distribution (i.e., the median, the upper 90th percentile, and the upper 95th percentile) are obtained from the transformed distribution as shown in Section 11.5.9. The 90% confidence estimate of the median is derived from Equation 11.43, with $z_\beta = 1.28$ (Table B.1), $T = R_p$, and $\mu_{\text{log-est}} = \Sigma \log R_p/N$. The result is Equation 11.25.

Equation 11.30 estimates the standard deviation of the transformed distribution:

$$\sigma_{\text{log-est}} = \sqrt{\frac{N}{(N-1)}\left[\frac{\sum (\log T_i)^2}{N} - (\mu_{\text{log-est}})^2\right]} \quad \textbf{(11.47)}$$

Referring to any text on statistical inference, you will see that the 90% confidence estimate of the standard deviation of the transformed distribution is

$$\sigma_{\text{log-est } 0.90} = \sigma_{\text{log-est}} \times \sqrt{\frac{N}{\chi^2_{0.1.0.N-1}}} \quad \textbf{(11.48)}$$

Since the number of tasks for a typical restorability demonstration test is 50 or more, let us take the worst case, that is, where $N = 50$, and

$$\begin{aligned}\sigma_{\text{log-est } 0.90} &= \sigma_{\text{log-est}} \times \sqrt{\frac{N}{\chi^2_{0.1.0.N-1}}} = \sigma_{\text{log-est}} \times \sqrt{\frac{50}{\chi^2_{0.10,49}}} \\ &= \sigma_{\text{log-est}} \times \sqrt{\frac{50}{37.7}} = 1.1\ \sigma_{\text{log-est}}\end{aligned} \quad \textbf{(11.49)}$$

where the $\chi^2_{0.10,49}$ value is taken from Table B.2.

Equations 11.26 and 11.27 are taken from Equation 11.44, with $z_\beta = 1.28$ once again and with $z_\alpha = 1.28$ for the upper 90th percentile (Equation 11.26) and $z_\alpha = 1.645$ for the upper 95th percentile (Equation 11.27). Equation 11.49 is applied by multiplying 1.1 by the z_α values to factor in the 90% confidence estimate of $\sigma_{\text{log-est}}$. Hence, the coefficients $1.1 \times 1.28 = 1.4$ and $1.1 \times 1.645 = 1.8$ are used to obtain Equations 11.26 and 11.27 from Equation 11.44:

$$\begin{aligned}Rp_{0.90}(\alpha) &= \log^{-1}[\mu_{\text{log-est}} + z_{0.90}\sigma_{\mu\text{-log}} + 1.1 z_\alpha \sigma_{\text{log-est}}] \\ &= \log^{-1}\left\{\frac{\sum (\log Rp_i)}{N} + 1.28\sqrt{\frac{1}{(N-1)}\left[\frac{\sum (\log Rp_i)^2}{N} - (\mu_{\text{log-est}})^2\right]}\right. \\ &\quad \left. + 1.1 z_\alpha \sqrt{\frac{N}{(N-1)}\left[\frac{\sum (\log Rp_i)^2}{N} - (\mu_{\text{log-est}})^2\right]}\right\} \\ &= \log^{-1}\left\{\frac{\sum (\log Rp)}{N} + \frac{1.28}{\sqrt{N}}\sqrt{\frac{N}{(N-1)}\left[\frac{\sum (\log Rp)^2}{N} - \left(\frac{\sum \log Rp}{N}\right)^2\right]}\right. \\ &\quad \left. + 1.1 z_\alpha \sqrt{\frac{N}{(N-1)}\left[\frac{\sum (\log Rp)^2}{N} - \left(\frac{\sum \log Rp}{N}\right)^2\right]}\right\} \\ &= \log^{-1}\left\{\frac{\sum (\log Rp)}{N} + \left(\frac{1.28}{\sqrt{N}} + 1.1 z_\alpha\right)\sqrt{\frac{N}{(N-1)}\left[\frac{\sum (\log Rp)^2}{N} - \left(\frac{\sum \log Rp}{N}\right)^2\right]}\right\} \\ &= \log^{-1}\left\{\frac{\sum (\log Rp}{N} + \left(\frac{1.28}{\sqrt{N}} + 1.1 z_\alpha\right)\sqrt{\left[\frac{N\sum (\log Rp)^2 - (\sum \log Rp)^2}{N(N-1)}\right]}\right\}\end{aligned} \quad \textbf{(11.50)}$$

BIBLIOGRAPHY

1. B. Epstein. 1960. "Tests for the Validity of the Assumption that the Underlying Distribution of Life is Exponential, I and II," *Technometrics* 83–102, 167–184.
2. E. E. Lewis. 1987. *Introduction to Reliability Engineering.* New York: John Wiley.
3. P. D. T. O'Connor. 1991. *Practical Reliability Engineering,* 3d ed. New York: John Wiley.
4. C. O. Smith. 1976. *Introduction to Reliability in Design.* New York: McGraw-Hill.
5. MIL-STD-105D. 1963. *Sampling Procedures and Tables for Inspecton by Attributes.* Arlington, Va.: Department of Defense.
6. MIL-STD-471A. 1973. *Maintainability Demonstration.* Arlington,Va.: Department of Defense.
7. MIL-STD-781D. 1986. *Reliability Testing for Equipment Development, Qualification and Production.* Arlington, Va.: Department of Defense.

EXERCISES

11.1. Use Table 11.3 to design a reliability demonstration test to demonstrate a system success probability of 0.999 in fewer than 200 cycles.

11.2. If the system subjected to the demonstration test of Exercise 11.1 actually has a success probability of 0.999, what is the probability that it will be accepted by the test?

11.3. If the system subjected to the demonstration test of Exercise 11.1 actually has a success probability of 0.982, what is the probability that it will be accepted by the test?

***11.4.** If the system subjected to the demonstration test of Exercise 11.1 actually has a success probability of 0.990, what is the probability that it will be accepted by the test?

11.5. If the system subjected to the demonstration test of Exercise 11.1 actually has a success probability of 0.999, what is the probability that during a test there will be no failures?

11.6. Use Table 11.3 to design a reliability demonstration test to demonstrate a system success probability of 0.990 in fewer than 100 cycles.

11.7. If the system subjected to the demonstration test of Exercise 11.6 actually has a success probability of 0.990, what is the probability that it will be accepted by the test?

11.8. If the system subjected to the demonstration test of Exercise 11.6 actually has a success probability of 0.923, what is the probability that it will be accepted by the test?

***11.9.** If the system subjected to the demonstration test of Exercise 11.6 actually has a success probability of 0.950, what is the probability that it will be accepted by the test?

11.10. Use Table 11.4 to develop a multiple reliability demonstration test plan equivalent to that of Exercise 11.6. What is the advantage of this plan over that of Exercise 11.6?

11.11. A surveillance system has been designed to meet its specified MTBF of 400 h. Using one prototype of this system, we wish to demonstrate achievement of the specified MTBF by operating continuously on a 24-h-a-day schedule. We want decision risks of no more than 10%, and we want the test to be completed in less than 6 months. Select a fixed-length test plan from among those listed in Table 11.6 that will satisfy these conditions.

11.12. If the MTBF of the system in Exercise 11.11 is actually 400 h, what is the probability of acceptance by the selected test plan?

11.13. If the MTBF of the system in Exercise 11.11 is actually 200 h, what is the probability of acceptance by the selected test plan?

***11.14.** If the MTBF of the system in Exercise 11.11 is actually 300 h, what is the probability of acceptance by the selected test plan?

11.15. Use Table 11.9 to develop a sequential MTBF demonstration test plan equivalent to that of Exercise 11.11. What is the advantage of this plan over that of Exercise 11.11? What is the longest possible test time for reaching a decision according to this plan? What is the shortest? Which plan do you recommend using, the fixed length (of Exercise 11.11) or the sequential (of this exercise). Why?

11.16. A specified MTBF of 750 h is to be demonstrated according to Test Plan 8 of Table 11.6.
- **a.** What is the required test time?
- **b.** What is the alternative test MTBF?
- **c.** What is the maximum allowable number of failures?
- **d.** What is the producer's risk?
- **e.** What is the consumer's risk?

11.17. A specified MTBF of 2000 h is to be demonstrated according to Test Plan 5 of Table 11.6.
- **a.** What is the required test time?
- **b.** What is the alternative test MTBF?
- **c.** What is the maximum allowable number of failures?
- **d.** What is the producer's risk?
- **e.** What is the consumer's risk?

11.18. A specified MTBF of 1200 h is to be demonstrated according to Test Plan 4 of Table 11.6.
- **a.** What is the required test time?
- **b.** What is the alternative test MTBF?
- **c.** What is the maximum allowable number of failures?
- **d.** What is the producer's risk?
- **e.** What is the consumer's risk?

11.19. If the system being tested to the test plan of Exercise 11.8 has an MTBF of 1200 h, what is the probability of test acceptance?

11.20. If the system being tested to the test plan of Exercise 11.8 has an MTBF of 600 h, what is the probability of test acceptance?

11.21. A specified MTBF of 1200 h is to be demonstrated according to Test Plan 4A of Table 11.9.
- **a.** What is the longest time possible for reaching a decision?
- **b.** What is the shortest time possible for reaching a decision?
- **c.** What is the alternative test MTBF?
- **d.** What is the producer's risk?
- **e.** What is the consumer's risk?

***11.22.** Design a fixed-length test plan to demonstrate an MTBF of 500 h, with decision risks $\alpha = \beta = 5\%$ and a discrimination ratio of 3.0.

11.23. A restorability demonstration test is required for a system composed of eight LRUs, with failure rates estimated as follows.

LRU	Failure Rate	LRU	Failure Rate
1	144.3×10^{-6}/h	5	475.6×10^{-6}/h
2	28.6×10^{-6}/h	6	71.2×10^{-6}/h
3	243.5×10^{-6}/h	7	837.4×10^{-6}/h
4	188.7×10^{-6}/h	8	265.1×10^{-6}/h

Suppose the test is to be composed of 100 restoration tasks. Allocate the tasks to the eight LRUs.

11.24. A system has a specified MTTR of 2.0 h and a specified maximum restoration time at the 95th percentile of 4.0 h. A restorability demonstration test composed of 100 tasks is conducted according to the practices of Section 11.5. The 100 restoration task times are recorded in minutes, yielding the following results:

$$\sum Rp = 8604 \text{ min} \qquad \sum(\log Rp) = 189.5443$$
$$\sum(Rp^2) = 917{,}500 \text{ min}^2 \qquad \sum(\log Rp)^2 = 362.2676$$
$$N = 100$$

Using 90% confidence estimates, determine whether compliance with the restorability specifications has been demonstrated.

11.25. Use Equation 11.26 to predict the maximum restoration time at the 90th percentile for the system of Exercise 11.24.

11.26. Why is it that restorability demonstration test results can be used to make preditions, as in Exercise 11.25, and MTBF demonstration test results cannot?

***11.27.** Use the demonstration test results of Exercise 11.24 to predict the percentage of restoration tasks expected to be acomplished in less than 2.0 h.

12

RELIABILITY GROWTH TESTING

12.1 INTENT

Some practitioners prefer what is known as **growth testing** as an approach for ascertaining a system's reliability and restorability achieement. Adherents of growth testing prefer it over demonstration testing for what they see as a lack of substantial "value added" of the latter. As seen in Chapter 11, demonstration testing, particularly MTBF demonstration testing, can be lengthy and expensive. Furthermore, MTBF or success probability demonstration tests are restricted to being accept-reject decision makers, lacking the power to make predictions or to lead the way toward significant reliability improvement. The growth-testing philosophy is to test with the objective of uncovering system failure modes and eliminating them through investigation and redesign. Of course, these tests are also subject to time restrictions and the randomness of incidence of particular failure modes in practice.

12.2 THE DUANE CURVE

During the 1950s, J. T. Duane, while maintaining MTBF data on helicopter engines, observed that newly released models appeared to experience an MTBF that was more frequent than predicted. With maturity the observed MTBF seemed to improve. Empirically, he developed the **Duane curve,** illustrated in Figure 12.1, which exhibits an approximately linear relationship between the logarithm of operating time and estimated MTBF.

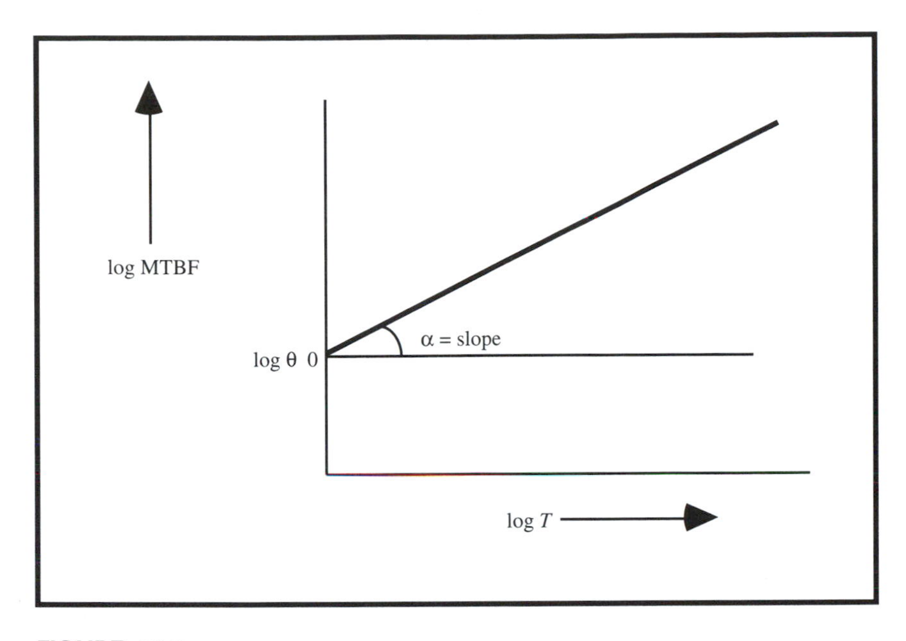

FIGURE 12.1
The Duane curve.

Followers of Duane's theory hypothesized that when a system is first introduced to field operation, its observed MTBF is only about 10% of what had been predicted. With experience observed MTBF ultimately improves according to the Duane curve. Empirical data seemed to indicate that the slope α of the linear relationship between the logarithm of MTBF and the logarithm of operating time was somewhere between 0.1 and 0.6. A slope at the lower end ($\alpha = 0.1$) was interpreted to mean very little effort was being put into quality and reliability improvement. On the other hand, a slope at the upper end ($\alpha = 0.6$) was interpreted as the result of a hard-hitting, ambitious effort to improve quality and reliability.

According to Duane theory, newly introduced systems were not to be expected to immediately experience their specified MTBF, but by estimating the slope α through an assessment of the level of quality and reliability improvement, a prediction of **time to maturity,** that is, the time until the system is expected to finally achieve its specified MTBF, was possible. According to this theory, the expected MTBF after a given accumulated system operating time was predictable according to the relation

$$\log \theta_c = \log \theta_0 + \alpha \log T \tag{12.1}$$

where
θ_c = the expected MTBF after T system operating hours
θ_0 = the observed MTBF at time $T_0 = 0$
α = the slope of the Duane curve

See Figure 12.2.

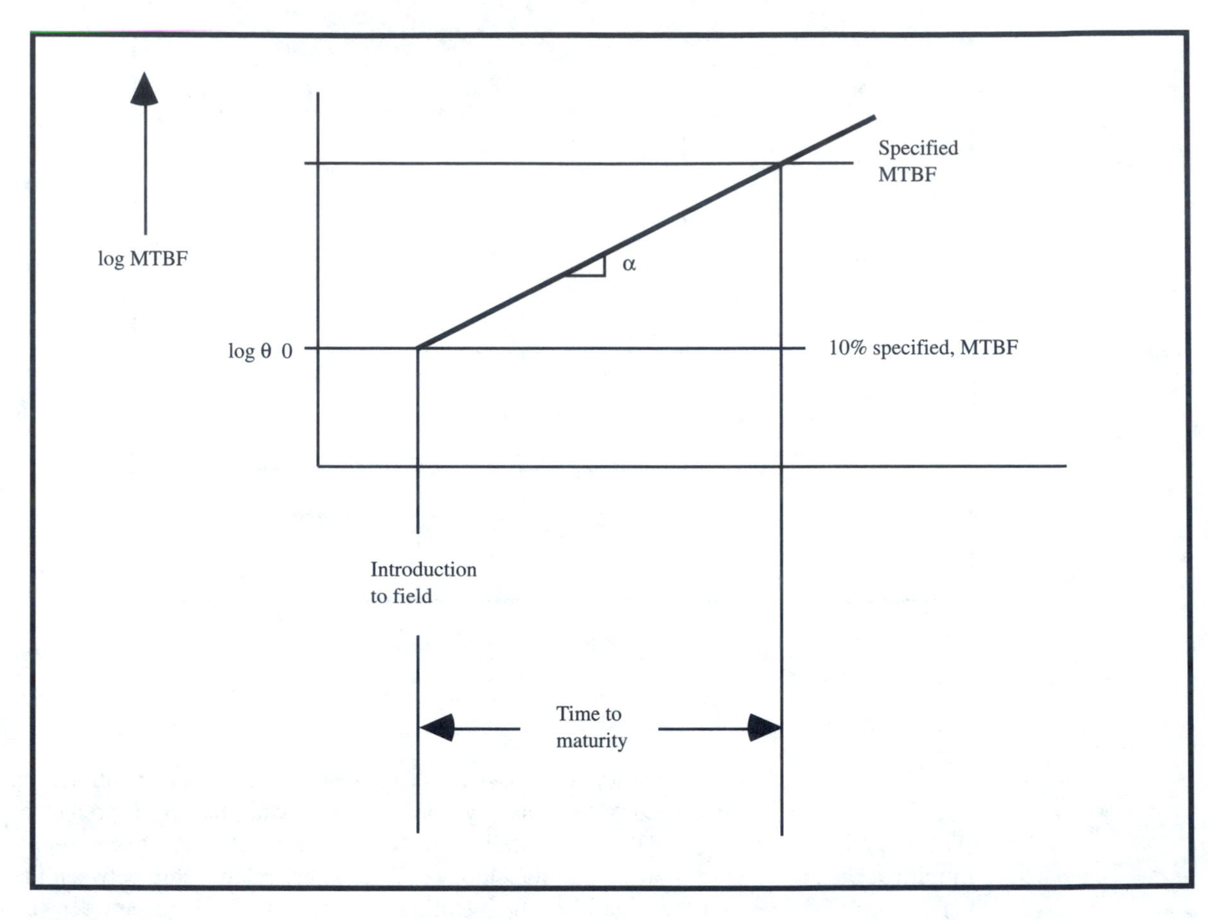

FIGURE 12.2
Duane theory.

12.3 TAAF

The **test, analyze and fix (TAAF)** program is a reliability growth alternative to demonstration testing that developed during the late 1960s from Duane theory. Similar to the MTBF demonstration procedure, TAAF requires the system or a set of systems to be put on test and operated in a field-simulated environment. Records are kept on the frequency of scorable failures, and the MTBF (the accumulated system operating time divided by the number of observed, scorable failures) is continuously plotted versus accumulated operating time, as illustrated in Figure 12.3. The restorability is demonstrated simultaneously by monitoring and recording the restoration times for the failures that happen to occur during the TAAF. At the end of the TAAF, confidence estimates of the MTTR and specified TTR percentile point can be computed as described in Chapter 11.

According to TAAF, if MTBF is truly improving, the cumulative MTBF as plotted in Figure 12.3 is presenting a pessimistic picture. It is, after all, computed by using all

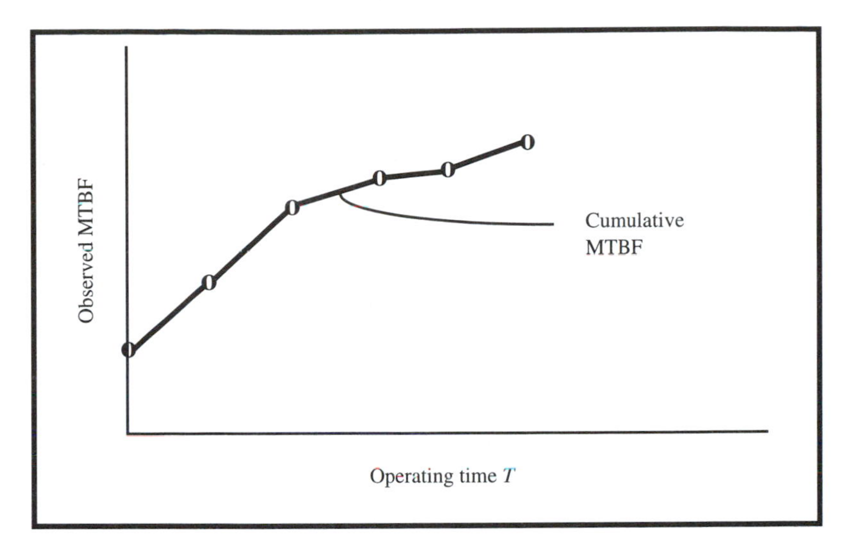

FIGURE 12.3
MTBF growth curve.

the observed failures right back to the beginning of the test, when the observed MTBF was supposedly lower. Hence, a moving average derived from the last 10 or fewer points can be expected to give a more current assessment.

Suppose we decide to plot a moving average of three. Then the *observed MTBF* will be computed by dividing the system operating time accumulated since the last three failures and dividing by 3. This new value is plotted versus accumulated operating time. Figure 12.4 illustrates how such a moving average growth curve might compare with a cumulative curve.

A feature of the TAAF is that every failure that occurs is analyzed to determine cause of failure and that every failure mode thus uncovered be eliminated through redesign or a manufacturing process adjustment. Those failure modes that are successfully eliminated can be expected to cause no additional failures. TAAF calls for the presentation of reliability growth as a consequence of this failure mode elimination. The *adjusted MTBF* is computed by dividing the total accumulated system operating time since the beginning of the test by only those failures caused by failure modes that have not yet been eliminated. Figure 12.5 presents comparative plots of the *cumulative, moving average,* and *adjusted MTBF* values versus cumulative operating time, to comprise what is known as a **reliability growth assessment.**

Some cautionary notes about the TAAF and reliability growth assessment approach concern

1. The possible appearance of growth that is really not taking place,
2. The likelihood that test incidents may direct us to solving low priority problems, and
3. The difficulty in meaningfully assessing time to maturity.

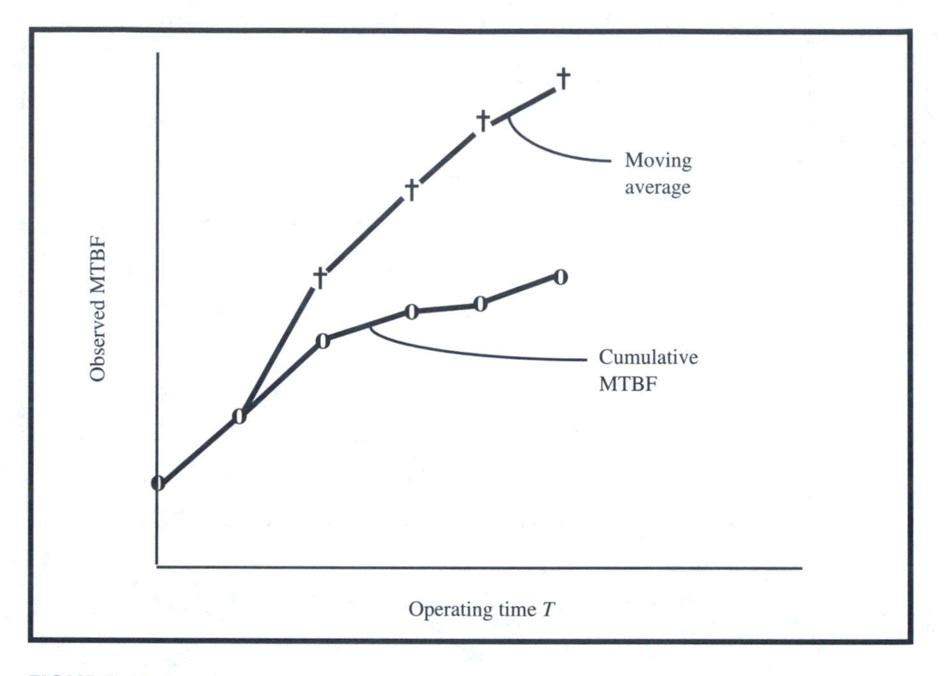

FIGURE 12.4
MTBF growth curves with moving average.

To illustrate the first point, suppose one prototype of a system with a constant failure rate is released to the field. And suppose that Duane's assumption about early releases of a new product or model not achieving the predicted MTBF is not true in this case. In fact, suppose that the actual MTBF is exactly as predicted. What is the probability that the prototype will actually run for a period equal to its MTBF without failure? Let us assume that wearout is not an issue here.

The probability of the prototype operating without failure for a period equal to its MTBF is the system reliability for a mission equal to the MTBF, which is

$$R(\text{MTBF}) = e^{-\lambda \cdot \text{MTBF}} = e^{-\text{MTBF}/\text{MTBF}} = e^{-1} = 0.368 \qquad \textbf{(12.2)}$$

This means that almost two out of three such prototypes will experience one or more failures during an initial operating period equal to its actual MTBF. During the early hours of accumulated field operation, MTBF computations on a new system are very likely to give the impression that the MTBF is less than it really is. When more units are released to the field and they have the chance to accumulate substantial operating time, the computed MTBF will start to approximate its actual value. New products are typically released in just this fashion, first one at a time, then a few of them, with operating time accumulating slowly, and finally a sizable fleet with a rapid accumulation of operating time. So even if Duane's assumption is not correct, it is highly likely that new products will initially appear on growth assessment curves to have a lower-than-predicted MTBF and that the MTBF will appear to be improving with age.

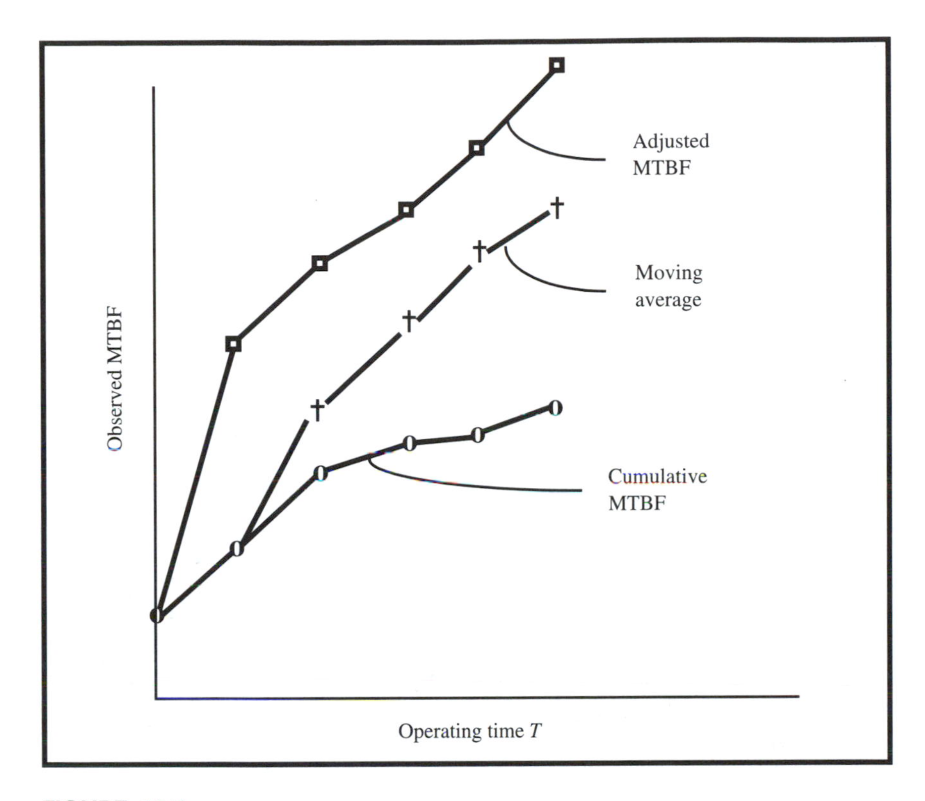

FIGURE 12.5
Cumulative, moving average, and adjusted MTBF growth curves.

We recognized at the beginning of this chapter that the incidence of particular failure modes in a short period is random. During a short operating period, which any MTBF test must be, out of necessity, there is no guarantee that what will eventually be the problems of highest priority will even make an appearance. The test period represents a very small statistical sample of the operating time for the product, and the few failure modes that do occur, purely at random, can initiate a long and expensive, in-depth investigation, leading us to eliminate failure modes that will not necessarily reduce the failure rate for the system.

Finally, the **time to maturity** is predicted through a rather loosely estimated slope on the logarithmic scale of Figure 12.2. Consider how sensitive this prediction is to the slightest change in the estimated angle of the slope. Also, we know that the MTBF will not continue to grow logarithmically.

12.4 R&M 2000

In 1985, the U.S. Air Force introduced a Reliability and Maintainability Action Plan called **R&M 2000.** Its stated purpose was to treat *reliability* as an objective in its own right, rather than as an input to a cost-benefit trade-off. Furthermore, this initiative emphasized *reliability improvement,* particularly with the introduction of new and updated systems. Among its features are minimum MTBF requirements for any electronic equipment of a given size, requirements to stress test all new equipment, assurance that each new generation of a system has a longer MTBF and a shorter MTTR than the system it is replacing, and assurance that all systems are designed for ease of operation and maintenance in their intended environments.

12.5 RELIABILITY IMPROVEMENT WARRANTIES

During the 1970s, reliability improvement warranty provisions began to appear in many government contracts. In one form or another these statements contained incentives for the supplier if reliability specifications were exceeded and penalties if reliability specifications were not met.

All reliability parameters (e.g., success probability, MTBF) are probabilistic. Therefore, basing a reliability warranty on, say, observed MTBF computed from early field experience, leaves the results of the warranty to chance. A reasonable and valid warranty takes the form of a service contract, whereby the supplier provides all maintenance, including parts and labor, for a given period at a cost based upon specified or predicted failure and restoration rates. There can also be a provision to compensate for lost time and loss of service to the consumer during system outage times. This approach has its own built-in incentives and penalties for the supplier and provides protection to the consumer.

BIBLIOGRAPHY

1. A. E. Green. 1983. *Safety Systems Analysis.* New York: John Wiley.
2. E. E. Lewis. 1987. *Introduction to Reliability Engineering.* New York: John Wiley.
3. P. D. T. O'Connor. 1991. *Practical Reliability Engineering,* 3d ed. New York: John Wiley.
4. MIL-STD-781D. 1978. *Reliability Testing for Equipment Development, Qualification and Production.* Arlington, Va.: Department of Defense.
5. MIL-STD-1635. 1978. *Test, Analyze and Fix Program.* Arlington, Va.: Department of Defense.
6. MIL-STD-2068. 1977. *Reliability Growth Assessment.* Arlington, Va.: Department of Defense.

EXERCISES

12.1 The specified MTBF for a new model of air-conditioning compressor is 2000 h. During the first year of production, field experience shows 83 failures reported during 4592 equipment-hours of operation. If we accept the Duane theory and assume an α value of 0.3, how long will it take the compressor to reach its specified MTBF?

12.2. Suppose that during the second year of production of the compressor of Exercise 12.1, there are 12 failures reported in 9872 equipment-hours of operation. Compute α based on the first two years of field experience.

12.3. Use the results of Exercise 12.2 to answer Exercise 12.1 based on the experience of the first 2 years.

12.4. A system is put on a TAAF type test and experiences failures at 40, 67, 214, 574, 1281, and 1680 h. Based on the initial 1680 h of testing, compute each value.

a. *The cumulative MTBF*

b. The MTBF based on a *moving average of three*

c. The *adjusted MTBF* if failure modes of two of the failures have been eliminated

12.5. Draw a set of observed MTBF-versus–operating time curves (similar to those of Figure 12.5) for the test results of Exercise 12.4. Assume that the failure modes that were eliminated were those of the second and fourth failures.

13

RISK ASSESSMENT

13.1 INTENT

For the first 12 chapters of this text we were concerned with what can go right with a system and how to maximize the likelihood that things will go right. But despite our best efforts, sometimes things will go wrong. We now address the question, "What happens if something goes wrong?" What are the risks inherent to the operation of our system, and how can we protect ourselves from the consequences of something going wrong?

Two common techniques for assessing risks are **failure modes and effects analysis** (FMEA) and the **fault tree analysis** (FTA). The former is a bottom-up approach, where we explore the details of our system design for all possible failure modes and their impact upon the system. The latter is a top-down approach, where we hypothesize all that can possibly go wrong for a system and determine what events or combinations of events within the system have to take place to produce each of these undesired outcomes. These techniques provide valuable design analysis information, and to maximize their effectiveness they ought to be conducted as soon as system design information becomes available and to be updated constantly as the design of a system evolves.

FMEA has the advantage of being absolutely thorough. For hardware systems, we can start with a complete parts list and explore all possible failure modes item by item. Similarly, for a service system we can explore every element of every procedural step for possible failure modes. FMEA exposes weaknesses within the system, components or elements that can lead to the system catastrophes. The disadvantage, however, is that the effect of combinations of malfunctions of unrelated components within the system can possibly be overlooked.

Conversely, FTA does consider the effects of combinations of failure modes, but it relies on the thoroughness of the analysts to produce an exhaustive set of undesired outcomes.

A derivative of FMEA is FMECA **(failure modes, effects, and criticality analysis),** which also evaluates the *criticality* of each effect to users and equipment as well as assessing the likelihood of the critical effects. FTA has a similar derivative, the **probabilistic risk assessment** (PRA), which discloses vulnerable areas within the system and assesses the likelihood of each undesired outcome.

The objective of both FMEA and FTA and their derivative analysis techniques (FMECA and PRA) is to improve system reliability and safety during the system design phase. They focus in on parts of a system that need additional protection to reduce the likelihood of accidents and to provide compensating measures to ameliorate the consequences of an accident. In other words, they address the question, "What happens if something goes wrong?"

13.2 FAILURE MODES AND EFFECTS ANALYSIS

FMEA is generally conducted on a form similar to the one shown in Figure 13.1. A separate form may be used for each subsystem or piece of equipment within an overall system (or, for that matter, for each process of a service-rendering system). In the Item column, every component or LRU (or every procedural step for a service) is listed. The function of each item is briefly described in the second column. In the Failure Modes column, every possible failure mode of each item is listed. For example, if the item is a gear, the failure modes can be a chip, a fracture, a crack, uneven wear, etc. If the item is a resister, possible failure modes are an open circuit, a short circuit, etc. If the item is a data-recording procedure, possible failure modes are omissions, erroneous entry, transposition of numerals, etc.

Figure 13.1 provides three columns for exploring the effects of each failure mode, from the immediate effect on the lowest assembly level to contain the item, to the effect upon the next-higher assembly level, up through the effect upon the system itself. Actually, many more columns may be necessary in order to trace the effects of a failure mode up through the system. We want to be especially aware of any single-mode failures—that is, any failure of an item in a particular mode that can by itself cause the entire system to go down. If such is the case we want to be able to protect the system through ready fault detection and through compensating design features.

If an item does fail in a critical mode, one that can jeopardize the system, we want to be certain that it is readily detected. In the Failure Detection column we describe how each of the failure modes announces itself, whether through a symptom or through a built-in audible sound from visual alarm feature. In this column we also describe the "testability" features that can facilitate the diagnosis. If FMEA reveals a single-mode failure (or a failure mode that is close to being so) with inadequate detection methods, design recommendations are in order.

Failure Modes and Effects Analysis

System: ____________________

Equipment: ____________________

Indenture level: ____________________

Date: ____________

Page: ______ of ______

Compiled by: ____________________

Item	Function	Failure Modes	Failure Effects			Failure-Detection Method, Testability	Compensating Provisions
			Local Effects	Next-higher Level	End Effects		

FIGURE 13.1
FMEA worksheet.

"Compensating provisions" for failure modes within a system may take many forms, such as fault tolerance, redundancy, and sparing. Again, if FMEA reveals that the system is not adequately protected from the effects of a critical failure mode through a compensating provision, we may wish to recommend further redundancy or some other form of design compensation. We do not need to address every single item failure mode within our system, only those that are likely to imperil our system. FMEA helps to prioritize those areas that need further design attention.

13.3 FAILURE MODES, EFFECTS, AND CRITICALITY ANALYSIS

FMECA, which is a refinement of FMEA, is conducted on a form similar to the one shown in Figure 13.2. FMECA, in addition to pointing out which failure modes contribute most heavily to system-level effects, assesses the severity of the consequences of the effects. It distinguishes among effects that endanger lives, effects that can permanently disable equipment, and effects that are less severe. FMECA can also quantify the likelihood of any of these occurrences. Hence, FMEA is an even more powerful tool for prioritizing areas of a system design that warrant further attention.

The last two columns of Figure 13.2 are for the Severity Class and the Criticality, the two features that distinguish the FMECA from the FMEA. Severity class deals with the consequences of the various system-level effects. Traditionally, the severity of a system effect is classified into five classes, as follows.

Class 1: Catastrophic This class is for irreversible and disastrous effects, such as loss of life or limb, enduring environmental damage, or permanent loss of property. With equipment, an explosion is an example of a catastrophic system failure. With a service-rendering system, a catastrophic failure might be the effect of an erroneous prescription of medication or an air-traffic controller error.

Class 2: Critical This category is for disastrous but restorable damage. It includes damage to equipment where nobody is hurt or where there is major but curable illness or injury. A service example might be a disastrous banking investment.

Class 3: Major This category is for serious malfunctions of the system where there is no physical injury to people or equipment. Included in this category might be erroneous purchasing orders or the breakdown of a road vehicle.

Class 4: Minor This category is reserved for faults that lead to marginal inconveniences to a system or its users. Examples might be a vending machine that momentarily cannot provide change or a bank's computer system that is down when a consumer requests a balance.

Class 5: Trivial This category is for inconsequential faults that cause no more than a nuisance to the users of the system. It may cause cosmetic imperfections to equipment. An example might be that an automobile dashboard light fails or a piece of mail gets torn in the mail slot.

Failure Modes, Effects and Criticality Analysis

System: ____________________ Date: __________

Equipment: ____________________ Page: ______ of ______

Indenture level: ____________________ Compiled by: ____________________

Item	Function	Failure Modes	Failure Effects			Failure-Detection Method, Testability	Compensating Provisions	Severity Class	Criticality
			Local Effects	Next-higher Level	End Effects				

FIGURE 13.2
An FMECA worksheet.

The criticality of a particular system effect is the probability that the effect will take place as the result of the failure mode in question. The criticality column, the last column in Figure 13.2, often appears as five columns, as shown in Figure 13.3. For each item, LRU, or function originally listed on the FMECA form, we list the failure rate λ, which we obtain from the failure rate estimates already determined as part of our reliability prediction activity. We next allocate the failure rate for each item to the various failure modes.

If, for example, a particular item has three failure modes, such as modes a, b, and c, we estimate through our experience the fraction of the time that the item is expected to fail in each of those modes. Suppose we expect that 50% of the time it fails it will fail in mode a, 30% of the time it will fail in mode b, and the remaining 20% of the time it will fail in mode c. Then the mode failure rates will be $\lambda_a = 0.50\lambda$, $\lambda_b = 0.30\lambda$, and $\lambda_c = 0.20\lambda$. The sum of the mode failure rates is, of course, the item failure rate.

The **at-risk time,** t, is either a specified mission or an interval of vulnerability during which the system effect due to the failure mode in question is possible.

Finally, we assign a **loss probability** β to the system effect. We know that just because a particular failure mode can cause an undesired system effect, it will not necessarily do so. For example, every time a gear tooth cracks, it will not necessarily cause a gear-train seizure. However, β is our assessment of the likelihood of the seizure.

The **criticality number,** C_m, for a failure mode is computed as

$$C_m = \beta(1 - e^{-\lambda_m t}) \tag{13.1}$$

Item	Failure Mode	Item Failure Rate (10^-6)	Mode Failure Rate	At-risk Time	Loss Probability	Criticality
VALVE	a)	12	12 x .15 = 1.8	8760	0.90	.016 x .9 = .0144
	b)	12	12 x .5 = .6	8760	.4	0.025
	c)	12	12 x .3 = 3.6	8760	1.0	.031
	d)	12	12 x .05 = .6	8760	.05	.00026

FIGURE 13.3
Criticality analysis form.

The criticality number, C, for an item is the sum of the criticality numbers for all its failure modes. The criticality number of an item expresses the likelihood that the item will cause an undesired system event.

These results can be used for generating a **critical items list,** which ranks all items within the system in descending order of their criticality numbers. Such information can also be presented through a **Pareto chart,** a bar graph that ranks the criticality numbers of the items within the system. The Pareto chart on item criticality numbers, combined with severity class numbers, are useful tools for prioritizing areas of the system that are candidates for redesign for additional compensating provisions and testability. Of specific interest to the designer or analyst at this point is what is known as a **single-mode-failure.** This is a failure mode that, by itself, can cause a system failure. We are especially concerned with a single-mode failure that has a high criticality number and a high-severity-class system-level effect.

EXAMPLE 13.1

Suppose that as part of an FMECA we have determined that a particular roller bearing has three failure modes:

1. Scored bearing
2. Scored bearing race
3. Cracked bearing race

Suppose that we estimate that 50% of the bearing failures are in the first mode, 40% in the second, and 10% in the third. We also estimate that the first mode will lead to the undesired system level effect 30% of the time, the second mode 40% of the time, and the third mode 90% of the time. If the bearing's failure rate is estimated to be 25×10^{-6} failures per hour and the at-risk time is 1000 hours, determine the criticality number for each failure mode and for the bearing.

The failure rate for the first failure mode, *scored bearing,* is $0.50 \times 25 \times 10^{-6} = 12.5 \times 10^{-6}$ failures per hour. The failure rate for the second failure mode, *scored bearing race,* is $0.40 \times 25 \times 10^{-6} = 10 \times 10^{-6}$ failures per hour. The failure rate for the third failure mode, *cracked bearing race,* is $0.10 \times 25 \times 10^{-6} = 2.5 \times 10^{-6}$ failures per hour.

For the first failure mode, the criticality number is

$$C_{m1} = \beta(1 - e^{-\lambda_{m}t}) = 0.30(1 - e^{-0.0125}) = 0.0037$$

For the second failure mode, the criticality number is

$$C_{m2} = \beta(1 - e^{-\lambda_{m}t}) = 0.40(1 - e^{-0.0100}) = 0.0040$$

For the third failure mode, the criticality number is

$$C_{m3} = \beta(1 - e^{-\lambda_{m}t}) = 0.90(1 - e^{-0.0025}) = 0.0022$$

Applying the table of Figure 13.3, we have the following:

Item	Failure Mode	Item Failure Rate	Mode Failure Rate	t	β	Criticality
Roller bearing	1. Scored bearing	25×10^{-6}	12.5×10^{-6}	1000	0.30	0.0037
	2. Scored race		10.0×10^{-6}	1000	0.40	0.0040
	3. Cracked race		2.5×10^{-6}	1000	0.90	0.0022

The total criticality number for the roller bearing is

$$C = 0.0037 + 0.0040 + 0.0022 = 0.0099$$

which assesses the likelihood that the bearing will cause an undesired system effect under the current system design.

13.4 THE FAULT TREE ANALYSIS

Using the symbols of Figure 13.4, a **fault tree analysis** (FTA) evaluates hypothesized **undesired events** in a system to expose their root causes. **Events** are connected through logic gates to describe how a system undesirable event is produced.

Figure 13.4 shows three types of *event symbols.* The rectangle is used where events are considered to be the result of other events within the system and where the relationship is going to be explored as part of the FTA. The circle is for a **basic event,** which is independent of all other events; hence, no further investigation in necessary. The

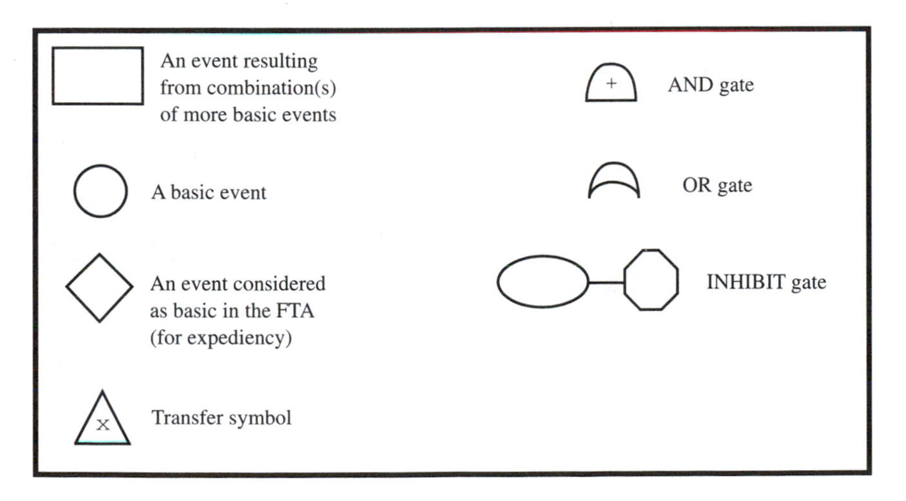

FIGURE 13.4
Fault tree analysis symbols.

rhombus is for events that are not basic but are considered to be so for the purpose of the FTA. We may choose to assign some lower-level events to this third category for the expediency of not exploring events of little consequence, or we may use this category for events that are analyzed on a separate fault tree.

The FTA is not nearly as compact as the FMEA and can go on for several pages. The **transfer symbol** is used for transferring a branch of the FTA onto another page. It consists of a triangle containing the page number of the continuation.

Three types of logic gates are shown in Figure 13.4. An event is connected to its causal events through an **AND gate** if *all* the causal events must happen in order for the resulting event to take place. An event is connected to its causal events through an **OR gate** if the resulting event can be produced by *any* of the causal events. For any other types of logic among events, an *INHIBIT gate* is used. The **INHIBIT gate** is composed of an octagon connected to an ellipse, in which the causal relationship is described.

The FTA technique is illustrated through the example in Figure 13.5, where the top undesired event is a structure burning down. For this to take place, both the connecting second-tier events must happen. Therefore, they are connected through an AND gate. The fire can be a chemical or electrical or other type of fire, so those three events

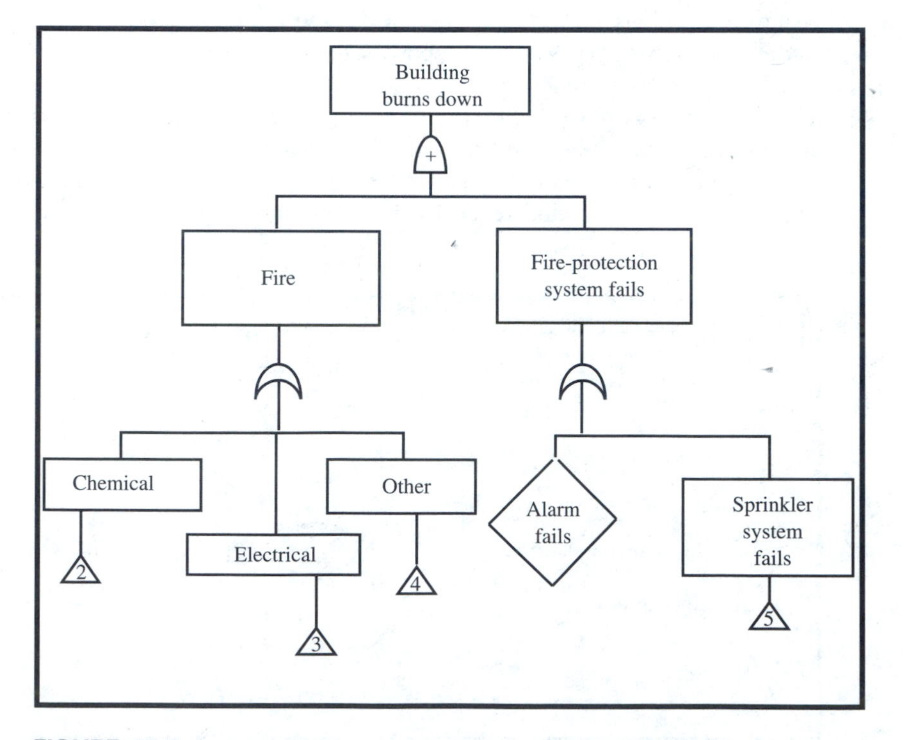

FIGURE 13.5
Example of a fault tree analysis.

are connected to the fire event through an OR gate. Since the types of fires are events to be further explored by the FTA, they are symbolized by rectangles, and their branches must continue. However, there is no room on the page shown, so they must be continued on other pages, as indicated by their transfer symbols. For example, the chemical fire branch continues on the second page of the FTA, the electrical fire on the third page, etc. The fire alarm failure, while not a basic event, is treated as such in this FTA. So the fire alarm branch does not continue.

The FTA technique is appropriate for examining events within a manufactured or service-rendering system that can lead to an accident or other undesired event. Examples of a system-level undesired event may be a vehicle crash, a space-shuttle accident, an infection received during a hospital stay, a chemical plant explosion, a power tool mishap, poisoning from a processed-food product, and a bank failure from an improper investment. To examine such events and assess the likelihood of their occurrence through the FTA technique requires an individual or, preferably, a team that is fully knowledgeable about the system. The FTA is an excellent vehicle for exposing single-mode failures. They would appear as *basic* events that connect to the top event through a series of OR gates only.

It is possible to assess the probability of a system-level undesired event from estimates of probabilities of the basic events in the fault tree. When a set of *mutually exclusive* events is connected to a higher-order event through an OR gate, we *add* the probabilities. When a set of *statistically independent* events is connected to a higher-order event through an AND gate, we multiply their probabilities. For situations where events that are not mutually exclusive are connected through an OR gate or dependent events are connected through an AND gate, we can follow Section A.1.2 for guidance on assessing probabilities. (See Section A.1.2 for definitions of *mutually exclusive* and *statistically independent,*) Also read Appendix A.1 for methods of dealing with inhibit-gate conditions.

EXAMPLE 13.2

Determine the probability of Event A in the fault tree of Figure 13.6 if all the basic events and events assumed to be basic are statistically independent and mutually exclusive of one another. The probabilities of the basic events are estimated to be the following:

$$P(D) = 5\% \qquad P(H) = 12\% \qquad P(L) = 5\%$$
$$P(F) = 15\% \qquad P(I) = 10\%$$
$$P(G) = 20\% \qquad P(K) = 5\%$$

Then,

$$P(J) = P(K) \times P(L) = (0.05)(0.05) = 0.0025$$
$$P(E) = P(I) + P(J) = 0.10 + 0.0025 = 0.1025$$
$$P(B) = P(D) + P(E) + P(F) = 0.05 + 0.1025 + 0.15 = 0.3025$$
$$P(C) = P(G) \times P(H) = (0.20)(0.12) = 0.0240$$
$$P(A) = P(B) \times P(C) = (0.3025)(0.0240) = 0.00726 = 0.73\%$$

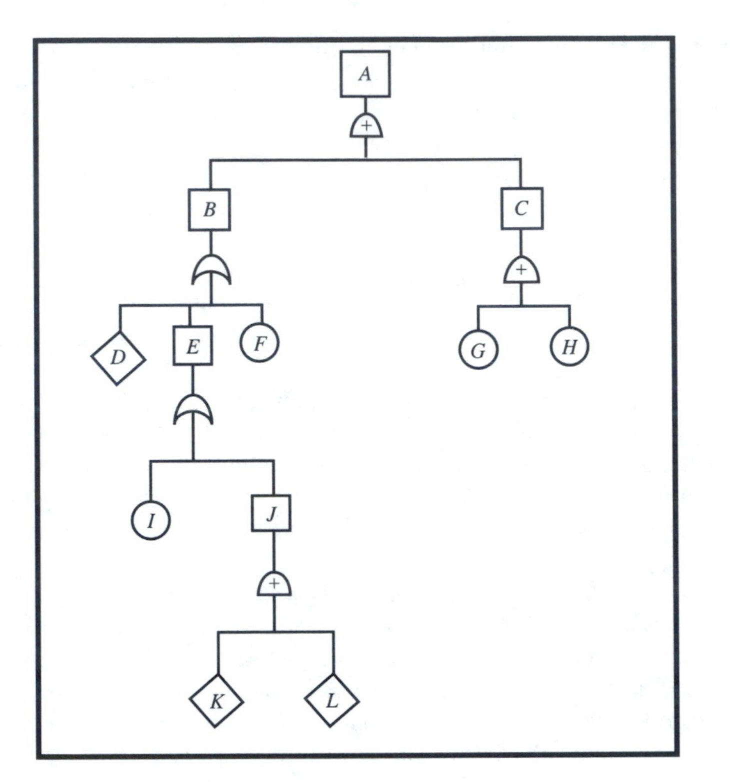

FIGURE 13.6
Fault tree for Example 13.2.

13.5 THE ISHIKAWA DIAGRAM

Because the FTA lacks the systematic bookkeeping approach of the FMEA, it is wise to use a team in attempting to develop an exhaustive set of *undesired events.* The cumulative knowledge of a team is often necessary to accurately develop the branches of the FTA. Often a multidisciplined team will be required to assemble all the cause-and-effect relationships that make up the fault tree. This means that both technical and nontechnical people will be cooperating on this effort.

As is certainly apparent at this point, the FTA structure requires an appreciation for schematic logic and a level of comfort with analytical techniques that all team participants may not possess. Therefore, in the early stages of an FTA evolution, it may be beneficial to work with a simplified version, which was developed by the Japanese industrial and quality consultant Kaoru Ishikawa during the 1960s.

In 1962, Ishikawa created **quality circles,** which were at that time composed of teams of production workers responsible for identifying and solving quality problems within their work areas. He then provided his quality circles with a set of what he referred to as *seven basic analytical tools* necessary to carry out these new responsibilities. All of Ishikawa's tools were actually uncomplicated adaptations of existing analytical techniques. Although Ishikawa's adaptations were more palatable to the quality team participants than the original versions, they remained quite effective in accomplishing the objectives.

Ishikawa identified the tool that he derived from the FTA as the **fishbone diagram** (because of its resemblance to a fish's skeleton). In the Ishikawa diagram, the top-level system effect (or undesired event) is placed in the "head of the fish," and all the causal events are placed in branches. Events in Ishikawa's fishbone diagram are not connected through logic gates. Hence, this technique is limited simply to revealing causes of effects, without considering the nature of their relationships or evaluating probabilities of occurrences. Such determinations can, of course, be made by an analyst at a more advanced stage of the FTA investigation.

Figure 13.7 presents a typical Ishikawa diagram. It corresponds to the fault tree of Figure 13.5, and it can be seen to provide cause-and-effect information without system logic. To facilitate its development, it is traditional to break the diagram into four major branches with the following titles:

1. People
2. Supplies or materials
3. Machinery, equipment, or procedures
4. Environment or policies

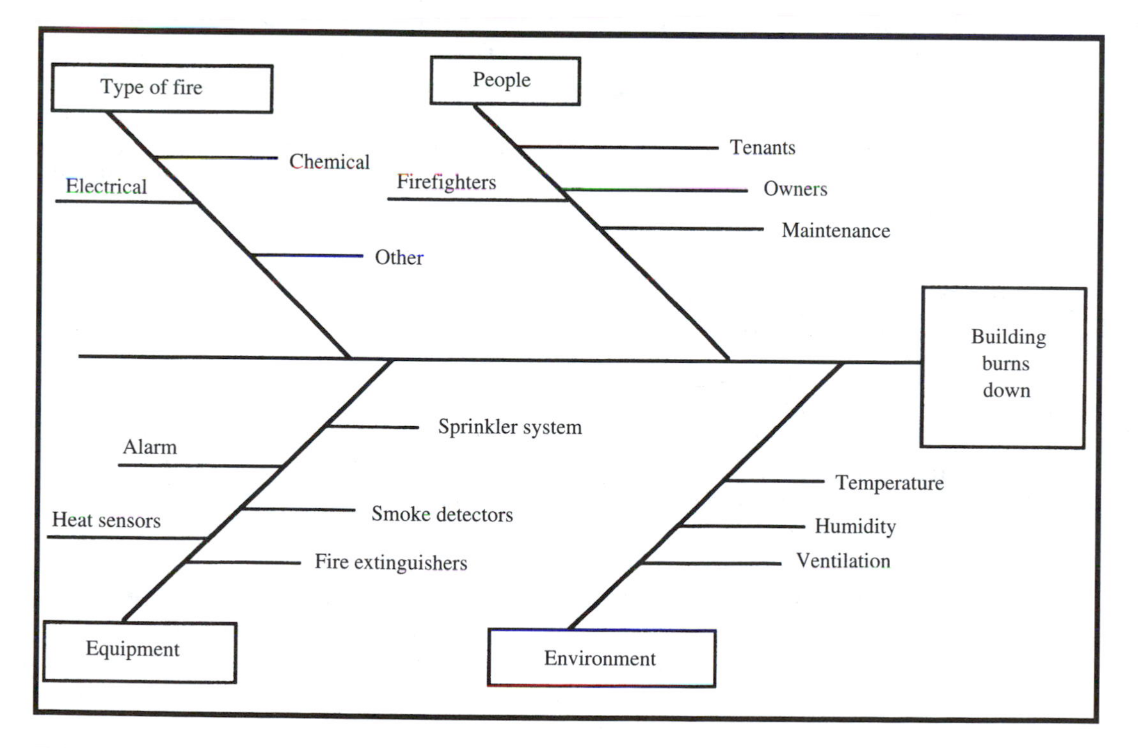

FIGURE 13.7
Ishikawa diagram.

In other words, the top level of the Ishikawa diagram asks the following questions: Who does it? To what? With what? Under what conditions?

The next level down contains branches to the four major branches. In Figure 13.7, for example, the people branch identifies all the people involved. Figure 13.7 stops at this point. However, the Ishikawa diagram can continue through many levels and sub-branches. Under each person identified, for instance, we can have a branch describing what that person can do to lead to the building burning. Under each type of fire we can have branches describing how that type of fire can start.

The Ishikawa diagram is far less formal than the rigorously structured fault tree. It permits a great deal of freedom, because its purpose is to do no more than to identify causes of system effects.

13.6 PROBABILISTIC RISK ASSESSMENT

The FTA technique has typically been used to evaluate system accident risks. But, as is apparent from the discussion thus far, it deals with the causal relationships among discrete events. Therefore, as presented in Section 13.2, the FTA cannot properly be applied to systems of time-dependent performance.

During the 1960s the FTA was, nevertheless, used on missile launch systems for NASA and DOD programs. But these systems had very short mission times (usually 30 min or less), and the error introduced by treating such missions as one-shot events was not significant. A decade later, when the NRC was using FTA to evaluate the safety of nuclear power plant systems, the technique had to be modified to accommodate time-dependent performance. The power plant systems were usually continuously operating or were safety-protection systems that had to be operable in a standby mode. The modification developed is known as **probabilistic risk assessment** (PRA), which identifies and solves the probabilities of *minimum cut sets.* The minimum cut set probabilities can be either discrete (cycle dependent) or continuous (time dependent).

13.7 MINIMUM CUT SET

A **cut set** is any set of basic events (or events considered to be basic) in the fault tree that in combination will create the top level undesired event. A **minimum cut set** is a cut set that is minimally sufficient for producing the top-level undesired event. In other words, a minimum cut set is any set of basic events that are *necessary* and *sufficient* for producing the top-level undesired event in a fault tree.

If a single basic event in a fault tree is, by itself, a minimum cut set, that event is a *single-mode failure event,* as described at the end of Section 13.1. As such, it is a candidate for system redesign to provide *compensating provisions* or to *improve testability.* Analysts are also interested in the revelation of minimum cut sets composed of too few basic events or minimum cut sets with high probabilities of occurrence, because such sets are points of vulnerability within the system. Consequently, the PRA technique can be a valuable catalyst for redesign considerations.

Assuming that the minimum cut sets within an FTA are mutually exclusive and exhaustive, the risk of the top-level system event can be determined by adding the probabilities of all minimum cut sets. Whatever method is appropriate can be used for predicting the probabilities of the minimum cut sets themselves. If the minimum cut sets are composed of statistically independent discrete events, the methods of Section 13.2 apply. If the minimum cut sets are composed of discrete events that are not necessarily statistically independent or are connected through inhibit-gate conditions, analytic techniques of Appendix A.1 are appropriate. If the minimum cut set is composed of time-dependent functions, the Markov model techniques of Appendix A.3 are appropriate. In this manner probabilistic risk assessment makes the FTA technique more universal.

EXAMPLE 13.3

Determine all the minimum cut sets of Example 13.2.

Notice that a minimum cut set must contain all events connected through an AND gate but only one event connected through an OR gate. For, example, event *B* will not produce event *A* unless event *C* takes place as well. For event *C* to take place, both events *G* and *H* must take place. So a minimum cut set must contain events *G* and *H* and events necessary for event *B* to take place. Since the subevents to event *B* are connected through an OR gate, any one of those subevents, event *D*, for instance, will cause event *B*. Hence, because *D*, *G*, and *H* are basic events, they comprise the first minimum cut set, as noted by the subscript 1 in Figure 13.8.

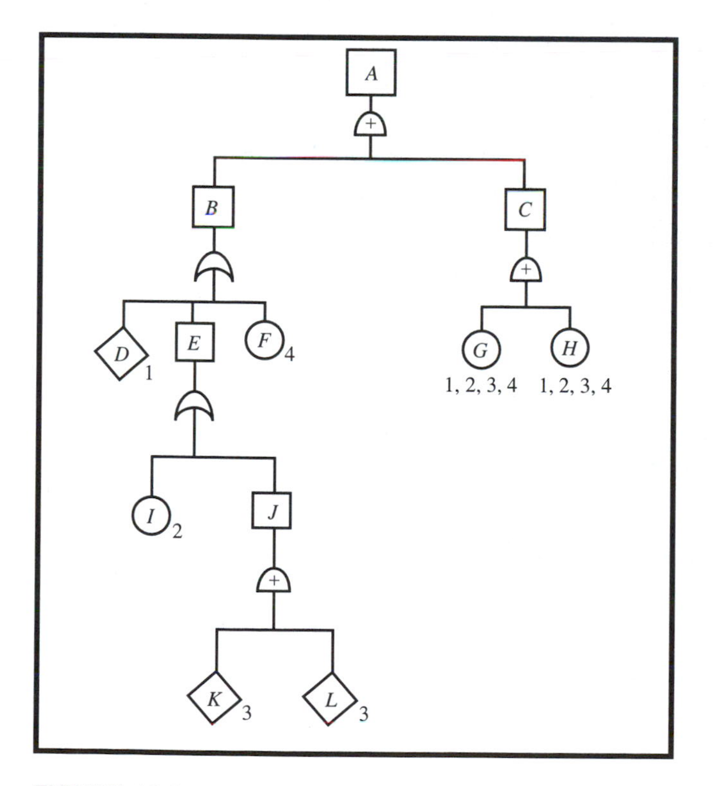

FIGURE 13.8
Fault tree for Example 13.3.

Similarly, event A will take place if events E, G, and H take place. But E is not a basic event. For it to happen, the basic event I or both K and L must take place. So the second and third minimum cut sets are events I, G, and H and events F, L, G, and H, respectively, as noted by the subscripts 2 and 3 in the illustration.

Finally, the fourth minimum cut set is composed of events F, G, and H, as noted by the subscript 4 in the illustration.

Notice that the four minimum cut sets identified,

1. $\{D, G, H\}$
2. $\{I, G, H\}$
3. $\{K, L, G, H\}$
4. $\{F, G, H\}$,

are mutually exclusive and exhaustive. Therefore, the probability of event A can be determined by summing up the probabilities of these minimum cut sets.

EXAMPLE 13.4 Determine the probabilities of each of the four minimum cut sets in Example 13.3.

Referring to Example 13.2,

$$P(D) = 5\% \qquad P(H) = 12\% \qquad P(L) = 5\%$$
$$P(F) = 15\% \qquad P(I) = 10\%$$
$$P(G) = 20\% \qquad P(K) = 5\%$$

Because all the basic events are statistically independent, the probability of the first minimum cut set is

$$P(DGH) = P(D) \times P(G) \times P(H) = (0.05)(0.20)(0.12) = 0.00120$$

The probability of the second minimum cut set is

$$P(IGH) = P(I) \times P(G) \times P(H) = (0.10)(0.20)(0.12) = 0.00240$$

The probability of the third minimum cut set is

$$P(KLGH) = P(K) \times P(L) \times P(G) \times P(H) = (0.05)(0.05)(0.20)(0.12) = 0.00006$$

The probability of the fourth minimum cut set is

$$P(FGH) = P(F) \times P(G) \times P(H) = (0.15)(0.20)(0.12) = 0.00360$$

EXAMPLE 13.5 Use the results of Example 13.4 to predict the risk of the top-level undesired event A of Example 13.3.

As concluded in Example 13.3, the risk of event A is the sum of the probabilities of the four minimum cut sets:

$$P(A) = P(DGH) + P(IGH) + P(KLGH) + P(FGH)$$
$$= 0.00120 + 0.00240 + 0.00006 + 0.00360 = 0.00726 \approx 0.73\%$$

Compare this result to that of Example 13.2.

BIBLIOGRAPHY

1. A. E. Green. 1983. *Safety Systems Analysis.* New York: John Wiley.
2. E. E. Lewis. 1987. *Introduction to Reliability Engineering.* New York: John Wiley.
3. P. D. T. O'onnor. 1991. *Practical Reliability Engineering,* 3d ed. New York: John Wiley.
4. C. O. Smith. 1976. *Introduction to Reliability in Design.* New York: McGraw-Hill.
5. MIL-STD-1629A. 1980. *Failure Mode and Effects Analysis.* Arlington, Va.: Department of Defense.

EXERCISES

13.1. Create a failure modes and effects analysis for a system with which you are familiar. Discuss any changes in design approach that you would suggest as a result of this analysis.

13.2. Create an Ishikawa diagram for an undesired event in your system of Exercise 13.1.

13.3. As part of an FMECA we have determined that a valve exhibits four failure modes: (a) cracks or ruptures, (b) seal leaks, (c) corrosion or freezing of moving parts, and (d) miscellaneous. Suppose we estimate that 15% of the failures are in mode a, 50% are in mode b, 30% are in mode c, and 5% are in mode d. We also estimate that mode a will lead to the undesired system level effect 90% of the time, mode b, 40% of the time, mode c, 100% of the time, and mode d, 5% of the time. If the valve's failure rate is estimated at 12×10^{-6} per hour and the at-risk time is 8760 h, determine the criticality number for each failure mode and for the valve.

13.4. Determine the probability of event A in the fault tree diagram of Figure 13.9. Basic events B, D, and E are statistically independent and mutually exclusive of one another.

$$P(B) = 20\% \qquad P(D) = 50\% \qquad P(E) = 30\%$$

13.5. Determine the probability of event A in the fault tree diagram of Figure 13.10. Basic events B, C, E, and F are statistically independent and mutually exclusive of one another.

$$P(B) = 25\% \qquad P(C) = 10\% \qquad P(E) = 30\% \qquad P(F) = 40\%$$

13.6. Determine the probability of event A in the fault tree diagram of Figure 13.11. Basic events B, C, E, G, and H are statistically independent and mutually exclusive of one another.

$$P(B) = 25\% \qquad P(C) = 10\% \qquad P(E) = 30\% \qquad P(G) = 40\% \qquad P(H) = 50\%$$

FIGURE 13.9
Fault tree for Exercise 13.4.

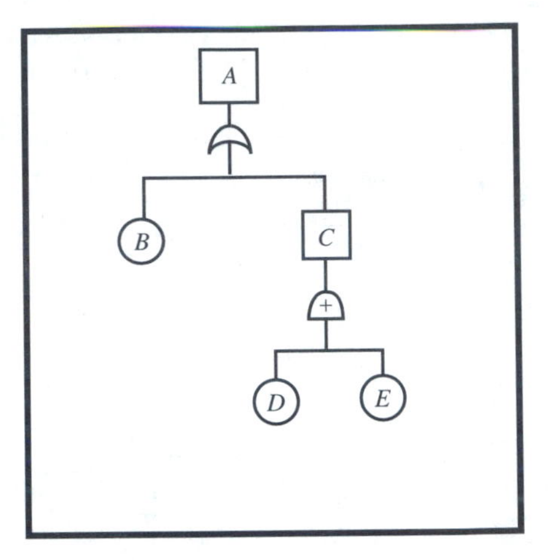

13.7. Determine all the minimum cut sets of Exercise 13.4. Compute the probabilities of each of them. Then use the probabilities of the minimum cut sets to compute the probability of event *A*. Compare your result with that of Exercise 13.4.

13.8. Determine all the minimum cut sets of Exercise 13.5. Compute the probabilities of each of them. Then use the probabilities of the minimum cut sets to compute the probability of event *A*. Compare your result with that of Exercise 13.5.

13.9. Determine all the minimum cut sets of Exercise 13.6. Compute the probabilities of each of them. Then use the probabilities of the minimum cut sets to compute the probability of event *A*. Compare your result with that of Exercise 13.6.

***13.10.** Can event *A* of Exercise 13.4 and event *A* of Exercise 13.6 be mutually exclusive events? Explain your answer.

FIGURE 13.10
Fault tree for Exercise 13.5.

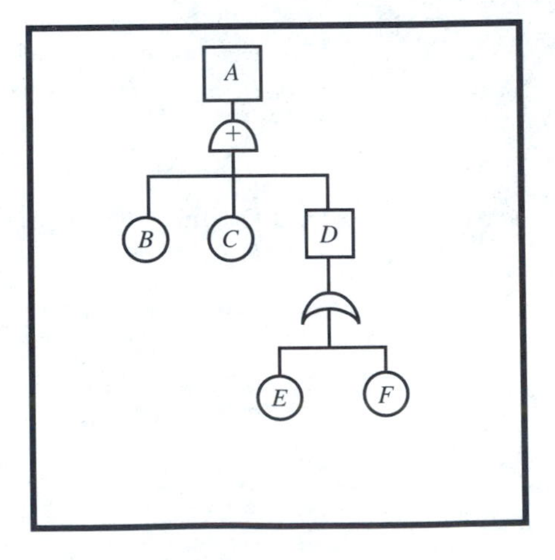

FIGURE 13.11
Fault tree for Exercise 13.6.

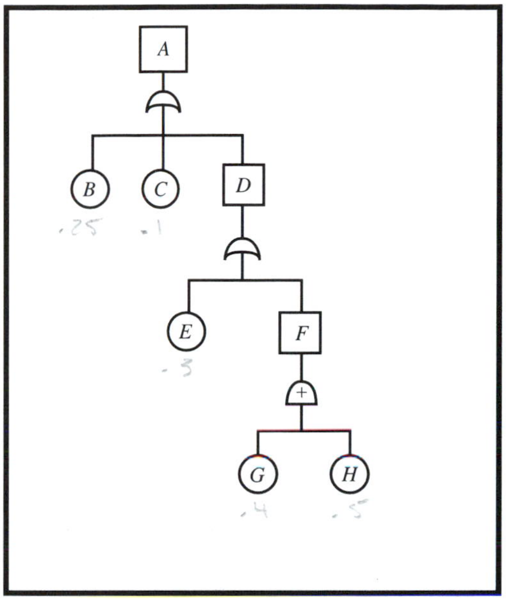

P(F) = .4 × .5 = .2
P(D) = .3 * .2 = .5
P(A) = .25 + .1 + .5 = .85

P = AB + AC + BC
.005 + .03 + .015

$m_1 + m_2 + m_3$
MIN CUT SETS

$I_{m_1} = \frac{A \cdot B}{P} = \%$

$I_{m_2} = \frac{A \cdot C}{P} = \%$ $I_{m_3} = \frac{B \cdot C}{P} = \%$

$I_A = \frac{.005 + .03}{P} = \%$ $I_B = \frac{.005 + .015}{P} = \%$ $I_C = \frac{.03 + .015}{P} = \%$

max terms F' = 0 + 0 + 0 F = 0 · 0 · 0 · 0
min terms = 1 + 1 + 1

14

EPILOGUE

During the four decades since the idea of system reliability was first introduced, changes in technology, world culture, and producer-consumer relations have expanded its definition and applications. From the days when the appearance of solid-state electronics raised concerns about complex equipment being able to perform when needed, through early aerospace development, through the emergence of nuclear power as a major energy source, and, finally, with the integration of artificial intelligence into system operation, the reliability engineering discipline has continually reinterpreted itself in order to best address the basic question, How are we to provide the user of our system *what* the user wants *when* the user wants it? We have learned that corollaries to this basic question are: What happens if something goes wrong? How can we best protect ourselves against the consequences of something going wrong? How do we quickly and efficiently *restore* our system when something does go wrong?

Over the past 15 years the concept of *total quality,* in one form or another, has infiltrated many organizations worldwide. The essence of total quality is focusing an entire organization on its prime objective to satisfy the needs of its "customers," the users of its products or services. Following this philosophy means that quality assurance efforts in a manufacturing plant are not restricted to its production operations. Rather, they address the entire way of doing business, including all service and support activities.

Indeed, total quality is applied to nonprofit, governmental, and other types of service organizations as well as to those that manufacture products.

System reliability and restorability practices are obvious extensions for organizations ready to go beyond their current total quality initiatives, because truly satisfying the needs of the customer means more than satisfying those needs at some delivery date. It means assuring that the customer gets what the customer needs when the customer needs it—every time that the customer needs it. It concentrates on the customer's *use* of the organization's product or service.

Reliability engineering practices during the four decades of their existence were limited to manufactured products and, as discussed in Chapter 1, often limited to products regulated by some government agency. Through total quality we can expect every form of organization to be incorporating these practices in the years to come. In anticipation, this text has broadened its definitions and techniques for service-system applications. The more general term *restorability* has been used instead of *maintainability* in order to have it pertain to the restoration of an interrupted service as well as to the repair of failed equipment. The employment of reliability and restorability practices for service-rendering systems were suggested in many of the examples presented.

Ultimately, it will be up to the reader and future users of these practices to expand them and to expose them to a variety of applications where there is a desire to have consumers get what they want from a system when they want it.

A

REVIEW OF PROBABILITY AND STATISTICS

A.1 HOW WE USE PROBABILITY AND STATISTICS

Probability theory, which had its origins in seventeenth-century France, is the study of the laws of chance that govern the occurrences or nonoccurrence of events. Until the twentieth century probability theory was used primarily as either a mathematical curiosity or for gambling. It entered industry through the scientific management techniques of Frederick W. Taylor and the introduction of statistical quality control by Walter Shewhart. A **statistic** is an event or data point. The **study of statistics** is the practice of combining a statistic with other statistics and applying them to an appropriate probability law in order to use past information to predict future events or to make a decision that will impact future events.

So, the study of statistics is used for making predictions or for making decisions, and as such it is useful in reliability and restorability in assessing the values of the parameters introduced in Chapter 2 and throughout Book 2 and in making the decisions for reliability and restorability demonstration in Book 3. Without the study of statistics, we generally make the predictions and decisions necessary to carry us through life (e.g., when to jam on our brakes on the highway or when and how to duck out of the way of a threatening object) by using our intuition. The device called intuition is actually the using of our experience in interpreting behavior of the world around us to make a prediction or decision on the basis of some observations. The study of statistics uses the same technique in an organized fashion. Instead of our personal experience, we use the applicable law(s) of probability to interpret behavior of a phenomenon, and instead of a few personal observations we deliberately gather appropriate data (or statistics).

The statistical analysis method consists of the following steps:

1. Determine the probability law that adequately describes the behavior of the phenomenon for which we are making a prediction or decision. The probability law is a mathematical expression in terms of one or more **statistical parameters.**
2. Gather data to estimate the statistical parameters. Notice here that the data are used to estimate the parameters of the probability law, *not* to actually make the prediction.
3. Use enough data to estimate the parameters closely enough to give a meaningful prediction from the statistical analysis. How close we need this estimate to be is dictated by the accuracy required in our prediction or the sensitivity of our decision. The closeness of the estimate is called **confidence.** Generally, the more data used, the greater the confidence, and thus the closer our estimate is to the actual value of the parameter. So if we make an 80% confidence estimate, there is an 80% chance that (as the case may be) the value of the parameter is greater than our estimate, less than our estimate, or within two (upper and lower) estimates. This description will be clearer later, but at this point try to accept that the term *confidence* always applies to the estimation of a statistical parameter of a probability law.
4. Sometimes we may want to use the data to see if the assumed probability law really does fit the phenomenon we are studying. This can be done through any number of different analytical or graphical goodness-of-fit tests. Here the statistical term **significance** applies. We accept our originally assumed probability law (because we had some very good reason for making that assumption in the first place) unless the data indicate a *significant* misfit—that is, unless there is overwhelming evidence that our probability law assumption was wrong.
5. We use the probability law with the estimated parameter values to calculate our predictions or to make the decision. Usually these calculations are facilitated through the use of commonly available tables, such as those in Appendix B, or by one of the many available computer programs.

A.2 PROBABILITY FUNCTIONS

A probability function $P(x)$ expresses the likelihood of an outcome x. For example, if we toss two coins, there are four possible outcomes: (1) both coins can be heads, (2) both can be tails, (3) the first can be a head and the second, a tail, or (4) the first can be a tail and the second, a head. Assuming that both are fair coins—that is, not weighted to the advantage of either head or tail, each of the four outcomes is equally likely; therefore, each has a probability of ¼:

$$P(HH) = P(HT) = P(TH) = P(TT) = 0.25$$

If we do not care to distinguish between the two coins (i.e., we do not bother to identify which coin turns up what), then outcomes 3 and 4 are indistinguishable, and we have only three possible outcomes: (1) two heads, (2) two tails, or (3) one of each. Assuming again that both are fair coins, the three outcomes are not equally likely. Although we choose not to distinguish between the two coins, the coins know the difference, and the third outcome is twice as likely as either of the other two. Hence,

$$P(2H) = 0.25, \qquad P(2T) = 0.25, \qquad P(1H + 1T) = 0.50$$

A.2.1 Properties of Probability Functions

You will be introduced to a variety of probability functions, and all of them possess the following five properties.

1. For *any* outcome x, $0 \leq P(x) \leq 1.0$.

2. *If* and *only if* $P(x) = 1.0$, the outcome x is a *certainty*.

3. *If* and *only if* $P(x) = 0$, the outcome x is an *impossibility*.

4. For any two outcomes x and y, the probability of the **union** of x and y—that is, the probability that the event x or y (or both) occurs—is

$$P(x \cup y) = P(x) + P(y) - P(xy)$$

where $P(xy)$ is the probability of the *intersection* of x and y, that is, the probability that both outcomes x and y occur together. In other words, the probability that at least the outcome x or y takes place is the sum of the probabilities that x and y occur minus the probability that they both occur together.

As a corollary to this property, *if* and *only if* x and y are *mutually exclusive*—that is, they cannot occur together (so $P(xy) = 0$)—then

$$P(x \cup y) = P(x) + P(y)$$

5. Bayes' theorem states that for any two outcomes x and y, the probability of the intersection of x and y—that is, the probability that the outcomes x and y occur together—is

$$P(xy) = P(x \mid y) \cdot P(y) = P(y \mid x) \cdot P(x)$$

where $P(x \mid y)$ is the probability of the outcome x conditional upon the outcome y—that is, the probability that x occurs whenever y does.

As a corollary to this property, *if* and *only if* x and y are statistically independent—that is, outcome x does not influence the probability of outcome y and outcome y does not influence the probability of outcome x (so $P(x \mid y) = P(x)$ and $P(y \mid x) = P(y)$)—then

$$P(xy) = P(x) \cdot P(y)$$

A.2.2 Some Definitions

In connection with the properties of probability functions, the following definitions of some of the terms are introduced.

Intersection The intersection of outcomes x and y means the occurrence of both x and y together. The probability of the intersection of x and y is written $P(xy)$. As an example, if x represents the event that we draw a spade from a deck of cards and y represents the event that we draw a 10, then $P(xy)$ represents the probability of drawing a 10 of spades. In our example, $P(xy) = 1/52$.

Union The union of outcomes x and y means the occurrence of at least x or y. The probability of the union of x and y—that is, the probability of outcome x or y—is written $P(x \cup y)$. As an example, if x represents the event that we draw a spade from a deck of cards and y represents the event that we draw a 10, then $P(x \cup y)$ represents the probability of drawing a card that is either a 10 or a spade. This means that the card drawn can be any of the 13 cards that are spades and any of the 4 cards that are 10s, including the 10 of spades and the other three 10s that are not spades. In our example, $P(x \cup y) = 16/52$.

Mutually exclusive The outcomes x and y are *mutually exclusive* if they cannot occur together; in other words, $P(xy) = 0$. For example, let x be the event that we draw a spade from a deck of cards and let y be the event that we draw a diamond. Obviously, the outcomes x and y are mutually exclusive, and $P(xy)$, the probability that a card is both a spade and a diamond, is 0.

Exhaustive The events x and y are called *exhaustive* if at least one of them must occur—that is, the outcome that at least one of the events x or y occurs is a certainty. Hence, if x and y are exhaustive, then $P(x \cup y) = 1.0$. An example is the outcomes x and y representing, respectively, the events that cards drawn from a deck are red or black.

Conditional outcomes The outcome $x \mid y$ represents the event that x occurs whenever y does. The probability $P(x \mid y)$ is the probability of the outcome x if we already know that the event y has taken place. For example, if x represents the outcome that we draw a king from a deck of cards and y represents the outcome that we draw a picture card, then $P(x \mid y)$ is the probability that we draw a king, given that we already know that it is a picture card. In our example, $P(x \mid y) = 1/3$.

Statistical independence The events x and y are considered to be *statistically independent* if the occurrence of one outcome does not influence the probability of the occurrence of the other. The outcomes x and y are then said to be statistically independent if $P(x \mid y) = P(x)$ and $P(y \mid x) = P(y)$. As examples, the outcome x that we draw a king and the outcome y that we draw a picture card are *not* statistically independent, whereas the outcome x that we draw a king and the outcome y that we draw a space *are* statistically independent. In the former example, $P(x) = 1/13$ and $P(x \mid y) = 1/3$. In the latter example, $P(x) = 1/13$ and $P(x \mid y) = 1/13$.

Notice that statistically independent outcomes are not necessarily mutually exclusive and that mutually exclusive outcomes are not necessarily exhaustive. These are common misapprehensions among beginning students in reliability classes.

A.3 NUMBERS

Referring to Figure A.1, numbers can be used for either coding or evaluating. Telephone numbers, street addresses, area codes, zip codes, social security numbers, part serial numbers, model numbers, and page numbers are all examples of numbers being used for the purpose of coding. Numbers are used for evaluating when they are used to count or measure something. Numbers used for counting are called **discrete** numbers; numbers used for measuring are called **continuous** numbers.

Continuous numbers must always be labeled (for example, feet, meters, pounds, grams, minutes, or revolutions per minute). Discrete numbers, on the other hand, have no labels; they are simply quantities. Whereas discrete numbers are expressed as exact and distinct quantities, a continuous number is always the result of a measurement taken with some instrument, meter, or other measuring device and, accordingly, is subject to the sensitivity of the instrument or the accuracy of the reading. The expression of a continuous number must, therefore, include an indication of the accuracy of the measurement or the implied error. Consequently, continuously numbers must always be

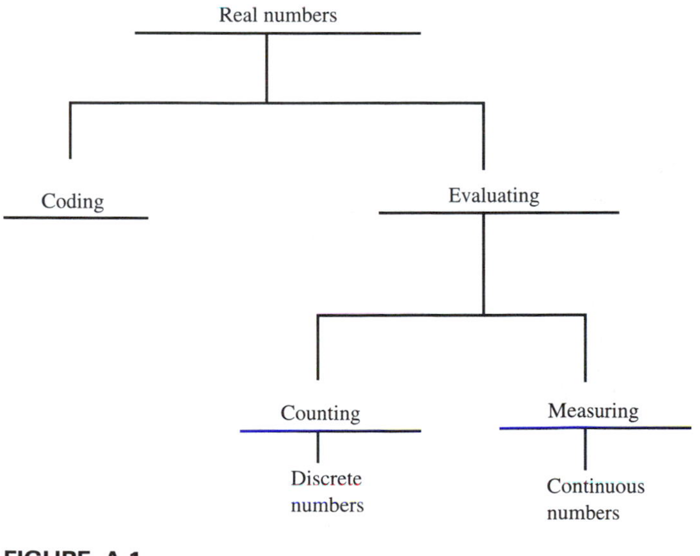

FIGURE A.1
Using numbers.

presented as a range or a nominal value plus a tolerance, such as between 24.5 and 25.5 in. or 250 $\pm$ 0.50 g.

With discrete numbers, the probability of the outcome of an exact and distinct value x can be expressed as some amount $p(x)$. For example, when tossing a pair of dice, the probability of the outcome 2 (which can be achieved only with a pair of 1s) is $P(x) = 1/36$. With continuous numbers, however, an exact and distinct value is impossible, and $P(x) = 0$ for any distinct value x. In other words, there is no such outcome as exactly 25 in. or exactly 250 g. With continuous numbers only ranges have probabilities. Thus, it is possible to express the probability of an outcome within the range of 24.5 and 25.5 in. or the probability of an outcome within 0.50 g of 250 g. With both discrete and continuous numbers it is possible to express the probablity of an outcome less than some value or of an outcome not more than some value: that is, $P(x < A)$ or $P(x \leq A)$.

A.4 FREQUENCY DISTRIBUTIONS

We can never predict what the actual outcome of a trial from a process is going to be. But we can state what all the possible outcomes are and we can determine the likelihood of each of those possible outcomes. For example, if we toss two coins (without distinguishing between the coins), as described before, there are three possible outcomes: no heads (thus, two tails), one head (thus, one tail), or two heads (thus, no tails). The probability of having no heads, as discussed earlier, is $P(0) = 0.25$. Also, as discussed earlier, the probabilities of having one head and two heads are, respectively, $P(1) = 0.50$ and $P(2) = 0.25$.

A frequency distribution is a graphical or analytical representation of the probabilities of all possible outcomes of a phenomenon or process. Figure A.2 is a frequency distribution of the example given, the tossing of two coins. In this figure, x represents the number of heads and $p(x)$ is the likelihood of x heads turning up in a random toss of two coins. Notice that there are only three possible outcomes: x can equal either 0, 1, or 2. And the likelihood of each of these outcomes is $p(0) = 0.25$, $p(1) = 0.50$, and $p(2) = 0.25$. For any other value of x, $p(x) = 0$.

Now notice how the properties of probability functions apply to Figure A.2: all the $p(x)$ values are between 0 and 1. Where an outcome x is impossible (i.e., outcomes other than 0, 1 or 2), $p(x) = 0$. The outcomes $x = 0$, 1, and 2 are mutually exclusive, so $p(0 \cup 1 \cup 2) = p(0) + p(1) + p(2)$. Also, because the outcomes $x = 0$, 1, and 2 are exhaustive, $p(0 \cup 1 \cup 2) = 1.0$.

The frequency distribution for any discrete phenomenon appears as a set of spikes, as in Figure A.2, which assign probability values to discrete integers x that represent all the possible outcomes (remember that we are always counting with discrete phenomena) and no value anyplace else. For a continuous phenomenon, on the other hand, a frequency distribution appears as a continuous curve, where there can be no probability value assigned to a discrete outcome, but probability values can be assigned to intervals along the measured continuum of x measurements. Figures A.3 and A.4 show typical discrete and continuous frequency distributions, respectively.

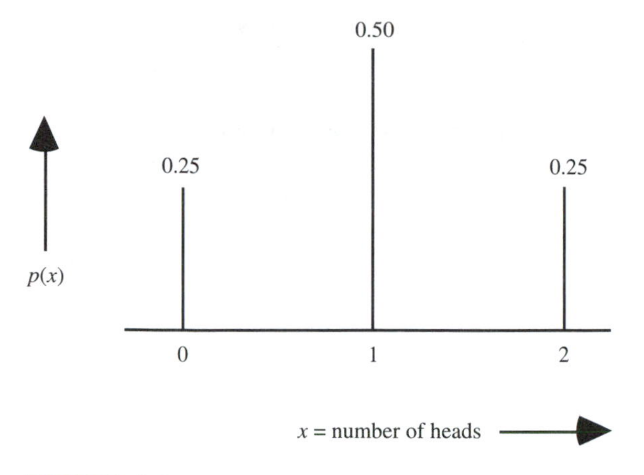

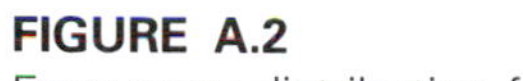

FIGURE A.2
Frequency distribution for the toss of two coins.

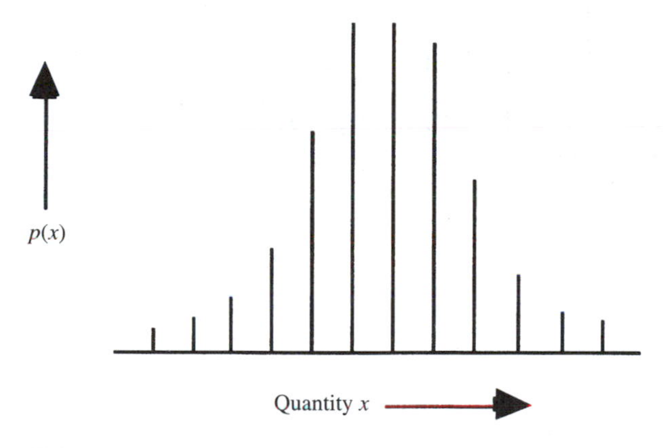

FIGURE A.3
Typical discrete frequency distribution.

FIGURE A.4
Typical continuous frequency distribution.

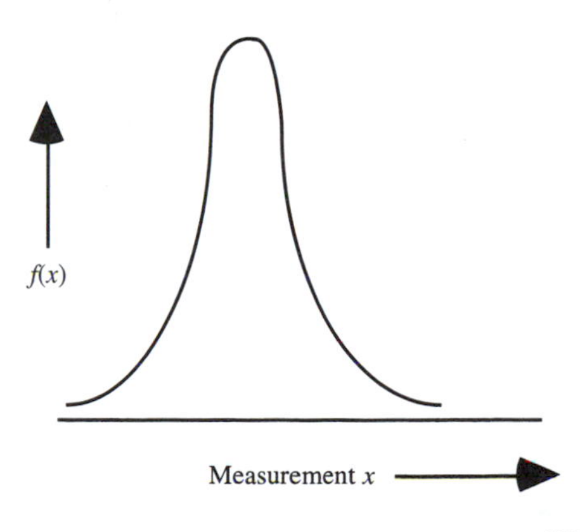

The mathematical function $p(x)$, which assigns probability values to outcomes x of discrete frequency distributions, is called the **probability mass function.** The mathematical function $f(x)$ for the continuous frequency distribution curves is called the **probability density function.** The probability of an outcome $x = A$ in a discrete distribution is the probability mass function evaluated at that quantity A: $P(x = A) = p(A)$. This situation is illustrated in Figure A.5. The probability of an outcome between the measurements A and B in a continuous distribution is the area under the probability density curve between A and B: $P(A < x < B) = \int_{x=A}^{B} f(x)\,dx$. This situation is illustrated in Figure A.6.

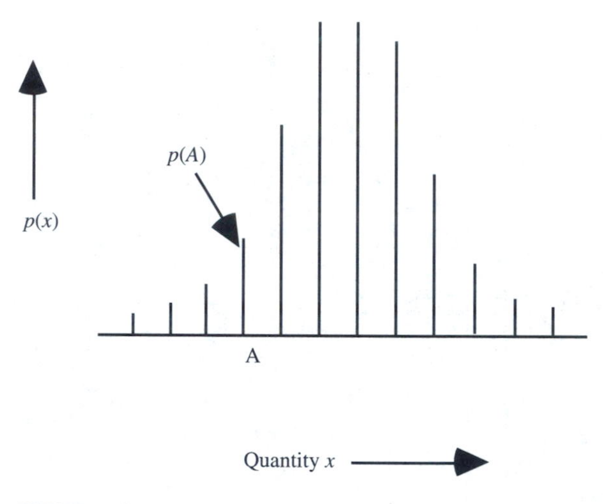

FIGURE A.5
A discrete frequency distribution showing the probability of the outcome *A*.

FIGURE A.6
A continuous frequency distribution showing the probability of an outcome between measurements *A* and *B*.

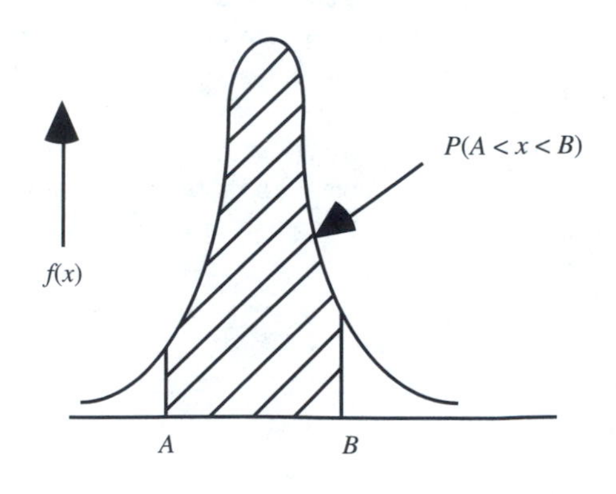

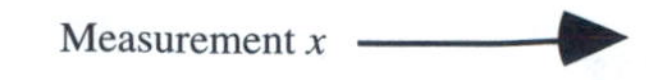

A.5 CUMULATIVE DISTRIBUTION FUNCTIONS

As we noted before, discrete phenomena can have probability values for distinct outcomes, whereas continuous phenomena cannot. In the two-coin-tossing example, there are probability values for the exact and distinct values $x = 0, 1,$ or 2 heads but for no other values of x. If, on the other hand, we are examining the distribution of overall lengths of pieces of stock produced from a given production process, the probability that a certain piece selected at random measures, say, exactly 5 in. is 0. The probability of that piece having an overall length that measures within a range of, say, between 4.75 and 5.25 in. or between 4.999 and 5.001 in. or even 5.0 ± 10^{-99} in. can be expressed. However, discrete and continuous phenomena do have the **cumulative distribution function** $F(x)$ in common. The function $F(x)$ expresses the probability of an outcome no greater than some value of x. In other words, the cumulative distribution function evaluated at A is the probability that x is no more than A:

$$F(A) = P(x \leq A)$$

For a discrete phenomenon, such as the coin-tossing example, the cumulative distribution function is

$$F(A) = P(x \leq A) = \sum_{x=0}^{x=A} p(x)$$

For a continuous phenomenon, such as the length-of-stock example, the cumulative distribution function is

$$F(A) = P(x \leq A) = \int_{-\infty}^{A} f(x)\, dx$$

Notice that since for continuous phenomena we cannot have probability values for exact and distinct outcomes x, $P(A) = 0$; consequently, the probability of x being no greater than A is the same as the probability that x is less than A: $P(x \leq A) = P(x < A)$. This statement cannot be made about discrete phenomena, where we must be careful about the inclusion or exclusion of our boundaries. For discrete phenomena, the probability that x is no greater than A is the probability that x is less than A plus the probability that x equals A: $P(x \leq A) = (x < A) + P(A)$.

For the example of the tossing of two coins, the cumulative distribution function $F(x)$ values are $F(0) = 0.25$, $F(1) = 0.75$, and $F(2) = 1.0$. For $x < 0$, $F(x) = 0$; for $0 < x < 1$, $F(x) = 0.25$; for $1 < x < 2$, $F(x) = 0.75$; and for $x > 2$, $F(x) = 1.0$. Hence, the cumulative distribution function for the tossing of two coins appears graphically as a set of stairs, with the largest step in the center, as shown in Figure A.7.

The cumulative distribution function for any discrete phenomenon, as a matter of fact, will always appear graphically as a staircase, where sudden changes take place at the discrete values x where an outcome is possible. Otherwise, it is flat. Discrete cumulative distribution functions evaluated at any point $x = A$ are always going to be the sum of all the $p(x)$ values from $x = 0$ to $x =$ A: $F(x) = \sum_{x=0}^{x=A} p(x)$. A typical discrete

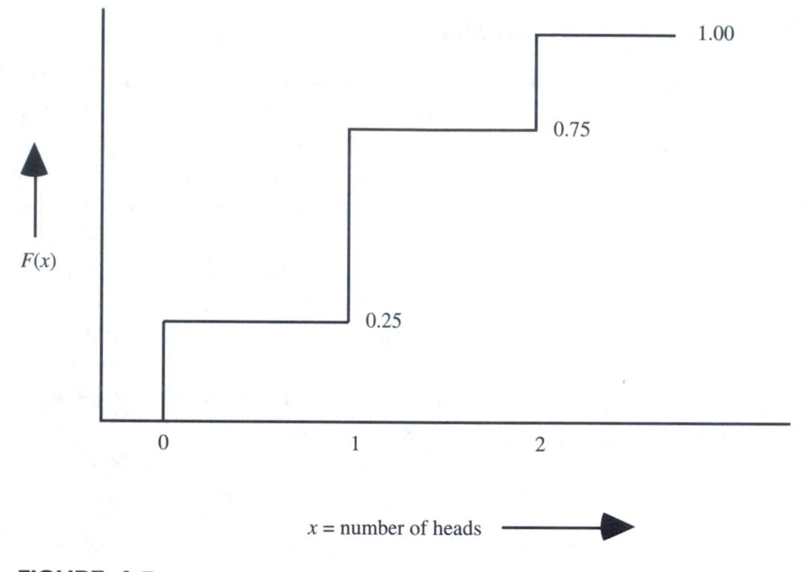

FIGURE A.7
Cumulative distribution function for the tossing of two coins.

cumulative distribution function appears in Figure A.8, whch also illustrates the value $F(A) = P(x \leq A)$.

The cumulative distribution function for a continuous phenomenon has, as expected, the continuous appearance of a staircase handrail, as shown in Figure A.9, which also illustrates the cumulative distribution functions evaluated at both $x = A$ and $x = B$: $F(A) = P(x < A) = \int_{x=-\infty}^{A} f(x)\,dx$, and $F(B) = P(x < B) = \int_{x=-\infty}^{B} f(x)\,dx$. Figure A.9 also illustrates the probability of an outcome between the measurements A and B: $P(A < x < B) = F(B) - F(A) = \int_{x=A}^{B} f(x)\,dx$.

Cumulative distribution functions have to possess all the properties of probability functions presented here. Because they are common to both continuous and discrete distributions, cumulative distribution functions are most convenient for our probability determinations. All statistical tables presented in Appendix B are x-versus-$F(x)$ tables, and these tables can be used for determining the probability of an outcome equal to, less than, or more than some value of x. The x-versus-$F(x)$ tables can also be used for determining the probability of an outcome between two values of x.

For discrete distributions, probabilities can be determined from the cumulative distribution function (hence, the tables in Appendix B) as follows:

$$P(x = A) = F(A) - F(A - 1)$$
$$P(x \leq A) = F(A)$$
$$P(x < A) = F(A - 1)$$
$$P(x > A) = 1 - F(A)$$
$$P(A < x \leq B) = F(B) - F(A)$$
$$P(A < x < B) = F(B - 1) - F(A)$$
$$P(A \leq x \leq B) = F(B) - F(A - 1)$$

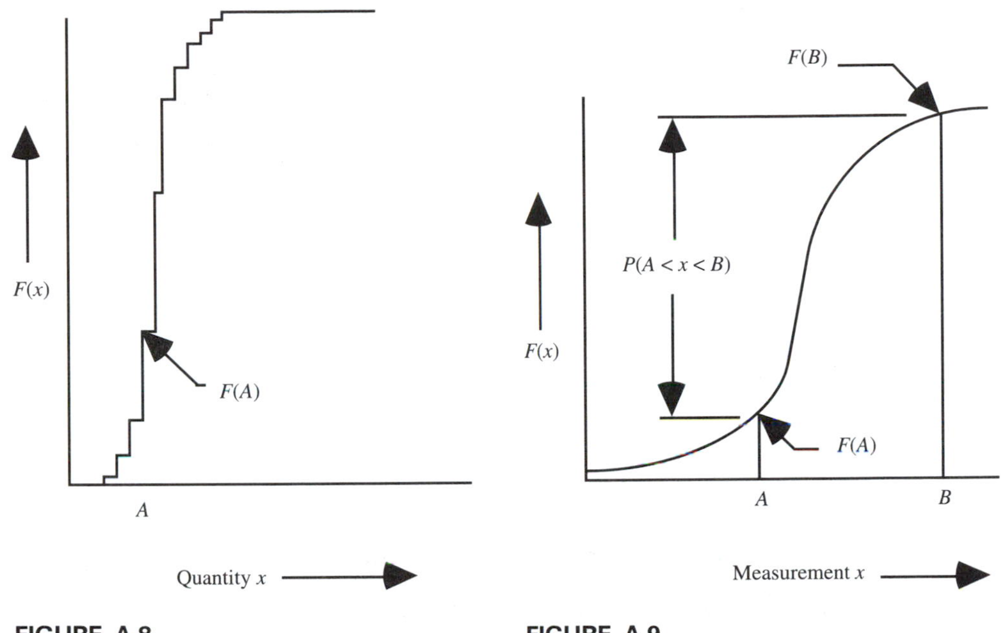

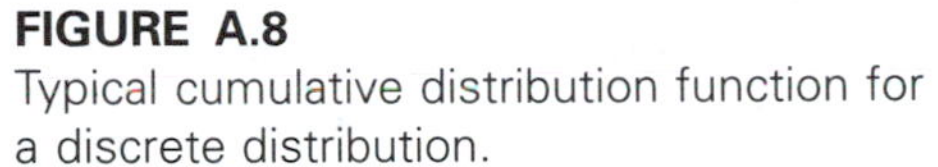
FIGURE A.8
Typical cumulative distribution function for a discrete distribution.

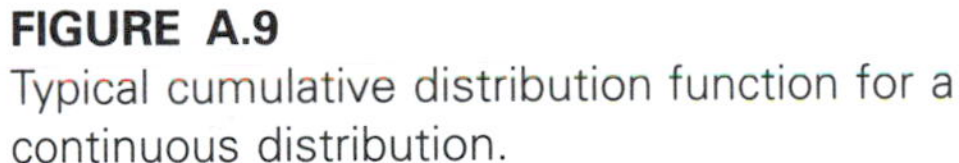
FIGURE A.9
Typical cumulative distribution function for a continuous distribution.

For continuous distributions, because $P(x) = 0$, the determinations of probabilities from the cumulative distribution function (hence, the tables in Appendix B) are simpler, because we do not have to distinguish between $x \leq A$ and $x < A$:

$$\begin{aligned} P(x = A) &= 0 \\ P(x < A) &= F(A) \\ P(x > A) &= 1 - F(A) \\ P(A < x < B) &= F(B) - F(A) \end{aligned}$$

A.6 HOW TO COUNT

Counting can always be done on your fingers, and, when you run out of fingers, counting can be accomplished by making marks on a piece of paper, as in Figure A.10, which contains a tally of 16. Figure A.11 also has a tally of 16; however, it is easier to count and less prone to error. The only difference between the tallies in Figures A.10 and A.11 is that the latter is organized. The counting techniques presented here are simply organized methods of doing what can be accomplished by counting on your fingers. These techniques will help you count things up faster, more accurately, and with more ease. In other words, they do analytically what Figure A.11 does graphically to improve the presentation of the information in Figure A.10.

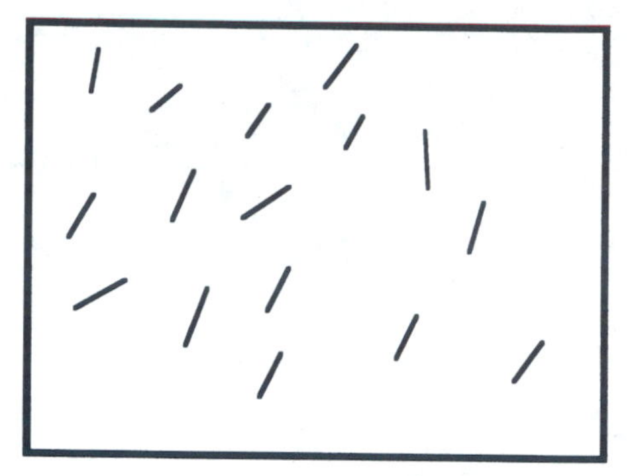

FIGURE A.10
Tally.

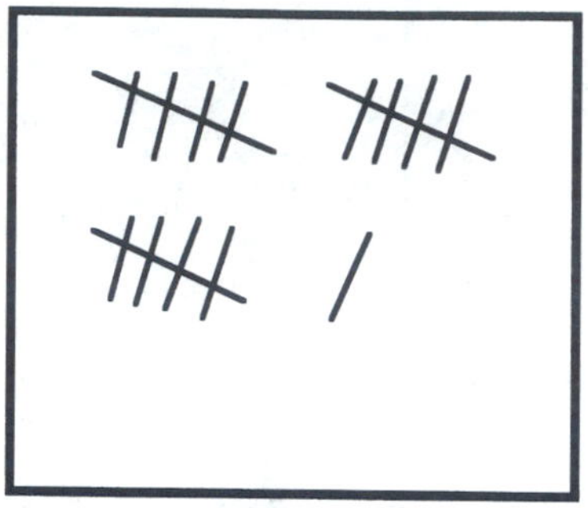

FIGURE A.11
Organized tally.

A.6.1 The Arithmetic Progression

There is a legend about a second-grade teacher in an eighteenth-century German classroom who kept an unruly class of boys after school one day and, as punishment, told them to add all the integers from 1 to 100. As soon as a pupil came up with the correct total, he would be allowed to go. Within a few minutes, according to the legend, the worst-behaved of the students, the 7-year-old Carl Frederick Gauss, presented the correct total. This could have been accomplished by making tally marks on a piece of paper as in the preceding figures. Or, more systematically, the numbers could have been added up one by one:

$$1 + 2 = 3, \quad 3 + 3 = 6, \quad 6 + 4 = 10, \quad 10 + 5 = 15, \quad 15 + 6 = 21, \ldots$$

However, the young Gauss recognizd that the first 100 integers can be joined as 50 top-and-bottom pairs, as follows:

$$1 + 100 = 101, \quad 2 + 99 = 101, \quad 3 + 98 = 101, \ldots, 50 + 51 = 101$$

There are 50 such pairs, each equaling 101. So the total is $50 \times 101 = 5050$. Notice that the only difference between young Gauss' approach and the other methods of adding up the first 100 integers is the way in which the numbers were organized to do the calculation. In other words, the computations were simplified and less prone to error through better counting organization, as with the tally examples in the preceding figures. The improved organization approach is one we recognize as the arithmetic progression that we all learned in high school. But in substance it is no different than counting on our fingers or making tally marks on a piece of paper. Similarly, the techniques presented next are nothing more than means of better organizing a count for simplicity and accuracy, counts that can always be accomplished by counting on your fingers.

A.6.2 Powers

Suppose we are at an event where we are among 500 guests holding tickets for a door prize. If three prizes are to be awarded, if there is a premium for the higher-order prizes (that is, first prize is more valuable than second prize, which is more valuable than third), and if the drawings are to be performed with replacement (that is, winning tickets are returned to the drum, so that winners are eligible for other prizes), how many outcomes are possible? All 500 guests are eligible for the first prize. All 500 guests, including the winner of the first prize, are eligible for the second prize. Similarly, all 500 guests are eligible for the third prize. So the total number of possible first-, second-, and third-place outcomes is $500 \times 500 \times 500 = 500^3$, which is 125,000,000. As a general case, if there are N guests and k prizes to be awarded, where order counts (premium for a higher order prize) and the drawings are with replacement, the number of possible outcomes is N^k. As before, we can reach these results by counting on our fingers, but the method of powers provides a better organized count.

A.6.3 Permutations

Now suppose that at the same event with the 500 guests and three prizes, order still counts (first prize is more valuable than second prize, which is more valuable than third), but the drawings are without replacement. In other words, this time the winner of one prize is not eligible for another. How many outcomes are possible here? Certainly, there are fewer than in the preceding example. All 500 guests are eligible for the first prize. But only 499 guests, all but the first-prize winner, are eligible for the second prize. And 498 guests, all but the first two prize winners, are eligible for the third prize. Hence, the total number of possible first-, second-, and third-place outcomes is $500 \times 499 \times 498 = 124{,}251{,}000$. This counting technique is known as a **permutation** and is symbolized by the notation $(500)_3 = 500 \times 499 \times 498$. As a general case, if there are N guests and k prizes to be awarded, where order counts and the drawings are *without* replacement, the number of possible outcomes is $(N)_k = N(N-1)(N-2)\cdots(N-k+1) = N!/(N-k)!$. Once again, these results can be reached with a lot of tedium by counting on our fingers, but the method of permutations provides a better organized count.

A.6.4 Combinations

This time let us suppose that the event with the 500 guests awards three prizes where order *does not count.* Three tickets are drawn and all three winners get the same prize. So we are not concerned with whose ticket is drawn when; all we care about is who the three prize winners are. Obviously, this is accomplished without replacement, as there will be three different prize winners. How many outcomes are possible here? Clearly, there are fewer than with the preceding example. In fact, there are fewer by the number of orders. That is, each outcome here would generate $3! = 3 \times 2 \times 1$ outcomes in the preceding example (any of the three winners could have won first prize, leaving two winners who could have won second prize, leaving one third-prize winner). So, the total number of possible outcomes of three prize winners with no distinction of the order of drawings is the number of possible outcomes where order counts divided by the number of orders: $(500)_3/3! = 124{,}251{,}000/3 \cdot 2 \cdot 1 = 20{,}708{,}500$. This is known as a **combination** and is symbolized by the notation

$$\binom{500}{0} = \frac{(500)_3}{3!} = \frac{124{,}251{,}000}{3 \cdot 2 \cdot 1} = 20{,}708{,}500$$

So, a combination is a permutation divided by the number of orders. As a general case, if there are N guests and k prizes to be awarded, where order does not count and the drawings are without replacement, the number of possible outcomes is

$$\binom{N}{k} = \frac{(N)_k}{k!} = \frac{N(N-1)(N-2)\cdots(N-k+1)}{k!} = \frac{N!}{k!(N-k)!}$$

As in all preceding cases, these results can be reached by finger-counting, but the method of combinations provides the benefit of an organized count.

A.7 PROBABILITY LAWS AND THEIR PARAMETERS

The statistical analysis method described at the beginning of this appendix recognizes that in nature every process or phenomenon produces not a single, predictable outcome, but an apparently random outcome of a predictable distribution of possible outcomes. So we select and use an appropriate probability law that best describes the distribution of outcomes for the phenomenon in question. Then, according to the described statistical analysis method, we gather enough data to estimate values of the law's parameters and use the probability law to make predictions or decisions.

Every probability law, continuous or discrete, has an associated frequency distribution, as illustrated in Figures A.3 and A.4. The distributions in those figures appear to be single-modal, symmetrical distributions. However, a frequency distribution can be multimodal or nonsymmetrical as well. Figures A.12 through A.14 are examples of such distributions. Figure A.12 shows a bimodal distribution; Figures A.13 and A.14 show

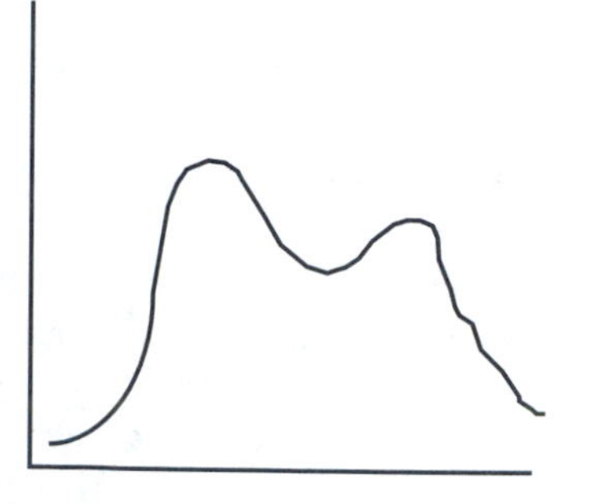

FIGURE A.12
Bimodal distribution.

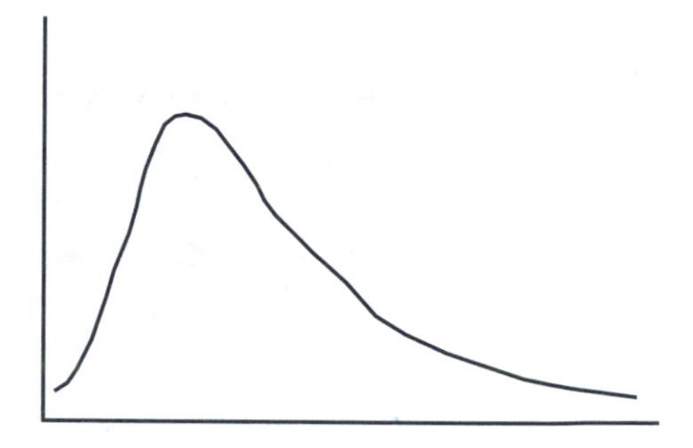

FIGURE A.13
Positively skewed distribution.

FIGURE A.14
Negatively skewed distribution.

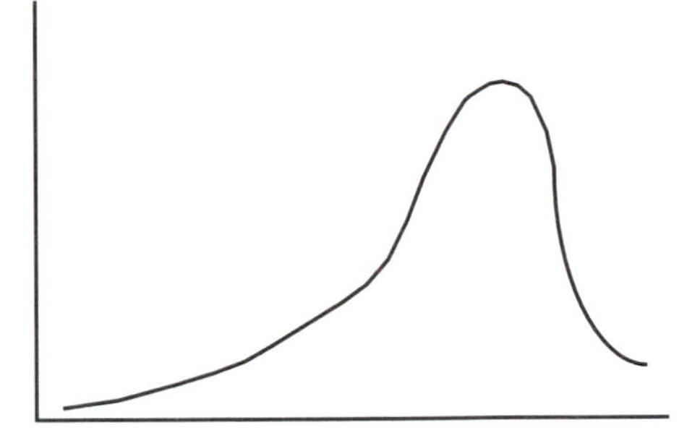

single-modal, nonsymmetrical distributions. The distribution in Figure A.13 skews to the right (positively skewed), and the distribution in Figure A.14 is negatively skewed (skews to the left).

Typical parameters for all types of distributions identify the center and spread of the distribution of possible outcomes. The center of a distribution is commonly identified by the mean, median, or, sometimes, by the mode for a single-modal distribution. The mean is the expected outcome or arithmetic average μ, which is $\mu = \int_{x=-\infty}^{\infty} xf(x)\,dx$ for a continuous distribution and $\mu = \sum_{x=0}^{\infty} xp(x)$ for a discrete distribution, where $f(x)$ and $p(x)$ are the probability density and mass functions, as defined earlier. The parameter μ is estimated from a set of randomly acquired data by the arithmetic mean $\bar{x} = (\Sigma x)/N$ of the data set, where N is the quantity of data items (measurements or tallies) being averaged. The median, $(x)_{0.5}$, is the 50th percentile point of the distribution, or that point such that half of the outcomes are expected to be above it and half are expectd to be below it. The median is estimated by the median of the data set, $\tilde{x} = x_{0.5}$. The mode of a distribution is its peak and is estimated from a data set by the mode, or most frequent outcome, $\hat{x}$, of the set of data.

In a single-modal, symmetrical distribution, the mean, median, and mode are equal to each other. However, in a positively skewed distribution, the mean is higher than the median, which is higher than the mode. The reverse is true of a negatively skewed distribution, as illustrated in Figure A.15.

The spread of a distribution is commonly identified by its **variance** or by the square root of the variance, called the **standard deviation.** The variance σ^2 of a distribution is its second moment about the mean, which is $\sigma^2 = \int_{x=-\infty}^{\infty} (x - \mu)^2 f(x)\,dx$ for a continuous distribution and $\sigma^2 = \sum_{x=0}^{\infty} (x - \mu)^2 p(x)$ for a discrete distribution. The standard deviation σ is the square root of the variance, and it is estimated from a set of randomly acquired data by the standard deviation estimator,

$$s = \sqrt{\frac{\Sigma(x - \bar{x})^2}{N - 1}} = \sqrt{\left(\frac{N}{N - 1}\right)\left(\frac{\Sigma x^2}{N} - \bar{x}^2\right)} = \sqrt{\frac{\Sigma x^2 - N\bar{x}^2}{N - 1}}$$

where N is the number of data points (measurements or tallies).

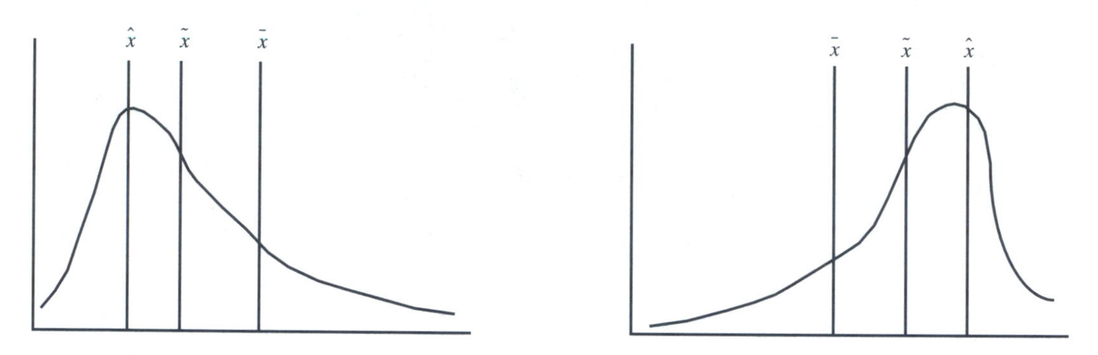

FIGURE A.15
Means, medians, and modes of positively and negatively skewed distributions.

Tables A.1 and A.2 present pertinent information on six discrete and six continuous distributions, respectively. The tables identify the probability laws and their parameters, provide the probability mass functions for the discrete laws and the probability density functions for the continuous laws, present solutions for the mean and standard deviation for each probability law, and supply examples of typical applications of each of the laws. The examples presented in the table are from both the world of quality assurance and the world in general. We conclude this section by describing some of the probability laws that are most important to the field of reliability and restorability.

A.8 DISCRETE PROBABILITY LAWS

Discrete probability laws, those of Table A.1, are for the distributions of outcomes of tallied phenomena, that is, where something is being counted. Those most commonly seen in reliability and restorability analysis are the Bernoulli Trial, the Binomial Probability Law, and the Poisson Probability law.

A.8.1 The Bernoulli Trial

A Bernoulli Trial is a single experiment or happening that can produce either of two possible mutually exclusive and exhaustive outcomes, such as the single toss of a coin. The coin toss can yield either a head or a tail. If we let x be the outcome *head,* the probability mass function $p(x) = \phi$ is simply the probability that the single toss of the coin results in the outcome *head.* A Bernoulli Trial has one parameter, ϕ, which is the probability that a single coin toss will produce a head. The statistical analysis method suggests that we now gather data to estimate the parameter ϕ. This is done by tossing the coin many times and estimating ϕ as the number of heads divided by the number of tosses. For example, if we toss the coin 1000 times and get 492 heads and 508 tails, our estimate of ϕ is 0.492. This now raises a question regarding the adequacy of our sample of 1000 tosses, or, in other words, the *confidence* in our estimate of the parameter provided by the 1000 tosses. If we are not satisfied with the achieved confidence, we will have to toss some more coins to get an estimate of ϕ. The issue of confidence estimates

TABLE A.1
Discrete probability laws.

Law and Parameters	Probability Mass Function	Mean	Standard Deviation	General Applications	Quality Assurance Applications
Bernoulli $0 \le \phi \le 1$	$p(x) = \phi$ if $x = 1$ $p(x) = (1 - \phi)$ if $x = 0$ $p(x) = 0$ otherwise	ϕ	$\sqrt{\phi(1 - \phi)}$	What is the probability of landing a head in a single coin toss?	What is the probability of a given production lot being accepted?
Binomial $n = 0, 1, 2,\ldots,$ $0 \le \phi \le 1$	$p(x) = \binom{n}{x}\phi^x(1 - \phi)^{n-x}$ if $x = 0, 1, 2,\ldots,n$ $p(x) = 0$ otherwise	$n\phi$	$\sqrt{n\phi(1 - \phi)}$	What is the probability of getting exactly x heads out of n consecutive tosses if ϕ is the probability of a head in a single toss?	What is the probability of getting exactly x defective units in a single sample of n units from an infinitely large production lot with defective rate ϕ?
Poisson $\mu > 0$	$p(x) = e^{-\mu}\left(\frac{\mu^x}{x!}\right)$ if $x = 0,1,2,\ldots$ $p(x) = 0$ otherwise	μ	$\sqrt{\mu}$	What is the probability of getting exactly x phone calls in a given hour if calls come in at a rate of μ calls per hour?	What is the probability of getting exactly x defects in a given sample of constant size from a production lot with a constant defect rate of μ per sample?
Hypergeometric $N = 1,2,3,\ldots,$ $n = 1,2,3,\ldots,$ $0 \le \phi \le 1$	$p(x) = \frac{\binom{N\phi}{x}\binom{N(1-\phi)}{(n-x)}}{\binom{N}{n}}$ if $x = 0,1,2,\ldots,n$ $p(x) = 0$ otherwise	$n\phi$	$\sqrt{n\phi(1 - \phi)}\sqrt{\frac{N - n}{N - 1}}$	What is the probability of being dealt a poker hand with four aces? $N = 52$ $x = 4$ $n = 5$ $\phi = 1/13$	What is the probability of getting exactly x defective units in a single sample of n units from a small production lot of N units with defective rate ϕ?
Geometric $0 \le \phi \le 1$	$p(x) = (\phi)(1 - \phi)^{x-1}$ if $x = 0,1,2,\ldots$ $p(x) = 0$ otherwise	$\frac{1}{\phi}$	$\frac{\sqrt{1 - \phi}}{\phi}$	What is the probability that you will have to toss a coin exactly x times to get your first head if ϕ is the probability of a head in a single toss?	What is the probability that exactly $x - 1$ units from an infinitely large production lot will be accepted before reaching the first reject (i.e., the xth unit inspected will be the first reject) if the reject rate is ϕ?
Negative Binomial $r > 0,$ $0 \le \phi \le 1$	$p(x) = \binom{r+x-1}{x}\phi^r(1 - \phi)^x$ if $x = 0,1,2,\ldots$ $p(x) = 0$ otherwise	$\frac{r - r\phi}{\phi}$	$\frac{\sqrt{r - r\phi}}{\phi}$	What is the probability that you will toss exactly x heads before you get r tails if ϕ is the probability of getting a tail in a single toss?	What is the probability of getting exactly x defective items from an infinitely large production lot before getting exactly r acceptable items, where $1 - \phi$ is the defective rate?

TABLE A.2
Continuous probability laws.

Law and Parameters	Probability Density Function	Mean	Standard Deviation	General Applications	Quality Assurance Applications
Uniform $-\infty < a < b < \infty$	$f(x) = \frac{1}{b-a}$ for $a < x < b$ $f(x) = 0$ otherwise	$\frac{a+b}{2}$	$\frac{b-a}{2\sqrt{3}}$	Roulette wheel: What is the probability of having a freely spun wheel land between a and b degrees from its original position?	What is the average undetected outage of an item whose performance is monitored every $(b - a)$ days?
Normal $-\infty < \mu < \infty$, $\sigma > 0$	$f(x) = \frac{1}{\sigma\sqrt{2\pi}} e^{-(x-\mu)^2/2\sigma^2}$	μ	σ	What is the probability that the next person who walks through the door will be between A and B inches tall?	What is the probability that a part inspected for a given measured criterion will be within its specification limits?
Exponential $\mu > 0$	$f(x) = \mu e^{-\mu x}$ if $x > 0$ $f(x) = 0$ otherwise	$\frac{1}{\mu}$	$\frac{1}{\mu}$	What is the probability that you will receive your next phone call between A and B minutes from now if calls arrive at a constant rate of μ per minute?	What is the probability that a system will experience its first failure in the time interval between A and B hours if it fails at a constant rate of μ per hour?
Gamma $\mu > 0$, $r > 0$	$f(x) = \frac{\mu}{\Gamma(r)}(\mu x)^{r-1}e^{-\mu x}$ if $x > 0$ $f(x) = 0$ otherwise	$\frac{r}{\mu}$	$\frac{\sqrt{r}}{\mu}$	What is the probability that you will receive your rth phone call between A and B minutes from now if calls arrive at a constant rate of μ per minute?	What is the probability that a system will experience its rth failure in the time interval between A and B hours if it fails at a constant rate of μ per hour?
Weibull $\mu > 0$ (scale), $\beta > 0$ (shape)	$f(x) = \beta\mu^{\beta}x^{\beta-1}e^{-(\mu x)^{\beta}}$ if $x > 0$ $f(x) = 0$ otherwise	$\frac{\beta}{\mu}$	$\frac{\beta}{\mu}$	Same as the exponential law, but the arrival rate is not necessarily constant with time: If $\beta < 1$, the rate decreases. If $\beta > 1$, the rate increases. If $\beta = 1$, the rate is constant.	Same as the exponential law, but the failure rate is not necessarily constant with time: If $\beta < 1$, the failure rate decreases. If $\beta > 1$, the failure rate increases. If $\beta = 1$, the failure rate is constant.
Chi-Square $\chi^2, \sigma > 0$, $n = 1,2,\ldots$	$f(x) = \frac{x^{(n/2)-1}e^{-x/2\sigma}}{2^{n/2}\sigma^n\Gamma(n/2)}$ if $x > 0$ $f(x) = 0$ otherwise	n	$\sqrt{2n}$	What is the probability that a given set of data is a random sample from a normally distributed population with variance σ^2?	

is discussed in further detail in the text. Another approach is to assume that we have a fair coin—that is, one that is equally weighted to turn up a head with the same frequency that it turns up a tail, in which case we assume that $\phi = 0.500$. An example of a Bernoulli Trial more pertinent to this text is the success probability or the fail probability of a one-shot device.

A.8.2 The Binomial Probability Law

A series of Bernoulli Trials produces a distribution of outcomes governed by the Binomial Probability Law. An example of such a series of Bernoulli trials is the simultaneous tossing of a number of coins. Using the example introduced earlier, the tossing of two coins (see Figures A.2 and A.7), the Binomial Law applies. The Binomial Law has two parameters, the sample size, n, and the probability of an outcome in a single Bernoulli Trial, ϕ. In this example, $n = 2$ (two tosses), and if x is the event *head* and we assume fairly weighted coins, $\phi = 0.500$. We can use the probability mass function $p(x) = \binom{x}{n}\phi^x(1 - \phi)^{n-x}$ to find the probability of getting 0, 1, or 2 heads, that is, the probability that $x = 1$, 2, or 3:

$$
\begin{aligned}
p(0) &= \tbinom{2}{0}(0.5)^0(1 - 0.5)^{2-0} = 1(0.5)^0(0.5)^2 = 0.25 \\
p(1) &= \tbinom{2}{1}(0.5)^1(1 - 0.5)^{2-1} = 2(0.5)(0.5) = 0.50 \\
p(2) &= \tbinom{2}{2}(0.5)^2(1 - 0.5)^{2-2} = 1(0.5)^2(0.5)^0 = 0.25
\end{aligned}
$$

As another example, suppose we are to test 100 one-shot devices and we wish to predict the probability of having no more than one failure. Suppose that we have sufficient data to estimate the fail probability of a single such device (a Bernoulli Trial) at 0.050. The Binomial Law applies with parameters $n = 100$ and $\phi = 0.050$. Hence, the probability mass function is

$$p(x) = \tbinom{100}{x}(0.050)^x(1 - 0.050)^{100-x}$$

The probability mass function can be used to predict the probability that the number of failures will be no more than one (that is, $x \leq 1$):

$$
\begin{aligned}
p(x \leq 1) &= p(x = 0) + p(x = 1) \\
&= \tbinom{100}{0}(0.050)^0(0.950)^{100} + \tbinom{100}{1}(0.050)^1(0.950)^{99} \\
&= 0.950^{100} + 100(0.050)(0.950)^{99} = 0.037
\end{aligned}
$$

A.8.3 The Poisson Probability Law

When we count features within our sample, the distribution of outcomes very likely is governed by the Poisson Probability Law with a single parameter, μ. The parameter μ is the quantity that you expect to get, on the average, in your sample. As an example of an application of the Poisson Probability Law, suppose we wish to predict the probability of getting no more than 25 phone calls in a given hour. To estimate the parameter μ, we have to get data on the rate of phone calls actually arriving. By gathering such data for a number of such 1-h periods, the average number of phone calls per hour is an estimate of μ. How good an estimate it is—that is, the confidence in the estimate—is discussed in the text. Suppose now that we have an estimate of 20 calls per hour. We can predict our probability from the probability mass function for the Poisson Probability law: $p(x) = e^{-\mu}(\mu^x/x!)$. Using our estimate $\mu = 20$ and solving for $p(x \leq 25)$,

$$p(x \leq 25) = \sum_{x=0}^{25} e^{-20}\left(\frac{20^x}{x!}\right) = e^{-20}\sum_{x=0}^{25}\frac{20^x}{x!} = 0.888$$

This result can also be derived from Table B.9 (Appendix B). All the probability distribution tables in Appendix B provide cumulative distribution functions $F(A) = P(x \leq A)$. So, if we look up the value corresponding to $\mu = 20$ and $A = 25$, we will find the probability that x is 25 or less where the parameter μ is 20. The table value is the same as our computed probability, 0.888.

Suppose we want to use the table to obtain the probability that x is a distinct quantity, say $p(x = 25)$. Recognizing that the tables provide only cumulative probabilities, we shall have to look up the probabilities corresponding to both $A = 25$ and $A = 24$ and subtract: $p(x = 25) = p(x \leq 25) - p(x \leq 24) = 0.888 - 0.843 = 0.045$. This result can also be computed through the probability mass function $p(x) = e^{-\mu}(\mu^x/x!) = e^{-20}(20^{25}/25!) = 0.045$.

To use an example more pertinent to reliability analysis, suppose a system that has a malfunction at an average rate of once every 100 h is put into operation for 30 days, running 24 h a day. What is the probability that in the 30 days of operation it experiences no more than 10 malfunctions? The distribution of malfunctions during a randomly selected 30-day period is expected to be governed by the Poisson Probability law with parameter μ. The value of μ is the quantity of malfunctions expected, on the average, during a 30-day (or $30 \times 24 = 720$-h) period. If a malfunction occurs on the average of once every 100 h, we expect an average of 7.2 in a 720-h period. Using Table B.9, with $\mu = 7.2$ and $A = 10$, $p(x \leq 10) = 0.887$. This can also be computed from the probability mass function for the Poisson Probability Law with $\mu = 7.2$: $p(x \leq 10) = \sum_{x=0}^{10} e^{-7.2}\frac{7.2^x}{x!} = 0.887$.

A.9 CONTINUOUS PROBABILITY LAWS

Continuous probability laws, those of Table A.2, are for the distributions of outcomes of measured phenomena. Those most commonly seen in reliability and restorability analysis are the Normal, Exponential, Weibull, and Chi-Square Distributions.

A.9.1 The Normal Probability Law

The Normal Probability Law governs the distributions of measured outcomes of many types of commonly observed natural phenomena. We can generally state that a normal distribution of measured outcomes is expected whenever there is no reason to believe otherwise, that is, when our observed phenomenon is untampered with or otherwise constrained. The Normal Probability Law has two parameters, the mean, μ, and the standard deviation, σ. Both parameters are estimated from data as described before: μ is estimated from $\bar{x} = \Sigma x/N$ and σ is estimated from

$$s = \sqrt{\frac{\Sigma(x - \bar{x})^2}{N - 1}} = \sqrt{\left(\frac{N}{N-1}\right)\left(\frac{\Sigma x^2}{N} - \bar{x}^2\right)} = \sqrt{\frac{\Sigma x^2 - N\bar{x}^2}{N - 1}}$$

As an example of an application of the Normal Distribution, suppose we know that a certain type of component has an average wearout life of 10,000 h and we have sufficient data to estimate the standard deviation at 750 h. What is the probability that one of these components will last for 8000 h without wearing out? In other words, we are looking for the probability that the wearout will take place beyond 8000 h (that the wearout time x is greater than 8000 h). This, of course, is a measured phenomenon (time), and there is no reason to believe that the distribution of wearout times is not *normal.* Hence, using the probability density function for the Normal Probability Law, the probability that the component wearout time will be beyond 8000 h can be predicted:

$$P(x > 8000) = \int_{x=8000}^{\infty} \frac{1}{\sigma\sqrt{2\pi}} e^{-(x-\mu)^2/2\sigma^2}$$

Table B.1, which is a table of cumulative probabilities for the Normal Probability Law, actually performs this integration for us. But Table B.1 will enable us to solve $P(x < 8000)$, and we can reach our desired result by subtracting from 1.0.

Table B.1 provides solutions for

$$P(x < \mu + z\sigma) = \int_{x=-\infty}^{\mu=z\sigma} \frac{1}{\sigma\sqrt{2\pi}} e^{-(x-\mu)^2/2\sigma^2}$$

To solve for $P(x < A)$ through Table B.1, we have to first set $A = \mu + z\sigma$; then solve for z in terms of A, μ, and σ; and, finally, look up the cumulative probability in the table that corresponds to the z value. In our example, we have estimated the parameters to be $\mu = 10{,}000$ h and $\sigma = 750$ h. To solve $P(x < 8000 \text{ h})$, we first set $8000 \text{ h} = \mu + z_\sigma = 10{,}000 \text{ h} + (750 \text{ h})z$. Then,

$$z = \frac{8000 \text{ h} - 10{,}000 \text{ h}}{750 \text{ h}} = -2.67$$

In Table B.1, the cumulative probability corresponding to $z = -2.67$ is 0.0038, which is $P(x < 8000 \text{ h})$. So, our prediction is $P(x > 8000 \text{ h}) = 1 - 0.0038 = 0.9962$. That is, we predict that 99.62% of these components will not wear out within 8000 h.

The Lognormal Distribution, used for making restorability predictions, is actually a logarithmic transformation of the Normal Distribution. As such, it is a positively skewed distribution that fits time-to-repair phenomena and is constrained by the fact that the most frequent malfunctions are most familiar to repair personnel and tend to be repaired quickly, whereas malfunctions that seldom occur are unfamiliar and tend to cause interminably long repair times. The derivation, suitability, and application of the Lognormal Distribution are detailed in the text with time-to-restore predictions.

A.9.2 The Exponential Probability Law

The Exponential Probability Law is the continuous cousin to the Poisson Probability Law. In discussing the Poisson Law we used an example in which an operating system has a malfunction at an average rate of once every 100 h and we asked the probability of having no more than a given number of malfunctions during a specified operating period (720 h in our example). In the Poisson, we sample a specified time period and count the number of malfunctions, but with the Exponential Law we do it the other way around: Our sample is the occurrence of the next malfunction, and we are measuring the time to that next occurrence. The Exponential Probability Law would apply if, for the system with a malfunction rate averaging once every 100 h, we want to determine the probability that the first malfunction occurs between 100 and 150 h into the operation. This can be computed by integrating the probability density function with parameter $\mu = 0.01$ malfunctions per hour:

$$P(100 < x < 150) = \mu \int_{x=100}^{150} e^{-\mu x}\, dx = 0.01 \int_{x=100}^{150} e^{-0.01x}\, dx$$

There are no tables for the Exponential Probability Law in this text, because graphical solutions are used within the text for exponential distribution solutions. This distribution is applicable to time-to-failure predictions when we are dealing with constant failure rates.

A.9.3 The Weibull Probability Law

The Weibull Probability Law is a more generalized form of the Exponential Law. It is used for predicting time to failure when we are dealing with a failure rate that is not necessarily constant. In comparing the Weibull Probability Law with the Exponential Law in Table A.2, notice that the Weibull has an additional parameter β. When $\beta < 1$, we are dealing with a failure rate that decreases with time; if $\beta > 1$, we are dealing with an increasing failure rate; and if $\beta = 1$, the failure rate is constant. Notice that if we set $\beta = 1$ in the Weibull Probability Density Function we get the density function for the Exponential Law.

A.9.4 The Chi-Square Probability Law

The Chi-Square Probability Law has a variety of applications throughout the text, from the determination of confidence estimates of parameters, as discussed before, to goodness-of-fit tests for ascertaining the suitability of an assumed probability law. Chi-Square Distribution probabilities are determined through Table B.2, and the use of this probability is discussed with its applications in the text.

A.10 GOODNESS OF FIT

The first step in the statistical analysis method presented at the start of this appendix is the selection of the probability law that adequately describes the distribution of outcomes of the phenomenon for which we are making predictions or decision. Generally speaking, we can choose the appropriate probability law from the descriptions of the laws' applications in the preceding paragraphs and in Tables A.1 and A.2. But we may want some additional reassurance that the selected probability law is, indeed, appropriate. This additional reassurance is accomplished by fitting a random set of actual data to the expected distribution suggested by the assumed probability law. If there is a *significant* difference between the distribution of the actual data and the distribution expected by the assumed law, we may want to reject our assumption and try to find a better-fitting probability law. Remembering that we had good reason to select the assumed probability law in the first place, we are willing to abandon this assumption only upon overwhelming evidence that there is a bad fit.

This text introduces three types of tests for goodness of fit of the assumed probability law: a graphical fit, the chi-square test, and the Kolmogorov-Smirnov test. All three are of the variety called *significance tests,* suggesting that we are willing to abandon our original assumption only if we see a *significant difference* between the expected and observed distribution of outcomes. The graphical-fit techniques are presented in Chapter 6. The two analytical techniques, chi-square and Kolmogorov-Smirnov, which are used in Chapter 5, are described in the paragraphs that follow.

A.10.1 The Chi-Square Test for Goodness of Fit

Consider a phenomenon that has a possibility of k different outcomes, $E_1, E_2, \ldots, E_k$. We can assume that the phenomenon is governed by some probability law. For n trials, the probability law assigns an expected frequency e_i to each event E_i. Where P_i is the probability of the outcome E_i, $e_i = nP_i$. So,

$e_1 = nP_1$ is the expected frequency for the event E_1
$e_2 = nP_2$ is the expected frequency for the event E_2
.
.
.
$e_k = nP_k$ is the expected frequency for the event E_k

For **discrete distributions,**

$$P_i = p_i$$

p_i being the *probability mass function.*

For **continuous distributions,** where the event E_i is an outcome between a and b,

$$P_i = P(a < x < b) = \int_a^b f(x)\,dx$$

$f(x)$ being the *probability density function.*

If a given test has n trials with k possible outcomes and

$$o_1 = \text{the observed frequency of the event } E_1$$
$$o_2 = \text{the observed frequency of the event } E_2$$
$$\vdots$$
$$o_k = \text{the observed frequency of the event } E_k$$

then a measure of the discrepancy of the assumption (that the phenomenon is governed by the assumed probability law) is the statistic

$$\chi^2 = \frac{(o_1 - e_1)^2}{e_1} + \frac{(o_2 - e_2)^2}{e_2} + \cdots + \frac{(o_k - e_k)^2}{e_k} = \sum_{j=1}^{k} \frac{(o_j - e_j)^2}{e_j}$$

Notice that $\sum_{j=1}^{k} o_j = \sum_{j=1}^{k} e_j = n$.

A theorem in probability states that *as n approaches infinity, the sampling distribution of the statistic* χ^2 *approaches the Chi-Square Distribution with f degrees of freedom, where*

$f = k - 1$	if the $e_i = nP_i$ values can be computed without having to estimate parameters from the sampling statistics
$f = k - 1 - m$	if the $e_i = nP_i$ values can be computed only by estimating m parameters from the sample statistics

Note: The 1 is subtracted from k because of the constraint $\sum_{j=1}^{k} o_j = \sum_{j-1}^{k} e_j = n$ (if $k - 1$ of the frequencies are known, we can determine the other one). Each parameter estimated from the sample statistics adds another constraint.

A.10.2 Application of the Chi-Square Test

The **chi-square test** is performed as follows:

1. List all the possible outcomes of the phenomenon you are studying.

2. Establish a hypothesis: The frequencies of occurrence of the various outcomes of the phenomenon you are studying are governed by a given probability law.

3. Compute the expected frequencies e_i for the possible outcomes E_i based on the hypothesized probability law.

4. Conduct the experiment and record the observed frequencies o_i for each outcome E_i.

5. Compute the statistic $\chi^2 = \sum_{j=1}^{k} \frac{(o_j - e_j)^2}{e_j}$.

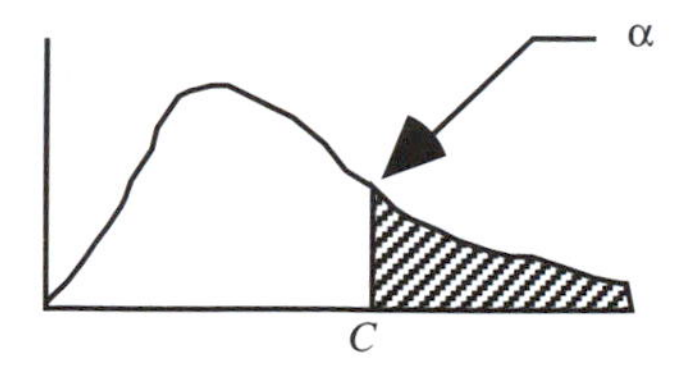

FIGURE A.16
The Chi-Square Test.

6. Select the significance level α. Usually, a significance level of 5% is used. A significance level of 10% is used for a stricter test, and 1% is used for a more lenient test.

7. Determine the degrees of freedom $f = k - 1 - m$, where k is the number of possible outcomes and m is the number of parameters you have to estimate from the sample statistics.

8. Determine the critical value $(\chi^2)_{1-\alpha,f}$ from Table B.2.

Analytically, what you are looking up in the table is a critical value $C = (\chi^2)_{1-\alpha,f}$ (see Figure A.16), such that

$$\alpha = \int_C^{\infty} f(\chi^2)\, d\chi^2$$

where

$$f(\chi^2) = \frac{1}{2\Gamma(f/2)}\left(\frac{\chi^2}{2}\right)^{f/2-1} e^{-(\chi^2/2)}$$

9. Compare the chi-square statistic determined in Step 5 with the critical chi-square value determined in Step 8.

10. If $\chi^2 \leq (\chi^2)_{1-\alpha,f}$, accept the hypothesized probability law as a good fit. Otherwise, reject it. A reject decision means that the difference between the distribution of the observed data and the distribution suggested by the hypothesized probability law is *significant* at the α significance level.

EXAMPLE A.1 Suppose five pennies are tossed 1000 times.

38 times there are 0 heads.

144 times there is 1 head.

342 times there are 2 heads.

287 times there are 3 heads.

164 times there are 4 heads.

25 times there are 5 heads.

At the 5% significance level, use the chi-square test to ascertain the fit of these data to the Binomial Probability Law.

If the Binomial fits, then the probability of getting x heads in a single toss of five pennies is

$$P(x) = \binom{5}{x}\phi^x(1 - \phi)^{5-x}$$

There is one parameter to estimate from the data, the parameter ϕ, which can be estimated by

$$\frac{\text{no. of heads}}{5 \times 1000} = \frac{144 + 2(342) + 3(287) + 4(164) + 5(25)}{5000}$$
$$= \frac{2470}{5000} = 0.494$$

Therefore, $m = 1$. So, the probability of tossing x heads is

$$P(x) = \binom{5}{x}(0.494)^x(0.506)^{5-x}$$

and the probability of event E_1, that $x = 0$ heads, is

$$P_1 = \binom{5}{0}(0.494)^0(0.506)^5$$

the probability of event E_2, that $x = 1$ head, is

$$P_2 = \binom{5}{1}(0.494)^1(0.506)^4$$

the probability of event E_3, that $x = 2$ heads, is

$$P_3 = \binom{5}{2}(0.494)^2(0.506)^3$$

the probability of event E_4, that $x = 3$ heads, is

$$P_4 = \binom{5}{3}(0.494)^3(0.506)^2$$

the probability of event E_5, that $x = 4$ heads, is

$$P_5 = \binom{5}{4}(0.494)^4(0.506)^1$$

the probability of the event E_6, that $x = 5$ heads, is

$$P_6 = \binom{5}{5}(0.494)^5(0.506)^0$$

There are six possible outcomes, so $k = 6$.

The expected frequencies are $e_i = nP_i = 1000 \times P_i$, and the observed frequencies o_i are as given. The computations for the chi-square statistic can be completed in tabular form as follows.

Event E_i	E_1	E_2	E_3	E_4	E_5	E_6	Total
x = no. of heads	$x = 0$	$x = 1$	$x = 2$	$x = 3$	$x = 4$	$x = 5$	
Probability P_i	0.0332	0.1619	0.3162	0.3087	0.1507	0.0293	1.0000
Expected frequency $e_i = 1000P_i$	33.2	161.9	316.2	308.7	150.7	29.3	1000
Observed frequency o_i	38	144	342	287	164	25	1000

The chi-square statistic is

$$\chi^2 = \frac{(38 - 33.2)^2}{33.2} + \frac{(144 - 161.9)^2}{161.9} + \frac{(342 - 316.2)^2}{316.2} + \frac{(287 - 308.7)^2}{308.7} + \frac{(164 - 150.7)^2}{150.7} + \frac{(25 - 29.3)^2}{29.3} = 7.51$$

Looking up the critical chi-square value from Table B.2 for $\alpha = 0.05$ and $f = k - 1 - m = 6 - 1 - 1 = 4$ degrees of freedom,

$$(\chi^2)_{1-a,f} = (\chi^2)_{0.95,4} = 9.49$$

Because the chi-square statistic, 7.51, is less than the critical value, 9.49, our assumption is accepted at the 5% significance level. Table A.1 suggests that the binomial probability ought to be the one that governs the distribution of outcomes of the tossing of five coins. According to Step 4 of the statistical analysis method described in Section A.1, we decided to subject the Binomial assumption to a goodness-of-fit test. The purpose of this test was to see if there is a significant difference between the distribution of our observed data and the Binomial distribution. In this case the test indicated no significance, meaning that there is no reason to reject the binomial assumption. Had the test indicated significance, we would be wise to develop another mathematical model for making our predictions.

In this example, we used our data to estimate the parameter $\phi = 0.494$. Suppose we had decided to test whether the coins tossed were fair coins, with the distribution of outcomes governed by the Binomial distribution. In other words, suppose we just assumed that the parameter ϕ was 0.500. In that case, we would not have used the data at all to estimate a parameter. Then, $m = 0$, and

$$f = k - 1 - m = 6 - 1 - 0 = 5$$

Also, the probabilities are

$$P_i = P(x) = \binom{5}{x}(0.500)^x(0.500)^{5-x} = P(x) = \binom{5}{x}(0.500)^5 = 0.03125\binom{5}{x}$$

and

$$e_i = 1000P_i = 31.25\binom{5}{x}$$

So

$$e_1 = 31.25 \times 1 = 31.25, \qquad e_2 = 31.25 \times 5 = 156.25$$
$$e_3 = 31.25 \times 10 = 312.5, \qquad e_4 = 31.25 \times 10 = 312.5$$
$$e_5 = 31.25 \times 5 = 156.25, \qquad e_6 = 31.25 \times 1 = 31.25$$

As before, the chi-square statistic is

$$\chi^2 = \frac{(38 - 31.25)^2}{31.25} + \frac{(144 - 156.25)^2}{156.25} + \frac{(342 - 312.5)^2}{312.5} + \frac{(287 - 312.5)^2}{312.5} + \frac{(164 - 156.25)^2}{156.25} + \frac{(25 - 31.25)^2}{31.25} = 8.92$$

This time the critical value from Table B.2 is taken for $\alpha = 0.05$ and $f = 5$ degrees of freedom:

$$(\chi^2)_{1-a,f} = (\chi^2)_{0.95,5} = 11.1$$

Again, the chi-square statistic is less than the critical value, and there is no significance; that is, we have no reason to reject an assumption of fair coins and a binomial distribution of outcomes.

EXAMPLE A.2 Forty units are taken at random from a production run.

4 of them weigh 26 pounds.

8 of them weigh 27 pounds.

16 of them weigh 28 pounds.

10 of them weigh 29 pounds.

2 of them weigh 30 pounds.

Table A.2 suggests that we can assume the weight of these units to be distributed according to the Normal probability law. Use the chi-square test for goodness of fit at the 5% significance level to determine whether there is significant evidence that we cannot make that assumption.

There are two parameters to be estimated from the sampling statistics given: the mean, μ, and the standard deviation, σ. Hence, $m = 2$. This being a continuous phenomenon, the outcomes can be anywhere from minus to plus infinity. The data given, however, are rounded off to the nearest pound (meaning that the precision of the measurements, as recorded, is 0.5 lb). The data presented can then be classified into one of five outcomes: less than 26.5 lb, between 26.5 and 27.5 lb, between 27.5 and 28.5 lb, between 28.5 and 29.5 lb, and more than 29.5 lb. So, $k = 5$. Therefore, the number of degrees of freedom for the critical chi-square value is

$$f = k - 1 - m = 5 - 1 - 2 = 2$$

As indicated in Section A.1.9, the parameter μ is estimated by

$$\bar{x} = \frac{\sum x}{40} = \frac{1118}{40} = 27.95$$

and the parameter σ is estimated by

$$\sqrt{\frac{40}{39}}\sqrt{\frac{\sum x^2}{40} - 27.95^2} = 1.04$$

For any value A, $P(x < A) = P(x < \mu + z\sigma)$, where $P(x < A)$ can be determined in Table B.1, corresponding to z. Applying our estimates for μ and σ,

$$P(x < A) = P(x < 27.95 + 1.04z),$$

where $z = (A - 27.95)/1.04$.

To compute the probabilities P_i we must determine $P(x < A)$, where $A = 26.5$, 27.5, 28.5, and 29.5.

Where $x = 26.5$, $z = (26.5 - 27.95)/1.04 = -1.39$.

Where $x = 27.5$, $z = (27.5 - 27.95)/1.04 = -0.43$.

Where $x = 28.5$, $z = (28.5 - 27.95)/1.04 = +0.55$.

Where $x = 29.5$, $z = (29.5 - 27.95)/1.04 = +1.49$.

From Table B.1, corresponding to $z = -1.39, -0.43, +0.55, +1.49$, respectively, we can determine

$$P(x < 26.5) = 0.0823, \quad P(x < 27.5) = 0.3336, \quad P(x < 28.5) = 0.7088$$
$$P(x < 29.5) = 0.9319, \quad P(x > 29.5) = 1.0000 - 0.9319 = 0.0681$$

So,

$$\begin{aligned} P_1 &= P(x < 26.5) = 0.082 \\ P_2 &= P(26.5 < x < 27.5) = 0.3336 - 0.0823 = 0.252 \\ P_3 &= P(27.5 < x < 28.5) = 0.7088 - 0.3336 = 0.375 \\ P_4 &= P(28.5 < x < 29.5) = 0.9319 - 0.7088 = 0.223 \\ P_5 &= P(x > 29.5) = 0.068 \end{aligned}$$

The expected frequencies $e_i = nP_i = 40P_i$, and the observed frequencies o_i are as given. The computations for the chi-square statistic can be completed in tabular form as follows.

Event E_i	E_1	E_2	E_3	E_4	E_5	Total
x = weight	< 26.5 lb	26.5–27.5 lb	27.5–28.5 lb	28.5–29.5 lb	> 29.5 lb	
Probability P_i	0.082	0.252	0.375	0.223	0.068	1.0000
Expected frequency $e_i = 40P_i$	3.28	10.08	15.00	8.92	2.72	40
Observed frequency o_i	4	8	16	10	2	40

The chi-square statistic is

$$\chi^2 = \frac{(4 - 3.28)^2}{3.28} + \frac{(8 - 10.08)^2}{10.08} + \frac{(16 - 15.0)^2}{15.0} + \frac{(10 - 8.92)^2}{8.92} + \frac{(2 - 2.72)^2}{2.72} = 0.975$$

From Table B.2, the critical chi-square value is

$$(\chi^2)_{1-\alpha,f} = (\chi^2)_{0.95,2} = 5.99$$

Because the chi-square statistic is less than the critical value, we have no significant evidence that the Normal assumption is inappropriate.

Chapter 5 contains examples of the Chi-Square Goodness-of-Fit Test being used to determine if a distribution of failure occurrence is governed by the exponential probability law. We also learned in Chapter 5 that the exponential law applies if and only if the failure rate is constant with operating time. So the goodness-of-fit test is actually used in Chapter 5 to substantiate an assumption of constant failure rate.

A.10.3 The Kolmogorov-Smirnov Test for Goodness of Fit

The **Kolmogorov-Smirnov Test** is less common but simpler to use than the Chi-Square Test. It is based on the cumulative distribution function $F(x)$ and uses the theorem

$$P\{\max|F(x) - F_n(x)| > K - S_\alpha\} = \alpha$$

FIGURE A.17
The Kolmogorov-Smirnov Test.

where $F(x)$ = the cumulative distribution function for the assumed probability law

$F_n(x)$ = the empirical distribution function from a sample of data distributed by that law

and K-S_α = the K-S statistic at the α significance level.

A.10.4 Application of the Kolmogorov-Smirnov Test

The Kolmogorov-Smirnov Test is performed as follows:

1. Create a table, listing all possible outcomes of the phenomenon being tested.

2. Establish a hypothesis: The frequencies of occurrence of the various outcomes of the phenomenon you are studying is governed by a given probability law.
3. Compute and tabulate the observed cumulative fraction F_n at each of the listed outcomes.
4. Use the hypothesized probability law to compute and tabulate the expected cumulative probability $F(x)$ for each possible outcome.
5. For each possible outcome listed in the table, compute and tabulate $|F(x) - F_n(x)|$.
6. Determine the maximum of the $|F(x) - F_n(x)|$ values, that is, $\max|F(x) - F_n(x)|$.
7. Use Table B.3 to look up the critical Kolmogorov-Smirnov value K-S$_\alpha$ for the applicable number of trials n used for the test.
8. If $\max|F(x) - F_n(x)| >$ K-S$_\alpha$, then there is less than α probability that the data is governed by the assumed probability law. In other words, there is *significance* at the α significance level. So we reject our assumption. We accept our assumption, on the other hand, if $\max|F(x) - F_n(x)| <$ K-S$_\alpha$.

EXAMPLE A.3 Use the Kolmogorov-Smirnov Test to evaluate the Binomial Probability Law assumption for the data provided in Example A.1.

The probabilities P_i from the table in Example A.1 are used in the following table. The cumulative probabilities $F(x)$ for each outcome E_i are computed by adding the probabilities of all outcomes through E_i. The observed cumulative frequencies $F_n(x)$ are computed by dividing the cumulative frequencies for each outcome by 1000 (the number of tossings):

$$\text{for outcome } E_i, \qquad F_n(x) = (o_1 + o_2 + \cdots + o_i)/1000$$

Event E_i	x = No. of Heads	P_i	$F(x)$	$F_n(x)$	$\lvert F(x) - F_n(x)\rvert$
E_1	0	0.0332	0.0332	0.038	0.005
E_2	1	0.1619	0.1951	0.182	0.013
E_3	2	0.3162	0.5113	0.524	0.013
E_4	3	0.3087	0.8200	0.811	0.009
E_5	4	0.1507	0.9707	0.975	0.004
E_6	5	0.0293	1.0000	1.0000	0.000

The largest of all the $|F(x) - F_n(x)|$ values is

$$\max|F(x) - F_n(x)| = 0.013$$

From Table B.3, for $n = 1000$, K-S$_{0.05}$ $= 1.36/(1000)^{1/2} = 0.043$. Because $\max|F(x) - F_n(x)| <$ K-S$_{0.05}$, we can accept the assumption that the distribution of the tossing of these coins is governed by the Binomial Probability Law.

EXAMPLE A.4 Use the Kolmogorov-Smirnov test to evaluate the Normal Probability Law assumption for the data provided in Example A.2.

The probabilities P_i from the table in Example A.2 are used in the following table. The cumulative probabilities $F(x)$ for each outcome E_i are again computed by adding the probabilities of all outcomes through E_i. The observed cumulative frequencies $F_n(x)$ are computed by dividing the cumulative frequencies for each outcome by 40 (the number of measurements):

$$\text{for outcome } E_i, \qquad F_n(x) = (o_1 + o_2 + \cdots + o_i)/40$$

Event E_i	x = Weight (in lbs)	P_i	$F(x)$	$F_n(x)$	$\lvert F(x) - F_n(x)\rvert$
E_1	< 26.5	0.082	0.082	0.100	0.018
E_2	26.5–27.5	0.252	0.334	0.300	0.034
E_3	27.5–28.5	0.375	0.709	0.700	0.009
E_4	28.5–29.5	0.223	0.932	0.950	0.018
E_5	> 29.5	0.068	1.000	1.000	0.000

The largest of all the $|F(x) - F_n(x)|$ values is

$$\max|F(x) - F_n(x)| = 0.034$$

From Table B.3, for $n = 40$, $\text{K-S}_{0.05} = 0.210$

Because $\max|F(x) - F_n(x)| < \text{K-S}_{0.05}$, we can accept the assumption that the distribution of weights of these units is governed by the Normal Probability Law.

Chapter 6 presented linear regresson analysis, determination of the best-fit straight line (in the least squares sense) through a set of data points plotted on an x-versus-$f(x)$ graph. This type of plot, which is the basis of the graphical evaluation techniques of Chapter 6, is the *least squares test* for goodness of fit.

BIBLIOGRAPHY

1. W. Feller. 1962. *An Introduction to Probability Theory and its Applications,* Vol. I, 2d ed. New York: John Wiley.
2. W. Feller. 1966. *An Introduction to Probability Theory and its Applications,* Vol. II, 2d ed. New York: John Wiley.
3. A. Hald. 1952. *Statistical Theory with Engineering Applications.* New York: John Wiley.
4. E. Parzen. 1960. *Modern Probability Theory and its Applications.* New York: John Wiley.
5. E. Parzen. 1962. *Stochastic Processes.* New York: John Wiley.

PROBABILITY TABLES

TABLE B.1
Normal distribution.

$P[x < (\mu + z\sigma)]$

z	0.00	0.01	0.02	0.03	0.04	0.05	0.06	0.07	0.08	0.09
−3.0	0.0014	0.0013	0.0013	0.0012	0.0012	0.0011	0.0011	0.0011	0.0010	0.0010
−2.9	0.0019	0.0018	0.0017	0.0017	0.0016	0.0016	0.0015	0.0015	0.0014	0.0014
−2.8	0.0026	0.0025	0.0014	0.0023	0.0023	0.0022	0.0021	0.0021	0.0020	0.0019
−2.7	0.0035	0.0034	0.0033	0.0032	0.0031	0.0030	0.0029	0.0028	0.0027	0.0026
−2.6	0.0047	0.0045	0.0044	0.0043	0.0041	0.0040	0.0039	0.0038	0.0037	0.0036
−2.5	0.0062	0.0060	0.0059	0.0057	0.0055	0.0054	0.0052	0.0051	0.0049	0.0048
−2.4	0.0082	0.0080	0.0078	0.0075	0.0073	0.0071	0.0069	0.0068	0.0066	0.0064
−2.3	0.0107	0.0104	0.0102	0.0099	0.0096	0.0094	0.0091	0.0089	0.0087	0.0084
−2.2	0.0139	0.0136	0.0132	0.0129	0.0125	0.0122	0.0119	0.0116	0.0113	0.0110
−2.1	0.0179	0.0174	0.0170	0.0166	0.0162	0.0158	0.0154	0.0150	0.0146	0.0143
−2.0	0.0228	0.0222	0.0217	0.0212	0.0207	0.0202	0.0197	0.0192	0.0188	0.0183
−1.9	0.0287	0.0281	0.0274	0.0268	0.0262	0.0256	0.0250	0.0244	0.0239	0.0233
−1.8	0.0359	0.0351	0.0344	0.0336	0.0329	0.0322	0.0314	0.0307	0.0301	0.0294
−1.7	0.0446	0.0436	0.0427	0.0418	0.0409	0.0401	0.0392	0.0384	0.0375	0.0367
−1.6	0.0548	0.0537	0.0526	0.0516	0.0505	0.0495	0.0485	0.0475	0.0465	0.0455
−1.5	0.0668	0.0655	0.0643	0.0630	0.0618	0.0606	0.0594	0.0582	0.0574	0.0559
−1.4	0.0808	0.0793	0.0778	0.0764	0.0749	0.0735	0.0721	0.0708	0.0694	0.0681
−1.3	0.0968	0.0951	0.0934	0.0918	0.0901	0.0885	0.0869	0.0853	0.0838	0.0823
−1.2	0.1151	0.1131	0.1112	0.1093	0.1075	0.1057	0.1038	0.1020	0.1003	0.0985
−1.1	0.1357	0.1335	0.1314	0.1292	0.1271	0.1251	0.1230	0.1210	0.1190	0.1170
−1.0	0.1587	0.1562	0.1539	0.1515	0.1492	0.1469	0.1446	0.1423	0.1401	0.1379
−0.9	0.1841	0.1814	0.1788	0.1762	0.1736	0.1711	0.1685	0.1660	0.1635	0.1611
−0.8	0.2119	0.2090	0.2061	0.2033	0.2005	0.1977	0.1949	0.1922	0.1894	0.1867
−0.7	0.2420	0.2389	0.2358	0.2327	0.2297	0.2266	0.2236	0.2207	0.2177	0.2148
−0.6	0.2743	0.2709	0.2676	0.2643	0.2611	0.2578	0.2546	0.2514	0.2483	0.2451
−0.5	0.3085	0.3050	0.3015	0.2981	0.2946	0.2912	0.2877	0.2843	0.2810	0.2776
−0.4	0.3446	0.3409	0.3372	0.3336	0.3300	0.3264	0.3228	0.3192	0.3156	0.3121
−0.3	0.3821	0.3783	0.3745	0.3707	0.3669	0.3632	0.3594	0.3557	0.3520	0.3483
−0.2	0.4207	0.4168	0.4129	0.4090	0.4052	0.4013	0.3974	0.3936	0.3897	0.3859
−0.1	0.4602	0.4562	0.4522	0.4483	0.4443	0.4404	0.4364	0.4325	0.4286	0.4247
−0.0	0.5000	0.4960	0.4920	0.4880	0.4840	0.4801	0.4761	0.4721	0.4681	0.4641
+0.0	0.5000	0.5040	0.5080	0.5120	0.5160	0.5199	0.5239	0.5279	0.5319	0.5359
+0.1	0.5398	0.5438	0.5478	0.5517	0.5557	0.5596	0.5636	0.5675	0.5714	0.5753
+0.2	0.5793	0.5832	0.5871	0.5910	0.5948	0.5987	0.6026	0.6064	0.6103	0.6141
+0.3	0.6179	0.6217	0.6255	0.6293	0.6331	0.6368	0.6406	0.6443	0.6480	0.6517
+0.4	0.6554	0.6591	0.6628	0.6664	0.6700	0.6736	0.6772	0.6808	0.6844	0.6879
+0.5	0.6915	0.6950	0.6985	0.7019	0.7054	0.7088	0.7123	0.7157	0.7190	0.7224
+0.6	0.7257	0.7291	0.7324	0.7357	0.7389	0.7422	0.7454	0.7486	0.7517	0.7549
+0.7	0.7580	0.7611	0.7642	0.7673	0.7704	0.7734	0.7764	0.7794	0.7823	0.7852
+0.8	0.7881	0.7910	0.7939	0.7967	0.7995	0.8023	0.8051	0.8079	0.8106	0.8133
+0.9	0.8159	0.8186	0.8212	0.8238	0.8264	0.8289	0.8315	0.8340	0.8365	0.8389
+1.0	0.8413	0.8438	0.8461	0.8485	0.8508	0.8531	0.8554	0.8577	0.8599	0.8621
+1.1	0.8643	0.8665	0.8686	0.8708	0.8729	0.8749	0.8770	0.8790	0.8810	0.8830
+1.2	0.8849	0.8869	0.8888	0.8907	0.8925	0.8944	0.8962	0.8980	0.8997	0.9015
+1.3	0.9032	0.9049	0.9066	0.9082	0.9099	0.9115	0.9131	0.9147	0.9162	0.9177
+1.4	0.9192	0.9207	0.9222	0.9236	0.9251	0.9265	0.9279	0.9292	0.9306	0.9319
+1.5	0.9332	0.9345	0.9357	0.9370	0.9382	0.9394	0.9406	0.9418	0.9429	0.9441
+1.6	0.9452	0.9463	0.9474	0.9484	0.9495	0.9505	0.9515	0.9525	0.9535	0.9545
+1.7	0.9554	0.9564	0.9573	0.9582	0.9591	0.9599	0.9608	0.9616	0.9625	0.9633
+1.8	0.9641	0.9649	0.9656	0.9664	0.9671	0.9678	0.9686	0.9693	0.9699	0.9706
+1.9	0.9713	0.9719	0.9726	0.9732	0.9738	0.9744	0.9750	0.9756	0.9761	0.9767
+2.0	0.9773	0.9778	0.9783	0.9788	0.9793	0.9798	0.9803	0.9808	0.9812	0.9817
+2.1	0.9821	0.9826	0.9830	0.9834	0.9838	0.9842	0.9846	0.9850	0.9854	0.9857
+2.2	0.9861	0.9864	0.9868	0.9871	0.9875	0.9878	0.9881	0.9884	0.9887	0.9890
+2.3	0.9893	0.9896	0.9898	0.9901	0.9904	0.9906	0.9909	0.9911	0.9913	0.9916
+2.4	0.9918	0.9920	0.9922	0.9925	0.9927	0.9929	0.9931	0.9932	0.9934	0.9936
+2.5	0.9938	0.9940	0.9941	0.9943	0.9945	0.9946	0.9948	0.9949	0.9951	0.9952
+2.6	0.9953	0.9955	0.9956	0.9957	0.9959	0.9960	0.9961	0.9962	0.9963	0.9964
+2.7	0.9965	0.9966	0.9967	0.9968	0.9968	0.9970	0.9971	0.9972	0.9973	0.9974
+2.8	0.9974	0.9975	0.9976	0.9977	0.9977	0.9978	0.9979	0.9979	0.9980	0.9981
+2.9	0.9981	0.9982	0.9983	0.9983	0.9984	0.9984	0.9985	0.9985	0.9986	0.9986
+3.0	0.9987	0.9987	0.9987	0.9988	0.9988	0.9989	0.9989	0.9989	0.9990	0.9990

TABLE B.2
Chi-square distribution.

Value χ^2 such that $P(x < \chi^2) = \alpha$

Degrees of Freedom μ:	α: 0.005	0.010	0.025	0.05	0.10	0.20	0.30	0.40	0.50	0.60	0.70	0.80	0.90	0.95	0.975	0.990	0.995
1	0.0	0.0	0.0	0.0	0.0	0.1	0.1	0.3	0.5	0.7	1.1	1.6	2.7	3.8	5.0	6.6	7.9
2	0.0	0.0	0.1	0.1	0.2	0.4	0.7	1.0	1.4	1.8	2.4	3.2	4.6	6.0	7.4	9.2	10.6
3	0.1	0.1	0.2	0.4	0.6	1.0	1.4	1.9	2.4	3.0	3.7	4.6	6.3	7.8	9.4	11.3	12.8
4	0.2	0.3	0.5	0.7	1.1	1.7	2.2	2.8	3.4	4.0	4.9	6.0	7.8	9.5	11.1	13.3	14.9
5	0.4	0.6	0.8	1.2	1.6	2.3	3.0	3.7	4.4	5.1	6.1	7.3	9.2	11.1	12.8	15.1	16.7
6	0.7	0.9	1.2	1.6	2.2	3.1	3.8	4.6	5.4	6.2	7.2	8.6	10.6	12.6	14.4	16.8	18.5
7	1.0	1.2	1.7	2.2	2.8	3.8	4.7	5.5	6.4	7.3	8.4	9.8	12.0	14.1	16.0	18.5	20.3
8	1.3	1.7	2.2	2.7	3.5	4.6	5.5	6.4	7.3	8.4	9.5	11.0	13.4	15.5	17.5	20.1	22.0
9	1.7	2.1	2.7	3.3	4.2	5.4	6.4	7.4	8.3	9.4	10.7	12.2	14.7	16.9	19.0	21.7	23.6
10	2.2	2.6	3.3	3.9	4.9	6.2	7.3	8.3	9.3	10.5	11.8	13.4	16.0	18.3	20.5	23.2	25.2
11	2.6	3.1	3.8	4.6	5.6	7.0	8.2	9.2	10.3	11.5	12.9	14.6	17.3	19.7	21.9	24.7	26.8
12	3.1	3.6	4.4	5.2	6.3	7.8	9.0	10.2	11.3	12.6	14.0	15.8	18.5	21.0	23.3	26.2	28.3
13	3.6	4.1	5.0	5.9	7.0	8.6	9.9	11.1	12.3	13.6	15.1	17.0	19.8	22.4	24.7	27.7	29.8
14	4.1	4.7	5.6	6.6	7.8	9.5	10.8	12.1	13.3	14.7	16.2	18.2	21.1	23.7	26.1	29.1	31.3
15	4.6	5.2	6.3	7.3	8.6	10.3	11.7	13.0	14.3	15.7	17.3	19.3	22.3	25.0	27.5	30.6	32.8
16	5.1	5.8	6.9	8.0	9.3	11.2	12.6	14.0	15.3	16.8	18.4	20.5	23.5	26.3	28.8	32.0	34.3
17	5.7	6.4	7.6	8.7	10.1	12.0	13.5	14.9	16.3	17.8	19.5	21.6	24.8	27.6	30.2	33.4	35.7
18	6.3	7.0	8.2	9.4	10.9	12.9	14.4	15.9	17.3	18.9	20.6	22.8	26.0	28.9	31.5	34.8	37.2
19	6.8	7.6	8.9	10.1	11.7	13.7	15.4	16.9	18.3	19.9	21.7	23.9	27.2	30.1	32.9	36.2	38.6
20	7.4	8.3	9.6	10.9	12.4	14.6	16.3	17.8	19.3	21.0	22.8	25.0	28.4	31.4	34.2	37.6	40.0
21	8.0	8.9	10.3	11.6	13.2	15.4	17.2	18.8	20.3	22.0	23.9	26.2	29.6	32.7	35.5	38.9	41.4
22	8.6	9.5	11.0	12.3	14.0	16.3	18.1	19.7	21.3	23.0	24.9	27.3	30.8	33.9	36.8	40.3	42.8
23	9.3	10.2	11.7	13.1	14.8	17.2	19.0	20.7	22.3	24.1	26.0	28.4	32.0	35.2	38.1	41.6	44.2
24	9.9	10.9	12.4	13.8	15.7	18.1	19.9	21.7	23.3	25.1	27.1	29.6	33.2	36.4	39.4	43.0	45.6
25	10.5	11.5	13.1	14.6	16.5	18.9	20.9	22.6	24.3	26.1	28.2	30.7	34.4	37.7	40.6	44.3	46.9
26	11.2	12.2	13.8	15.4	17.3	19.8	21.8	23.6	25.3	27.2	29.2	31.8	35.6	38.9	41.9	45.6	48.3
27	11.8	12.9	14.6	16.2	18.1	20.7	22.7	24.5	26.3	28.2	30.3	32.9	36.7	40.1	43.2	47.0	49.6
28	12.5	13.6	15.3	16.9	18.9	21.6	23.6	25.5	27.3	29.2	31.4	34.0	37.9	41.3	44.5	48.3	51.0
29	13.1	14.3	16.0	17.7	19.8	22.5	24.6	26.5	28.3	30.3	32.5	35.1	39.1	42.6	45.7	49.6	52.3
30	13.8	15.0	16.8	18.5	20.6	23.4	25.5	27.4	29.3	31.3	33.5	36.3	40.3	43.8	47.0	50.9	53.7
35	17.2	18.5	20.6	22.5	24.8	27.8	30.2	32.3	34.3	36.5	38.9	41.8	46.1	49.8	53.2	57.3	60.3
40	20.7	22.2	24.4	26.5	29.1	32.3	34.9	37.1	39.3	41.6	44.2	47.3	51.8	55.8	59.3	63.7	66.8
45	24.3	25.9	28.4	30.6	33.4	36.9	39.6	42.0	44.3	46.8	49.5	52.7	57.5	61.7	65.4	70.0	73.2
50	28.0	29.7	32.4	34.8	37.7	41.4	44.3	46.9	49.3	51.9	54.7	58.2	63.2	67.5	71.4	76.2	79.5
75	47.2	49.5	52.9	56.1	59.8	64.5	68.1	71.3	74.3	77.5	80.9	85.1	91.1	96.2	101	106	110
100	67.3	70.1	74.2	77.9	82.4	87.9	92.1	95.8	99.3	103	107	112	119	124	130	136	140

TABLE B.3
Kolmogorov-Smirnov table.
Values K-S such that $P[\max | F^n(x) - F(x) |]\text{K-S} = \alpha$

Number of Trials, n	Level of Significance, α			
	0.10	0.05	0.02	0.01
1	0.950	0.975	0.990	0.995
2	0.776	0.842	0.900	0.929
3	0.636	0.708	0.785	0.829
4	0.565	0.624	0.689	0.734
5	0.509	0.563	0.627	0.669
6	0.468	0.519	0.577	0.617
7	0.436	0.483	0.538	0.576
8	0.410	0.454	0.507	0.542
9	0.387	0.430	0.480	0.513
10	0.369	0.409	0.457	0.489
11	0.352	0.391	0.437	0.468
12	0.338	0.375	0.419	0.449
13	0.325	0.361	0.404	0.432
14	0.314	0.349	0.390	0.418
15	0.304	0.338	0.377	0.404
16	0.295	0.327	0.366	0.392
17	0.286	0.318	0.355	0.381
18	0.279	0.309	0.346	0.371
19	0.271	0.301	0.337	0.361
20	0.265	0.294	0.329	0.352
21	0.259	0.287	0.321	0.344
22	0.253	0.281	0.314	0.337
23	0.247	0.275	0.307	0.330
24	0.242	0.269	0.301	0.323
25	0.238	0.264	0.295	0.317
26	0.233	0.259	0.290	0.311
27	0.229	0.254	0.284	0.305
28	0.225	0.250	0.279	0.300
29	0.221	0.246	0.275	0.295
30	0.218	0.242	0.270	0.290
31	0.214	0.238	0.266	0.285
32	0.211	0.234	0.262	0.281
33	0.208	0.231	0.258	0.277
34	0.205	0.227	0.254	0.273
35	0.202	0.224	0.251	0.269
36	0.199	0.221	0.247	0.265
37	0.196	0.218	0.244	0.262
38	0.194	0.215	0.241	0.258
39	0.191	0.213	0.238	0.255
40	0.189	0.210	0.235	0.252
>40	$1.22/n^{1/2}$	$1.36/n^{1/2}$	$1.51/n^{1/2}$	$1.63/n^{1/2}$

TABLE B.4
Binomial distribution.

$$P(x \leq k) = \sum_{x=0}^{k} \binom{n}{k} \phi^x (1 - \phi)^{n-x}$$

n	k	ϕ: 0.05	0.10	0.15	0.20	0.25	0.30	0.35	0.40	0.45	0.50
2	0	0.9025	0.8100	0.7225	0.6400	0.5625	0.4900	0.4225	0.3600	0.3025	0.2500
	1	0.9975	0.9900	0.9775	0.9600	0.9375	0.9100	0.8775	0.8400	0.7975	0.7500
3	0	0.8574	0.7290	0.6141	0.5120	0.4219	0.3430	0.2746	0.2160	0.1664	0.1250
	1	0.9927	0.9720	0.9393	0.8960	0.8438	0.7840	0.7183	0.6480	0.5748	0.5000
	2	0.9999	0.9990	0.9966	0.9920	0.9844	0.9730	0.9571	0.9360	0.9089	0.8750
4	0	0.8145	0.6561	0.5220	0.4096	0.3164	0.2401	0.1785	0.1296	0.0915	0.0625
	1	0.9860	0.9477	0.8905	0.8192	0.7383	0.6517	0.5630	0.4752	0.3910	0.3125
	2	0.9995	0.9963	0.9880	0.9728	0.9492	0.9163	0.8735	0.8208	0.7585	0.6875
	3	1.0000	0.9999	0.9995	0.9984	0.9961	0.9919	0.9850	0.9744	0.9590	0.9375
5	0	0.7738	0.5905	0.4437	0.3277	0.2373	0.1681	0.1160	0.0778	0.0503	0.0313
	1	0.9774	0.9185	0.8352	0.7373	0.6328	0.5282	0.4284	0.3370	0.2562	0.1875
	2	0.9988	0.9914	0.9734	0.9421	0.8965	0.8369	0.7648	0.6826	0.5931	0.5000
	3	1.0000	0.9995	0.9978	0.9933	0.9844	0.9692	0.9460	0.9130	0.8688	0.8125
	4	1.0000	1.0000	0.9999	0.9997	0.9990	0.9976	0.9947	0.9898	0.9815	0.9688
6	0	0.7351	0.5314	0.3771	0.2621	0.1780	0.1176	0.0754	0.0467	0.0277	0.0156
	1	0.9672	0.8857	0.7765	0.6554	0.5339	0.4202	0.3191	0.2333	0.1636	0.1094
	2	0.9978	0.9841	0.9527	0.9011	0.8306	0.7443	0.6471	0.5443	0.4415	0.3438
	3	0.9999	0.9987	0.9941	0.9830	0.9624	0.9295	0.8826	0.8208	0.7447	0.6563
	4	1.0000	0.9999	0.9996	0.0084	0.9954	0.9891	0.9777	0.9590	0.9308	0.8906
	5	1.0000	1.0000	1.0000	0.9999	0.9998	0.9993	0.9982	0.9959	0.9917	0.9844
7	0	0.6983	0.4783	0.3206	0.2097	0.1335	0.0824	0.0490	0.0280	0.0152	0.0078
	1	0.9556	0.8503	0.7166	0.5767	0.4449	0.3294	0.2338	0.1586	0.1024	0.0625
	2	0.9962	0.9743	0.9262	0.8520	0.7564	0.6471	0.5323	0.4199	0.3164	0.2266
	3	0.9998	0.9973	0.9879	0.9667	0.9294	0.8740	0.8002	0.7102	0.6983	0.5000
	4	1.0000	0.9998	0.9988	0.9953	0.9871	0.9712	0.9444	0.9037	0.8471	0.7734
	5	1.0000	1.0000	0.9999	0.9996	0.9987	0.9962	0.9910	0.9812	0.9643	0.9375
	6	1.0000	1.0000	1.0000	1.0000	0.9999	0.9998	0.9994	0.9984	0.9963	0.9922
8	0	0.6634	0.4305	0.2725	0.1678	0.1001	0.0567	0.0319	0.0168	0.0084	0.0039
	1	0.9428	0.8131	0.6572	0.5033	0.3671	0.2553	0.1691	0.1064	0.0632	0.0352
	2	0.9942	0.9619	0.8948	0.7969	0.6785	0.5518	0.4278	0.3154	0.2201	0.1445
	3	0.9996	0.9950	0.9786	0.9437	0.8862	0.8059	0.7064	0.5941	0.4770	0.3633
	4	1.0000	0.9996	0.9971	0.9896	0.9727	0.9420	0.8939	0.8263	0.7396	0.6367
	5	1.0000	1.0000	0.9998	0.9988	0.9958	0.9887	0.9747	0.9502	0.9115	0.8555
	6	1.0000	1.0000	1.0000	0.9999	0.9996	0.9987	0.9964	0.9915	0.9819	0.9648
	7	1.0000	1.0000	1.0000	1.0000	1.0000	0.9999	0.9998	0.9993	0.9983	0.9961
9	0	0.6302	0.3874	0.2316	0.1342	0.0751	0.0404	0.0207	0.0101	0.0046	0.0020
	1	0.9299	0.7748	0.5995	0.4362	0.3003	0.1960	0.1211	0.0705	0.0385	0.0195
	2	0.9916	0.9470	0.8591	0.7382	0.6007	0.4628	0.3373	0.2318	0.1495	0.0898
	3	0.9994	0.9917	0.9661	0.9144	0.8343	0.7297	0.6089	0.4826	0.3614	0.2539
	4	1.0000	0.9991	0.9944	0.9804	0.9511	0.9012	0.8283	0.7334	0.6214	0.5000
	5	1.0000	0.9999	0.9994	0.9969	0.9900	0.9747	0.9464	0.9006	0.8342	0.7461
	6	1.0000	1.0000	1.0000	0.9997	0.9987	0.9957	0.9888	0.9750	0.9502	0.9102
	7	1.0000	1.0000	1.0000	1.0000	0.9999	0.9996	0.9986	0.9962	0.9909	0.9805
	8	1.0000	1.0000	1.0000	1.0000	1.0000	1.0000	0.9999	0.9997	0.9992	0.9980
10	0	0.5987	0.3487	0.1969	0.1074	0.0563	0.0282	0.0135	0.0060	0.0025	0.0010
	1	0.9139	0.7361	0.5443	0.3758	0.2440	0.1493	0.0860	0.0464	0.0233	0.0107
	2	0.9885	0.9298	0.8202	0.6778	0.5256	0.3828	0.2616	0.1673	0.0996	0.0547
	3	0.9990	0.9872	0.9500	0.8791	0.7759	0.6496	0.5138	0.3823	0.2660	0.1719
	4	0.9999	0.9984	0.9901	0.9672	0.9219	0.8497	0.7515	0.6331	0.5044	0.3770
	5	1.0000	0.9999	0.9986	0.9936	0.9803	0.9527	0.9051	0.8338	0.7384	0.6230
	6	1.0000	1.0000	0.9999	0.9991	0.9965	0.9894	0.9740	0.9452	0.8980	0.8281

TABLE B.4
Binomial distribution (continued).

n	k	ϕ: 0.05	0.10	0.15	0.20	0.25	0.30	0.35	0.40	0.45	0.50
10	7	1.0000	1.0000	1.0000	0.9999	0.9996	0.9984	0.9952	0.9877	0.9726	0.9453
	8	1.0000	1.0000	1.0000	1.0000	1.0000	0.9999	0.9995	0.9983	0.9955	0.9893
	9	1.0000	1.0000	1.0000	1.0000	1.0000	1.0000	1.0000	0.9999	0.9997	0.9990
11	0	0.5688	0.3138	0.1673	0.0859	0.0422	0.0198	0.0088	0.0036	0.0014	0.0005
	1	0.8981	0.6974	0.4922	0.3221	0.1971	0.1130	0.0606	0.0302	0.0139	0.0059
	2	0.9848	0.9104	0.7788	0.6174	0.4552	0.3127	0.2001	0.1189	0.0652	0.0327
	3	0.9984	0.9815	0.9306	0.8389	0.7133	0.5696	0.4256	0.2963	0.1911	0.1133
	4	0.9999	0.9972	0.9841	0.9496	0.8854	0.7897	0.6683	0.5328	0.3971	0.2744
	5	1.0000	0.9997	0.9973	0.9883	0.9657	0.9218	0.8513	0.7535	0.6331	0.5000
	6	1.0000	1.0000	0.9997	0.9980	0.9924	0.9784	0.9499	0.9006	0.8262	0.7256
	7	1.0000	1.0000	1.0000	0.9998	0.9988	0.9957	0.9878	0.9707	0.9390	0.8867
	8	1.0000	1.0000	1.0000	1.0000	0.9999	0.9994	0.9980	0.9941	0.9852	0.9673
	9	1.0000	1.0000	1.0000	1.0000	1.0000	1.0000	0.9998	0.9993	0.9978	0.9941
	10	1.0000	1.0000	1.0000	1.0000	1.0000	1.0000	1.0000	1.0000	0.9998	0.9995
12	0	0.5404	0.2824	0.1422	0.0687	0.0317	0.0138	0.0057	0.0022	0.0008	0.0002
	1	0.8816	0.6590	0.4435	0.2749	0.1584	0.0850	0.0424	0.0196	0.0083	0.0032
	2	0.9804	0.8891	0.7358	0.5583	0.3907	0.2528	0.1513	0.0834	0.0421	0.0193
	3	0.9978	0.9744	0.9078	0.7946	0.6488	0.4925	0.3467	0.2253	0.1345	0.0730
	4	0.9998	0.9957	0.9761	0.9274	0.8424	0.7237	0.5833	0.4382	0.3044	0.1938
	5	1.0000	0.9995	0.9954	0.9806	0.9456	0.8822	0.7873	0.6652	0.5269	0.3872
	6	1.0000	0.9999	0.9993	0.9961	0.9857	0.9614	0.9154	0.8418	0.7393	0.6128
	7	1.0000	1.0000	0.9999	0.9994	0.9972	0.9905	0.9745	0.9427	0.8883	0.8062
	8	1.0000	1.0000	1.0000	0.9999	0.9996	0.9983	0.9944	0.9847	0.9644	0.9270
	9	1.0000	1.0000	1.0000	1.0000	1.0000	0.9998	0.9992	0.9972	0.9921	0.9807
	10	1.0000	1.0000	1.0000	1.0000	1.0000	1.0000	0.9999	0.9997	0.9989	0.9968
	11	1.0000	1.0000	1.0000	1.0000	1.0000	1.0000	1.0000	1.0000	0.9999	0.9998
13	0	0.5133	0.2542	0.1209	0.0550	0.0238	0.0097	0.0037	0.0013	0.0004	0.0001
	1	0.8646	0.6213	0.3983	0.2336	0.1267	0.0637	0.0296	0.0126	0.0049	0.0017
	2	0.9755	0.8661	0.6920	0.5017	0.3326	0.2025	0.1132	0.0579	0.0268	0.0112
	3	0.9969	0.9658	0.8820	0.7473	0.5843	0.4206	0.2783	0.1686	0.0929	0.0461
	4	0.9997	0.9935	0.9658	0.9009	0.7940	0.6543	0.5005	0.3530	0.2279	0.1334
	5	1.0000	0.9991	0.9925	0.9700	0.9198	0.8346	0.7159	0.5744	0.4268	0.2905
	6	1.0000	0.9999	0.9987	0.9930	0.9757	0.9376	0.8705	0.7712	0.6437	0.5000
	7	1.0000	1.0000	0.9998	0.9988	0.9944	0.9818	0.9538	0.9023	0.8212	0.7095
	8	1.0000	1.0000	1.0000	0.9998	0.9990	0.9960	0.9874	0.9679	0.9302	0.8666
	9	1.0000	1.0000	1.0000	1.0000	0.9999	0.9993	0.9975	0.9922	0.9797	0.9539
	10	1.0000	1.0000	1.0000	1.0000	1.0000	0.9999	0.9997	0.9987	0.9959	0.9888
	11	1.0000	1.0000	1.0000	1.0000	1.0000	1.0000	1.0000	1.0000	1.0000	1.0000
14	0	0.4877	0.2288	0.1028	0.0440	0.0178	0.0068	0.0024	0.0008	0.0002	0.0001
	1	0.8470	0.5846	0.3567	0.1979	0.1010	0.0475	0.0205	0.0081	0.0029	0.0009
	2	0.9699	0.8416	0.6479	0.4481	0.2811	0.1608	0.0838	0.0398	0.0170	0.0065
	3	0.9958	0.9559	0.8535	0.6982	0.5213	0.3552	0.2205	0.1243	0.0632	0.0287
	4	0.9996	0.9908	0.9533	0.8702	0.7415	0.5842	0.4227	0.2793	0.1672	0.0898
	5	1.0000	0.9985	0.9885	0.9561	0.8883	0.7805	0.6405	0.4859	0.3373	0.2120
	6	1.0000	0.9998	0.9978	0.9884	0.9617	0.9067	0.8164	0.6925	0.5461	0.3953
	7	1.0000	1.0000	0.9997	0.9976	0.9897	0.9685	0.9247	0.8499	0.7414	0.6047
	8	1.0000	1.0000	1.0000	0.9996	0.9978	0.9917	0.9757	0.9417	0.8811	0.7880
	9	1.0000	1.0000	1.0000	1.0000	0.9997	0.9983	0.9940	0.9825	0.9574	0.9102
	10	1.0000	1.0000	1.0000	1.0000	1.0000	0.9998	0.9989	0.9961	0.9886	0.9713
	11	1.0000	1.0000	1.0000	1.0000	1.0000	1.0000	0.9999	0.9994	0.9978	0.9935
	12	1.0000	1.0000	1.0000	1.0000	1.0000	1.0000	1.0000	1.0000	1.0000	1.0000
15	0	0.4633	0.2059	0.0874	0.0352	0.0134	0.0047	0.0016	0.0005	0.0001	0.0000
	1	0.8290	0.5490	0.3186	0.1671	0.0802	0.0353	0.0142	0.0052	0.0017	0.0005
	2	0.9638	0.8159	0.6042	0.3980	0.2361	0.1268	0.0617	0.0271	0.0107	0.0037
	3	0.9945	0.9444	0.8227	0.6482	0.4613	0.2969	0.1729	0.0905	0.0424	0.0176
	4	0.9994	0.9873	0.9383	0.8358	0.6865	0.5155	0.3519	0.2173	0.1204	0.0592
	5	0.9999	0.9977	0.9832	0.9389	0.8516	0.7216	0.5643	0.4032	0.2608	0.1509

TABLE B.4
Binomial distribution (continued).

n	k	ϕ: 0.05	0.10	0.15	0.20	0.25	0.30	0.35	0.40	0.45	0.50
15	6	1.0000	0.9997	0.9964	0.9819	0.9434	0.8689	0.7548	0.6098	0.4522	0.3036
	7	1.0000	1.0000	0.9994	0.9958	0.9827	0.9500	0.8868	0.7869	0.6535	0.5000
	8	1.0000	1.0000	0.9999	0.9992	0.9958	0.9848	0.9578	0.9050	0.8182	0.6964
	9	1.0000	1.0000	1.0000	0.9999	0.9992	0.9963	0.9876	0.9662	0.9231	0.8491
	10	1.0000	1.0000	1.0000	1.0000	0.9999	0.9993	0.9972	0.9907	0.9745	0.9408
	11	1.0000	1.0000	1.0000	1.0000	1.0000	0.9999	0.9995	0.9981	0.9937	0.9824
	12	1.0000	1.0000	1.0000	1.0000	1.0000	1.0000	0.9999	0.9997	0.9989	0.9963
	13	1.0000	1.0000	1.0000	1.0000	1.0000	1.0000	1.0000	1.0000	1.0000	1.0000
16	0	0.4401	0.1853	0.0743	0.0281	0.0100	0.0033	0.0010	0.0003	0.0001	0.0000
	1	0.8108	0.5147	0.2839	0.1407	0.0635	0.0261	0.0098	0.0033	0.0010	0.0003
	2	0.9571	0.7892	0.5614	0.3518	0.1971	0.0994	0.0451	0.0183	0.0066	0.0021
	3	0.9930	0.9316	0.7899	0.5981	0.4050	0.2459	0.1339	0.0651	0.0281	0.0106
	4	0.9991	0.9830	0.9209	0.7982	0.6302	0.4499	0.2892	0.1666	0.0853	0.0384
	5	0.9999	0.9967	0.9765	0.9183	0.8103	0.6598	0.4900	0.3288	0.1976	0.1051
	6	1.0000	0.9995	0.9944	0.9733	0.9204	0.8247	0.6881	0.5272	0.3660	0.2272
	7	1.0000	0.9999	0.9989	0.9930	0.9729	0.9256	0.8406	0.7161	0.5629	0.4018
	8	1.0000	1.0000	0.9998	0.9985	0.9925	0.9743	0.9329	0.8577	0.7441	0.5982
	9	1.0000	1.0000	1.0000	0.9998	0.9984	0.9929	0.9771	0.9417	0.8759	0.7728
	10	1.0000	1.0000	1.0000	1.0000	0.9997	0.9984	0.9938	0.9809	0.9514	0.8949
	11	1.0000	1.0000	1.0000	1.0000	1.0000	0.9997	0.9987	0.9951	0.9851	0.9616
	12	1.0000	1.0000	1.0000	1.0000	1.0000	1.0000	0.9998	0.9991	0.9965	0.9894
	13	1.0000	1.0000	1.0000	1.0000	1.0000	1.0000	1.0000	0.9999	0.9994	0.9979
	14	1.0000	1.0000	1.0000	1.0000	1.0000	1.0000	1.0000	1.0000	1.0000	1.0000
17	0	0.4181	0.1668	0.0631	0.0225	0.0075	0.0023	0.0007	0.0002	0.0000	0.0000
	1	0.7922	0.4818	0.2525	0.1182	0.0501	0.0193	0.0067	0.0021	0.0006	0.0001
	2	0.9497	0.7618	0.5198	0.3096	0.1637	0.0774	0.0327	0.0123	0.0041	0.0012
	3	0.9912	0.9174	0.7556	0.5489	0.3530	0.2019	0.1028	0.0464	0.0184	0.0064
	4	0.9988	0.9779	0.9013	0.7582	0.5739	0.3887	0.2348	0.1260	0.0596	0.0245
	5	0.9999	0.9953	0.9681	0.8943	0.7653	0.5968	0.4197	0.2639	0.1471	0.0717
	6	1.0000	0.9992	0.9917	0.9623	0.8929	0.7752	0.6188	0.4478	0.2902	0.1662
	7	1.0000	0.9999	0.9983	0.9891	0.9598	0.8954	0.7872	0.6405	0.4743	0.3145
	8	1.0000	1.0000	0.9997	0.9974	0.9876	0.9597	0.9006	0.8011	0.6626	0.5000
	9	1.0000	1.0000	1.0000	0.9995	0.9969	0.9873	0.9617	0.9081	0.8166	0.6855
	10	1.0000	1.0000	1.0000	0.9999	0.9994	0.9968	0.9880	0.9652	0.9174	0.8338
	11	1.0000	1.0000	1.0000	1.0000	0.9999	0.9993	0.9970	0.9894	0.9699	0.9283
	12	1.0000	1.0000	1.0000	1.0000	1.0000	0.9999	0.9994	0.9975	0.9914	0.9755
	13	1.0000	1.0000	1.0000	1.0000	1.0000	1.0000	0.9999	0.9995	0.9981	0.9936
	14	1.0000	1.0000	1.0000	1.0000	1.0000	1.0000	1.0000	1.0000	1.0000	1.0000
18	0	0.3972	0.1501	0.0536	0.0180	0.0056	0.0016	0.0004	0.0001	0.0000	0.0000
	1	0.7735	0.4503	0.2241	0.0991	0.0395	0.0142	0.0046	0.0013	0.0003	0.0001
	2	0.9419	0.7338	0.4797	0.2713	0.1353	0.0600	0.0236	0.0082	0.0025	0.0007
	3	0.9891	0.9018	0.7202	0.5010	0.3057	0.1646	0.0783	0.0328	0.0120	0.0038
	4	0.9985	0.9718	0.8794	0.7164	0.5187	0.3327	0.1886	0.0942	0.0411	0.0154
	5	0.9998	0.9936	0.9581	0.8671	0.7175	0.5344	0.3550	0.2088	0.1077	0.0481
	6	1.0000	0.9988	0.9882	0.9487	0.8610	0.7217	0.5491	0.3743	0.2258	0.1189
	7	1.0000	0.9998	0.9973	0.9837	0.9431	0.8593	0.7283	0.5634	0.3915	0.2403
	8	1.0000	1.0000	0.9995	0.9957	0.9807	0.9404	0.8609	0.7368	0.5778	0.4073
	9	1.0000	1.0000	0.9999	0.9991	0.9946	0.9790	0.9403	0.8653	0.7473	0.5927
	10	1.0000	1.0000	1.0000	0.9998	0.9988	0.9939	0.9788	0.9424	0.8720	0.7597
	11	1.0000	1.0000	1.0000	1.0000	0.9998	0.9986	0.9938	0.9797	0.9463	0.8811
	12	1.0000	1.0000	1.0000	1.0000	1.0000	0.9997	0.9986	0.9942	0.9817	0.9519
	13	1.0000	1.0000	1.0000	1.0000	1.0000	1.0000	0.9997	0.9987	0.9951	0.9846
	14	1.0000	1.0000	1.0000	1.0000	1.0000	1.0000	1.0000	0.9998	0.9990	0.9962
	15	1.0000	1.0000	1.0000	1.0000	1.0000	1.0000	1.0000	1.0000	1.0000	1.0000
19	0	0.3774	0.1351	0.0456	0.0144	0.0042	0.0011	0.0003	0.0001	0.0000	0.0000
	1	0.7547	0.4203	0.1985	0.0829	0.0310	0.0104	0.0031	0.0008	0.0002	0.0000
	2	0.9335	0.7054	0.4413	0.2369	0.1113	0.0462	0.0170	0.0055	0.0015	0.0004

TABLE B.4
Binomial distribution (continued).

n	k	ϕ: 0.05	0.10	0.15	0.20	0.25	0.30	0.35	0.40	0.45	0.50
19	3	0.9868	0.8850	0.6841	0.4551	0.2631	0.1332	0.0591	0.0230	0.0077	0.0022
	4	0.9980	0.9648	0.8556	0.6733	0.4654	0.2822	0.1500	0.0696	0.0280	0.0096
	5	0.9998	0.9914	0.9463	0.8369	0.6678	0.4739	0.2968	0.1629	0.0777	0.0318
	6	1.0000	0.9983	0.9837	0.9324	0.8251	0.6655	0.4812	0.3081	0.1727	0.0835
	7	1.0000	0.9997	0.9959	0.9767	0.9225	0.8180	0.6656	0.4878	0.3169	0.1796
	8	1.0000	1.0000	0.9992	0.9933	0.9713	0.9161	0.8145	0.6675	0.4940	0.3238
	9	1.0000	1.0000	0.9999	0.9984	0.9911	0.9674	0.9125	0.8139	0.6710	0.5000
	10	1.0000	1.0000	1.0000	0.9997	0.9977	0.9895	0.9653	0.9115	0.8159	0.6762
	11	1.0000	1.0000	1.0000	1.0000	0.9995	0.9972	0.9886	0.9648	0.9129	0.8204
	12	1.0000	1.0000	1.0000	1.0000	0.9999	0.9994	0.9969	0.9884	0.9658	0.9165
	13	1.0000	1.0000	1.0000	1.0000	1.0000	0.9999	0.9993	0.9969	0.9891	0.9682
	14	1.0000	1.0000	1.0000	1.0000	1.0000	1.0000	0.9999	0.9994	0.9972	0.9904
	15	1.0000	1.0000	1.0000	1.0000	1.0000	1.0000	1.0000	0.9999	0.9995	0.9978
	16	1.0000	1.0000	1.0000	1.0000	1.0000	1.0000	1.0000	1.0000	1.0000	1.0000
20	0	0.3585	0.1216	0.1388	0.0115	0.0032	0.0008	0.0002	0.0000	0.0000	0.0000
	1	0.7358	0.3917	0.1756	0.0692	0.0243	0.0076	0.0021	0.0005	0.0001	0.0000
	2	0.9245	0.6769	0.4049	0.2061	0.0913	0.0355	0.0121	0.0036	0.0009	0.0002
	3	0.9841	0.8670	0.6477	0.4114	0.2252	0.1071	0.0444	0.0160	0.0049	0.0013
	4	0.9974	0.9568	0.8298	0.6296	0.4148	0.2375	0.1182	0.0510	0.0189	0.0059
	5	0.9997	0.9887	0.9327	0.8042	0.6172	0.4164	0.2454	0.1256	0.0553	0.0207
	6	1.0000	0.9976	0.9781	0.9133	0.7858	0.6080	0.4166	0.2500	0.1299	0.0577
	7	1.0000	0.9996	0.9941	0.9679	0.8982	0.7723	0.6010	0.4159	0.2520	0.1316
	8	1.0000	0.9999	0.9987	0.9900	0.9591	0.8867	0.7624	0.5956	0.4143	0.2517
	9	1.0000	1.0000	0.9998	0.9974	0.9861	0.9520	0.8782	0.7553	0.5914	0.4119
	10	1.0000	1.0000	1.0000	0.9994	0.9961	0.9829	0.9468	0.8725	0.7507	0.5881
	11	1.0000	1.0000	1.0000	0.9999	0.9991	0.9949	0.9804	0.9435	0.8692	0.7483
	12	1.0000	1.0000	1.0000	1.0000	0.9998	0.9987	0.9940	0.9790	0.9420	0.8684
	13	1.0000	1.0000	1.0000	1.0000	1.0000	0.9997	0.9985	0.9935	0.9786	0.9423
	14	1.0000	1.0000	1.0000	1.0000	1.0000	1.0000	0.9997	0.9984	0.9936	0.9793
	15	1.0000	1.0000	1.0000	1.0000	1.0000	1.0000	1.0000	0.9997	0.9985	0.9941
	16	1.0000	1.0000	1.0000	1.0000	1.0000	1.0000	1.0000	1.0000	0.9997	0.9987
	17	1.0000	1.0000	1.0000	1.0000	1.0000	1.0000	1.0000	1.0000	1.0000	1.0000

TABLE B.5
Binomial 50% confidence levels.

$$\text{Values } R_{0.50} \text{ such that } P(x > F) = 0.50 = 1 - P(x \le F) = 1.0 - \sum_{x=0}^{F} \binom{N}{x} R_{0.50}{}^{N-x}(1 - R_{0.50})^{x}$$

Sample Size	Number of Failures in Sample = F										
N	0	1	2	3	4	5	6	7	8	9	10
1	0.500										
2	0.707	0.293									
3	0.794	0.500	0.206								
4	0.841	0.614	0.386	0.159							
5	0.871	0.686	0.500	0.314	0.129						
6	0.891	0.736	0.579	0.421	0.264	0.109					
7	0.906	0.772	0.636	0.500	0.364	0.228	0.094				
8	0.917	0.799	0.679	0.560	0.440	0.321	0.201	0.083			
9	0.926	0.820	0.714	0.607	0.500	0.393	0.286	0.180	0.074		
10	0.933	0.838	0.741	0.646	0.548	0.452	0.355	0.259	0.162	0.067	
11	0.939	0.852	0.764	0.676	0.589	0.500	0.412	0.324	0.235	0.148	0.061
12	0.944	0.864	0.783	0.702	0.621	0.541	0.460	0.379	0.298	0.217	0.136
13	0.948	0.874	0.800	0.725	0.650	0.575	0.500	0.425	0.350	0.275	0.200
14	0.952	0.883	0.814	0.744	0.674	0.605	0.535	0.466	0.395	0.326	0.256
15	0.955	0.891	0.826	0.761	0.695	0.630	0.565	0.500	0.436	0.370	0.305
16	0.958	0.897	0.836	0.775	0.714	0.653	0.592	0.531	0.469	0.409	0.347
17	0.960	0.903	0.846	0.788	0.731	0.673	0.615	0.558	0.500	0.442	0.385
18	0.962	0.908	0.854	0.800	0.745	0.691	0.636	0.582	0.527	0.473	0.418
19	0.964	0.913	0.862	0.810	0.758	0.707	0.655	0.603	0.552	0.500	0.448
20	0.966	0.917	0.869	0.819	0.770	0.721	0.672	0.623	0.574	0.525	0.475
21	0.968	0.921	0.875	0.828	0.781	0.734	0.687	0.641	0.594	0.547	0.500
22	0.969	0.925	0.880	0.836	0.791	0.746	0.701	0.657	0.612	0.567	0.522
23	0.970	0.928	0.885	0.843	0.800	0.757	0.714	0.671	0.629	0.586	0.543
24	0.972	0.931	0.890	0.849	0.808	0.767	0.726	0.685	0.644	0.603	0.562
25	0.973	0.934	0.894	0.855	0.816	0.776	0.737	0.697	0.658	0.618	0.579
26	0.974	0.936	0.898	0.861	0.823	0.785	0.747	0.709	0.671	0.633	0.595
27	0.975	0.939	0.902	0.866	0.829	0.793	0.756	0.719	0.683	0.646	0.610
28	0.976	0.941	0.906	0.870	0.835	0.800	0.765	0.729	0.694	0.659	0.624
29	0.976	0.943	0.909	0.875	0.841	0.807	0.773	0.739	0.704	0.670	0.636
30	0.977	0.945	0.912	0.879	0.846	0.813	0.780	0.747	0.714	0.681	0.648
31	0.978	0.946	0.915	0.883	0.851	0.819	0.787	0.755	0.723	0.691	0.660
32	0.979	0.948	0.917	0.886	0.856	0.825	0.794	0.762	0.732	0.701	0.670
33	0.979	0.950	0.920	0.890	0.860	0.830	0.800	0.770	0.740	0.710	0.680
34	0.980	0.951	0.922	0.893	0.864	0.835	0.806	0.777	0.748	0.718	0.689
35	0.980	0.953	0.924	0.896	0.868	0.840	0.811	0.783	0.755	0.726	0.698
36	0.981	0.954	0.926	0.899	0.871	0.844	0.816	0.789	0.761	0.734	0.706
37	0.981	0.955	0.928	0.902	0.875	0.848	0.821	0.795	0.768	0.741	0.714
38	0.982	0.956	0.930	0.904	0.878	0.852	0.826	0.800	0.774	0.748	0.722
39	0.982	0.957	0.932	0.907	0.881	0.856	0.830	0.805	0.780	0.754	0.729
40	0.983	0.958	0.934	0.909	0.884	0.859	0.835	0.810	0.785	0.760	0.736
41	0.983	0.959	0.935	0.911	0.887	0.863	0.839	0.814	0.790	0.766	0.742
42	0.984	0.960	0.937	0.913	0.890	0.866	0.842	0.819	0.795	0.772	0.748
43	0.984	0.961	0.938	0.915	0.892	0.869	0.846	0.823	0.800	0.777	0.754
44	0.984	0.962	0.940	0.917	0.895	0.872	0.850	0.827	0.804	0.782	0.759
45	0.985	0.963	0.941	0.919	0.897	0.875	0.853	0.831	0.809	0.787	0.765
46	0.985	0.964	0.942	0.921	0.899	0.878	9.856	0.834	0.813	0.791	0.770
47	0.985	0.965	0.944	0.922	0.901	0.880	0.859	0.838	0.817	0.796	0.775
48	0.986	0.965	0.945	0.924	0.903	0.883	0.862	0.841	0.821	0.800	0.779
49	0.986	0.966	0.946	0.926	0.905	0.885	0.865	0.845	0.824	0.804	0.784
50	0.986	0.967	0.947	0.927	0.907	0.887	0.868	0.848	0.828	0.808	0.788

TABLE B.5
Binomial 50% confidence levels (continued).

Sample Size N	Number of Failures in Sample = F										
	0	1	2	3	4	5	6	7	8	9	10
51	0.987	0.967	0.948	0.928	0.909	0.890	0.870	0.851	0.831	0.812	0.792
52	0.987	0.968	0.949	0.930	0.911	0.982	0.873	0.853	0.834	0.815	0.796
53	0.987	0.969	0.950	0.931	0.912	0.894	0.875	0.856	0.837	0.819	0.800
54	0.987	0.969	0.951	0.932	0.914	0.896	0.877	0.859	0.841	0.822	0.804
55	0.987	0.970	0.952	0.934	0.916	0.898	0.879	0.861	0.843	0.825	0.807
56	0.9877	0.9702	0.9525	0.9348	0.9171	0.8994	0.8816	0.8639	0.8461	0.8284	0.8106
57	0.9879	0.9707	0.9534	0.9360	0.9185	0.9011	0.8837	0.8662	0.8488	0.8314	0.8139
58	0.9881	0.9712	0.9542	0.9371	0.9199	0.9028	0.8857	0.8685	0.8514	0.8343	0.8171
59	0.9883	0.9717	0.9549	0.9381	0.9213	0.9044	0.8876	0.8708	0.8539	0.8371	0.8202
60	0.9885	0.9722	0.9557	0.9391	0.9226	0.9060	0.8895	0.8729	0.8563	0.8398	0.8232
61	0.9887	0.9726	0.9564	0.9401	0.9239	0.9076	0.8913	0.8750	0.8587	0.8424	0.8261
62	0.9889	0.9731	0.9571	0.9411	0.9251	0.9090	0.8930	0.8770	0.8609	0.8449	0.8289
63	0.9891	0.9735	0.9578	0.9420	0.9263	0.9105	0.8947	0.8789	0.8631	0.8473	0.8316
64	0.9892	0.9739	0.9584	0.9429	0.9274	0.9119	0.8963	0.8808	0.8653	0.8497	0.8342
65	0.9894	0.9743	0.9591	0.9438	0.9285	0.9132	0.8979	0.8826	0.8673	0.8520	0.8367
66	0.9896	0.9747	0.9597	0.9446	0.9296	0.9145	0.8995	0.8843	0.8693	0.8542	0.8392
67	0.9897	0.9751	0.9603	0.9455	0.9306	0.9158	0.9010	0.8861	0.8713	0.8564	0.8416
68	0.9899	0.9754	0.9609	0.9463	0.9317	0.9170	0.9024	0.8878	0.8731	0.8585	0.8439
69	0.9900	0.9758	0.9614	0.9470	0.9326	0.9182	0.9038	0.8894	0.8750	0.8606	0.8461
70	0.9902	0.9761	0.9620	0.9478	0.9336	0.9194	0.9052	0.8910	0.8768	0.8625	0.8483
71	0.9903	0.9765	0.9625	0.9485	0.9345	0.9205	0.9065	0.8925	0.8785	0.8645	0.8505
72	0.9904	0.9768	0.9630	0.9492	0.9354	0.9216	0.9079	0.8940	0.8802	0.8663	0.8525
73	0.9906	0.9771	0.9635	0.9499	0.9363	0.9227	0.9091	0.8954	0.8818	0.8682	0.8545
74	0.9907	0.9774	0.9640	0.9506	0.9372	0.9237	0.9103	0.8968	0.8834	0.8699	0.8565
75	0.9908	0.9778	0.9645	0.9513	0.9380	0.9247	0.9115	0.8982	0.8849	0.8717	0.8584
76	0.9909	0.9780	0.9650	0.9519	0.9388	0.9257	0.9126	0.8995	0.8864	0.8733	0.8602
77	0.9910	0.9783	0.9654	0.9525	0.9396	0.9267	0.9138	0.9008	0.8879	0.8750	0.8621
78	0.9912	0.9786	0.9659	0.9531	0.9404	0.9276	0.9149	0.9021	0.8893	0.8766	0.8638
79	0.9913	0.9789	0.9663	0.9537	0.9411	0.9285	0.9159	0.9033	0.8907	0.8781	0.8655
80	0.9914	0.9791	0.9667	0.9543	0.9419	0.9294	0.9170	0.9045	0.8921	0.8796	0.8672
81	0.9915	0.9794	0.9671	0.9549	0.9426	0.9303	0.9180	0.9057	0.8934	0.8811	0.8688
82	0.9916	0.9796	0.9675	0.9554	0.9433	0.9311	0.9190	0.9069	0.8947	0.8826	0.8704
83	0.9917	0.9799	0.9679	0.9559	0.9440	0.9320	0.9200	0.9080	0.8960	0.8840	0.8720
84	0.9918	0.9801	0.9683	0.9565	0.9446	0.9323	0.9209	0.9091	0.8972	0.8854	0.8735
85	0.9919	0.9803	0.9687	0.9570	0.9453	0.9336	0.9218	0.9101	0.8984	0.8867	0.8750
86	0.9920	0.9806	0.9690	0.9575	0.9459	0.9343	0.9228	0.9112	0.8996	0.8880	0.8764
87	0.9921	0.9808	0.9694	0.9580	0.9465	0.9351	0.9236	0.9122	0.9007	0.8893	0.8774
88	0.9922	0.9810	0.9697	0.9584	0.9471	0.9358	0.9245	0.9132	0.9019	0.8905	0.8792
89	0.9922	0.9812	0.9701	0.9589	0.9477	0.9365	0.9253	0.9142	0.9030	0.8918	0.8806
90	0.9923	0.9814	0.9704	0.9594	0.9483	0.9372	0.9262	0.9151	0.9040	0.8930	0.8819
91	0.9924	0.9816	0.9707	0.9598	0.9489	0.9379	0.9270	0.9160	0.9051	0.8941	0.8832
92	0.9925	0.9818	0.9710	0.9602	0.9494	0.9386	0.9278	0.9169	0.9061	0.8953	0.8845
93	0.9926	0.9820	0.9714	0.9607	0.9500	0.9393	0.9285	0.9178	0.9071	0.8964	0.8857
94	0.9927	0.9822	0.9717	0.9611	0.9505	0.9399	0.9293	0.9187	0.9081	0.8975	0.8869
95	0.9927	0.9824	0.9720	0.9615	0.9510	0.9405	0.9300	0.9196	0.9091	0.8986	0.8881
96	0.9928	0.9826	0.9722	0.9619	0.9515	0.9411	0.9308	0.9204	0.9100	0.8996	0.8893
97	0.9929	0.9828	0.9725	0.9623	0.9520	0.9418	0.9315	0.9212	0.9109	0.9007	0.8904
98	0.9930	0.9829	0.9728	0.9627	0.9525	0.9423	0.9322	0.9220	0.9118	0.9017	0.8915
99	0.9930	0.9831	0.9731	0.9630	0.9530	0.9429	0.9329	0.9228	0.9127	0.9027	0.8926
100	0.9931	0.9833	0.9734	0.9634	0.9535	0.9435	0.9335	0.9236	0.9136	0.9036	0.8937
101	0.9932	0.9834	0.9736	0.9638	0.9539	0.9441	0.9342	0.9243	0.9145	0.9046	0.8947
102	0.9932	0.9836	0.9739	0.9641	0.9544	0.9446	0.9348	0.9251	0.9153	0.9055	0.8958
103	0.9933	0.9838	0.9741	0.9645	0.9548	0.9451	0.9355	0.9258	0.9161	0.9064	0.8968
104	0.9934	0.9839	0.9744	0.9648	0.9552	0.9457	0.9361	0.9265	0.9169	0.9073	0.8978
105	0.9934	0.9841	0.9746	0.9651	0.9557	0.9462	0.9367	0.9272	0.9177	0.9082	0.8987

TABLE B.5
Binomial 50% confidence levels (continued).

Sample Size N	Number of Failures in Sample = F										
	0	1	2	3	4	5	6	7	8	9	10
106	0.9935	0.9842	0.9749	0.9655	0.9561	0.9467	0.9373	0.9279	0.9185	0.9091	0.8997
107	0.9935	0.9844	0.9751	0.9658	0.9565	0.9472	0.9379	0.9286	0.9192	0.9099	0.9006
108	0.9936	0.9845	0.9753	0.9661	0.9569	0.9477	0.9384	0.9292	0.9200	0.9108	0.9052
109	0.9937	0.9847	0.9755	0.9664	0.9573	0.9481	0.9390	0.9229	0.9207	0.9116	0.9024
110	0.9937	0.9848	0.9758	0.9667	0.9577	0.9486	0.9396	0.9305	0.9214	0.9124	0.9033
111	0.9938	0.9849	0.9760	0.9670	0.9581	0.9491	0.9401	0.9311	0.9221	0.9132	0.9042
112	0.9938	0.9851	0.9762	0.9673	0.9584	0.9495	0.9406	0.9317	0.9228	0.9139	0.9050
113	0.9939	0.9852	09764	0.9676	0.9588	0.9500	0.9412	0.9323	0.9235	0.9147	0.9059
114	0.9939	0.9853	0.9766	0.9679	0.9592	0.9504	0.9417	0.9329	0.9242	0.9154	0.9067
115	0.9940	0.9855	0.9768	0.9682	0.9595	0.9508	0.9422	0.9335	0.9248	0.9162	0.9075
116	0.9940	0.9856	0.9770	0.9684	0.9599	0.9513	0.9427	0.9341	0.9255	0.9169	0.9083
117	0.9941	0.9857	0.9772	0.9687	0.9602	0.9517	0.9432	0.9346	0.9261	0.9176	0.9091
118	0.9941	0.9858	0.9774	0.9690	0.9605	0.9521	0.9436	0.9352	0.9267	0.9183	0.9098
119	0.9942	0.9859	0.9776	0.9692	0.9609	0.9525	0.9441	0.9357	0.9274	0.9190	0.9106
120	0.9942	0.9861	0.9778	0.9695	0.9612	0.9529	0.9446	0.9363	0.9280	0.9197	0.9113
121	0.9943	0.9862	0.9780	0.9697	0.9615	0.9533	0.9450	0.9368	0.9286	0.9203	0.9121
122	0.9943	0.9863	0.9781	0.9700	0.9618	0.9537	0.9455	0.9373	0.9291	0.9210	0.9128
123	0.9944	0.9864	0.9783	0.9702	0.9621	0.9540	0.9459	0.9378	0.9297	0.9216	0.9135
124	0.9944	0.9865	0.9785	0.9705	0.9624	0.9544	0.9464	0.9383	0.9303	0.9222	0.9142
125	0.9945	0.9866	0.9787	0.9707	0.9627	0.9548	0.9468	0.9388	0.9308	0.9229	0.9149
126	0.9945	0.9867	0.9788	0.9709	0.9630	0.9551	0.9472	0.9393	0.9314	0.9235	0.9156
127	0.9946	0.9868	0.9790	0.9712	0.9633	0.9555	0.9476	0.9398	0.9319	0.9241	0.9162
128	0.9946	0.9869	0.9792	0.9714	0.9636	0.9558	0.9480	0.9402	0.9324	0.9247	0.9169
129	0.9946	0.9870	0.9793	0.9716	0.9639	0.9562	0.9484	0.9407	0.9330	0.9252	0.9175
130	0.9947	0.9871	0.9795	0.9718	0.9642	0.9565	0.9488	0.9412	0.9335	0.9258	0.9181
131	0.9947	0.9872	0.9796	0.9720	0.9644	0.9568	0.9492	0.9416	0.9340	0.9264	0.9188
132	0.9948	0.9873	0.9798	0.9723	0.9647	0.9572	0.9496	0.9421	0.9345	0.9269	0.9194
133	0.9948	0.9874	0.9800	0.9725	0.9650	0.9575	0.9500	0.9425	0.9350	0.9275	0.9200
134	0.9948	0.9875	0.9801	0.9727	0.9652	0.9578	0.9504	0.9429	0.9355	0.9280	0.9206
135	0.9949	0.9876	0.9802	0.9729	0.9655	0.9581	0.9507	0.9433	0.9359	0.9286	0.9212
136	0.9949	0.9877	0.9804	0.9731	0.9657	0.9584	0.9511	0.9438	0.9364	0.9291	0.9218
137	0.9950	0.9878	0.9805	0.9733	0.9660	0.9587	0.9514	0.9442	0.9369	0.9296	0.9223
138	0.9950	0.9879	0.9807	0.9735	0.9662	0.9590	0.9518	0.9446	0.9373	0.9301	0.9229
139	0.9950	0.9880	0.9808	0.9737	0.9665	0.9593	0.9521	0.9450	0.9378	0.9306	0.9234
140	0.9951	0.9880	0.9810	0.9738	0.9667	0.9596	0.9525	0.9454	0.9382	0.9311	0.9240
141	0.9951	0.9881	0.9811	0.9740	0.9670	0.9599	0.9528	0.9457	0.9387	0.9316	0.9245
142	0.9951	0.9882	0.9812	0.9742	0.9672	0.9602	0.9531	0.9461	0.9391	0.9321	0.9251
143	0.9952	0.9883	0.9813	0.9744	0.9674	0.9604	0.9535	0.9465	0.9395	0.9325	0.9256
144	0.9952	0.9884	0.9815	0.9746	0.9676	0.9607	0.9538	0.9469	0.9400	0.9330	0.9261
145	0.9952	0.9885	0.9816	0.9747	0.9679	0.9610	0.9541	0.9472	0.9404	0.9335	0.9266
146	0.9952	0.9885	0.9817	0.9749	0.9681	0.9613	0.9544	0.9476	0.9408	0.9339	0.9271
147	0.9953	0.9886	0.9819	0.9751	0.9683	0.9615	0.9547	0.9480	0.9412	0.9344	0.9276
148	0.9953	0.9887	0.9820	0.9752	0.9685	0.9618	0.9550	0.9483	0.9416	0.9348	0.9281
149	0.9954	0.9888	0.9821	0.9754	0.9687	0.9620	0.9553	0.9486	0.9420	0.9353	0.9286
150	0.9954	0.9888	0.9822	0.9756	0.9689	0.9623	0.9556	0.9490	0.9423	0.9357	0.9290
151	0.9954	0.9889	0.9823	0.9757	0.9691	0.9625	0.9559	0.9493	0.9427	0.9361	0.9295
152	0.9955	0.9890	0.9824	0.9759	0.9693	0.9628	0.9562	0.9497	0.9431	0.9365	0.9300
153	0.9955	0.9891	0.9826	0.9761	0.9695	0.9630	0.9565	0.9500	0.9435	0.9369	0.9304
154	0.9955	0.9891	0.9827	0.9762	0.9697	0.9633	0.9568	0.9503	0.9438	0.9374	0.9309
155	0.9955	0.9892	0.9828	0.9764	0.9699	0.9635	0.9571	0.9506	0.9442	0.9378	0.9313

TABLE B.5
Binomial 50% confidence levels (continued).

Sample Size N	Number of Failures in Sample = F										
	0	1	2	3	4	5	6	7	8	9	10
156	0.9956	0.9893	0.9829	0.9765	0.9701	0.9637	0.9573	0.9509	0.9446	0.9382	0.9318
157	0.9956	0.9893	0.9830	0.9767	0.9703	0.9640	0.9576	0.9513	0.9449	0.9386	0.9322
158	0.9956	0.9894	0.9831	0.9768	0.9705	0.9642	0.9579	0.9516	0.9453	0.9389	0.9326
159	0.9957	0.9895	0.9832	0.9770	0.9707	0.9644	0.9581	0.9519	0.9456	0.9393	0.9330
160	0.9957	0.9895	0.9833	0.9771	0.9709	0.9646	0.9584	0.9522	0.9459	0.9397	0.9335
161	0.9957	0.9896	0.9834	0.9772	0.9711	0.9649	0.9587	0.9525	0.9463	0.9401	0.9339
162	0.9957	0.9897	0.9835	0.9774	0.9712	0.9651	0.9589	0.9528	0.9466	0.9404	0.9343
163	0.9958	0.9897	0.9836	0.9775	0.9714	0.9653	0.9592	0.9531	0.9469	0.9408	0.9347
164	0.9958	0.9898	0.9837	0.9777	0.9716	0.9655	0.9594	0.9533	0.9473	0.9412	0.9351
165	0.9958	0.9899	0.9838	0.9778	0.9718	0.9657	0.9597	0.9536	0.9476	0.9415	0.9355
166	0.9958	0.9899	0.9839	0.9779	0.9719	0.9659	0.9599	0.9539	0.9479	0.9419	0.9359
167	0.9959	0.9900	0.9840	0.9781	0.9721	0.9661	0.9601	0.9542	0.9482	0.9422	0.9362
168	0.9959	0.9900	0.9841	0.9782	0.9723	0.9663	0.9604	0.9544	0.9485	0.9426	0.9366
169	0.9959	0.9901	0.9842	0.9783	0.9724	0.9665	0.9606	0.9547	0.9488	0.9429	0.9370
170	0.9959	0.9902	0.9843	0.9784	0.9726	0.9667	0.9608	0.9550	0.9491	0.9432	0.9374
171	0.9960	0.9902	0.9844	0.9786	0.9727	0.9669	0.9611	0.9552	0.9494	0.9436	0.9377
172	0.9960	0.9903	0.9845	0.9787	0.9729	0.9671	0.9613	0.9555	0.9497	0.9439	0.9381
173	0.9960	0.9903	0.9846	0.9788	0.9731	0.9673	0.9615	0.9558	0.9500	0.9442	0.9385
174	0.9960	0.9904	0.9847	0.9789	0.9732	0.9675	0.9617	0.9560	0.9503	0.9445	0.9388
175	0.9961	0.9904	0.9848	0.9791	0.9734	0.9677	0.9620	0.9563	0.9506	0.9449	0.9392
176	0.9961	0.9905	0.9848	0.9792	0.9735	0.9678	0.9622	0.9565	0.9508	0.9452	0.9395
177	0.9961	0.9905	0.9849	0.9793	0.9737	0.9680	0.9624	0.9568	0.9511	0.9455	0.9398
178	0.9961	0.9906	0.9850	0.9794	0.9738	0.9682	0.9626	0.9570	0.9514	0.9458	0.9402
179	0.9961	0.9906	0.9851	0.9795	0.9740	0.9684	0.9628	0.9572	0.9517	0.9461	0.9405
180	0.9962	0.9907	0.9852	0.9796	0.9741	0.9686	0.9630	0.9575	0.9519	0.9464	0.9408
181	0.9962	0.9907	0.9853	0.9798	0.9742	0.9687	0.9632	0.9577	0.9522	0.9467	0.9412
182	0.9962	0.9908	0.9853	0.9799	0.9744	0.9689	0.9634	0.9579	0.9525	0.9470	0.9415
183	0.9962	0.9909	0.9854	0.9800	0.9745	0.9691	0.9636	0.9582	0.9527	0.9473	0.9418
184	0.9962	0.9909	0.9855	0.9801	0.9747	0.9692	0.9638	0.9584	0.9530	0.9476	0.9421
185	0.9963	0.9909	0.9856	0.9802	0.9748	0.9694	0.9640	0.9586	0.9532	0.9478	0.9424
186	0.9963	0.9910	0.9857	0.9803	0.9749	0.9696	0.9642	0.9588	0.9535	0.9481	0.9428
187	0.9963	0.9910	0.9857	0.9804	0.9751	0.9697	0.9644	0.9591	0.9537	0.9484	0.9431
188	0.9963	0.9911	0.9858	0.9805	0.9752	0.9699	0.9646	0.9593	0.9540	0.9487	0.9433
189	0.9963	0.9911	0.9859	0.9806	0.9753	0.9701	0.9648	0.9595	0.9542	0.9489	0.9437
190	0.9964	0.9912	0.9860	0.9807	0.9755	0.9702	0.9650	0.9597	0.9545	0.9492	0.9440
191	0.9964	0.9912	0.9860	0.9808	0.9756	0.9704	0.9651	0.9599	0.9547	0.9495	0.9442
192	0.9964	0.9913	0.9861	0.9809	0.9757	0.9705	0.9653	0.9601	0.9549	0.9497	0.9445
193	0.9964	0.9913	0.9862	0.9810	0.9758	0.9707	0.9655	0.9603	0.9552	0.9500	0.9448
194	0.9964	0.9914	0.9862	0.9811	0.9760	0.9708	0.9657	0.9605	0.9554	0.9503	0.9451
195	0.9965	0.9914	0.9863	0.9812	0.9761	0.9710	0.9659	0.9607	0.9556	0.9505	0.9454
196	0.9965	0.9915	0.9864	0.9813	0.9762	0.9711	0.9660	0.9609	0.9559	0.9508	0.9457
197	0.9965	0.9915	0.9865	0.9814	0.9763	0.9713	0.9662	0.9611	0.9561	0.9510	0.9459
198	0.9965	0.9915	0.9865	0.9815	0.9765	0.9714	0.9664	0.9613	0.9563	0.9513	0.9462
199	0.9965	0.9916	0.9866	0.9816	0.9766	0.9716	0.9665	0.9615	0.9565	0.9515	0.9465
200	0.9965	0.9916	0.9867	0.9817	0.9767	0.9717	0.9667	0.9617	0.9567	0.9517	0.9468
201	0.9966	0.9917	0.9867	0.9818	0.9768	0.9718	0.9669	0.9619	0.9569	0.9520	0.9470
202	0.9966	0.9917	0.9868	0.9819	0.9769	0.9720	0.9670	0.9621	0.9572	0.9522	0.9473
203	0.9966	0.9918	0.9869	0.9819	0.9770	0.9721	0.9672	0.9623	0.9574	0.9525	0.9475
204	0.9966	0.9918	0.9869	0.9820	0.9771	0.9723	0.9674	0.9625	0.9576	0.9527	0.9478
205	0.9966	0.9918	0.9870	0.9921	0.9773	0.9724	0.9675	0.9627	0.9578	0.9529	0.9480

TABLE B.5

Binomial 50% confidence levels (continued).

Sample Size N	Number of Failures in Sample = F: 0	1	2	3	4	5	6	7	8	9	10
206	0.9966	0.9919	0.9870	0.9822	0.9774	0.9725	0.9677	0.9628	0.9580	0.9531	0.9483
207	0.9967	0.9919	0.9871	0.9823	0.9775	0.9727	0.9678	0.9630	0.9582	0.9534	0.9485
208	0.9967	0.9919	0.9872	0.9824	0.9776	0.9728	0.9680	0.9632	0.9584	0.9536	0.9488
209	0.9967	0.9920	0.9872	0.9825	0.9777	0.9729	0.9681	0.9634	0.9586	0.9538	0.9490
210	0.9967	0.9920	0.9873	0.9825	0.9778	0.9730	0.9683	0.9635	0.9588	0.9540	0.9493
211	0.9967	0.9921	0.9874	0.9826	0.9779	0.9732	0.9684	0.9637	0.9590	0.9543	0.9495
212	0.9967	0.9921	0.9874	0.9827	0.9780	0.9733	0.9686	0.9639	0.9592	0.9545	0.9498
213	0.9968	0.9921	0.9875	0.9828	0.9781	0.9734	0.9687	0.9641	0.9594	0.9547	0.9500
211	0.9968	0.9922	0.9875	0.9829	0.9782	0.9736	0.9689	0.9642	0.9596	0.9549	0.9502
215	0.9968	0.9922	0.9876	0.9830	0.9783	0.9737	0.9690	0.9644	0.9597	0.9551	0.9505
216	0.9968	0.9922	0.9876	0.9830	0.9784	0.9738	0.9692	0.9646	0.9599	0.9553	0.9507
217	0.9968	0.9923	0.9877	0.9831	0.9785	0.9739	0.9693	0.9647	0.9601	0.9555	0.9509
218	0.9968	0.9923	0.9878	0.9832	0.9786	0.9740	0.9695	0.9649	0.9603	0.9557	0.9511
219	0.9968	0.9924	0.9878	0.9833	0.9787	0.9742	0.9696	0.9650	0.9605	0.9559	0.9514
220	0.9969	0.9924	0.9879	0.9833	0.9788	0.9743	0.9697	0.9652	0.9607	0.9561	0.9516
221	0.9969	0.9924	0.9879	0.9834	0.9789	0.9744	0.9699	0.9654	0.9608	0.9563	0.9518
222	0.9969	0.9925	0.9880	0.9835	0.9790	0.9745	0.9700	0.9655	0.9610	0.9565	0.9520
223	0.9969	0.9925	0.9880	0.9836	0.9791	0.9746	0.9701	0.9657	0.9612	0.9567	0.9522
224	0.9969	0.9925	0.9881	0.9836	0.9792	0.9747	0.9703	0.9658	0.9614	0.9569	0.9524
225	0.9969	0.9926	0.9881	0.9837	0.9793	0.9748	0.9704	0.9660	0.9615	0.9571	0.9527
226	0.9969	0.9926	0.9882	0.9838	0.9794	0.9750	0.9705	0.9661	0.9617	0.9573	0.9529
227	0.9970	0.9926	0.9882	0.9839	0.9795	0.9751	0.9707	0.9663	0.9619	0.9575	0.9531
228	0.9970	0.9927	0.9883	0.9839	0.9795	0.9752	0.9708	0.9664	0.9620	0.9577	0.9533
229	0.9970	0.9927	0.9883	0.9840	0.9796	0.9753	0.9709	0.9666	0.9622	0.9578	0.9535
230	0.9970	0.9927	0.9884	0.9841	0.9797	0.9754	0.9710	0.9667	0.9624	0.9580	0.9537
235	0.9971	0.9929	0.9886	0.9844	0.9802	0.9759	0.9717	0.9674	0.9632	0.9589	0.9547
240	0.9971	0.9930	0.9889	0.9847	0.9806	0.9764	0.9723	0.9681	0.9639	0.9598	0.9556
245	0.9972	0.9932	0.9891	0.9850	0.9810	0.9769	0.9728	0.9687	0.9647	0.9606	0.9565
250	0.9972	0.9933	0.9893	0.9853	0.9813	0.9774	0.9734	0.9694	0.9654	0.9614	0.9574

TABLE B.6
Binomial 75% confidence levels.

Values $R_{0.75}$ such that $P(x > F) = 0.75 = 1 - P(x \leq F) = 1.0 - \sum_{x=0}^{F} \binom{N}{x} R_{0.75}{}^{N-x}(1 - R_{0.75})^{x}$

Sample Size N	Number of Failures in Sample = F: 0	1	2	3	4	5	6	7	8	9	10
1	0.250										
2	0.500	0.134									
3	0.630	0.326	0.091								
4	0.707	0.456	0.243	0.069							
5	0.758	0.546	0.359	0.194	0.056						
6	0.794	0.610	0.447	0.297	0.161	0.047					
7	0.821	0.659	0.514	0.379	0.253	0.138	0.040				
8	0.841	0.699	0.567	0.445	0.329	0.221	0.121	0.035			
9	0.857	0.728	0.612	0.498	0.392	0.291	0.196	0.107	0.031		
10	0.871	0.753	0.645	0.545	0.445	0.351	0.261	0.176	0.096	0.028	
11	0.882	0.773	0.674	0.580	0.491	0.402	0.317	0.236	0.159	0.088	0.026
12	0.891	0.791	0.699	0.611	0.527	0.448	0.366	0.290	0.216	0.146	0.080
13	0.899	0.806	0.720	0.639	0.560	0.483	0.412	0.337	0.267	0.199	0.134
14	0.906	0.819	0.739	0.662	0.588	0.516	0.447	0.381	0.312	0.247	0.185
15	0.912	0.830	0.755	0.683	0.613	0.546	0.480	0.415	0.354	0.290	0.230
16	0.917	0.840	0.769	0.702	0.636	0.572	0.509	0.448	0.388	0.331	0.271
17	0.922	0.849	0.782	0.718	0.656	0.595	0.535	0.477	0.420	0.364	0.311
18	0.926	0.857	0.794	0.733	0.673	0.616	0.559	0.504	0.449	0.395	0.343
19	0.930	0.864	0.804	0.746	0.690	0.635	0.581	0.528	0.475	0.424	0.374
20	0.933	0.871	0.813	0.758	0.704	0.652	0.600	0.549	0.499	0.450	0.402
21	0.936	0.877	0.822	0.769	0.717	0.667	0.618	0.569	0.521	0.474	0.428
22	0.939	0.882	0.829	0.779	0.730	0.681	0.634	0.587	0.542	0.496	0.451
23	0.942	0.887	0.837	0.788	0.741	0.694	0.649	0.604	0.560	0.516	0.473
24	0.944	0.892	0.843	0.796	0.751	0.706	0.663	0.620	0.577	0.535	0.494
25	0.946	0.896	0.849	0.804	0.760	0.718	0.675	0.634	0.593	0.552	0.512
26	0.948	0.900	0.855	0.811	0.769	0.728	0.687	0.647	0.608	0.568	0.530
27	0.950	0.903	0.860	0.818	0.777	0.738	0.698	0.660	0.621	0.583	0.546
28	0.952	0.907	0.865	0.824	0.785	0.746	0.709	0.671	0.634	0.597	0.561
29	0.953	0.910	0.869	0.830	0.792	0.755	0.718	0.682	0.646	0.611	0.575
30	0.955	0.913	0.873	0.836	0.799	0.763	0.727	0.692	0.657	0.623	0.589
31	0.956	0.915	0.877	0.841	0.805	0.770	0.735	0.701	0.668	0.634	0.601
32	0.958	0.918	0.881	0.846	0.811	0.777	0.743	0.710	0.678	0.645	0.613
33	0.959	0.920	0.885	0.850	0.816	0.783	0.751	0.719	0.687	0.655	0.624
34	0.960	0.923	0.888	0.854	0.822	0.790	0.758	0.727	0.696	0.665	0.635
35	0.961	0.925	0.891	0.858	0.827	0.795	0.765	0.734	0.704	0.674	0.645
36	0.962	0.927	0.894	0.862	0.831	0.801	0.771	0.741	0.712	0.683	0.654
37	0.963	0.929	0.897	0.866	0.836	0.806	0.777	0.748	0.719	0.691	0.663
38	0.964	0.931	0.899	0.869	0.840	0.811	0.782	0.754	0.726	0.699	0.671
39	0.965	0.932	0.902	0.873	0.844	0.816	0.788	0.760	0.733	0.706	0.680
40	0.966	0.934	0.904	0.876	0.848	0.820	0.793	0.766	0.740	0.713	0.687
41	0.967	0.936	0.907	0.879	0.851	0.824	0.798	0.772	0.746	0.720	0.695
42	0.968	0.937	0.909	0.881	0.855	0.828	0.803	0.777	0.752	0.726	0.701
43	0.968	0.939	0.911	0.884	0.858	0.832	0.807	0.782	0.757	0.733	0.708
44	0.969	0.940	0.913	0.887	0.861	0.836	0.811	0.787	0.763	0.738	0.715
45	0.970	0.941	0.915	0.889	0.864	0.840	0.815	0.791	0.768	0.744	0.721
46	0.970	0.943	0.917	0.891	0.867	0.843	0.819	0.796	0.772	0.749	0.727
47	0.971	0.944	0.918	0.894	0.870	0.846	0.823	0.800	0.777	0.755	0.732
48	0.972	0.945	0.920	0.896	0.872	0.849	0.827	0.804	0.782	0.759	0.738
49	0.972	0.946	0.922	0.898	0.875	0.852	0.830	0.808	0.786	0.764	0.743
50	0.973	0.947	0.923	0.900	0.877	0.855	0.833	0.812	0.790	0.769	0.748

TABLE B.6
Binomial 75% confidence levels (continued).

Sample Size N	Number of Failures in Sample = F										
	0	1	2	3	4	5	6	7	8	9	10
51	0.973	0.948	0.925	0.902	0.880	0.858	0.836	0.815	0.794	0.773	0.752
52	0.974	0.949	0.926	0.904	0.882	0.861	0.840	0.819	0.798	0.777	0.757
53	0.974	0.950	0.927	0.906	0.884	0.863	0.842	0.822	0.802	0.782	0.762
54	0.975	0.951	0.929	0.907	0.886	0.966	0.845	0.825	0.805	0.785	0.766
55	0.975	0.952	0.930	0.909	0.888	0.868	0.848	0.828	0.809	0.789	0.770
56	0.976	0.953	0.931	0.910	0.890	0.870	0.851	0.831	0.812	0.973	0.774
57	0.976	0.953	0.932	0.912	0.892	0.873	0.853	0.834	0.815	0.796	0.778
58	0.976	0.954	0.934	0.914	0.894	0.875	0.856	0.837	0.818	0.800	0.782
59	0.977	0.955	0.935	0.915	0.896	0.877	0.858	0.840	0.821	0.803	0.785
60	0.977	0.956	0.936	0.916	0.897	0.879	0.860	0.842	0.824	0.806	0.789
61	0.978	0.956	0.937	0.918	0.899	0.881	0.863	0.845	0.827	0.809	0.792
62	0.978	0.957	0.938	0.919	0.901	0.883	0.865	0.847	0.830	0.812	0.795
63	0.978	0.958	0.939	0.920	0.902	0.884	0.867	0.850	0.832	0.815	0.798
64	0.979	0.958	0.940	0.921	0.904	0.886	0.869	0.852	0.835	0.818	0.801
65	0.979	0.959	0.941	0.923	0.905	0.888	0.871	0.854	0.837	0.821	0.804
66	0.979	0.960	0.941	0.924	0.907	0.890	0.873	0.856	0.840	0.824	0.807
67	0.980	0.960	0.942	0.925	0.908	0.891	0.875	0.858	0.842	0.826	0.810
68	0.980	0.961	0.943	0.926	0.909	0.893	0.877	0.860	0.844	0.829	0.813
69	0.980	0.961	0.944	0.927	0.911	0.894	0.878	0.862	0.847	0.831	0.816
70	0.980	0.962	0.945	0.928	0.912	0.896	0.880	0.864	0.849	0.833	0.818
71	0.981	0.963	0.946	0.929	0.913	0.897	0.882	0.866	0.851	0.836	0.821
72	0.981	0.963	0.946	0.930	0.914	0.899	0.883	0.868	0.853	0.838	0.823
73	0.981	0.964	0.947	0.931	0.915	0.900	0.885	0.870	0.855	0.840	0.825
74	0.981	0.964	0.948	0.932	0.917	0.901	0.886	0.872	0.857	0.842	0.828
75	0.982	0.965	0.948	0.933	0.918	0.903	0.888	0.873	0.859	0.844	0.830
76	0.982	0.965	0.959	0.934	0.919	0.904	0.889	0.875	0.861	0.846	0.832
77	0.982	0.965	0.950	0.935	0.920	0.905	0.891	0.876	0.862	0.848	0.834
78	0.982	0.966	0.950	0.935	0.921	0.906	0.892	0.878	0.864	0.850	0.836
79	0.983	0.966	0.951	0.936	0.922	0.907	0.893	0.880	0.866	0.852	0.838
80	0.983	0.967	0.952	0.937	0.923	0.909	0.895	0.881	0.867	0.854	0.840
81	0.983	0.967	0.952	0.938	0.924	0.910	0.896	0.882	0.869	0.856	0.842
82	0.983	0.968	0.953	0.938	0.925	0.911	0.897	0.884	0.871	0.857	0.844
83	0.983	0.968	0.953	0.939	0.295	0.912	0.899	0.885	0.872	0.859	0.846
84	0.934	0.968	0.954	0.940	0.926	0.913	0.900	0.887	0.874	0.861	0.848
85	0.984	0.969	0.954	0.941	0.927	0.914	0.901	0.888	0.875	0.862	0.850
86	0.984	0.969	0.955	0.941	0.928	0.915	0.902	0.889	0.877	0.864	0.851
87	0.984	0.969	0.955	0.942	0.929	0.916	0.903	0.890	0.878	0.865	0.853
88	0.984	0.970	0.956	0.943	0.930	0.917	0.904	0.892	0.879	0.867	0.855
89	0.985	0.970	0.956	0.943	0.930	0.918	0.905	0.893	0.881	0.868	0.856
90	0.985	0.970	0.957	0.944	0.931	0.918	0.906	0.894	0.882	0.870	0.858
91	0.985	0.971	0.957	0.945	0.932	0.920	0.907	0.895	0.883	0.871	0.859
92	0.985	0.971	0.958	0.945	0.933	0.920	0.908	0.896	0.884	0.873	0.861
93	0.985	0.971	0.958	0.946	0.933	0.921	0.909	0.897	0.886	0.874	0.862
94	0.985	0.972	0.959	0.946	0.934	0.922	0.910	0.898	0.887	0.875	0.864
95	0.985	0.972	0.959	0.947	0.935	0.923	0.911	0.900	0.888	0.877	0.865
96	0.986	0.972	0.960	0.947	0.935	0.924	0.912	0.901	0.889	0.878	0.867
97	0.986	0.972	0.960	0.948	0.936	0.924	0.913	0.902	0.890	0.879	0.868
98	0.986	0.973	0.960	0.948	0.937	0.925	0.914	0.903	0.891	0.880	0.869
99	0.986	0.973	0.961	0.949	0.937	0.926	0.915	0.904	0.892	0.881	0.871
100	0.986	0.973	0.961	0.949	0.938	0.927	0.916	0.904	0.894	0.883	0.872

TABLE B.6
Binomial 75% confidence levels (continued).

Sample Size N	Number of Failures in Sample = F										
	0	1	2	3	4	5	6	7	8	9	10
101	0.986	0.974	0.962	0.950	0.939	0.927	0.916	0.905	0.895	0.884	0.873
102	0.987	0.974	0.962	0.950	0.939	0.928	0.917	0.906	0.896	0.885	0.874
103	0.987	0.974	0.962	0.951	0.940	0.929	0.918	0.907	0.897	0.886	0.875
104	0.987	0.974	0.963	0.951	0.940	0.929	0.919	0.908	0.898	0.887	0.877
105	0.987	0.975	0.963	0.952	0.941	0.930	0.920	0.909	0.899	0.888	0.878
106	0.987	0.975	0.963	0.952	0.941	0.931	0.920	0.910	0.899	0.889	0.879
107	0.987	0.975	0.964	0.953	0.942	0.931	0.921	0.911	0.900	0.890	0.880
108	0.987	0.975	0.964	0.953	0.943	0.932	0.922	0.911	0.901	0.891	0.881
109	0.987	0.975	0.964	0.954	0.943	0.933	0.922	0.912	0.902	0.892	0.882
110	0.987	0.976	0.965	0.954	0.944	0.933	0.923	0.913	0.903	0.893	0.883
111	0.988	0.976	0.965	0.954	0.944	0.934	0.924	0.914	0.904	0.984	0.884
112	0.988	0.976	0.965	0.955	0.945	0.934	0.924	0.915	0.905	0.895	0.885
113	0.988	0.976	0.966	0.955	0.945	0.935	0.925	0.915	0.906	0.896	0.886
114	0.988	0.977	0.966	0.956	0.946	0.936	0.926	0.916	0.906	0.897	0.887
115	0.988	0.977	0.966	0.956	0.946	0.936	0.926	0.917	0.907	0.898	0.888
116	0.9881	0.9770	0.9665	0.9564	0.9464	0.9367	0.9271	0.9175	0.9081	0.8986	0.8893
117	0.9882	0.9772	0.9668	0.9567	0.9469	0.9372	0.9277	0.9182	0.9088	0.8995	0.8902
118	0.9883	0.9773	0.9671	0.9571	0.9473	0.9377	0.9283	0.9189	0.9096	0.9003	0.8911
119	0.9884	0.9775	0.9673	0.9574	0.9478	0.9383	0.9289	0.9196	0.9103	0.9011	0.8920
120	0.9885	0.9777	0.9676	0.9578	0.9482	0.9388	0.9295	0.9202	0.9111	0.9019	0.8929
121	0.9886	0.9779	0.9679	0.9581	0.9486	0.9393	0.9300	0.9209	0.9118	0.9027	0.8938
122	0.9887	0.9781	0.9681	0.9585	0.9491	0.9398	0.9306	0.9215	0.9125	0.9035	0.8946
123	0.9888	0.9783	0.9684	0.9588	0.9495	0.9403	0.9312	0.9222	0.9132	0.9043	0.8955
124	0.9889	0.9784	0.9686	0.9591	0.9499	0.9407	0.9317	0.9228	0.9139	0.9051	0.8963
125	0.9890	0.9786	0.9689	0.9595	0.9503	0.9412	0.9323	0.9234	0.9146	0.9058	0.8971
126	0.9891	0.9788	0.9691	0.9598	0.9507	0.9417	0.9328	0.9240	0.9153	0.9066	0.8979
127	0.9891	0.9789	0.9694	0.9601	0.9510	0.9421	0.9333	0.9246	0.9159	0.9073	0.8987
128	0.9892	0.9791	0.9696	0.9604	0.9514	0.9426	0.9338	0.9252	0.9166	0.9080	0.8995
129	0.9893	0.9793	0.9698	0.9607	0.9518	0.9430	0.9343	0.9257	0.9172	0.9087	0.9003
130	0.9894	0.9794	0.9701	0.9610	0.9522	0.9434	0.9348	0.9263	0.9178	0.9094	0.9010
131	0.9895	0.9796	0.9703	0.9613	0.9525	0.9439	0.9353	0.9269	0.9185	0.9101	0.9018
132	0.9896	0.9797	0.9705	0.9616	0.9529	0.9443	0.9358	0.9274	0.9191	0.9108	0.9025
133	0.9896	0.9799	0.9707	0.9619	0.9532	0.9447	0.9363	0.9280	0.9197	0.9114	0.9033
133	0.9897	0.9800	0.9710	0.9622	0.9536	0.9451	0.9368	0.9285	0.9203	0.9121	0.9040
135	0.9898	0.9802	0.9712	0.9625	0.9539	0.9455	0.9372	0.9290	0.9209	0.9127	0.9047
136	0.9899	0.9803	0.9714	0.9627	0.9543	0.9459	0.9377	0.9295	0.9214	0.9134	0.9054
137	0.9899	0.9805	0.9716	0.9630	0.9546	0.9463	0.9381	0.9300	0.9220	0.9140	0.9060
138	0.9900	0.9806	0.9718	0.9633	0.9549	0.9467	0.9386	0.9305	0.9226	0.9146	0.9067
139	0.9901	0.9808	0.9720	0.9635	0.9552	0.9471	0.9390	0.9310	0.9231	0.9152	0.9074
140	0.9902	0.9809	0.9722	0.9638	0.9556	0.9474	0.9394	0.9315	0.9237	0.9158	0.9080
141	0.9902	0.9810	0.9724	0.9640	0.9559	0.9478	0.9399	0.9320	0.9242	0.9164	0.9087
142	0.9903	0.9812	0.9726	0.9643	0.9562	0.9482	0.9403	0.9325	0.9247	0.9170	0.9093
143	0.9904	0.9813	0.9728	0.9645	0.9565	0.9485	0.9407	0.9329	0.9252	0.9175	0.9099
144	0.9904	0.9814	0.9730	0.9648	0.9568	0.9489	0.9411	0.9334	0.9258	0.9181	0.9106
145	0.9905	0.9815	0.9731	0.9650	0.9571	0.9492	0.9415	0.9339	0.9263	0.9187	0.9112
146	0.9906	0.9817	0.9733	0.9653	0.9574	0.9496	0.9419	0.9343	0.9268	0.9192	0.9118
147	0.9906	0.9818	0.9735	0.9655	0.9577	0.9499	0.9423	0.9348	0.9273	0.9198	0.9124
148	0.9907	0.9819	0.9737	0.9657	0.9579	0.9503	0.9427	0.9352	0.9277	0.9203	0.9130
149	0.9907	0.9820	0.9739	0.9660	0.9582	0.9506	0.9431	0.9356	0.9282	0.9209	0.9135
150	0.9908	0.9822	0.9740	0.9662	0.9585	0.9509	0.9435	0.9360	0.9287	0.9214	0.9141

TABLE B.6
Binomial 75% confidence levels (continued).

Sample Size N	Number of Failures in Sample = F										
	0	1	2	3	4	5	6	7	8	9	10
151	0.9909	0.9823	0.9742	0.9664	0.9588	0.9512	0.9438	0.9365	0.9292	0.9219	0.9147
152	0.9909	0.9824	0.9744	0.9666	0.9590	0.9516	0.9442	0.9369	0.9296	0.9224	0.9152
153	0.9910	0.9825	0.9745	0.9668	0.9593	0.9519	0.9446	0.9373	0.9301	0.9229	0.9158
154	0.9910	0.9826	0.9747	0.9671	0.9596	0.9522	0.9449	0.9377	0.9305	0.9234	0.9163
155	0.9911	0.9827	0.9749	0.9673	0.9598	0.9525	0.9453	0.9381	0.9310	0.9239	0.9168
156	0.9912	0.9828	0.9750	0.9675	0.9601	0.9528	0.9456	0.9385	0.9314	0.9244	0.9174
157	0.9912	0.9829	0.9752	0.9677	0.9603	0.9531	0.9460	0.9389	0.9318	0.9248	0.9179
158	0.9913	0.9831	0.9753	0.9679	0.9606	0.9534	0.9463	0.9393	0.9323	0.9253	0.9184
159	0.9913	0.9832	0.9755	0.9681	0.9608	0.9537	0.9466	0.9396	0.9327	0.9258	0.9189
160	0.9914	0.9833	0.9756	0.9683	0.9611	0.9540	0.9470	0.9400	0.9331	0.9262	0.9194
161	0.9914	0.9834	0.9758	0.9685	0.9613	0.9542	0.9473	0.9404	0.9335	0.9267	0.9200
162	0.9915	0.9835	0.9759	0.9687	0.9615	0.9545	0.9476	0.9407	0.9339	0.9271	0.9204
163	0.9915	0.9836	0.9761	0.9689	0.9618	0.9548	0.9479	0.9411	0.9343	0.9276	0.9209
164	0.9916	0.9837	0.9762	0.9690	0.9620	0.9551	0.9482	0.9415	0.9347	0.9280	0.9214
165	0.9916	0.9838	0.9764	0.9692	0.9622	0.9553	0.9486	0.9418	0.9351	0.9284	0.9218
166	0.9917	0.9839	0.9765	0.9694	0.9625	0.9556	0.9489	0.9422	0.9355	0.9289	0.9223
167	0.9917	0.9840	0.9767	0.9696	0.9627	0.9559	0.9492	0.9425	0.9359	0.9293	0.9228
168	0.9918	0.9841	0.9768	0.9698	0.9629	0.9561	0.9495	0.9428	0.9363	0.9297	0.9232
169	0.9918	0.9842	0.9769	0.9700	0.9631	0.9564	0.9498	0.9432	0.9366	0.9301	0.9237
170	0.9919	0.9842	0.9771	0.9701	0.9633	0.9567	0.9501	0.9435	0.9370	0.9305	0.9241
171	0.9919	0.9843	0.9772	0.9703	0.9636	0.9569	0.9503	0.9438	0.9374	0.9309	0.9246
172	0.9920	0.9844	0.9773	0.9705	0.9638	0.9572	0.9506	0.9442	0.9378	0.9313	0.9250
173	0.9920	0.9845	0.9775	0.9706	0.9640	0.9574	0.9509	0.9445	0.9381	0.9317	0.9254
174	0.9921	0.9846	0.9776	0.9708	0.9642	0.9576	0.9512	0.9448	0.9385	0.9321	0.9258
175	0.9921	0.9847	0.9777	0.9710	0.9644	0.9579	0.9515	0.9451	0.9388	0.9325	0.9263
176	0.9922	0.9848	0.9779	0.9711	0.9646	0.9581	0.9517	0.9454	0.9391	0.9329	0.9267
177	0.9922	0.9849	0.9780	0.9713	0.9648	0.9584	0.9520	0.9457	0.9395	0.9333	0.9271
178	0.9922	0.9849	0.9781	0.9715	0.9650	0.9586	0.9523	0.9460	0.9398	0.9336	0.0275
179	0.9923	0.9850	0.9782	0.9716	0.9652	0.9588	0.9525	0.9463	0.9402	0.9340	0.9279
180	0.9923	0.9851	0.9783	0.9718	0.9654	0.9590	0.9528	0.9466	0.9405	0.9344	0.9283
181	0.9924	0.9852	0.9785	0.9719	0.9656	0.9593	0.9531	0.9469	0.9408	0.9347	0.9287
182	0.9924	0.9853	0.9786	0.9721	0.9657	0.9595	0.9533	0.9472	0.9411	0.9351	0.9291
183	0.9925	0.9854	0.9787	0.9722	0.9659	0.9597	0.9536	0.9475	0.9415	0.9354	0.9295
184	0.9925	0.9854	0.9788	0.9724	0.9661	0.9599	0.9538	0.9478	0.9418	0.9358	0.9298
185	0.9925	0.9855	0.9789	0.9725	0.9663	0.9601	0.9541	0.9481	0.9421	0.9361	0.9302
186	0.9926	0.9856	0.9790	0.9727	0.9665	0.9604	0.9543	0.9483	0.9424	0.9365	0.9306
187	0.9926	0.9857	0.9791	0.9728	0.9667	0.9606	0.9546	0.9486	0.9427	0.9368	0.9310
188	0.9927	0.9857	0.9793	0.9730	0.9668	0.9608	0.9548	0.9489	0.9430	0.9371	0.9313
189	0.9927	0.9858	0.9794	0.9731	0.9670	0.9610	0.9550	0.9491	0.9433	0.9375	0.9317
190	0.9927	0.9859	0.9795	0.9733	0.9672	0.9612	0.9553	0.9494	0.9436	0.9378	0.9320
191	0.9928	0.9860	0.9796	0.9734	0.9673	0.9614	0.9555	0.9497	0.9439	0.9381	0.9324
192	0.9928	0.9860	0.9797	0.9735	0.9675	0.9616	0.9557	0.9499	0.9442	0.9384	0.9327
193	0.9928	0.9861	0.9798	0.9737	0.9677	0.9618	0.9560	0.9502	0.9445	0.9387	0.9331
194	0.9929	0.9862	0.9799	0.9738	0.9679	0.9620	0.9562	0.9504	0.9448	0.9391	0.9334
195	0.9929	0.9863	0.9800	0.9739	0.9680	0.9622	0.9564	0.9507	0.9450	0.9394	0.9338
196	0.9930	0.9863	0.9801	0.9741	0.9681	0.9624	0.9566	0.9510	0.9453	0.9397	0.9341
197	0.9930	0.9864	0.9802	0.9742	0.9683	0.9626	0.9569	0.9512	0.9456	0.9400	0.9344
198	0.9930	0.9865	0.9803	0.9743	0.9685	0.9627	0.9571	0.9514	0.9459	0.9403	0.9348
199	0.9931	0.9865	0.9804	0.9745	0.9687	0.9629	0.9573	0.9517	0.9461	0.9406	0.9351
200	0.9931	0.9866	0.9805	0.9746	0.9688	0.9631	0.9575	0.9519	0.9464	0.9409	0.9354

TABLE B.6
Binomial 75% confidence levels (continued).

Sample Size N	Number of Failures in Sample = F										
	0	1	2	3	4	5	6	7	8	9	10
201	0.9931	0.9867	0.9806	0.9747	0.9689	0.9633	0.9577	0.9522	0.9467	0.9412	0.9357
202	0.9932	0.9867	0.9807	0.9748	0.9691	0.9635	0.9579	0.9524	0.9469	0.9415	0.9361
203	0.9932	0.9868	0.9808	0.9750	0.9693	0.9637	0.9581	0.9526	0.9472	0.9417	0.9364
204	0.9932	0.9869	0.9809	0.9751	0.9694	0.9638	0.9583	0.9529	0.9474	0.9420	0.9367
205	0.9933	0.9869	0.9810	0.9752	0.9696	0.9640	0.9585	0.9531	0.9477	0.9423	0.9370
206	0.9933	0.9870	0.9811	0.9753	0.9697	0.9642	0.9587	0.9533	0.9479	0.9426	0.9373
207	0.9933	0.9871	0.9812	0.9754	0.9699	0.9644	0.9589	0.9535	0.9482	0.9429	0.9376
208	0.9934	0.9871	0.9812	0.9756	0.9700	0.9645	0.9591	0.9538	0.9484	0.9431	0.9379
209	0.9934	0.9872	0.9813	0.9757	0.9701	0.9647	0.9593	0.9540	0.9487	0.9434	0.9382
210	0.9934	0.9872	0.9814	0.9758	0.9703	0.9649	0.9595	0.9542	0.9489	0.9437	0.9385
211	0.9935	0.9873	0.9815	0.9759	0.9704	0.9650	0.9597	0.9544	0.9492	0.9439	0.9388
212	0.9935	0.9874	0.9816	0.9760	0.9706	0.9652	0.9599	0.9546	0.9494	0.9442	0.9390
213	0.9935	0.9874	0.9817	0.9761	0.9707	0.9654	0.9601	0.9548	0.9496	0.9445	0.9393
214	0.9935	0.9875	0.9818	0.9762	0.9708	0.9655	0.9603	0.9551	0.9499	0.9447	0.9396
215	0.9936	0.9875	0.9819	0.9764	0.9710	0.9657	0.9604	0.9553	0.9501	0.9450	0.9399
216	0.9936	0.9876	0.9819	0.9765	0.9711	0.9658	0.9606	0.9555	0.9503	0.9452	0.9402
217	0.9936	0.9876	0.9820	0.9766	0.9712	0.9660	0.9608	0.9557	0.9506	0.9455	0.9404
218	0.9937	0.9877	0.9821	0.9767	0.9714	0.9661	0.9610	0.9559	0.9508	0.9457	0.9407
219	0.9937	0.9878	0.9822	0.9768	0.9715	0.9663	0.9612	0.9561	0.9510	0.9460	0.9410
220	0.9937	0.9878	0.9823	0.9769	0.9716	0.9665	0.9613	0.9563	0.9512	0.9462	0.9412
221	0.9938	0.9879	0.9823	0.9770	0.9718	0.9666	0.9615	0.9565	0.9515	0.9465	0.9415
222	0.9938	0.9879	0.9824	0.9771	0.9719	0.9668	0.9617	0.9567	0.9517	0.9467	0.9418
223	0.9938	0.9880	0.9825	0.9772	0.9720	0.9669	0.9619	0.9569	0.9519	0.9469	0.9420
224	0.9938	0.9880	0.9826	0.9773	0.9721	0.9670	0.9620	0.9570	0.9521	0.9472	0.9423
225	0.9939	0.9881	0.9827	0.9774	0.9723	0.9672	0.9622	0.9572	0.9523	0.9474	0.9425
226	0.9939	0.9881	0.9827	0.9775	0.9724	0.9673	0.9624	0.9574	0.9525	0.9476	0.9428
227	0.9939	0.9882	0.9828	0.9776	0.9725	0.9675	0.9625	0.9576	0.9527	0.9479	0.9430
228	0.9939	0.9882	0.9829	0.9777	0.9726	0.9676	0.9627	0.9578	0.9529	0.9481	0.9433
229	0.9940	0.9883	0.9830	0.9778	0.9727	0.9678	0.9629	0.9580	0.9531	0.9483	0.9435
230	0.9940	0.9883	0.9830	0.9779	0.9729	0.9679	0.9630	0.9582	0.9533	0.9485	0.9438
235	0.9941	0.9886	0.9834	0.9784	0.9734	0.9686	0.9638	0.9590	0.9543	0.9496	0.9450
240	0.9942	0.9888	0.9837	0.9788	0.9740	0.9692	0.9645	0.9599	0.9553	0.9507	0.9461
245	0.9944	0.9891	0.9841	0.9792	0.9745	0.9699	0.9652	0.9607	0.9562	0.9517	0.9472
250	0.9945	0.9893	0.9844	0.9797	0.9750	0.9705	0.9660	0.9615	0.9571	0.9526	0.9483

TABLE B.7
Binomial 95% confidence levels.

Values $R_{0.95}$ such that $P(x > F) = 0.95 = 1 - P(x \le F) = 1.0 - \sum_{x=0}^{F} \binom{N}{x} R_{0.95}^{N-x}(1 - R_{0.95})^{x}$

Sample Size *N*	Number of Failures in Sample = *F*										
	0	1	2	3	4	5	6	7	8	9	10
1	0.050										
2	0.224	0.025									
3	0.368	0.135	0.017								
4	0.473	0.249	0.098	0.013							
5	0.549	0.343	0.189	0.076	0.010						
6	0.607	0.418	0.271	0.153	0.063	0.009					
7	0.655	0.479	0.341	0.225	0.129	0.053	0.007				
8	0.688	0.534	0.400	0.289	0.193	0.111	0.046	0.006			
9	0.717	0.571	0.455	0.345	0.251	0.169	0.098	0.041	0.006		
10	0.741	0.606	0.493	0.399	0.306	0.222	0.150	0.087	0.037	0.005	
11	0.762	0.636	0.530	0.436	0.355	0.271	0.200	0.135	0.079	0.033	0.005
12	0.779	0.661	0.562	0.473	0.391	0.320	0.245	0.181	0.123	0.072	0.030
13	0.794	0.684	0.590	0.505	0.427	0.355	0.292	0.224	0.166	0.113	0.066
14	0.807	0.703	0.615	0.534	0.460	0.390	0.325	0.268	0.206	0.153	0.104
15	0.819	0.721	0.637	0.560	0.489	0.423	0.360	0.300	0.248	0.191	0.142
16	0.829	0.736	0.656	0.583	0.516	0.452	0.391	0.333	0.279	0.231	0.178
17	0.838	0.750	0.674	0.604	0.539	0.478	0.420	0.364	0.311	0.260	0.216
18	0.847	0.762	0.690	0.623	0.561	0.502	0.446	0.392	0.341	0.291	0.244
19	0.854	0.774	0.704	0.641	0.581	0.524	0.470	0.418	0.368	0.320	0.274
20	0.861	0.784	0.717	0.656	0.599	0.544	0.492	0.442	0.394	0.347	0.302
21	0.867	0.793	0.729	0.671	0.616	0.563	0.513	0.464	0.417	0.372	0.328
22	0.873	0.802	0.741	0.684	0.631	0.580	0.532	0.485	0.439	0.395	0.353
23	0.878	0.810	0.751	0.696	0.645	0.596	0.549	0.504	0.460	0.417	0.375
24	0.883	0.817	0.760	0.708	0.658	0.611	0.565	0.521	0.479	0.437	0.397
25	0.887	0.824	0.769	0.718	0.670	0.625	0.581	0.538	0.496	0.456	0.417
26	0.891	0.830	0.777	0.728	0.682	0.637	0.595	0.553	0.513	0.473	0.436
27	0.895	0.836	0.785	0.737	0.692	0.649	0.608	0.568	0.529	0.490	0.453
28	0.899	0.841	0.792	0.746	0.702	0.661	0.620	0.581	0.543	0.506	0.470
29	0.902	0.847	0.798	0.754	0.712	0.671	0.632	0.594	0.557	0.521	0.486
30	0.905	0.851	0.805	0.761	0.720	0.681	0.643	0.606	0.570	0.535	0.501
31	0.908	0.856	0.811	0.768	0.729	0.690	0.653	0.617	0.582	0.548	0.515
32	0.911	0.860	0.816	0.775	0.736	0.699	0.663	0.628	0.594	0.561	0.528
33	0.913	0.864	0.821	0.781	0.744	0.707	0.672	0.638	0.605	0.572	0.540
34	0.916	0.868	0.826	0.787	0.751	0.715	0.681	0.648	0.615	0.584	0.552
35	0.918	0.871	0.831	0.793	0.757	0.723	0.690	0.657	0.625	0.594	0.564
36	0.920	0.875	0.835	0.798	0.764	0.730	0.697	0.666	0.635	0.604	0.575
37	0.922	0.878	0.839	0.804	0.769	0.737	0.705	0.674	0.644	0.614	0.585
38	0.924	0.881	0.843	0.808	0.775	0.743	0.712	0.682	0.652	0.623	0.595
39	0.926	0.884	0.847	0.813	0.780	0.749	0.719	0.689	0.660	0.632	0.604
40	0.928	0.887	0.851	0.817	0.786	0.755	0.723	0.696	0.668	0.640	0.613
41	0.930	0.889	0.854	0.822	0.790	0.761	0.732	0.703	0.676	0.648	0.622
42	0.931	0.892	0.858	0.826	0.795	0.766	0.737	0.710	0.683	0.656	0.630
43	0.933	0.894	0.861	0.829	0.800	0.771	0.743	0.716	0.689	0.663	0.638
44	0.934	0.897	0.864	0.833	0.804	0.776	0.749	0.722	0.696	0.670	0.645
45	0.936	0.899	0.867	0.837	0.808	0.780	0.754	0.728	0.702	0.677	0.652
46	0.937	0.901	0.869	0.840	0.812	0.785	0.759	0.733	0.708	0.683	0.659
47	0.938	0.903	0.872	0.843	0.816	0.789	0.764	0.378	0.714	0.690	0.666
48	0.939	0.905	0.875	0.846	0.819	0.793	0.768	0.744	0.719	0.696	0.672
49	0.941	0.907	0.877	0.849	0.823	0.797	0.773	0.748	0.725	0.701	0.678
50	0.942	0.909	0.879	0.852	0.826	0.801	0.777	0.753	0.730	0.707	0.684

TABLE B.7
Binomial 95% confidence levels (continued).

Sample Size N	Number of Failures in Sample = F										
	0	1	2	3	4	5	6	7	8	9	10
51	0.943	0.910	0.882	0.855	0.829	0.805	0.781	0.758	0.735	0.712	0.690
52	0.944	0.912	0.884	0.858	0.833	0.808	0.785	0.762	0.740	0.717	0.696
53	0.945	0.914	0.886	0.860	0.836	0.812	0.789	0.766	0.744	0.722	0.701
54	0.946	0.915	0.888	0.863	0.838	0.815	0.792	0.770	0.749	0.727	0.706
55	0.947	0.917	0.890	0.865	0.841	0.818	0.796	0.774	0.753	0.732	0.711
56	0.948	0.918	0.892	0.867	0.844	0.821	0.800	0.778	0.757	0.736	0.716
57	0.949	0.919	0.894	0.870	0.847	0.824	0.803	0.782	0.761	0.741	0.721
58	0.950	0.921	0.895	0.872	0.849	0.827	0.806	0.785	0.765	0.745	0.725
59	0.950	0.922	0.897	0.874	0.852	0.830	0.809	0.789	0.769	0.749	0.730
60	0.951	0.923	0.899	0.876	0.854	0.833	0.812	0.792	0.772	0.753	0.734
61	0.952	0.925	0.900	0.878	0.856	0.835	0.815	0.795	0.776	0.757	0.738
62	0.953	0.926	0.902	0.880	0.858	0.838	0.818	0.798	0.779	0.760	0.742
63	0.954	0.927	0.903	0.882	0.861	0.840	0.821	0.801	0.783	0.764	0.746
64	0.954	0.928	0.905	0.883	0.863	0.843	0.823	0.804	0.786	0.767	0.749
65	0.955	0.929	0.906	0.885	0.865	0.845	0.826	0.807	0.789	0.771	0.753
66	0.956	0.930	0.908	0.887	0.867	0.847	0.829	0.810	0.792	0.774	0.757
67	0.956	0.931	0.909	0.888	0.869	0.849	0.831	0.813	0.795	0.777	0.760
68	0.957	0.932	0.910	0.890	0.870	0.852	0.833	0.815	0.798	0.780	0.763
69	0.958	0.933	0.912	0.891	0.872	0.854	0.836	0.818	0.801	0.783	0.767
70	0.958	0.954	0.913	0.893	0.874	0.856	0.838	0.820	0.803	0.786	0.770
71	0.959	0.935	0.914	0.894	0.876	0.858	0.840	0.823	0.806	0.789	0.773
72	0.959	0.936	0.915	0.896	0.877	0.860	0.842	0.825	0.809	0.792	0.776
73	0.960	0.937	0.916	0.897	0.879	0.861	0.844	0.828	0.811	0.795	0.779
74	0.960	0.937	0.917	0.899	0.881	0.863	0.846	0.830	0.813	0.797	0.782
75	0.961	0.938	0.918	0.900	0.882	0.865	0.848	0.832	0.816	0.800	0.784
76	0.961	0.939	0.919	0.901	0.884	0.867	0.850	0.834	0.818	0.802	0.787
77	0.962	0.940	0.920	0.902	0.885	0.868	0.852	0.836	0.820	0.805	0.790
78	0.962	0.941	0.921	0.904	0.886	0.870	0.854	0.838	0.823	0.807	0.792
79	0.963	0.941	0.922	0.905	0.888	0.872	0.856	0.840	0.825	0.810	0.795
80	0.963	0.942	0.923	0.906	0.889	0.873	0.857	0.842	0.827	0.812	0.797
81	0.964	0.943	0.924	0.907	0.891	0.875	0.859	0.844	0.829	0.814	0.800
82	0.964	0.943	0.925	0.908	0.892	0.876	0.861	0.846	0.831	0.816	0.802
83	0.965	0.944	0.926	0.909	0.893	0.878	0.862	0.847	0.833	0.818	0.804
84	0.965	0.945	0.927	0.910	0.894	0.879	0.864	0.849	0.835	0.820	0.806
85	0.965	0.945	0.928	0.911	0.896	0.880	0.865	0.851	0.837	0.823	0.809
86	0.966	0.946	0.929	0.912	0.897	0.882	0.867	0.853	0.839	0.824	0.811
87	0.966	0.947	0.929	0.913	0.898	0.883	0.868	0.854	0.840	0.826	0.813
88	0.967	0.947	0.930	0.914	0.899	0.884	0.870	0.856	0.842	0.828	0.815
89	0.967	0.948	0.931	0.915	0.900	0.885	0.871	0.857	0.844	0.830	0.817
90	0.967	0.948	0.932	0.916	0.901	0.887	0.873	0.859	0.845	0.832	0.819
91	0.968	0.949	0.932	0.917	0.902	0.888	0.874	0.860	0.847	0.833	0.821
92	0.968	0.949	0.933	0.918	0.903	0.889	0.875	0.862	0.849	0.835	0.823
93	0.968	0.950	0.934	0.919	0.904	0.890	0.877	0.863	0.850	0.837	0.824
94	0.969	0.951	0.935	0.920	0.905	0.891	0.878	0.865	0.852	0.839	0.826
95	0.969	0.951	0.935	0.920	0.906	0.893	0.879	0.866	0.853	0.841	0.828
96	0.969	0.952	0.936	0.921	0.907	0.894	0.880	0.867	0.855	0.842	0.830
97	0.970	0.952	0.937	0.922	0.908	0.895	0.882	0.869	0.856	0.844	0.831
98	0.970	0.953	0.937	0.923	0.909	0.896	0.883	0.870	0.858	0.845	0.833
99	0.970	0.953	0.938	0.924	0.910	0.897	0.884	0.871	0.859	0.847	0.835
100	0.970	0.953	0.938	0.924	0.911	0.898	0.885	0.873	0.860	0.848	0.836

TABLE B.7

Binomial 95% confidence levels (continued).

Sample Size N	Number of Failures in Sample = F										
	0	1	2	3	4	5	6	7	8	9	10
101	0.971	0.954	0.939	0.925	0.912	0.899	0.886	0.874	0.862	0.850	0.838
102	0.971	0.955	0.940	0.926	0.913	0.900	0.887	0.875	0.863	0.851	0.839
103	0.971	0.955	0.940	0.926	0.913	0.901	0.888	0.876	0.864	0.852	0.841
104	0.972	0.955	0.941	0.927	0.914	0.902	0.889	0.877	0.866	0.854	0.842
105	0.972	0.956	0.941	0.928	0.915	0.902	0.890	0.878	0.867	0.855	0.844
106	0.972	0.956	0.942	0.928	0.916	0.903	0.891	0.880	0.868	0.856	0.845
107	0.972	0.956	0.942	0.929	0.916	0.904	0.892	0.881	0.869	0.858	0.847
108	0.973	0.957	0.943	0.930	0.917	0.905	0.893	0.882	0.870	0.859	0.848
109	0.973	0.957	0.943	0.930	0.918	0.906	0.894	0.883	0.872	0.860	0.849
110	0.973	0.957	0.944	0.931	0.919	0.907	0.895	0.884	0.873	0.862	0.851
111	0.973	0.958	0.944	0.932	0.919	0.908	0.896	0.885	0.874	0.863	0.852
112	0.974	0.858	0.945	0.932	0.920	0.908	0.897	0.886	0.875	0.864	0.853
113	0.974	0.959	0.945	0.933	0.921	0.909	0.898	0.887	0.876	0.865	0.855
114	0.974	0.959	0.946	0.933	0.922	0.910	0.899	0.888	0.877	0.866	0.856
115	0.974	0.959	0.946	0.934	0.922	0.911	0.900	0.889	0.878	0.867	0.857
116	0.975	0.960	0.947	0.935	0.923	0.912	0.901	0.890	0.879	0.869	0.858
117	0.975	0.960	0.947	0.935	0.923	0.912	0.901	0.891	0.880	0.870	0.859
118	0.975	0.960	0.948	0.936	0.924	0.913	0.902	0.892	0.881	0.871	0.861
119	0.975	0.961	0.948	0.936	0.925	0.914	0.903	0.892	0.882	0.872	0.862
120	0.975	0.961	0.948	0.937	0.925	0.914	0.904	0.893	0.883	0.873	0.863
121	0.976	0.961	0.949	0.937	0.926	0.915	0.905	0.894	0.884	0.874	0.864
122	0.976	0.962	0.949	0.938	0.927	0.916	0.905	0.895	0.885	0.875	0.865
123	0.976	0.962	0.950	0.938	0.927	0.916	0.906	0.896	0.886	0.876	0.866
124	0.976	0.962	0.950	0.939	0.928	0.917	0.907	0.897	0.887	0.877	0.867
125	0.976	0.963	0.950	0.939	0.928	0.918	0.907	0.897	0.887	0.878	0.868
126	0.977	0.963	0.951	0.940	0.929	0.918	0.908	0.898	0.888	0.879	0.869
127	0.977	0.963	0.951	0.940	0.929	0.919	0.909	0.899	0.889	0.880	0.870
128	0.977	0.963	0.952	0.941	0.930	0.920	0.910	0.900	0.890	0.881	0.871
129	0.977	0.964	0.952	0.941	0.930	0.920	0.910	0.901	0.891	0.881	0.872
130	0.977	0.964	0.952	0.941	0.931	0.921	0.911	0.901	0.892	0.882	0.873
131	0.977	0.964	0.953	0.942	0.931	0.921	0.912	0.902	0.893	0.883	0.874
132	0.978	0.965	0.953	0.942	0.932	0.922	0.912	0.903	0.893	0.884	0.875
133	0.978	0.965	0.953	0.943	0.932	0.923	0.913	0.903	0.894	0.885	0.876
134	0.978	0.965	0.954	0.943	0.933	0.923	0.914	0.904	0.895	0.886	0.877
135	0.978	0.965	0.954	0.944	0.933	0.924	0.914	0.905	0.896	0.887	0.878
136	0.978	0.966	0.954	0.944	0.934	0.924	0.915	0.906	0.896	0.887	0.879
137	0.978	0.966	0.955	0.944	0.934	0.925	0.915	0.906	0.897	0.888	0.879
138	0.979	0.966	0.955	0.945	0.935	0.925	0.916	0.907	0.898	0.889	0.880
139	0.979	0.966	0.955	0.945	0.935	0.926	0.917	0.908	0.899	0.890	0.881
140	0.979	0.967	0.956	0.946	0.936	0.926	0.917	0.908	0.899	0.891	0.882
141	0.979	0.967	0.956	0.946	0.936	0.927	0.918	0.909	0.900	0.891	0.883
142	0.979	0.967	0.956	0.946	0.937	0.927	0.918	0.909	0.901	0.892	0.884
143	0.979	0.967	0.957	0.947	0.937	0.928	0.919	0.910	0.901	0.893	0.884
144	0.979	0.967	0.957	0.947	0.938	0.928	0.919	0.911	0.902	0.893	0.885
145	0.980	0.968	0.957	0.947	0.938	0.929	0.920	0.911	0.903	0.894	0.886
146	0.980	0.968	0.958	0.948	0.938	0.929	0.921	0.912	0.903	0.895	0.887
147	0.980	0.968	0.958	0.948	0.939	0.930	0.921	0.912	0.904	0.896	0.887
148	0.980	0.968	0.958	0.948	0.939	0.930	0.922	0.913	0.905	0.896	0.888
149	0.980	0.969	0.958	0.949	0.940	0.931	0.922	0.914	0.905	0.897	0.889
150	0.980	0.969	0.959	0.949	0.940	0.931	0.923	0.914	0.906	0.898	0.890

TABLE B.7
Binomial 95% confidence levels (continued).

Sample Size N	Number of Failures in Sample = F										
	0	1	2	3	4	5	6	7	8	9	10
151	0.980	0.969	0.959	0.949	0.940	0.932	0.923	0.915	0.907	0.898	0.890
152	0.980	0.969	0.959	0.950	0.941	0.932	0.924	0.915	0.907	0.899	0.891
153	0.981	0.969	0.959	0.950	0.941	0.933	0.924	0.916	0.908	0.900	0.892
154	0.981	0.970	0.960	0.950	0.942	0.933	0.925	0.916	0.908	0.900	0.892
155	0.981	0.970	0.960	0.951	0.942	0.933	0.925	0.917	0.909	0.901	0.893
156	0.981	0.970	0.960	0.951	0.942	0.934	0.926	0.917	0.909	0.901	0.894
157	0.981	0.970	0.960	0.951	0.943	0.934	0.926	0.918	0.910	0.902	0.894
158	0.981	0.970	0.961	0.952	0.943	0.935	0.926	0.918	0.911	0.903	0.895
159	0.981	0.971	0.961	0.952	0.943	0.935	0.927	0.919	0.911	0.903	0.896
160	0.981	0.971	0.961	0.952	0.944	0.935	0.927	0.919	0.912	0.904	0.896
161	0.982	0.971	0.961	0.953	0.944	0.936	0.928	0.920	0.912	0.904	0.897
162	0.982	0.971	0.962	0.953	0.944	0.936	0.928	0.920	0.913	0.905	0.898
163	0.982	0.971	0.962	0.953	0.945	0.937	0.929	0.921	0.913	0.906	0.898
164	0.982	0.971	0.962	0.953	0.945	0.937	0.929	0.921	0.914	0.906	0.899
165	0.982	0.972	0.962	0.954	0.945	0.937	0.930	0.922	0.914	0.907	0.899
166	0.982	0.972	0.963	0.954	0.946	0.938	0.930	0.922	0.915	0.907	0.900
167	0.982	0.972	0.963	0.954	0.946	0.938	0.930	0.923	0.915	0.907	0.901
168	0.982	0.972	0.963	0.954	0.946	0.938	0.931	0.923	0.916	0.908	0.901
169	0.982	0.972	0.963	0.955	0.947	0.939	0.931	0.924	0.916	0.909	0.902
170	0.983	0.972	0.963	0.955	0.947	0.939	0.932	0.924	0.917	0.909	0.902
171	0.983	0.973	0.964	0.955	0.947	0.940	0.932	0.925	0.917	0.910	0.903
172	0.983	0.973	0.964	0.956	0.948	0.940	0.932	0.925	0.918	0.910	0.903
173	0.983	0.973	0.964	0.956	0.948	0.940	0.933	0.925	0.918	0.911	0.904
174	0.983	0.973	0.964	0.956	0.948	0.941	0.933	0.926	0.919	0.911	0.904
175	0.983	0.973	0.964	0.956	0.948	0.941	0.933	0.926	0.919	0.912	0.905
176	0.983	0.973	0.965	0.957	0.949	0.941	0.934	0.927	0.920	0.912	0.906
177	0.983	0.973	0.965	0.957	0.949	0.942	0.934	0.927	0.920	0.913	0.906
178	0.983	0.974	0.965	0.957	0.949	0.942	0.935	0.927	0.920	0.913	0.907
179	0.983	0.974	0.965	0.957	0.950	0.942	0.935	0.928	0.921	0.913	0.907
180	0.983	0.974	0.965	0.957	0.950	0.943	0.935	0.928	0.921	0.914	0.908
181	0.984	0.974	0.966	0.958	0.950	0.943	0.936	0.929	0.922	0.915	0.908
182	0.984	0.974	0.966	0.958	0.950	0.943	0.936	0.929	0.922	0.915	0.908
183	0.984	0.974	0.966	0.958	0.951	0.943	0.936	0.929	0.923	0.916	0.909
184	0.984	0.974	0.966	0.958	0.951	0.944	0.937	0.930	0.923	0.916	0.910
185	0.984	0.975	0.966	0.959	0.951	0.944	0.937	0.930	0.923	0.917	0.910
186	0.984	0.975	0.967	0.959	0.951	0.944	0.937	0.931	0.924	0.917	0.911
187	0.984	0.975	0.967	0.959	0.952	0.945	0.938	0.931	0.924	0.918	0.911
188	0.984	0.975	0.967	0.959	0.952	0.945	0.938	0.931	0.925	0.918	0.911
189	0.984	0.975	0.967	0.959	0.952	0.945	0.938	0.932	0.925	0.918	0.912
190	0.984	0.975	0.967	0.960	0.952	0.945	0.939	0.932	0.925	0.919	0.912
191	0.984	0.976	0.967	0.960	0.953	0.946	0.939	0.932	0.926	0.919	0.913
192	0.985	0.976	0.968	0.960	0.953	0.946	0.939	0.933	0.926	0.920	0.913
193	0.985	0.976	0.968	0.960	0.953	0.946	0.940	0.933	0.927	0.920	0.914
194	0.985	0.976	0.968	0.961	0.953	0.947	0.940	0.933	0.927	0.920	0.914
195	0.985	0.976	0.968	0.961	0.954	0.947	0.940	0.934	0.927	0.921	0.915
196	0.985	0.976	0.968	0.961	0.954	0.947	0.941	0.934	0.928	0.921	0.915
197	0.985	0.976	0.968	0.961	0.954	0.947	0.941	0.934	0.928	0.922	0.916
198	0.985	0.976	0.969	0.961	0.954	0.948	0.941	0.935	0.928	0.922	0.916
199	0.985	0.976	0.969	0.962	0.955	0.948	0.941	0.935	0.929	0.922	0.916
200	0.985	0.977	0.969	0.962	0.955	0.948	0.942	0.935	0.929	0.923	0.917

TABLE B.7
Binomial 95% confidence levels (continued).

Sample Size N	Number of Failures in Sample = F										
	0	1	2	3	4	5	6	7	8	9	10
201	0.985	0.977	0.969	0.962	0.955	0.948	0.942	0.936	0.929	0.923	0.917
202	0.985	0.977	0.969	0.962	0.955	0.949	0.942	0.936	0.930	0.924	0.918
203	0.985	0.977	0.969	0.962	0.955	0.949	0.943	0.936	0.930	0.924	0.918
204	0.985	0.977	0.969	0.962	0.956	0.949	0.943	0.937	0.930	0.924	0.918
205	0.985	0.977	0.970	0.963	0.956	0.949	0.943	0.937	0.931	0.925	0.919
206	0.986	0.977	0.970	0.963	0.956	0.950	0.943	0.937	0.931	0.925	0.919
207	0.986	0.977	0.970	0.963	0.956	0.950	0.944	0.937	0.931	0.925	0.919
208	0.986	0.977	0.970	0.963	0.957	0.950	0.944	0.938	0.932	0.926	0.920
209	0.986	0.978	0.970	0.963	0.957	0.950	0.944	0.938	0.932	0.926	0.920
210	0.986	0.978	0.970	0.963	0.957	0.951	0.944	0.938	0.932	0.926	0.921
211	0.986	0.978	0.970	0.964	0.957	0.951	0.945	0.939	0.933	0.927	0.921
212	0.986	0.978	0.971	0.964	0.957	0.951	0.945	0.939	0.933	0.927	0.921
213	0.986	0.978	0.971	0.964	0.958	0.951	0.945	0.939	0.933	0.927	0.922
214	0.986	0.978	0.971	0.964	0.958	0.952	0.945	0.939	0.934	0.928	0.922
215	0.986	0.978	0.971	0.964	0.958	0.952	0.946	0.940	0.934	0.928	0.922
216	0.986	0.978	0.971	0.964	0.958	0.952	0.946	0.940	0.934	0.928	0.923
217	0.986	0.978	0.971	0.965	0.958	0.952	0.946	0.940	0.935	0.929	0.923
218	0.986	0.978	0.971	0.965	0.959	0.952	0.946	0.941	0.935	0.929	0.923
219	0.986	0.979	0.972	0.965	0.959	0.953	0.947	0.941	0.935	0.929	0.924
220	0.986	0.979	0.972	0.965	0.959	0.953	0.947	0.941	0.935	0.930	0.924
221	0.987	0.979	0.972	0.965	0.959	0.953	0.947	0.941	0.936	0.930	0.924
222	0.987	0.979	0.972	0.965	0.959	0.953	0.947	0.942	0.936	0.930	0.925
223	0.987	0.979	0.972	0.966	0.959	0.953	0.948	0.942	0.936	0.931	0.925
224	0.987	0.979	0.972	0.966	0.960	0.954	0.948	0.942	0.937	0.931	0.925
225	0.987	0.979	0.972	0.966	0.960	0.954	0.948	0.942	0.937	0.931	0.926
226	0.987	0.979	0.972	0.966	0.960	0.954	0.948	0.943	0.937	0.932	0.926
227	0.987	0.979	0.973	0.966	0.960	0.954	0.949	0.943	0.937	0.932	0.926
228	0.987	0.979	0.973	0.966	0.960	0.954	0.949	0.943	0.938	0.932	0.927
229	0.987	0.979	0.973	0.966	0.960	0.955	0.949	0.943	0.938	0.932	0.927
230	0.987	0.980	0.973	0.967	0.961	0.955	0.949	0.944	0.938	0.932	0.928
235	0.987	0.980	0.973	0.967	0.961	0.956	0.950	0.945	0.939	0.934	0.929
240	0.988	0.980	0.974	0.968	0.962	0.957	0.951	0.946	0.941	0.935	0.930
245	0.988	0.981	0.975	0.969	0.963	0.958	0.952	0.947	0.942	0.937	0.932
250	0.988	0.981	0.975	0.969	0.964	0.958	0.953	0.948	0.943	0.938	0.933

TABLE B.8
Table of $e^{-\lambda t}$.

λt	$e^{-\lambda t}$	λt	$e^{-\lambda t}$	λt	$e^{-\lambda t}$
<0.002	1.0−λt	0.051	0.950	0.11	0.896
0.002	0.998	0.052	0.949	0.12	0.887
0.003	0.997	0.053	0.948	0.13	0.878
0.004	0.996	0.054	0.947	0.14	0.869
0.005	0.995	0.055	0.946	0.15	0.861
0.006	0.994	0.056	0.946	0.16	0.852
0.007	0.993	0.057	0.945	0.17	0.844
0.008	0.992	0.058	0.944	0.18	0.835
0.009	0.991	0.059	0.943	0.19	0.827
0.010	0.990	0.060	0.942	0.20	0.819
0.011	0.989	0.061	0.941	0.21	0.811
0.012	0.988	0.062	0.940	0.22	0.803
0.013	0.987	0.063	0.939	0.23	0.795
0.014	0.986	0.064	0.938	0.24	0.787
0.015	0.985	0.065	0.937	0.25	0.779
0.016	0.984	0.066	0.936	0.26	0.771
0.017	0.983	0.067	0.935	0.27	0.763
0.018	0.982	0.068	0.934	0.28	0.756
0.019	0.981	0.069	0.933	0.29	0.748
0.020	0.980	0.070	0.932	0.30	0.741
0.021	0.979	0.071	0.931	0.31	0.733
0.022	0.978	0.072	0.931	0.32	0.726
0.023	0.977	0.073	0.930	0.33	0.719
0.024	0.976	0.074	0.929	0.34	0.712
0.025	0.975	0.075	0.928	0.35	0.705
0.026	0.974	0.076	0.927	0.36	0.698
0.027	0.973	0.077	0.926	0.37	0.691
0.028	0.972	0.078	0.925	0.38	0.684
0.029	0.971	0.079	0.924	0.39	0.677
0.030	0.970	0.080	0.923	0.40	0.670
0.031	0.969	0.081	0.922	0.41	0.664
0.032	0.969	0.082	0.921	0.42	0.657
0.033	0.968	0.083	0.920	0.43	0.651
0.034	0.967	0.084	0.919	0.44	0.644
0.035	0.966	0.085	0.919	0.45	0.638
0.036	0.965	0.086	0.918	0.46	0.631
0.037	0.964	0.087	0.917	0.47	0.625
0.038	0.963	0.088	0.916	0.48	0.619
0.039	0.962	0.089	0.915	0.49	0.613
0.040	0.961	0.090	0.914	0.50	0.607
0.041	0.960	0.091	0.913	0.55	0.577
0.042	0.959	0.092	0.912	0.60	0.549
0.043	0.958	0.093	0.911	0.65	0.522
0.044	0.957	0.094	0.910	0.70	0.497
0.045	0.956	0.095	0.909	0.75	0.472
0.046	0.955	0.096	0.908	0.80	0.449
0.047	0.954	0.097	0.908	0.85	0.427
0.048	0.953	0.098	0.907	0.90	0.407
0.049	0.952	0.099	0.906	0.95	0.387
0.050	0.951	0.100	0.905	1.00	0.368

TABLE B.9
Poisson distribution.

$$P(x \leq c) = \sum_{x=0}^{c} e^{-\mu}\left(\frac{\mu^x}{x!}\right)$$

μ	$c =$ 0	1	2	3	4	5	6	7	8	9
0.02	0.980	1.000								
0.04	0.961	0.999	1.000							
0.06	0.942	0.998	1.000							
0.08	0.923	0.977	1.000							
0.10	0.905	0.995	1.000							
0.15	0.861	0.990	0.999	1.000						
0.20	0.819	0.982	0.999	1.000						
0.25	0.779	0.974	0.998	1.000						
0.30	0.741	0.963	0.996	1.000						
0.35	0.705	0.951	0.994	1.000						
0.40	0.670	0.938	0.992	0.999	1.000					
0.45	0.638	0.925	0.989	0.999	1.000					
0.50	0.607	0.910	0.986	0.998	1.000					
0.55	0.577	0.894	0.982	0.998	1.000					
0.60	0.549	0.878	0.977	0.997	1.000					
0.65	0.522	0.861	0.972	0.996	0.999	1.000				
0.70	0.497	0.844	0.966	0.994	0.999	1.000				
0.75	0.472	0.827	0.959	0.993	0.999	1.000				
0.80	0.449	0.809	0.953	0.991	0.999	1.000				
0.85	0.427	0.791	0.945	0.989	0.998	1.000				
0.90	0.407	0.772	0.937	0.987	0.998	1.000				
0.95	0.387	0.754	0.929	0.984	0.997	1.000				
1.00	0.368	0.736	0.920	0.981	0.996	0.999	1.000			
1.10	0.333	0.699	0.900	0.974	0.995	0.999	1.000			
1.20	0.301	0.663	0.879	0.966	0.992	0.998	1.000			
1.30	0.273	0.627	0.857	0.957	0.989	0.998	1.000			
1.40	0.247	0.592	0.833	0.946	0.986	0.997	0.999	1.000		
1.50	0.223	0.558	0.809	0.934	0.981	0.996	0.999	1.000		
1.60	0.202	0.525	0.783	0.921	0.976	0.994	0.999	1.000		
1.70	0.183	0.493	0.757	0.907	0.970	0.992	0.998	1.000		
1.80	0.165	0.463	0.731	0.891	0.964	0.990	0.997	0.999	1.000	
1.90	0.150	0.434	0.704	0.875	0.956	0.987	0.997	0.999	1.000	
2.00	0.135	0.406	0.677	0.857	0.947	0.983	0.995	0.999	1.000	
2.20	0.111	0.355	0.623	0.819	0.928	0.975	0.993	0.998	1.000	
2.40	0.091	0.308	0.570	0.779	0.904	0.964	0.988	0.997	0.999	1.000
2.60	0.074	0.267	0.518	0.736	0.877	0.951	0.983	0.995	0.999	1.000
2.80	0.061	0.231	0.469	0.692	0.848	0.935	0.976	0.992	0.998	0.999
3.00	0.050	0.199	0.423	0.647	0.815	0.916	0.966	0.988	0.996	0.999

TABLE B.9
Poisson distribution (continued).

μ	c = 0	1	2	3	4	5	6	7	8	9
3.2	0.041	0.171	0.380	0.603	0.781	0.895	0.955	0.983	0.994	0.998
3.4	0.033	0.147	0.340	0.558	0.744	0.871	0.942	0.977	0.992	0.997
3.6	0.027	0.126	0.303	0.515	0.706	0.844	0.927	0.969	0.988	0.996
3.8	0.022	0.107	0.269	0.473	0.668	0.816	0.909	0.960	0.984	0.994
4.0	0.018	0.092	0.238	0.433	0.629	0.785	0.889	0.949	0.979	0.992
4.2	0.015	0.078	0.210	0.395	0.590	0.753	0.867	0.936	0.972	0.989
4.4	0.012	0.066	0.185	0.359	0.551	0.720	0.844	0.921	0.964	0.985
4.6	0.010	0.056	0.163	0.326	0.513	0.686	0.818	0.905	0.955	0.980
4.8	0.008	0.048	0.143	0.294	0.476	0.651	0.791	0.887	0.944	0.975
5.0	0.007	0.040	0.125	0.265	0.440	0.616	0.762	0.867	0.932	0.968
5.2	0.006	0.034	0.109	0.238	0.406	0.581	0.732	0.845	0.918	0.960
5.4	0.005	0.029	0.095	0.213	0.373	0.546	0.702	0.822	0.903	0.951
5.6	0.004	0.024	0.082	0.191	0.342	0.512	0.670	0.797	0.886	0.941
5.8	0.003	0.021	0.072	0.170	0.313	0.478	0.638	0.771	0.867	0.929
6.0	0.002	0.017	0.062	0.151	0.285	0.446	0.606	0.744	0.847	0.916
6.2	0.002	0.015	0.054	0.134	0.259	0.414	0.574	0.716	0.826	0.902
6.4	0.002	0.012	0.046	0.119	0.235	0.384	0.542	0.687	0.803	0.886
6.6	0.001	0.010	0.040	0.105	0.213	0.355	0.511	0.658	0.780	0.869
6.8	0.001	0.009	0.034	0.093	0.192	0.327	0.480	0.628	0.755	0.850
7.0	0.001	0.007	0.030	0.082	0.173	0.301	0.450	0.599	0.729	0.830

μ	c = 10	11	12	13	14	15	16	17	18	19
2.8	1.000									
3.0	1.000									
3.2	1.000									
3.4	0.999	1.000								
3.6	0.999	1.000								
3.8	0.998	0.999	1.000							
4.0	0.997	0.999	1.000							
4.2	0.996	0.999	1.000							
4.4	0.994	0.998	0.999	1.000						
4.6	0.992	0.997	0.999	1.000						
4.8	0.990	0.996	0.999	1.000						
5.0	0.986	0.995	0.998	0.999	1.000					
5.2	0.982	0.993	0.997	0.999	1.000					
5.4	0.977	0.990	0.996	0.999	1.000					
5.6	0.972	0.988	0.995	0.998	0.999	1.000				
5.8	0.965	0.984	0.993	0.997	0.999	1.000				
6.0	0.957	0.980	0.991	0.996	0.999	0.999	1.000			
6.2	0.949	0.975	0.989	0.995	0.998	0.999	1.000			
6.4	0.939	0.969	0.986	0.994	0.997	0.999	1.000			
6.6	0.972	0.963	0.982	0.992	0.997	0.999	0.999	1.000		
6.8	0.915	0.955	0.978	0.990	0.996	0.998	0.999	1.000		
7.0	0.901	0.947	0.973	0.987	0.994	0.998	0.999	1.000		

TABLE B.9

Poisson distribution (continued).

μ	$c =$ 0	1	2	3	4	5	6	7	8	9
7.2	0.001	0.006	0.025	0.072	0.156	0.276	0.420	0.569	0.703	0.810
7.4	0.001	0.005	0.022	0.063	0.140	0.253	0.392	0.539	0.676	0.788
7.6	0.001	0.004	0.019	0.055	0.125	0.231	0.365	0.510	0.648	0.765
7.8	0.000	0.004	0.016	0.048	0.112	0.210	0.338	0.481	0.620	0.741
8.0	0.000	0.003	0.014	0.042	0.100	0.191	0.313	0.453	0.593	0.717
8.5	0.000	0.002	0.009	0.030	0.074	0.150	0.256	0.386	0.523	0.653
9.0	0.000	0.001	0.006	0.021	0.055	0.116	0.207	0.324	0.456	0.587
9.5	0.000	0.001	0.004	0.015	0.040	0.089	0.165	0.269	0.392	0.522
10.0	0.000	0.000	0.003	0.010	0.029	0.067	0.130	0.220	0.333	0.458
11.0	0.000	0.000	0.001	0.005	0.015	0.038	0.079	0.143	0.232	0.341
12.0	0.000	0.000	0.001	0.002	0.008	0.020	0.046	0.090	0.155	0.242
13.0	0.000	0.000	0.000	0.001	0.004	0.011	0.026	0.054	0.100	0.166
14.0	0.000	0.000	0.000	0.000	0.002	0.006	0.014	0.032	0.062	0.109
15.0	0.000	0.000	0.000	0.000	0.001	0.003	0.008	0.018	0.037	0.070

μ	$c =$ 10	11	12	13	14	15	16	17	18	19
7.2	0.887	0.937	0.967	0.984	0.993	0.997	0.999	0.999	1.000	
7.4	0.871	0.926	0.961	0.980	0.991	0.996	0.998	0.999	1.000	
7.6	0.854	0.915	0.954	0.976	0.989	0.995	0.998	0.999	1.000	
7.8	0.835	0.902	0.945	0.971	0.986	0.993	0.997	0.999	1.000	
8.0	0.816	0.888	0.936	0.966	0.983	0.992	0.996	0.998	0.999	1.000
8.5	0.763	0.849	0.909	0.949	0.973	0.986	0.993	0.997	0.999	0.999
9.0	0.706	0.803	0.876	0.926	0.959	0.978	0.989	0.995	0.998	0.999
9.5	0.645	0.752	0.836	0.898	0.940	0.967	0.982	0.991	0.996	0.998
10.0	0.583	0.697	0.792	0.864	0.917	0.951	0.973	0.986	0.993	0.997
11.0	0.460	0.579	0.689	0.781	0.854	0.907	0.944	0.968	0.982	0.991
12.0	0.347	0.462	0.576	0.682	0.772	0.844	0.899	0.937	0.963	0.979
13.0	0.252	0.353	0.463	0.573	0.675	0.764	0.835	0.890	0.930	0.957
14.0	0.176	0.260	0.358	0.464	0.570	0.669	0.756	0.827	0.883	0.923
15.0	0.118	0.185	0.268	0.363	0.466	0.568	0.664	0.749	0.819	0.875

μ	$c =$ 20	21	22	23	24	25	26	27	28	29
8.5	1.000									
9.0	1.000									
9.5	0.999	1.000								
10.0	0.998	0.999	1.000							
11.0	0.995	0.998	0.999	1.000						
12.0	0.988	0.994	0.997	0.999	0.999	1.000				
13.0	0.975	0.986	0.992	0.996	0.998	0.999	1.000			
14.0	0.952	0.971	0.983	0.991	0.995	0.997	0.999	0.999	1.000	
15.0	0.917	0.947	0.967	0.981	0.989	0.994	0.997	0.998	0.999	1.000

TABLE B.9
Poisson distribution (continued).

μ	c = 0	1	2	3	4	5	6	7	8	9	10
16	0.000	0.000	0.000	0.000	0.000	0.001	0.004	0.010	0.022	0.043	0.077
17	0.000	0.000	0.000	0.000	0.000	0.001	0.002	0.005	0.013	0.026	0.049
18	0.000	0.000	0.000	0.000	0.000	0.000	0.001	0.003	0.007	0.015	0.030
19	0.000	0.000	0.000	0.000	0.000	0.000	0.001	0.002	0.004	0.009	0.018
20	0.000	0.000	0.000	0.000	0.000	0.000	0.000	0.001	0.002	0.005	0.011
21	0.000	0.000	0.000	0.000	0.000	0.000	0.000	0.000	0.001	0.003	0.006
22	0.000	0.000	0.000	0.000	0.000	0.000	0.000	0.000	0.001	0.002	0.004
23	0.000	0.000	0.000	0.000	0.000	0.000	0.000	0.000	0.000	0.001	0.002
24	0.000	0.000	0.000	0.000	0.000	0.000	0.000	0.000	0.000	0.000	0.001
25	0.000	0.000	0.000	0.000	0.000	0.000	0.000	0.000	0.000	0.000	0.001

μ	c = 11	12	13	14	15	16	17	18	19	20	21
16	0.127	0.193	0.275	0.368	0.467	0.566	0.659	0.742	0.812	0.868	0.911
17	0.085	0.135	0.201	0.281	0.371	0.468	0.564	0.655	0.736	0.805	0.861
18	0.055	0.092	0.143	0.208	0.287	0.375	0.469	0.562	0.651	0.731	0.799
19	0.035	0.061	0.098	0.150	0.215	0.292	0.378	0.469	0.561	0.647	0.725
20	0.021	0.039	0.066	0.105	0.157	0.221	0.297	0.381	0.470	0.559	0.644
21	0.013	0.025	0.043	0.072	0.111	0.163	0.227	0.302	0.384	0.471	0.558
22	0.008	0.015	0.028	0.048	0.077	0.117	0.169	0.232	0.306	0.387	0.472
23	0.004	0.009	0.017	0.031	0.052	0.082	0.123	0.175	0.238	0.310	0.389
24	0.003	0.005	0.011	0.020	0.034	0.056	0.087	0.128	0.180	0.243	0.314
25	0.001	0.003	0.006	0.012	0.022	0.038	0.060	0.092	0.134	0.185	0.247

μ	c= 22	23	24	25	26	27	28	29	30	31	32
16	0.942	0.963	0.978	0.987	0.993	0.996	0.998	0.999	0.999	1.000	
17	0.905	0.937	0.959	0.975	0.985	0.991	0.995	0.997	0.999	0.999	1.000
18	0.855	0.899	0.932	0.955	0.972	0.983	0.990	0.994	0.997	0.998	0.999
19	0.793	0.849	0.893	0.927	0.951	0.969	0.980	0.988	0.993	0.996	0.998
20	0.721	0.787	0.843	0.888	0.922	0.948	0.966	0.978	0.987	0.992	0.995
21	0.640	0.716	0.782	0.838	0.883	0.917	0.944	0.963	0.976	0.985	0.991
22	0.556	0.637	0.712	0.777	0.832	0.877	0.913	0.940	0.959	0.973	0.983
23	0.472	0.555	0.635	0.708	0.772	0.827	0.873	0.908	0.936	0.956	0.971
24	0.392	0.473	0.554	0.632	0.704	0.768	0.823	0.868	0.904	0.932	0.953
25	0.318	0.394	0.473	0.553	0.629	0.700	0.763	0.818	0.963	0.900	0.929

μ	c = 33	34	35	36	37	38	39	40	41	42	43
19	0.999	0.999	1.000								
20	0.997	0.999	0.999	1.000							
21	0.994	0.997	0.998	0.999	0.999	1.000					
22	0.989	0.994	0.996	0.998	0.999	0.999	1.000				
23	0.981	0.988	0.993	0.996	0.997	0.999	0.999	1.000			
24	0.969	0.979	0.987	0.992	0.995	0.997	0.998	0.999	0.999	1.000	
25	0.950	0.966	0.978	0.985	0.991	0.994	0.997	0.998	0.999	0.999	1.000

TABLE B.10
Correlation coefficient tables.

Values r, such that if $\{x, y\}$ represent a random pairing (i.e., no correlation), then $P(r_{x,y} < r) = \alpha$, where $r_{x,y}$ is as defined by Equation 6.13.

Degrees of Freedom	Significance Level				
	0.10	0.05	0.02	0.01	0.001
1	0.988	0.997	0.999	1.000	1.000
2	0.900	0.950	0.980	0.990	0.999
3	0.805	0.878	0.934	0.959	0.992
4	0.729	0.811	0.882	0.917	0.974
5	0.669	0.754	0.833	0.874	0.951
6	0.621	0.707	0.789	0.834	0.925
7	0.582	0.666	0.750	0.798	0.898
8	0.549	0.632	0.716	0.765	0.872
9	0.521	0.602	0.685	0.735	0.847
10	0.497	0.576	0.658	0.708	0.823
11	0.476	0.553	0.634	0.684	0.801
12	0.457	0.532	0.612	0.661	0.780
13	0.441	0.514	0.592	0.641	0.760
14	0.426	0.497	0.574	0.623	0.742
15	0.412	0.482	0.558	0.606	0.725
16	0.400	0.468	0.543	0.590	0.708
17	0.389	0.456	0.528	0.575	0.693
18	0.378	0.444	0.516	0.561	0.679
19	0.369	0.433	0.503	0.549	0.665
20	0.360	0.423	0.492	0.537	0.652
25	0.323	0.381	0.445	0.487	0.597
30	0.296	0.349	0.409	0.449	0.554
35	0.275	0.325	0.381	0.418	0.519
40	0.257	0.304	0.358	0.393	0.490
45	0.243	0.287	0.338	0.372	0.465
50	0.231	0.273	0.322	0.354	0.443
60	0.211	0.250	0.295	0.325	0.408
70	0.195	0.232	0.274	0.302	0.380
80	0.183	0.217	0.256	0.283	0.357
90	0.173	0.205	0.242	0.267	0.337
100	0.164	0.195	0.230	0.254	0.321

Condensed from R. A. Fisher and F. Yates, *Statistical Tables for Biological, Agricultural and Medical Research*, London: Oliver and Boyd, Table VI.

C

SYSTEM MTBF AND AVAILABILITY TABLES

TABLE C.1
MTBF solutions for active, redundant system, all items on-line.
One restoration at a time, item failure rate λ, and item restoration rate μ.

Number of Items N	Number Required k	System Effective Failure Rate λ_{sys}	System MTBF
2	1	$\frac{2\lambda^2}{\mu + 3\lambda}$	$\frac{\mu + 3\lambda}{2\lambda^2}$
3	2	$\frac{6\lambda^2}{\mu + 5\lambda}$	$\frac{\mu + 5\lambda}{6\lambda^2}$
4	3	$\frac{12\lambda^2}{\mu + 7\lambda}$	$\frac{\mu + 7\lambda}{12\lambda^2}$
5	4	$\frac{20\lambda^2}{\mu + 9\lambda}$	$\frac{\mu + 9\lambda}{20\lambda^2}$
N	$N - 1$	$\frac{(N)(N - 1)\lambda^2}{\mu + (2N - 1)\lambda}$	$\frac{\mu + (2N - 1)\lambda}{(N)(N - 1)\lambda^2}$
3	1	$\frac{6\lambda^3}{\mu^2 + 4\mu\lambda + 11\lambda^2}$	$\frac{\mu^2 + 4\mu\lambda + 11\lambda^2}{6\lambda^3}$

TABLE C.1 (continued).

Number of Items N	Number Required k	System Effective Failure Rate λ_{sys}	System MTBF
4	2	$\dfrac{24\lambda^3}{\mu^2 + 6\mu\lambda + 26\lambda^2}$	$\dfrac{\mu^2 + 6\mu\lambda + 26\lambda^2}{24\lambda^3}$
5	3	$\dfrac{60\lambda^3}{\mu^2 + 8\mu\lambda + 47\lambda^2}$	$\dfrac{\mu^2 + 8\mu\lambda + 47\lambda^2}{60\lambda^3}$
N	$N - 2$	$\dfrac{N(N-1)(N-2)\lambda^3}{\mu^2 + 2N\mu\lambda - 2\mu\lambda + 3N^2\lambda^2 - 6N\lambda^2 + 2\lambda^2}$	$\dfrac{\mu^2 + 2N\mu\lambda - 2\mu\lambda + 3N^2\lambda^2 - 6N\lambda^2 + 2\lambda^2}{N(N-1)(N-2)\lambda^3}$
N	$<(N - 2)$	Negligible	>>>>>>>>

TABLE C.2
MTBF solutions for active, redundant system, all items on-line. One restoration at a time, item failure rates are λ_A with all items up and λ_B with one item down, item restoration rate μ.

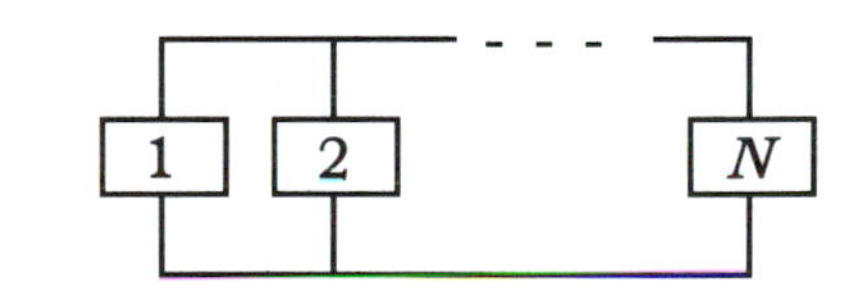

Number of Items N	Number Required k	System Effective Failure Rate λ_{sys}	System MTBF
2	1	$\dfrac{2\lambda_A\lambda_B}{\mu + 2(\lambda_A + \lambda_B) - \lambda_B}$	$\dfrac{\mu + 2(\lambda_A + \lambda_B) - \lambda_B}{2\lambda_A\lambda_B}$
3	2	$\dfrac{6\lambda_A\lambda_B}{\mu + 3(\lambda_A + \lambda_B) - \lambda_B}$	$\dfrac{\mu + 3(\lambda_A + \lambda_B) - \lambda_B}{6\lambda_A\lambda_B}$
4	3	$\dfrac{12\lambda_A\lambda_B}{\mu + 4(\lambda_A + \lambda_B) - \lambda_B}$	$\dfrac{\mu + 4(\lambda_A + \lambda_B) - \lambda_B}{12\lambda_A\lambda_B}$
5	4	$\dfrac{20\lambda_A\lambda_B}{\mu + 5(\lambda_A + \lambda_B) - \lambda_B}$	$\dfrac{\mu + 5(\lambda_A + \lambda_B) - \lambda_B}{20\lambda_A\lambda_B}$
6	5	$\dfrac{30\lambda_A\lambda_B}{\mu + 6(\lambda_A + \lambda_B) - \lambda_B}$	$\dfrac{\mu + 6(\lambda_A + \lambda_B) - \lambda_B}{30\lambda_A\lambda_B}$
N	$N - 1$	$\dfrac{N(N-1)\lambda_A\lambda_B}{\mu + N(\lambda_A + \lambda_B) - \lambda_B}$	$\dfrac{\mu + N(\lambda_A + \lambda_B) - \lambda_B}{N(N-1)\lambda_A\lambda_B}$
N	$<(N - 1)$	Use Table C.1.	Use Table C.1.

TABLE C.3

MTBF solutions for active, redundant system, two items ($N = 2$), both on line.

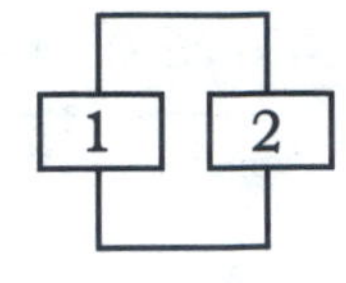

At least one item required to operate ($k = 1$). Item 1 has failure rate λ_1 and restoration rate μ_1. Item 2 has failure rate λ_2 and restoration rate μ_2.

If $\lambda_1 > 0.001$ or $\lambda_2 > 0.001$,

$$\lambda_{sys} = \frac{\lambda_1\lambda_2[(\mu_1 + \lambda_2) + (\mu_2 + \lambda_1)]}{(\mu_1 + \lambda_2)(\mu_2 + \lambda_1) + \lambda_1(\mu_2 + \lambda_1) + \lambda_2(\mu_1 + \lambda_2)}$$

$$\text{MTBF} = \frac{(\mu_1 + \lambda_2)(\mu_2 + \lambda_1) + \lambda_1(\mu_2 + \lambda_1) + \lambda_2(\mu_1 + \lambda_2)}{\lambda_1\lambda_2[(\mu_1 + \lambda_2) + (\mu_2 + \lambda_1)]}$$

If $\lambda_1 < 0.001$ and $\lambda_2 < 0.001$,

$$\lambda_{sys} = \frac{\lambda_1\lambda_2[(\mu_1 + \mu_2) + (\lambda_1 + \lambda_2)]}{\mu_1\mu_2 + [(\mu_1 + \mu_2)(\lambda_1 + \lambda_2)]}$$

$$\text{MTBF} = \frac{\mu_1\mu_2 + [(\mu_1 + \mu_2)(\lambda_1 + \lambda_2)]}{\lambda_1\lambda_2[(\mu_1 + \mu_2) + (\lambda_1 + \lambda_2)]}$$

TABLE C.4

MTBF solutions for active, redundant system, all items on line, no item on-line restoration.

Item failure rate λ.

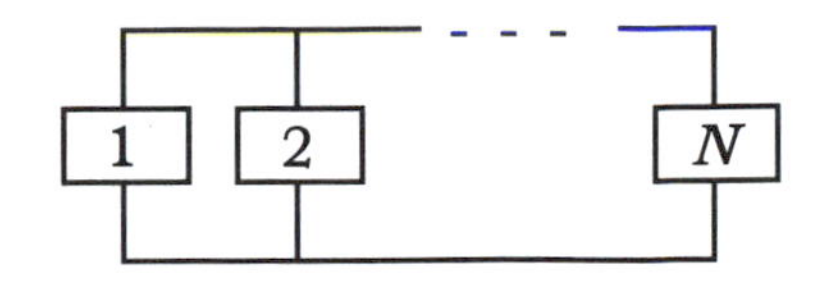

N = number of redundant items in system

k = minimum number of items required for system success

λ_{sys} = system effective failure rate

MTBF = system MTBF

N	k	λ_{sys}	MTBF	N	k	λ_{sys}	MTBF
1	1	λ	$1/\lambda$	8	1	$(280/761)\lambda$	$761/(280\lambda)$
2	1	$(2/3)\lambda$	$3/(2\lambda)$	8	2	$(280/481)\lambda$	$481/(280\lambda)$
2	2	2λ	$1/(2\lambda)$	8	3	0.821λ	$1.218/\lambda$
3	1	$(6/11)\lambda$	$11/(6\lambda)$	8	4	1.131λ	$0.884/\lambda$
3	2	$(6/5)\lambda$	$5/(6\lambda)$	8	5	1.576λ	$0.635/\lambda$
3	3	3λ	$1/(3\lambda)$	8	6	2.301λ	$0.435/\lambda$
4	1	$(12/25)\lambda$	$25/(12\lambda)$	8	7	3.733λ	$0.268/\lambda$
4	2	$(12/13)\lambda$	$13/(12\lambda)$	8	8	8.000λ	$0.125/\lambda$
4	3	$(12/7)\lambda$	$7/(12\lambda)$	9	1	0.353λ	$2.833/\lambda$
4	4	4λ	$1/(4\lambda)$	9	2	0.547λ	$1.828/\lambda$
5	1	$(60/137)\lambda$	$137/(60\lambda)$	9	3	0.752λ	$1.330/\lambda$
5	2	$(60/77)\lambda$	$77/(60\lambda)$	9	4	1.004λ	$0.996/\lambda$
5	3	$(60/47)\lambda$	$47/(60\lambda)$	9	5	1.341λ	$0.746/\lambda$
5	4	$(60/27)\lambda$	$27/(60\lambda)$	9	6	1.833λ	$0.546/\lambda$
5	5	5λ	$1/(5\lambda)$	9	7	2.639λ	$0.379/\lambda$
6	1	$(60/147)\lambda$	$147/(60\lambda)$	9	8	4.235λ	$0.236/\lambda$
6	2	$(60/87)\lambda$	$87/(60\lambda)$	9	9	9.000λ	$0.111/\lambda$
6	3	$(60/57)\lambda$	$57/(60\lambda)$	10	1	0.341λ	$2.933/\lambda$
6	4	$(60/37)\lambda$	$37/(60\lambda)$	10	2	0.518λ	$1.931/\lambda$
6	5	$(30/11)\lambda$	$11/(30\lambda)$	10	3	0.700λ	$0.143/\lambda$
6	6	6λ	$1/(6\lambda)$	10	4	0.913λ	$1.095/\lambda$
7	1	$(140/363)\lambda$	$363/(140\lambda)$	10	5	1.183λ	$0.845/\lambda$
7	2	$(140/223)\lambda$	$223/(140\lambda)$	10	6	1.549λ	$0.646/\lambda$
7	3	$(140/153)\lambda$	$153/(140\lambda)$	10	7	2.088λ	$0.479/\lambda$
7	4	$(420/319)\lambda$	$319/(420\lambda)$	10	8	2.975λ	$0.336/\lambda$
7	5	$(210/107)\lambda$	$107/(210\lambda)$	10	9	4.737λ	$0.211/\lambda$
7	6	$(42/13)\lambda$	$13/(42\lambda)$	10	10	10.000λ	$0.100/\lambda$
7	7	7λ	$1/(7\lambda)$	N	$N-1$	$\frac{N(N-1)}{2N-1}\lambda$	$\frac{2N-1}{N(N-1)\lambda}$

For any K out of N,

$$\lambda_{sys} = \frac{\lambda}{\sum_{i=k}^{N}(1/i)} \quad \text{and} \quad \text{MTBF} = \frac{\sum_{i=k}^{N}(1/i)}{\lambda}$$

The summation $\sum_{i=k}^{N}(1/i)$ can be solved through Table C.4a from the equation

$$\sum_{i=k}^{N}(1/i) = \frac{1}{k} + \frac{1}{k+1} + \cdots + \frac{1}{N} = \psi(N+1) - \psi(k)$$

where $\psi(n)$ is the *digamma function* of the integer n.

TABLE C.4a
The Digamma Function ψ (n).

Note: $\psi(z) = \frac{d}{dz}\ln \Gamma(z) = \frac{1}{\Gamma(z)}\frac{d}{dz}\Gamma(z)$, where $\Gamma(z)$ is the gamma function.

n	$\psi(n)$	n	$\psi(n)$	n	$\psi(n)$	n	$\psi(n)$
1	−0.57722	26	3.23874	51	3.92199	76	4.32414
2	+0.42278	27	3.27720	52	3.94160	77	4.33730
3	0.92278	28	3.31424	53	3.96083	78	4.35028
4	1.25612	29	3.34996	54	3.97970	79	4.36311
5	1.50612	30	3.38444	55	3.99821	80	4.37576
6	1.70612	31	3.41777	56	4.01640	81	4.38826
7	1.87278	32	3.45003	57	4.03425	82	4.40061
8	2.01564	33	3.48128	58	4.05180	83	4.41280
9	2.14064	34	3.51158	59	4.06904	84	4.42485
10	2.25175	35	3.54099	60	4.08599	85	4.43676
11	2.35175	36	3.56957	61	4.10265	86	4.44852
12	2.44266	37	3.59734	62	4.11905	87	4.46015
13	2.52600	38	3.62437	63	4.13518	88	4.47164
14	2.60292	39	3.65068	64	4.15105	89	4.48301
15	2.67435	40	3.67633	65	4.16668	90	4.49424
16	2.74101	41	3.70133	66	4.18206	91	4.50535
17	2.80351	42	3.72572	67	4.19721	92	4.51634
18	2.86234	43	3.74953	68	4.21214	93	4.52721
19	2.91789	44	3.77278	69	4.22684	94	4.53797
20	2.97052	45	3.79551	70	4.24134	95	4.54860
21	3.02052	46	3.81773	71	4.25562	96	4.55913
22	3.06814	47	3.83947	72	4.26971	97	4.56955
23	3.11359	48	3.86075	73	4.28359	98	4.57986
24	3.15708	49	3.88158	74	4.29729	99	4.59006
25	3.19874	50	3.90199	75	4.31081	100	4.60016

For $n > 100$, use $\psi(n) \approx \ln(n)$.

This table can be used for solving $\sum_{i=k}^{N}(1/i) = \psi(N + 1) - \psi(k)$.

EXAMPLE $(1/3) + (1/4) + (1/5) = 0.333 + 0.250 + 0.200 = 0.783$.
According to the digamma equation,

$$\begin{aligned}(1/3) + (1/4) + (1/5) &= \psi(N + 1) - \psi(k) = \psi(6) - \psi(3) \\ &= 1.70612 - 0.92278 = 0.783\end{aligned}$$

From *Handbook of Mathematical Functions*, U.S. Department of Commerce, 1964, Table 6.3.

TABLE C.5
MTBF solutions for active, redundant system, all items on-line.
Item failure rates are λ_0 with all items up, λ_1 with one item down, λ_2 with two items down, $\cdots$, λ_n with n items down.

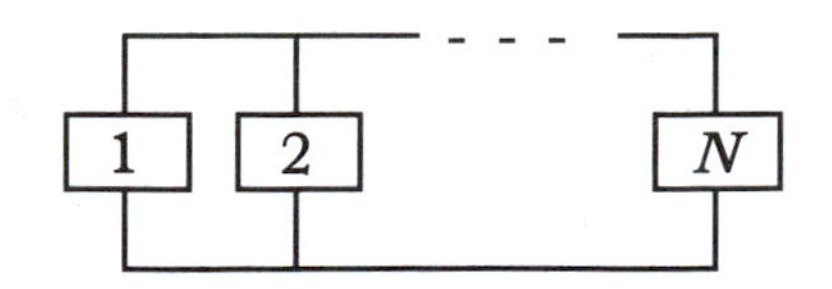

Number of Items N	Number Required k	System Effective Failure Rate λ_{sys}	System MTBF
2	1	$\dfrac{2\lambda_0\lambda_1}{2\lambda_0 + \lambda_1}$	$\dfrac{2\lambda_0 + \lambda_1}{2\lambda_0\lambda_1}$
3	1	$\dfrac{6\lambda_0\lambda_1\lambda_2}{6\lambda_0\lambda_1 + 3\lambda_0\lambda_2 + 2\lambda_1\lambda_2}$	$\dfrac{6\lambda_0\lambda_1 + 3\lambda_0\lambda_2 + 2\lambda_1\lambda_2}{6\lambda_0\lambda_1\lambda_2}$
3	2	$\dfrac{6\lambda_0\lambda_1}{3\lambda_0 + 2\lambda_1}$	$\dfrac{3\lambda_0 + 2\lambda_1}{6\lambda_0\lambda_1}$
4	2	$\dfrac{12\lambda_0\lambda_1\lambda_2}{6\lambda_0\lambda_1 + 4\lambda_0\lambda_2 + 3\lambda_1\lambda_2}$	$\dfrac{6\lambda_0\lambda_1 + 4\lambda_0\lambda_2 + 3\lambda_1\lambda_2}{12\lambda_0\lambda_1\lambda_2}$
4	3	$\dfrac{12\lambda_0\lambda_1}{4\lambda_0 + 3\lambda_1}$	$\dfrac{4\lambda_0 + 3\lambda_1}{12\lambda_0\lambda_1}$
N	$(N - 1)$	$\dfrac{N(N-1)\lambda_0\lambda_1}{N\lambda_0 + (N-1)\lambda_1}$	$\dfrac{N\lambda_0 + (N-1)\lambda_1}{N(N-1)\lambda_0\lambda_1}$
N	A	$\dfrac{1}{\sum_{i=0}^{N-A}\left(\dfrac{1}{\lambda_i[N-i]}\right)}$	$\sum_{i=0}^{N-A}\left(\dfrac{1}{\lambda_i[N-i]}\right)$

TABLE C.6
MTBF solutions for active, redundant system, no on-line item restoration, 2 items ($N = 2$), both on line.
At least one item required to operate ($k = 1$). Item 1 has failure rate λ_1. Item 2 has failure rate λ_2.

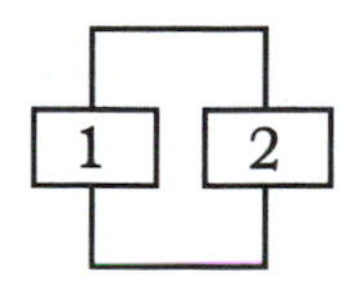

$$\text{System effective failure rate } \lambda_{sys} = \frac{\lambda_1^2\lambda_2 + \lambda_1\lambda_2^2}{\lambda_1^2 + \lambda_2^2 + \lambda_1\lambda_2}$$

$$\text{MTBF} = \frac{\lambda_1^2 + \lambda_2^2 + \lambda_1\lambda_2}{\lambda_1^2\lambda_2 + \lambda_1\lambda_2^2}$$

TABLE C.7
MTBF solutions for standby, redundant system, one restoration at a time, N items, perfect sensing and switching assumed.
k items on-line, $N - k$ items initially in standby, all k on-line items required to operate, on-line items have failure rate λ, standby items have negligible failure rates, item restoration rate is μ.

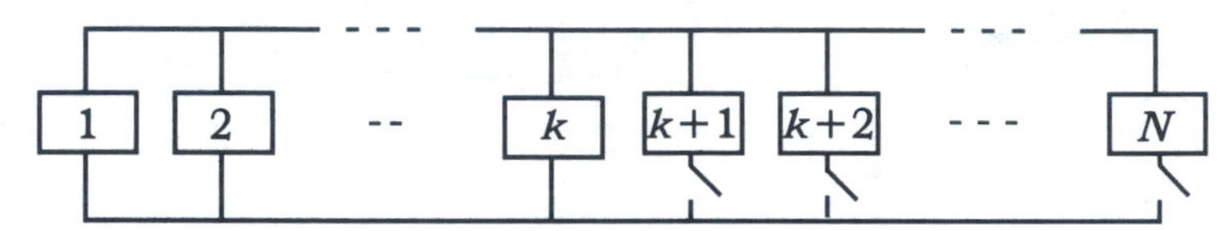

Number of Items N	Number Required k	System Effective Failure Rate λ_{sys}	System MTBF
2	1	$\frac{\lambda^2}{\mu + 2\lambda}$	$\frac{\mu + 2\lambda}{\lambda^2}$
3	2	$\frac{4\lambda^2}{\mu + 4\lambda}$	$\frac{\mu + 4\lambda}{4\lambda^2}$
4	3	$\frac{9\lambda^2}{\mu + 6\lambda}$	$\frac{\mu + 6\lambda}{9\lambda^2}$
5	4	$\frac{16\lambda^2}{\mu + 8\lambda}$	$\frac{\mu + 8\lambda}{16\lambda^2}$
6	5	$\frac{25\lambda^2}{\mu + 10\lambda}$	$\frac{\mu + 10\lambda}{25\lambda^2}$
7	6	$\frac{36\lambda^2}{\mu + 12\lambda}$	$\frac{\mu + 12\lambda}{36\lambda^2}$
8	7	$\frac{49\lambda^2}{\mu + 14\lambda}$	$\frac{\mu + 14\lambda}{49\lambda^2}$
9	8	$\frac{64\lambda^2}{\mu + 16\lambda}$	$\frac{\mu + 16\lambda}{64\lambda^2}$
10	9	$\frac{81\lambda^2}{\mu + 18\lambda}$	$\frac{\mu + 18\lambda}{81\lambda^2}$
11	10	$\frac{100\lambda^2}{\mu + 20\lambda}$	$\frac{\mu + 20\lambda}{100\lambda^2}$
12	11	$\frac{121\lambda^2}{\mu + 22\lambda}$	$\frac{\mu + 22\lambda}{121\lambda^2}$
13	12	$\frac{144\lambda^2}{\mu + 24\lambda}$	$\frac{\mu + 24\lambda}{144\lambda^2}$
14	13	$\frac{169\lambda^2}{\mu + 26\lambda}$	$\frac{\mu + 26\lambda}{169\lambda^2}$
15	14	$\frac{196\lambda^2}{\mu + 28\lambda}$	$\frac{\mu + 28\lambda}{196\lambda^2}$

TABLE C.7 (continued).

Number of Items N	Number Required k	System Effective Failure Rate λ_{sys}	System MTBF
16	15	$\frac{225\lambda^2}{\mu + 30\lambda}$	$\frac{\mu + 30\lambda}{225\lambda^2}$
17	16	$\frac{256\lambda^2}{\mu + 32\lambda}$	$\frac{\mu + 32\lambda}{256\lambda^2}$
N	$(N - 1)$	$\frac{(N - 1)^2\lambda^2}{\mu + 2(N - 1)\lambda}$	$\frac{\mu + 2(N - 1)\lambda}{(N - 1)^2\lambda^2}$
N	$< (N - 1)$	Use Table C.1	Use Table C.1

TABLE C.8

MTBF solutions for standby, redundant system, one restoration at a time, N items. k items on-line, $N - k$ items initially in standby, all k on-line items required to operate, on-line items have failure rate λ, standby items have negligible failure rates, item restoration rate is μ, and p_s = probability of successful switchover upon demand.

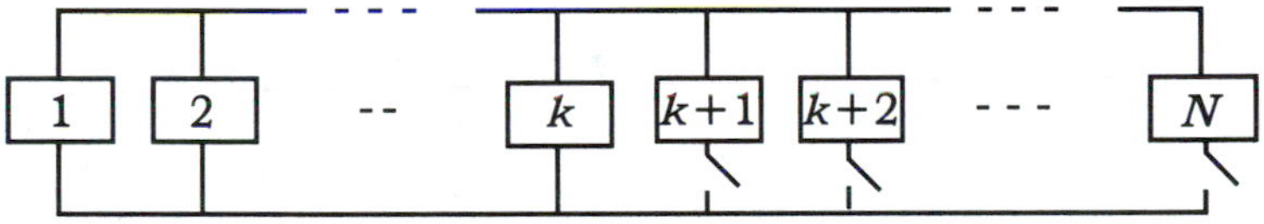

1 2 -- k $k+1$ $k+2$ --- N

Number of Items N	Number Required k	System Effective Failure Rate λ_{sys}	System MTBF
2	1	$\frac{[\lambda + (1 - p_s)\mu]\lambda}{\mu + (1 + p_s)\lambda}$	$\frac{\mu + (1 + p_s)\lambda}{[\lambda + (1 - p_s)\mu]\lambda}$
3	2	$\frac{2[2\lambda + (1 - p_s)\mu]\lambda}{\mu + 2(1 + p_s)\lambda}$	$\frac{\mu + 2(1 + p_s)\lambda}{2[2\lambda + (1 - p_s)\mu]\lambda}$
4	3	$\frac{3[3\lambda + (1 - p_s)\mu]\lambda}{\mu + 3(1 + p_s)\lambda}$	$\frac{\mu + 3(1 + p_s)\lambda}{3[3\lambda + (1 - p_s)\mu]\lambda}$
5	4	$\frac{4[4\lambda + (1 - p_s)\mu]\lambda}{\mu + 4(1 + p_s)\lambda}$	$\frac{\mu + 4(1 + p_s)\lambda}{4[4\lambda + (1 - p_s)\mu]\lambda}$
6	5	$\frac{5[5\lambda + (1 - p_s)\mu]\lambda}{\mu + 5(1 + p_s)\lambda}$	$\frac{\mu + 5(1 + p_s)\lambda}{5[5\lambda + (1 - p_s)\mu]\lambda}$
7	6	$\frac{6[6\lambda + (1 - p_s)\mu]\lambda}{\mu + 6(1 + p_s)\lambda}$	$\frac{\mu + 6(1 + p_s)\lambda}{6[6\lambda + (1 - p_s)\mu]\lambda}$
N	$(N - 1)$	$\frac{(N - 1)[(N - 1)\lambda + (1 - p_s)\mu]\lambda}{\mu + (N - 1)(1 + p_s)\lambda}$	$\frac{\mu + (N - 1)(1 + p_s)\lambda}{(N - 1)[(N - 1)\lambda + (1 - p_s)\mu]\lambda}$
N	$< (N - 1)$	Use Table C.1	Use Table C.1

TABLE C.9

MTBF solutions for standby, redundant system, no on-line restoration, N items.

k items on-line, $N - k$ items initially in standby, all k on-line items required to operate. On-line items have failure rate λ, standby items have negligible failure rates, perfect sensing and switching assumed.

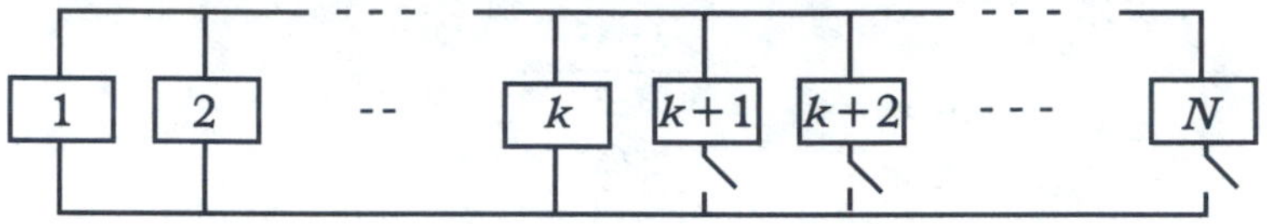

Number of Items N	Number Required k	System Effective Failure Rate λ_{sys}	System MTBF
2	1	$\frac{\lambda}{2}$	$\frac{2}{\lambda}$
3	2	λ	$\frac{1}{\lambda}$
4	3	$\frac{3\lambda}{2}$	$\frac{2}{3\lambda}$
5	4	2λ	$\frac{1}{2\lambda}$
6	5	$\frac{5\lambda}{2}$	$\frac{2}{5\lambda}$
7	6	3λ	$\frac{1}{3\lambda}$
8	7	$\frac{7\lambda}{2}$	$\frac{2}{7\lambda}$
9	8	4λ	$\frac{1}{4\lambda}$
10	9	$\frac{9\lambda}{2}$	$\frac{2}{9\lambda}$
11	10	5λ	$\frac{1}{5\lambda}$
12	11	$\frac{11\lambda}{2}$	$\frac{2}{11\lambda}$
13	12	6λ	$\frac{1}{6\lambda}$
14	13	$\frac{13\lambda}{2}$	$\frac{2}{13\lambda}$
15	14	7λ	$\frac{1}{7\lambda}$
16	15	$\frac{15\lambda}{2}$	$\frac{2}{15\lambda}$

TABLE C.9 (continued).

Number of Items N	Number Required k	System Effective Failure Rate λ_{sys}	System MTBF
17	16	8λ	$\frac{1}{8\lambda}$
N	$(N-1)$	$\frac{(N-1)\lambda}{2}$	$\frac{2}{(N-1)\lambda}$
N	$< (N-1)$	Use Table C.4	Use Table C.4

TABLE C.10

MTBF solutions for standby, redundant system, no on-line restoration, N items.

k items on-line, $N - k$ items initially in standby, all k on-line items required to operate, on-line items have failure rate λ, standby items have negligible failure rates, p_s = probability of successful switchover upon demand.

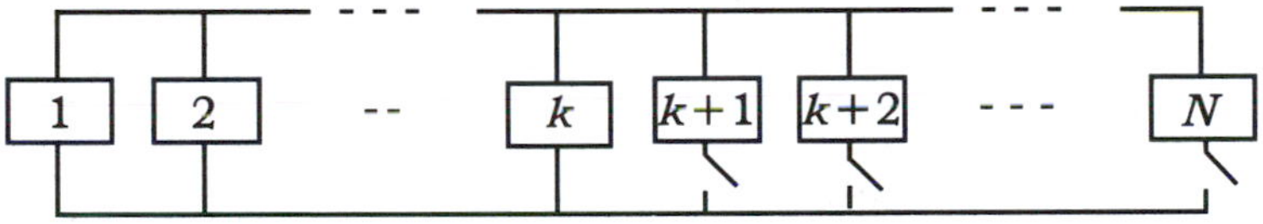

Number of Items N	Number Required k	System Effective Failure Rate λ_{sys}	System MTBF
2	1	$\frac{\lambda}{p_s + 1}$	$\frac{p_s + 1}{\lambda}$
3	2	$\frac{2\lambda}{p_s + 1}$	$\frac{p_s + 1}{2\lambda}$
4	3	$\frac{3\lambda}{p_s + 1}$	$\frac{p_s + 1}{3\lambda}$
5	4	$\frac{4\lambda}{p_s + 1}$	$\frac{p_s + 1}{4\lambda}$
6	5	$\frac{5\lambda}{p_s + 1}$	$\frac{p_s + 1}{5\lambda}$
7	6	$\frac{6\lambda}{p_s + 1}$	$\frac{p_s + 1}{6\lambda}$
N	$(N-1)$	$\frac{(N-1)\lambda}{p_s + 1}$	$\frac{p_s + 1}{(N-1)\lambda}$
N	$< (N-1)$	Use Table C.4.	Use Table C.4.

TABLE C.11

Availability solutions for active, redundant system, all items on line.
One restoration at a time, item failure rate λ, item restoration rate μ.

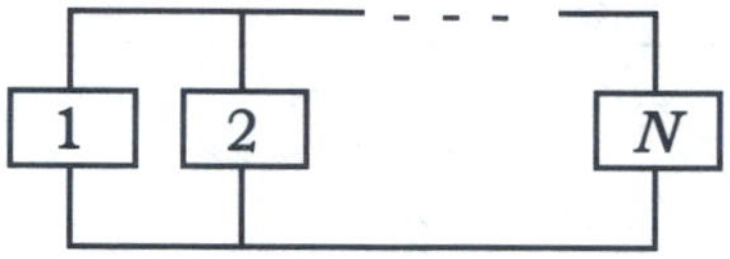

Number of Items N	Number Required k	System Availability A
1	1	$\dfrac{\mu}{\mu + \lambda}$
2	1	$\dfrac{\mu^2 + 2\mu\lambda}{\mu^2 + 2\mu\lambda + 2\lambda^2}$
2	2	$\dfrac{\mu^2}{\mu^2 + 2\mu\lambda + 2\lambda^2}$
3	1	$\dfrac{\mu^3 + 3\mu^2\lambda + 6\mu\lambda^2}{\mu^3 + 3\mu^2\lambda + 6\mu\lambda^2 + 6\lambda^3}$
3	2	$\dfrac{\mu^3 + 3\mu^2\lambda}{\mu^3 + 3\mu^2\lambda + 6\mu\lambda^2 + 6\lambda^3}$
3	3	$\dfrac{\mu^3}{\mu^3 + 3\mu^2\lambda + 6\mu\lambda^2 + 6\lambda^3}$
4	1	$\dfrac{\mu^4 + 4\mu^3\lambda + 12\mu^2\lambda^2 + 24\mu\lambda^3}{\mu^4 + 4\mu^3\lambda + 12\mu^2\lambda^4 + 24\mu\lambda^3 + 24\lambda^4}$
4	2	$\dfrac{\mu^4 + 4\mu^3\lambda + 12\mu^2\lambda^2}{\mu^4 + 4\mu^3\lambda + 12\mu^2\lambda^2 + 24\mu\lambda^3 + 24\lambda^4}$
4	3	$\dfrac{\mu^4 + 4\mu^3\lambda}{\mu^4 + 4\mu^3\lambda + 12\mu^2\lambda^2 + 24\mu\lambda^3 + 24\lambda^4}$
4	4	$\dfrac{\mu^4}{\mu^4 + 4\mu^3\lambda + 12\mu^2\lambda^2 + 24\mu\lambda^3 + 24\lambda^4}$
5	1	$\dfrac{\mu^5 + 5\mu^4\lambda + 20\mu^3\lambda^2 + 60\mu^2\lambda^3 + 120\mu\lambda^4}{\mu^5 + 5\mu^4\lambda + 20\mu^3\lambda^2 + 60\mu^2\lambda^3 + 120\mu\lambda^4 + 120\lambda^5}$
5	2	$\dfrac{\mu^5 + 5\mu^4\lambda + 20\mu^3\lambda^2 + 60\mu^2\lambda^3}{\mu^5 + 5\mu^4\lambda + 20\mu^3\lambda^2 + 60\mu^2\lambda^3 + 120\mu\lambda^4 + 120\lambda^5}$
5	3	$\dfrac{\mu^5 + 5\mu^4\lambda + 20\mu^3\lambda^2}{\mu^5 + 5\mu^4\lambda + 20\mu^3\lambda^2 + 60\mu^2\lambda^3 + 120\mu\lambda^4 + 120\lambda^5}$
5	4	$\dfrac{\mu^5 + 5\mu^4\lambda}{\mu^5 + 5\mu^4\lambda + 20\mu^3\lambda^2 + 60\mu^2\lambda^3 + 120\mu\lambda^4 + 120\lambda^5}$
5	5	$\dfrac{\mu^5}{\mu^5 + 5\mu^4\lambda + 20\mu^3\lambda^2 + 60\mu^2\lambda^3 + 120\mu\lambda^4 + 120\lambda^5}$

TABLE C.11 (continued).

Number of Items N	Number Required k	System Availability A
N	$N - 1$	$\dfrac{\sum_{i=N-1}^{N} (\mu^i / i!\lambda_i)}{\sum_{i=0}^{N} (\mu^i / i!\lambda_i)}$
N	k	$\dfrac{\sum_{i=k}^{N} (\mu^i / i!\lambda^i)}{\sum_{i=0}^{N} (\mu^i / i!\lambda^i)}$

TABLE C.12
Availability solutions for active, redundant system, all items on line.
One restoration at a time, item failure rate with all items up is λ_A and with one item down is λ_B; item restoration rate is μ.

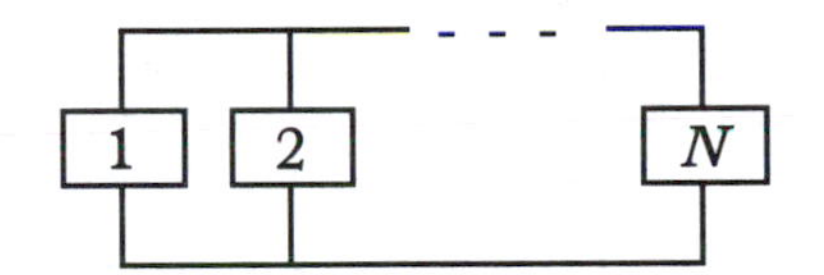

Number of Items N	Number Required k	System Availability A
N	$(N - 1)$	$\dfrac{\dfrac{\mu^{N-1}}{(N-1)!\lambda_B^{N-1}} + \dfrac{\mu^N}{N!\lambda_A\lambda_B^{N-1}}}{\dfrac{\mu^N}{N!\lambda_A\lambda_B^{N-1}} + \sum_{i=0}^{N-1} \dfrac{\mu^i}{i!\lambda_B^i}}$
N	$< (N - 1)$	Use Table C.11

TABLE C.13
Availability solutions for active, redundant system, two items ($N = 2$), both on line.
At least one item required to operate ($k = 1$). Item 1 has failure rate λ_1 and restoration rate μ_1. Item 2 has failure rate λ_2 and restoration rate μ_2.

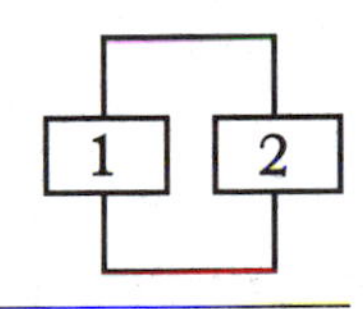

$$A = \frac{\mu_1\mu_2 + \mu_2\lambda_1 + \mu_1\lambda_2}{\mu_1\mu_2 + \mu_2\lambda_1 + \mu_1\lambda_2 + \lambda_1\lambda_2}$$

TABLE C.14

Availability solutions for active, redundant system, all items on line, no item on-line restoration. Item failure rate λ, system restoration rate μ.

N = number of redundant items in system
k = minimum number of items required for system success
A = system availability

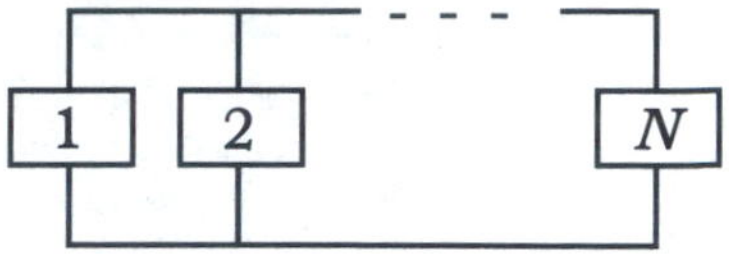

N	k	A	N	k	A
1	1	$\frac{\mu}{\mu + \lambda}$	6	2	$\frac{29\mu}{29\mu + 20\lambda}$
2	1	$\frac{3\mu}{3\mu + 2\lambda}$	6	3	$\frac{19\mu}{19\mu + 20\lambda}$
2	2	$\frac{\mu}{\mu + 2\lambda}$	6	4	$\frac{37\mu}{37\mu + 60\lambda}$
3	1	$\frac{11\mu}{11\mu + 6\lambda}$	6	5	$\frac{11\mu}{11\mu + 30\lambda}$
3	2	$\frac{5\mu}{5\mu + 6\lambda}$	6	6	$\frac{\mu}{\mu + 6\lambda}$
3	3	$\frac{\mu}{\mu + 3\lambda}$	7	1	$\frac{363\mu}{363\mu + 140\lambda}$
4	1	$\frac{25\mu}{25\mu + 12\lambda}$	7	2	$\frac{223\mu}{223\mu + 140\lambda}$
4	2	$\frac{13\mu}{13\mu + 12\lambda}$	7	3	$\frac{140\mu}{140\mu + 153\lambda}$
4	3	$\frac{7\mu}{7\mu + 12\lambda}$	7	4	$\frac{319\mu}{319\mu + 420\lambda}$
4	4	$\frac{\mu}{\mu + 4\lambda}$	7	5	$\frac{107\mu}{107\mu + 210\lambda}$
5	1	$\frac{137\mu}{137\mu + 60\lambda}$	7	6	$\frac{\mu}{\mu + 3.231\lambda}$
5	2	$\frac{77\mu}{77\mu + 60\lambda}$	7	7	$\frac{\mu}{\mu + 7\lambda}$
5	3	$\frac{47\mu}{47\mu + 60\lambda}$	8	2	$\frac{\mu}{\mu + 0.582\lambda}$
5	4	$\frac{9\mu}{9\mu + 20\lambda}$	8	3	$\frac{\mu}{\mu + 0.821\lambda}$
5	5	$\frac{\mu}{\mu + 5\lambda}$	8	4	$\frac{\mu}{\mu + 1.13\lambda}$
6	1	$\frac{49\mu}{49\mu + 20\lambda}$	8	5	$\frac{\mu}{\mu + 1.51\lambda}$

TABLE C.14 (continued).

N	k	A	N	k	A
8	6	$\dfrac{\mu}{\mu + 2.30\lambda}$	9	9	$\dfrac{\mu}{\mu + 9\lambda}$
8	7	$\dfrac{\mu}{\mu + 3.73\lambda}$	10	8	$\dfrac{\mu}{\mu + 2.98\lambda}$
8	8	$\dfrac{\mu}{\mu + 8\lambda}$	10	9	$\dfrac{\mu}{\mu + 4.74\lambda}$
9	7	$\dfrac{\mu}{\mu + 2.64\lambda}$	10	10	$\dfrac{\mu}{\mu + 10\lambda}$
9	8	$\dfrac{\mu}{\mu + 4.24\lambda}$	N	k	$\dfrac{\mu\sum_{i=k}^{N}(1/i)}{\lambda + \mu\sum_{i=k}^{N}(1/i)}$ (See Table C.4a)

TABLE C.15
Availability solutions for active, redundant system, no on-line item restoration, two items ($N = 2$), both on-line.
At least one item required to operate ($k = 1$). Item 1 has failure rate λ_1. Item 2 has failure rate λ_2. System restoration rate is μ.

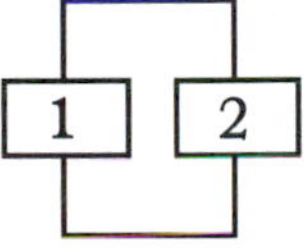

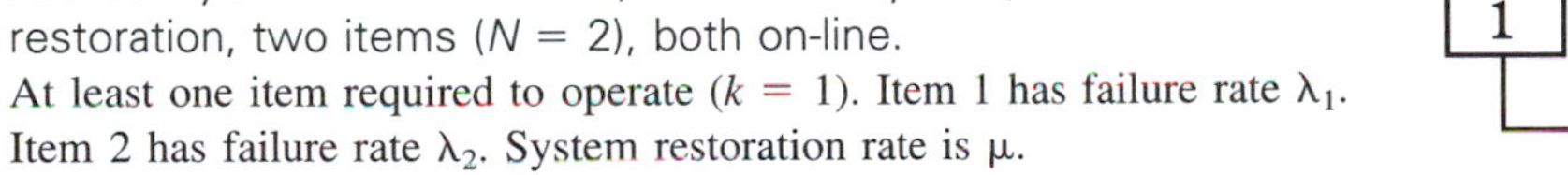

$$A = \frac{\mu(\lambda_1^2 + \lambda_2^2 + \lambda_1\lambda_2)}{\mu(\lambda_1^2 + \lambda_2^2 + \lambda_1\lambda_2) + \lambda_1^2\lambda_2 + \lambda_2^2\lambda_1}$$

TABLE C.16

Availability solutions for standby, redundant system, N items.

One restoration at a time, k items on-line, $N - k$ items initially in standby, all k on-line items required to operate, on-line items have failure rate λ, standby items have negligible failure rates, item restoration rate is μ, perfect sensing and switching assumed.

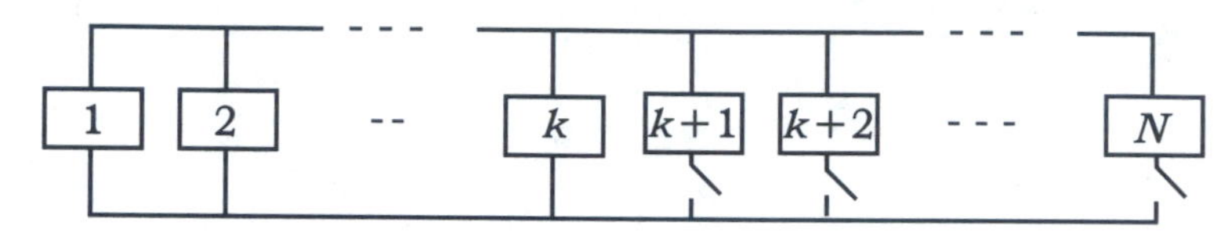

Number of Items N	Number Required k	System Availability A
1	1	$\frac{\mu}{\mu + \lambda}$
2	1	$\frac{\mu^2 + \mu\lambda}{\mu^2 + \mu\lambda + \lambda^2}$
2	2	$\frac{\mu^2}{\mu^2 + 2\mu\lambda + 2\lambda^2}$
3	1	$\frac{\mu^3 + \mu^2\lambda + \mu\lambda^2}{\mu^3 + \mu^2\lambda + \mu\lambda^2 + \lambda^3}$
3	2	$\frac{\mu^3 + 2\mu^2\lambda}{\mu^3 + 2\mu^2\lambda + 4\mu\lambda^2 + 4\lambda^3}$
3	3	$\frac{\mu^3}{\mu^3 + 3\mu^2\lambda + 6\mu\lambda^2 + 6\lambda^3}$
4	1	$\frac{\mu^4 + \mu^3\lambda + \mu^2\lambda^2 + \mu\lambda^3}{\mu^4 + \mu^3\lambda + \mu^2\lambda^2 + \mu\lambda^3 + \lambda^4}$
4	2	$\frac{\mu^4 + 2\mu^3\lambda + 4\mu^2\lambda^2}{\mu^4 + 2\mu^3\lambda + 4\mu^2\lambda^2 + 8\mu\lambda^3 + 8\lambda^4}$
4	3	$\frac{\mu^4 + 3\mu^3\lambda}{\mu^4 + 3\mu^3\lambda + 9\mu^2\lambda^2 + 18\mu\lambda^3 + 18\lambda^4}$
4	4	$\frac{\mu^4}{\mu^4 + 4\mu^3\lambda + 12\mu^2\lambda^2 + 24\mu\lambda^3 + 24\lambda^4}$
5	1	$\frac{\mu^5 + \mu^4\lambda + \mu^3\lambda^2 + \mu^2\lambda^3 + \mu\lambda^4}{\mu^5 + \mu^4\lambda + \mu^3\lambda^2 + \mu^2\lambda^3 + \mu\lambda^4 + \lambda^5}$
5	2	$\frac{\mu^5 + 2\mu^4\lambda + 4\mu^3\lambda^2 + 8\mu^2\lambda^3}{\mu^5 + 2\mu^4\lambda + 4\mu^3\lambda^2 + 8\mu^2\lambda^3 + 16\mu\lambda^4 + 16\lambda^5}$
5	3	$\frac{\mu^5 + 3\mu^4\lambda + 9\mu^3\lambda^2}{\mu^5 + 3\mu^4\lambda + 9\mu^3\lambda^2 + 27\mu^2\lambda^3 + 54\mu\lambda^4 + 54\lambda^5}$
5	4	$\frac{\mu^5 + 4\mu^4\lambda}{\mu^5 + 4\mu^4\lambda + 16\mu^3\lambda^2 + 48\mu^2\lambda^3 + 96\mu\lambda^4 + 96\lambda^5}$

TABLE C.16 (continued).

Number of Items N	Number Required k	System Availability A
5	5	$\dfrac{\mu^5}{\mu^5 + 5\mu^4\lambda + 20\mu^3\lambda^2 + 60\mu^2\lambda^3 + 120\mu\lambda^4 + 120\lambda^5}$
N	k	$\dfrac{\sum_{i=0}^{N-k} (\mu/k\lambda)^i}{\sum_{i=0}^{N-k} (\mu/k\lambda)^i + k!\sum_{i=1}^{k} (\lambda^i/(k-i)!\mu^i)}$

INDEX

A

B

C

D

E

F

G

H

I

K

L

M

N

O

P

Q

R

S

T

U

V

W